Environmental Science

Environmental Science

Edited by **Anne Offit**

SYRAWOOD
PUBLISHING HOUSE
New York

Published by Syrawood Publishing House,
750 Third Avenue, 9th Floor,
New York, NY 10017, USA
www.syrawoodpublishinghouse.com

Environmental Science
Edited by Anne Offit

International Standard Book Number: 978-1-68286-305-3 (Hardback)

Printed in China.

Contents

Preface

Every book is initially just a concept; it takes months of research and hard work to give it the final shape in which the readers receive it. In its early stages, this book also went through rigorous reviewing. The notable contributions made by experts from across the globe were first molded into patterned chapters and then arranged in a sensibly sequential manner to bring out the best results.

Environmental science is a discipline that assimilates concepts and information from several fields such as atmospheric science, mineralogy, oceanology, environmental engineering, geosciences, etc. to help solve the current environmental problems and assess various environmental impacts. This innovative and comprehensive book integrates the well-developed theory and practical applications of environmental science with the help of topics like pollution control, toxicology, environmental management, natural resource management, etc. This book, with its detailed analyses and data, will prove immensely beneficial to professionals and students involved in this area at various levels.

It has been my immense pleasure to be a part of this project and to contribute my years of learning in such a meaningful form. I would like to take this opportunity to thank all the people who have been associated with the completion of this book at any step.

Editor

Radioactivity of local tobacco and some selected brands of cigarettes

Agba Hemen Emmanuel[1], Kungur Samuel Terhemba[2]* and Akaagerger Nguvan Becky[1]

[1]Department of Physics, Benue State University, Makurdi, Nigeria.
[2]Department of Physics, College of Education, Katsina-Ala, Benue, Nigeria.

This study investigated the radioactivity levels of locally consumed tobacco and three (3) popular brands of cigarettes. The Ortec X-ray fluorescence spectroscopy (XRF) with model number XL12170 was used to determine the radioactivity and elemental composition of each of the samples. All the samples were found to be radioactive with the following elements, including-K, Ca, Ti, Cr, Mn, Fe, Pb, Rb, Sr, Zr and Nb. From the analysis, it was observed that the locally consumed tobaccos have a higher radioactive nature than the selected brands of cigarettes and this was ascribed to their different methods of processing/techniques.

Key words: Radioactivity, background radiation, X-ray fluorescence spectroscopy, radioactive elements, alpha radiation.

INTRODUCTION

Tobacco is an agricultural product processed from fresh leaves of plants in the *genus Nicotania*. It_contains minute quantities of radioactive isotopes that pose a radiation exposure hazard to those who intentionally or passively inhale it. Smokers inhale the Polonium (a radioactive component of tobacco), which settles in "hot spots" in the lungs and can cause cancer (Naomi and Erik, 2010). This is because, Po-210 emits alpha particles upon its decay and this radiation has a very destructive effect on tissues because virtually all of its very high ionizing energy is expended within the tissue. Due to its double positive charge, limited range in tissue and enormously high energy, an alpha particle can produce huge numbers of ion pairs in substances with which it interacts (Winickoff et al., 2009). Its effects may lead to thousands of deaths a year.

Basically, the most common usage of tobacco and its products is for smoking in the form of a cigarette or cigar.

Tobacco is commercially available in dried, cured and natural forms. In addition to been consumed as cigarettes and cigars, it can be smoked in a stem pipe, water pipe or hookah. Tobacco can also be chewed "dipped", or sniffed into the nose as finely powdered snuff.

According to the World Health Organization, radiation from tobacco smoke is the second biggest cause of death worldwide, just between hunger and malaria, having killed 100 million people in the 20th century (Kilthau, 1996).

All methods of tobacco consumption results in varying quantities of radiation to be absorbed into the smokers bloodstream. This exposes the smoker to a level of radiation that over time leads to several ailments like cancer, ulcer, leukemia and many other diseases (Ponte, 1986). Thus many countries set a minimum smoking age, regulating the purchase and use of tobacco products. Therefore, the basic aims and objectives of this work include the following:

(i) To determine the level of radioactivity in local dried tobacco and some brands of cigarettes.
(ii) To ascertain the elemental composition of all samples of tobacco chosen.
(iii) To obtain the radiation dose levels from the dried tobacco and brands of cigarettes.

*Corresponding author. E-mail: kungurs@yahoo.com.

Table 1. The elemental composition of sample A using XRF.

Element(s)	Energy (keV)	Intensity (C/S)	Sensitivity	Concentration (Fraction)
K	3.312	0.039	6.31E+01	6.63±1.02
Ca	3.690	0.024	7.49E+01	4.99±1.12
Cr	5.411	0.014	3.35E+02	0.32±LOD
Mn	5.895	0.014	4.26E+02	0.20±00.69
Fe	6.400	0.013	5.27E+02	0.13±00.48
Pb	10.540	0.009	9.55E+02	00.25±LOD
Sr	14.142	0.019	2.69E+03	00.16±000.42
Y	14.933	0.012	2.91E+03	000.85±LOD
Zr	15.746	0.016	3.74E+03	000.93±LOD
Nb	16.584	0.014	4.06E+03	000.70±LOD

(v) To draw conclusions on the potential hazards of radioactivity in tobacco and its products.

MATERIALS AND METHODS

The X-ray fluorescence spectroscopy was used to determine the radioactive materials in the samples. Also, the following brands of cigarettes that are commonly consumed in these localities and other materials were used for the research:

1. Sample A - ST Moritz (Dunhill)
2. Sample B - Benson and Hedges
3. Sample C - Rothmans king size
4. Samples D, E and F- Locally consumed tobacco from three different towns (Makurdi, Gboko and Katsina-Ala).
5. Handling forceps
6. Disposable plastic container.

Experimental methods

Radiation measurements using X-ray fluorescence (XRF) spectroscopy

The measurements using XRF spectrometer were carried out at Center for Energy and Research Technology (CERT), Ahmadu Bello University (ABU), Zaria to measure the elemental composition and the radioactive levels of the isotopes within the samples.

Sample preparation and analysis

All the samples were concealed in cellophane bags to prevent cross contamination. They were ground manually to powder with an agate mortar and pestle to grain size of less than 125 μm at CERT. Pellets of 19 mm diameter were prepared from 0.3 to 0.5 g powder mixed with three drops of organic liquid binder and pressed afterwards at 10 tons with a hydraulic press.

Measurements were performed using an annular 25 mCi ^{109}Cd as the excitation source that emits Ag-K X-ray (22.1 keV) in which all elements with lower characteristic excitation energies were accessible for detection in the samples. The system consists furthermore of a Si(Li) detector coupled to a computer.

Quantitative analysis of the samples was carried out using the Emission-Transmission (E-T) method and it involves the use of pure target material (Mo) to measure the absorption factors in the sample. The spectra for the samples were collected for 7000 s. with the ^{109}Cd source and then evaluated using the AXIL-QXAS program for the results.

EXPERIMENTAL RESULTS AND DISCUSSION

The research has investigated the radioactivity of local tobacco and some brands of cigarettes; these locally consumed tobaccos were collected from katsina-Ala, Gboko and Makurdi towns in Benue State, Nigeria. Also, commonly consumed cigarettes brands were investigated. These brands were Benson and Hedges, St. Moritz and Rothmans.

In sample A (Table 1), K, Ca and Sr has an activity higher than Cr, Mn, Fe, Pb, Y, Zr and Nb as shown in the bar chart, Figure 1. In sample B (Table 2), K, Ca, Mn, Rb, Sr, Zr and Nb has an activity higher than Ti, Cr, Fe and Pb as shown in Figure 2. Furthermore, in sample C (Table 3), K, Ca, Fe, Rb and Sr has an activity level higher than Ti, Cr, Mn, Pb, Zr and Nb as shown in Figure 3. It can therefore be inferred from samples A, B and C, that the activity of lead in sample C is higher than that in samples A and B. In sample D (Table 4), K, Ca, Fe, Rb and Sr are more in activity compared to Ti, Cr, Mn, Pb, Zr and Nb as shown in Figure 4. Also, in sample E (Table 5), K, Ca, Fe, Rb, Sr and Zr are more in activity than Ti, Cr, Mn, Pb, and Nb as shown in Figure 5. In sample F (Table 6), K, Ca, Fe, Sr and Zr are more in activity than Ti, Cr, Mn, Pb, Rb and Nb as shown in Figure 6. The isotope of lead which is seventh in the line spectrum has an activity which is quite small compared to the other radioisotopes in all the samples (Tahir and Alaamar, 2008). However, as a beta emitter, the isotopes of lead is an emitting precursor of Po-210 which is highly concentrated in tobacco trichomes and the trichome combustion in burning cigarettes produces insoluble, Pb-210-enriched particles in mainstream smoke (Monique et al., 2008). Hence the high concentration of Po- 210 observed at segmental bifurcations is as a result of the persistence of insoluble isotopes of lead enriched

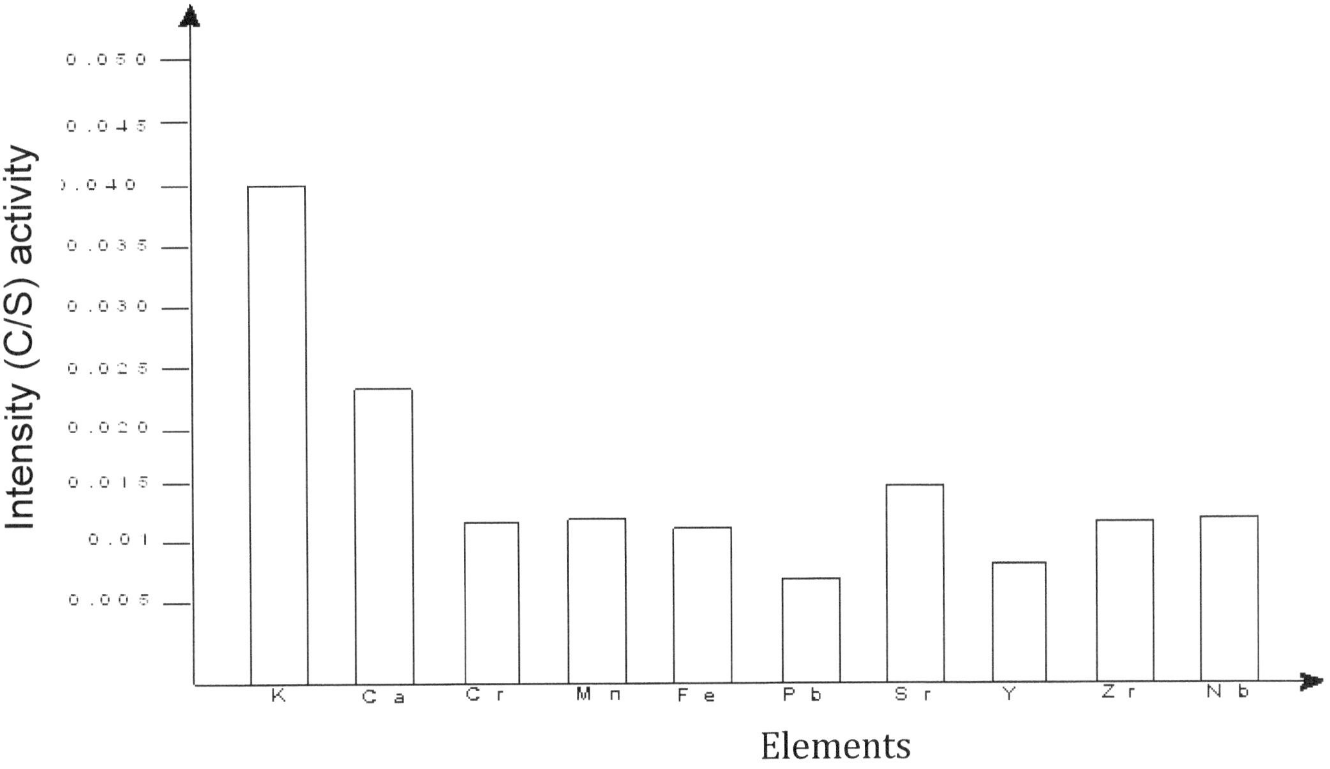

Figure 1. The bar chart of elemental composition for sample.

Table 2. The elemental composition of sample B using XRF.

Element(s)	Energy (keV)	Intensity (C/S)	Sensitivity	Concentration (Fraction)
K	3.312	0.025	6.60E+01	3.97±0.84
Ca	3.690	0.027	7.84E+01	4.24±0.86
Ti	4.508	0.010	1.80E+02	0.58±LOD
Cr	5.411	0.009	3.51E+02	0.18±LOD
Mn	5.895	0.015	4.46E+02	0.18±00.54
Fe	6.400	0.013	5.52E+02	0.11±LOD
Pb	10.540	0.011	9.99E+02	00.26±LOD
Rb	13.375	0.025	2.58E+03	00.21±LOD
Sr	14.142	0.029	2.81E+03	00.21±000.38
Zr	15.746	0.016	3.91E+03	000.80±LOD
Nb	16.584	0.015	4.24E+03	000.69±LOD

particles deposited at segmental bifurcations and by ingrowths of Po-210 in these particles after a long period of time (Purkayastha and Bhattacharyya, 1975).

Furthermore, from the brands of cigarettes selected, even though the isotope of lead are at a low limit of detection (LOD), a long period of exposure will bring about a malignant transformation by alpha interaction with basal cells leading to the cause of the several ailments mentioned previously. Now, considering the locally consumed tobacco and the selected brands of cigarettes, the level of radiation in the locally consumed tobacco in Katsina-Ala town is slightly higher due to high levels of K, Ca, Fe, Rb, Sr and Zr compared to the others from Gboko and Makurdi towns. Also, sample C (Rothmans king size) has a higher level of radiation compared to samples B and D (St. Moritzs and Benson Hedges). This could be attributed to differences in the application and use of chemical fertilizers and non-toxic

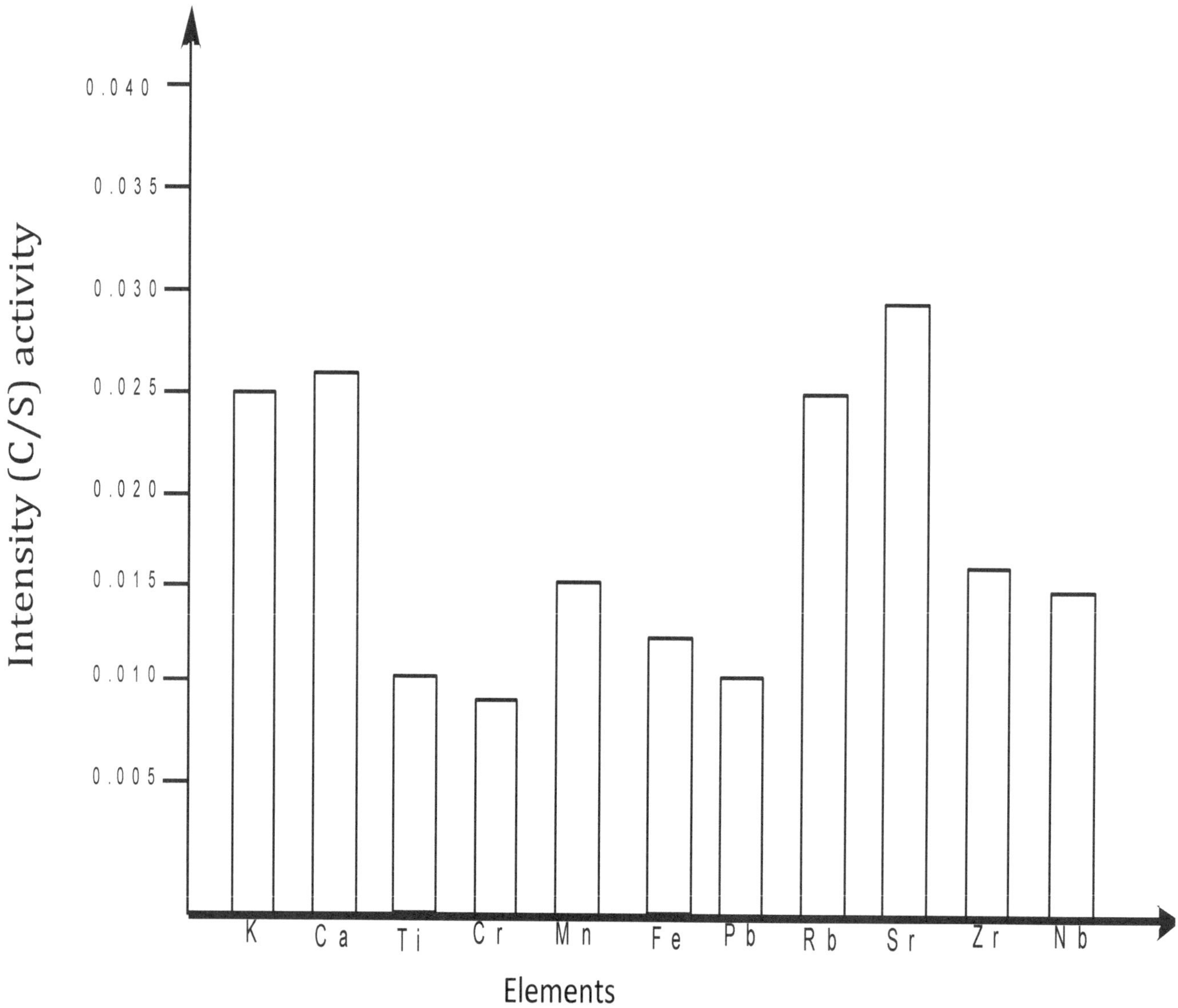

Figure 2. The bar chart of elemental composition for sample B.

Table 3. The elemental composition of sample C using XRF.

Element(s)	Energy (keV)	Intensity (C/S)	Sensitivity	Concentration (Fraction)
K	3.312	0.033	6.60E+01	5.22±0.92
Ca	3.690	0.028	7.84E+01	4.99±0.99
Ti	4.508	0.010	1.80E+02	0.66±LOD
Cr	5.411	0.011	3.51E+02	0.23±LOD
Mn	5.895	0.013	4.46E+02	0.18±LOD
Fe	6.400	0.015	5.52E+02	0.15±00.44
Pb	10.540	0.010	9.99E+02	00.28±LOD
Rb	13.375	0.019	2.58E+03	00.17±LOD
Sr	14.142	0.020	2.81E+03	00.65±000.42
Zr	15.746	0.014	3.91E+03	000.82±LOD
Nb	16.584	0.011	4.24E+03	000.61±LOD

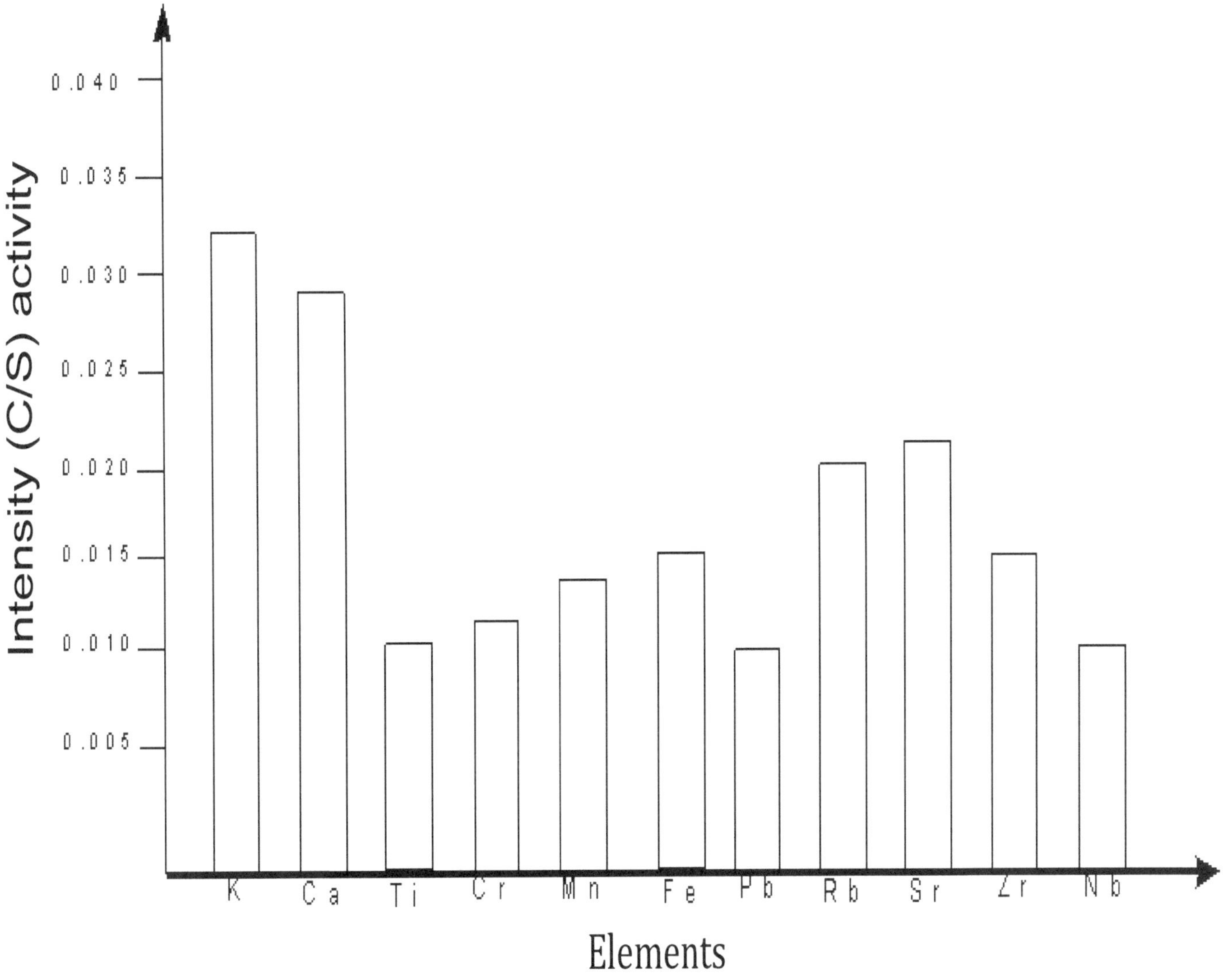

Figure 3. The bar chart of elemental composition for sample C.

Table 4. The elemental composition of sample D using XRF.

Element(s)	Energy (keV)	Intensity (C/S)	Sensitivity	Concentration (Fraction)
K	3.312	0.025	6.91E+01	3.71±0.99
Ca	3.690	0.050	8.21E+01	7.58±1.12
Ti	4.508	0.017	1.89E+02	1.12±0.41
Cr	5.411	0.013	3.67E+02	0.29±LOD
Mn	5.895	0.015	4.67E+02	0.21±LOD
Fe	6.400	0.054	5.78E+02	0.54±06.81
Pb	10.540	0.010	1.05E+03	02.68±LOD
Rb	13.375	0.025	2.71E+03	02.11±LOD
Sr	14.142	0.032	2.94E+03	02.45±004.90
Zr	15.746	0.015	4.10E+03	008.13±0032.27
Nb	16.584	0.016	4.44E+03	00.748±LOD

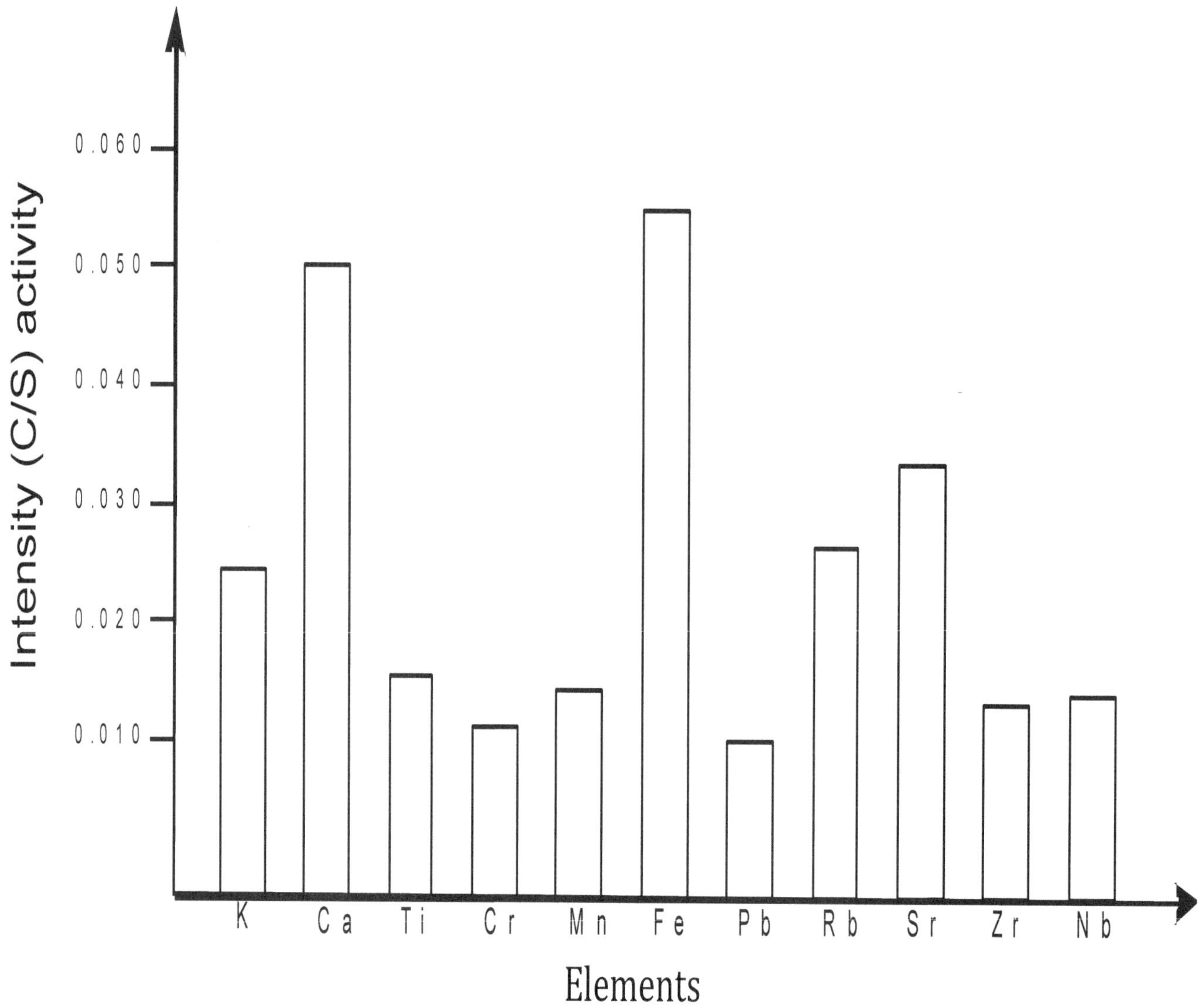

Figure 4. The bar chart of elemental composition for sample D.

Table 5. The elemental composition of sample E using XRF.

Element(s)	Energy (keV)	Intensity (C/S)	Sensitivity	Concentration (Fraction)
K	3.312	0.022	6.60E+01	3.52±0.86
Ca	3.690	0.034	7.84E+01	5.28±0.89
Ti	4.508	0.013	1.80E+02	0.75±LOD
Cr	5.411	0.013	3.51E+02	0.26±LOD
Mn	5.895	0.013	4.46E+02	0.16±LOD
Fe	6.400	0.052	5.52E+02	0.47±00.52
Pb	10.540	0.009	9.99E+02	00.24±LOD
Rb	13.375	0.028	2.58E+03	00.25±LOD
Sr	14.142	0.027	2.81E+03	00.21±000.42
Zr	15.746	0.029	3.91E+03	00.16±000.31
Nb	16.584	0.012	4.24E+03	000.60±LOD

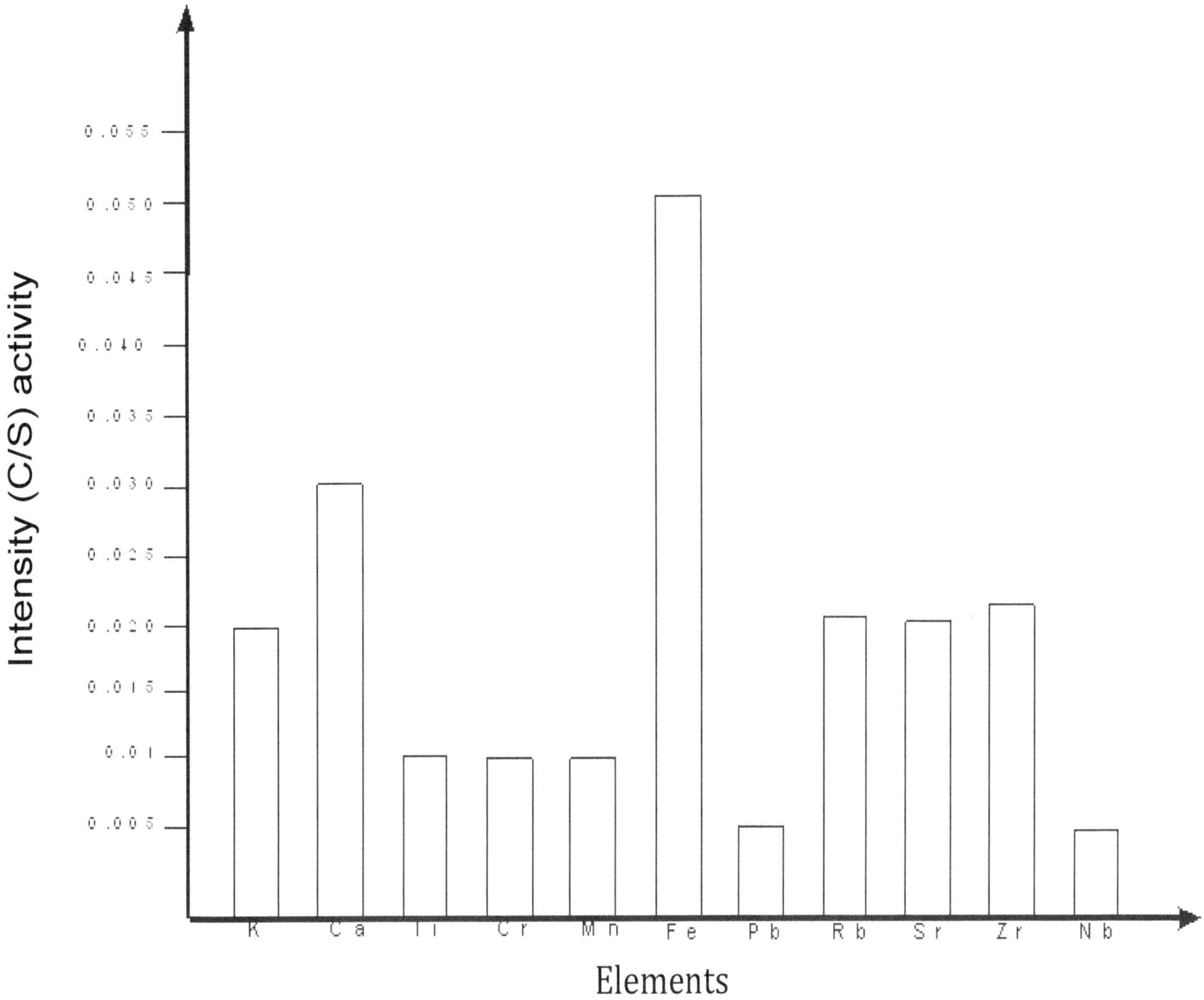

Figure 5. The bar chart of elemental composition for sample E.

Table 6. The elemental composition of sample F using XRF.

Element(s)	Energy (keV)	Intensity (C/S)	Sensitivity	Concentration (Fraction)
K	3.312	0.030	6.60E+01	4.92±0.88
Ca	3.690	0.030	7.84E+01	5.21±0.97
Ti	4.508	0.015	1.80E+02	0.94±0.32
Cr	5.411	0.010	3.51E+02	0.23±LOD
Mn	5.895	0.010	4.46E+02	0.15±LOD
Fe	6.400	0.057	5.52E+02	0.56±00.54
Pb	10.540	0.012	9.99E+02	00.34±LOD
Rb	13.375	0.020	2.58E+03	00.19±LOD
Sr	14.142	0.024	2.81E+03	00.19±000.43
Zr	15.746	0.110	3.91E+03	00.63±000.41
Nb	16.584	0.012	4.24E+03	000.64±LOD

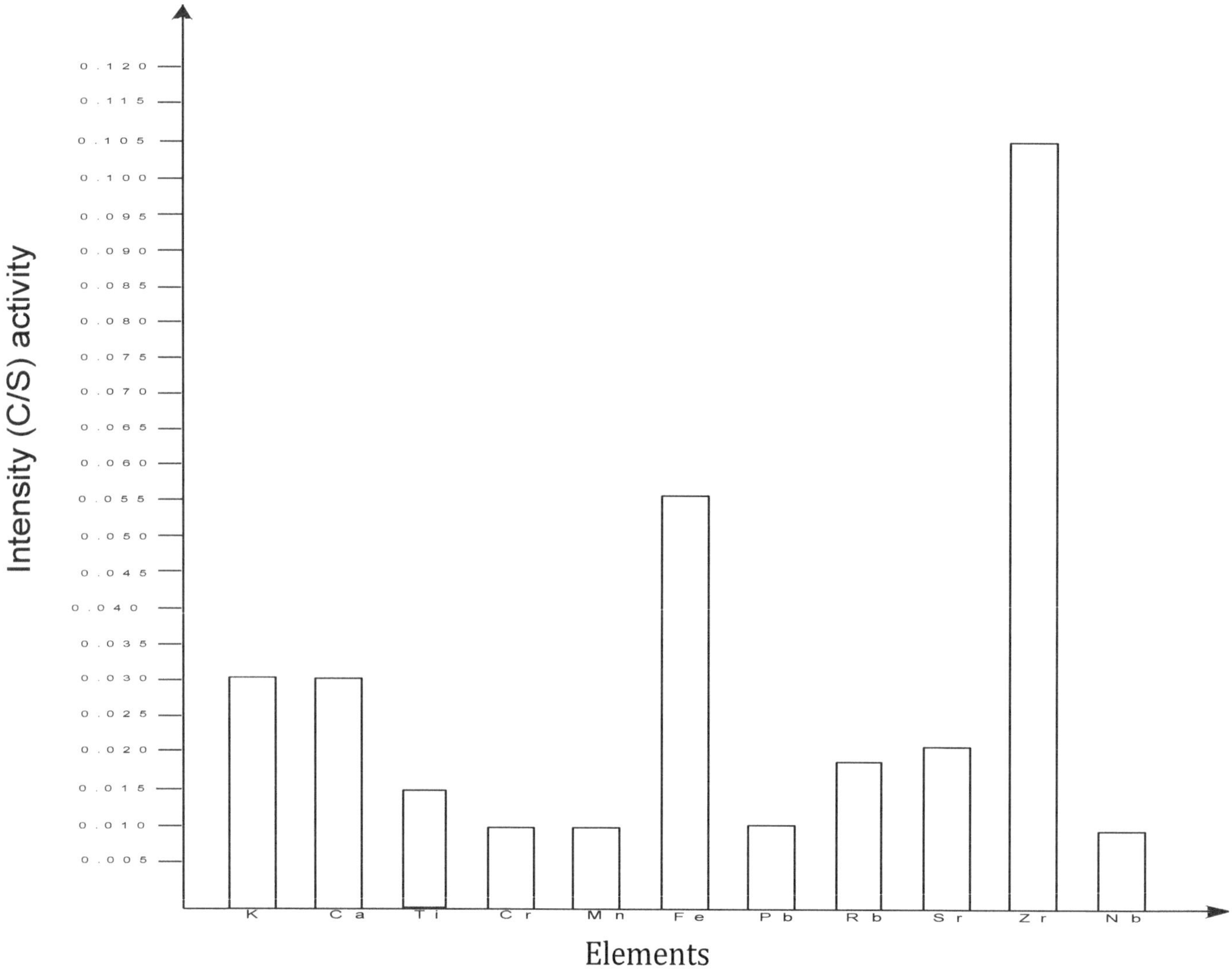

Figure 6. The bar chart of elemental composition for sample F.

pesticides, but an enhanced use of organic fertilizers which appears to emits less alpha radiation than a chemical fertilizer (Kilthau, 1996; Colgrove et al., 2011).

Since it has been proved that all the brands contain isotopes of lead (an alpha-emitter) which is a precursor to the production of polonium-210 at segmental bifurcations, then all the samples are radioactive (Martell, 1983).

Conclusion

Based on the findings of the research, it can be concluded that:

(i) All the selected brands of cigarettes and the locally consumed tobacco from the three different towns were found to be radioactive.

(ii) The mean radioactive levels of the locally consumed tobacco are found to be higher than the selected brands of cigarettes. This could be as a result of the differences in their processing techniques or methods.

REFERENCES

Colgrove J, Bayer R, Bachynski K (2011). No Where Left To Hide? The Banishment of Smoking from Public Places. New England. J. Med., 364: 2375- 2377.

Kilthau GF (1996). Cancer risk in relation to radioactivity in tobacco. Radiologic Technol., p. 67.

Martell EA (1983). Radiation Dose at Bronchial Bifurcations of Smokers from Indoor Exposure to Radon progeny. Harwood Pub., Iowa, pp. 5-13.

Monique EM, Jon OE, Channing R, Richard DH, Bogdan W, Barbara P (2008). Waking a Sleeping Giant: The Tobaccos Industry's Response to the Polonium-210 Issue. Am. J. Public Health, 98(9): 1643-1650.

Naomi O, Erik MC (2010). Merchants of Doubts: How a Handful of

Scientists Obscured the Truth on Issues from Tobacco Smoke to Global Warming. Bloombury Press, London, pp. 34-41.
Ponte L (1986). Radioactivity: The New-Found Danger in Cigarettes. pp. 123-127.
Purkayastha BC, Bhattacharyya DK (1975). Estimation of rare and radioactive constituents in samples of Indian tobacco with the aid of low-level beta-counter; J. Radioanalyt. Nuclear Chem., 27(2): 21-26.
Tahir SNA, Alaamer R (2008). Radiation Protection Dosimetry: Pb-210 concentrations in cigarettes & radiation doses to the smokers.130 (3): 389-391.
Winickoff J, Muggli E, Harrison J, Friebely J, Tanski S, Sherrod C, Matt G, Hovell M, McMillen R (2009). Beliefs about the Health Effects of Thirdhand Smoke and and Home Smoking Bans. Pediatrics, 13(1): 74-79.

Relationship between magnetospheric interplanetary parameters and radio refractivity for quiet and disturbed days at Minna

G. A. Agbo, O. N. Okoro and J. E. Ekpe

Department of Industrial Physics, Ebonyi State University, P. M. B. 053, Abakaliki, Nigeria.

The relationship of magnetospheric interplanetary parameters with radio refractivity is investigated from hourly measurement of solar wind energy and interplanetary parameters for quiet and disturbed days during dry and rainy season in Minna (Latitude 9° 36' 50" N and Longitude 6° 33' 24" E) North central, Nigeria for the year 2008. The solar wind energy, magnetospheric interplanetary parameters and radio waves are characterized by large-scale interactions. These introduce both random and systematic variations in radio refractivity which virtually causes variations in radio propagation. Data used for this work is made up of hourly interval record of solar wind energy and interplanetary parameters calculated from data obtained from the Center for Basic Space Science (CBSS), University of Nigeria, Nsukka. The data was obtained using Vantage PRO II Automatic Weather station, and also through http://omniweb.gsfc.nasa.gov/ for each day in all the months of the year in 2008. The result showed that variations of solar wind energy, radio refractivity and interplanetary parameters with time are almost an hourly cycle each following almost the same trend line for both quiet and disturbed days during dry and rainy season. The result of the correlation analysis of solar wind energy and radio refractivity also showed that solar wind energy has square of the correlation coefficients, $R^2 = 0.176$, $R^2 = 0.040$, $R^2 = 0.319$ and $R^2 = 0.016$ for both quiet and disturbed days during dry and rainy seasons which reviews weak correlation. Similarly, proton density (PD) and Dst Index, showed high correlation with square of the correlation coefficients, $R^2 = 0.598$ and $R^2 = 0.5949$, respectively for quiet day during dry season, and for disturbed day during dry season only PD showed high degree of correlation with square of the correlation coefficient, $R^2 = 0.543$. But during rainy season, the result indicated that there is no significant correlation existing between any of the interplanetary parameters with radio refractivity for both quiet and disturbed days.

Key words: Magnetospheric interplanetary parameters, radio refractivity.

INTRODUCTION

Magnetospheric structures and boundaries are ever changing with solar wind conditions, the interplanetary magnetic field (IMF), and substorm activity (Russell et al., 1998). In order to understand and quantify magnetospheric processes, one must understand the electromagnetic field and plasma topology which, in spite of many years of *in situ* observations, has not been mapped on a global scale (Fuselier et al., 1991).

According to some researchers, magnetospheric boundaries are in constant motion during solar wind magnetosphere coupling processes (Russell, 2000), this constant motion is responsible for the overall shape of

Earth's magnetosphere, and fluctuations in its speed, density, direction, and entrained magnetic field which strongly affect Earth's local space environment. For example, the levels of ionizing radiation and radio interference can vary by factors of hundreds to thousands, and the shape and location of the magnetopause and bow shock wave upstream of it can change by several Earth radii, exposing geosynchronous satellites to the direct solar wind (Fuselier et al., 1997).

Enhanced solar wind flow velocities and densities, such as those that can occur in coronal mass ejection (CME) events, can easily distort the dayside magnetopause and push it inside its normal location at about ten Earth radii distance (Phan et al., 1994). During large solar wind disturbances, the magnetopause can be pushed inside the geosynchronous orbit (Phan et al., 1994). At such times, the magnetic field increases to as much as twice its "quiescent" value. In addition, the magnetic field outside the magnetopause will have a polarity that is predominantly opposite to that inside the magnetosphere (Phan et al., 1994). CMEs generally form a large magnetic flux rope that propagates into interplanetary space.

The large flux rope contains very low density and a strong field that is observed to slowly rotate as the rope passes an observer. The ICME flux rope is often referred to as a magnetic cloud and if the cloud is oriented properly, it will produce a long steady period of slowly varying southward IMF at the magnetopause. If the flux rope is propagating faster than the Alfven velocity through the ambient solar wind plasma, a shock front and sheath will form in front of the cloud. The sheath may also contain large southward IMF parameters for a substantial period, prior to the encounter with the magnetic cloud. Some of the most severe magnetic storms, the so-called 'double hit storms' develop in response to these two subsequent intervals of sustained strong southward IMF parameters at the magnetopause leading to a complex radio refractivity changes within such period. Okoro et al. (2012) showed that solar wind-magnetosphere coupling can cause serious effect on radio refractivity in atmosphere. Thus, this work investigates the relationship between such Earth-magnetospheric interplanetary parameters with radio refractivity during solar events within the atmosphere.

Theoretical analysis

The study of the solar wind interaction with the planetary bodies is one of the main topics of space weather. During a geomagnetic storm, the sun and the magnetosphere are connected, giving rise to severe changes both in interplanetary space and terrestrial environment (Baker et al., 1998). Some examples are the acceleration of charged particles, enhancement of electric currents, auroras and magnetic field variations on the earth surface. These changes can produce important damages in electric power supplier, radio communications and spacecrafts. It is assumed that sun-earth interaction depends on solar wind. In fact, in-tense geomagnetic storms seem to be related to intense IMF with a southern component for a long time (Tsurutani and Meng, 1972). Several papers about geomagnetic storms (Dungey, 1963) have pointed out that reconnection between a southern IMF and the magnetospheric magnetic field as the physical mechanism responsible for sun-earth connection. Although several aspects on this mechanism are still open questions, it is accepted that reconnection in the day-side of magnetosphere produces a transference of magnetic flux to the magnetotail (Rickett, 1990). Then energetic particles of solar wind can go into the magnetosphere, along magnetic field lines, yielding an injection of plasma in the night-side of the magnetosphere.

The radiation belts are regions of terrestrial environment where charged particles become trapped on closed geomagnetic field lines. These particles show drifts due to magnetic field gradient and curvature as well as to gyration orbit effects. As drifts depend on the sign of charge, ions travel to West and electrons to east, giving rise to a ring current (RC) which extends from 4 to 8 terrestrial radii. Variations on this current result in variations of the magnetic field on the earth surface.

The main mechanism by which the magnetized solar wind powers the magnetosphere was first proposed by Dungey (1963). In his proposal, he stated that solar wind and magnetosphere are connected by IMF southward, and becomes connected to the terrestrial magnetic field at the sub solar point in a process known as reconnection. Reconnection produces open field lines with one end connected to the earth, by combining interplanetary field lines, not connected to the earth, with closed field lines, connected to the earth on both ends. The resulting V-shaped magnetic fields attempt to straighten and accelerate plasma (whose origin is both in the solar wind and the magnetosphere).

The earth's magnetosphere is a very dynamical system, whose configuration depends on many internal and external factors. The first factor is orientation of the earth's magnetic axis with respect to the direction of the incoming solar wind flow, which varies with time because of (i) Earth's diurnal rotation and its yearly orbital motion around sun, and (ii) frequent "side gusts" of the solar wind. The second is the state of the solar wind, in particular, the orientation and strength of the IMF, carried to the earth's orbit from sun due to the high electrical conductivity of the solar wind plasma. Interaction between the terrestrial and interplanetary fields becomes much more effective when the IMF turns ant-parallel to the earth's field on the dayside boundary of the magnetosphere. In this case, geomagnetic and interplanetary field lines connect across the magnetospheric boundary, which greatly enhances the

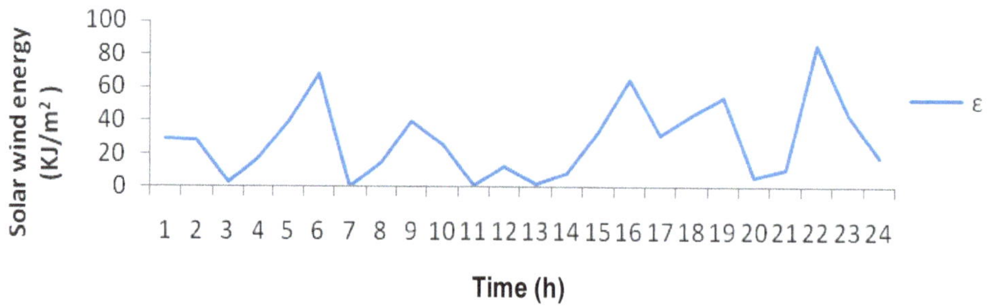

Figure 1. Variations of solar wind energy with time for quiet day during dry season.

transfer of the solar wind mass, energy, and electric field inside the magnetosphere. As a result, the magnetospheric field and plasma become involved in convection.

MATERIALS AND METHODS

The data used for this work covered the period of 1 year starting from January to December, 2008. The data are in two forms. First one is meteorological parameters which are raw recorded 5 min data of temperature, humidity and pressure. These data were provided by the Centre for Basic Space Science (CBSS), University of Nigeria Nsukka.

The second type of data is the interplanetary magnetic data obtained from OMINI website (www.ominiweb.com). This data consists of 5 min data value of solar wind velocity and IMF components of Bz(nT), Bx(nT), By(nT), PD (Ncm^{-3}), Dst Index (nT) AE Index (nT) and solar wind pressure.

From the data, hourly averages of solar wind IMF parameters for each day of the month were calculated using excel package. Solar wind energy (ε) of the magnetosphere will be calculated using:

$$\varepsilon = V_{sv} B^2 sin^4 \left(\frac{\theta}{2}\right) / l_o^2 \tag{1}$$

Where V_{sv} = solar wind velocity; |*B*| = magnitude of the IMF parameter; $\theta = \arctan(B_y / B_z) \; for B_z > 0 \; and$ $\theta = 180^0 - \arctan(B_y / B_z) \; for B_z \leq 0$; l_o = dayside magnetopause scale length which is equal to $7R_E$; B_z = z-component of IMF parameter; B_y = y-component of IMF parameter. Also, the hourly averages of meteorological parameters for each day of the month were calculated using excel package. From the values obtained, partial pressure of water (e) was determined from the equation as follows:

$$e = e_s . \frac{H}{100} \tag{2}$$

Where H is the relative humidity, and e_s is the saturation vapour pressure determined by Clausius-Clapeyron equation given as:

$$e_s = 6.11 exp\,[17.5(T - 273.16)/(T - 35.87)] \tag{3}$$

Employing the values of meteorological parameters (temperature, pressure and relative humidity) computed, radio refractivity was calculated using;

$$N = 77.6 \frac{P}{T} + 3.37 \; x \; 10^5 \; e/T^2 \tag{4}$$

Where P = atmospheric pressure (hPa), e = water vapour pressure (hPa), T = absolute temperature (K).

Equation 4 may be employed for the propagation of radio frequencies up to 100 GHz (Willoughby et al., 2002).

The two results (solar wind coupling and radio refractivity) were compared to actually deduce solar-wind magnetosphere coupling effect on radio refractivity at Minna in the year 2008.

RESULTS AND DISCUSSION

The graphical representation of hourly records of variations of solar wind energy (ε), radio refractivity (N), proton density (PD) and Dst index are presented in the figures for both quiet and disturbed days during dry and rainy seasons at Minna, Nigeria. Although the research was carried out in all the months of the year, the quiet and disturbed days of the months of February and August, respectively were considered. In the same manner, the relationship of solar wind energy (ε), PD and Dst index with radio refractivity (N) are the main focus of this work. This relationship is clearly demonstrated in the plots of variations and correlations of radio refractivity with magnetospheric interplanetary parameters as can be observed in Figures 1 to 32, respectively.

Figures 1 to 8 represent variations of solar wind energy and radio refractivity with time for quiet and disturbed days during dry and rainy seasons at Minna in 2008. Figures 9 to 12 represent correlation of solar wind energy with radio refractivity for quiet and disturbed days during dry and rainy seasons.

In Figure 1, 2, 3 and 4, there is a corresponding fall and rise in the peaks of variations of solar wind energy as well as radio refractivity with time almost throughout the early hours of the day, but they seem to exhibit maximum increase between 0600 to 2200 h for both quiet and disturbed days during dry and rainy seasons. This is caused by diurnal variations of solar wind energy, in which case, the IMF turns anti-parallel to the earth's field on the day side of the magnetosphere. This period is the time when there is maximum solar wind activity (solar maximum) caused by the high temperature released by

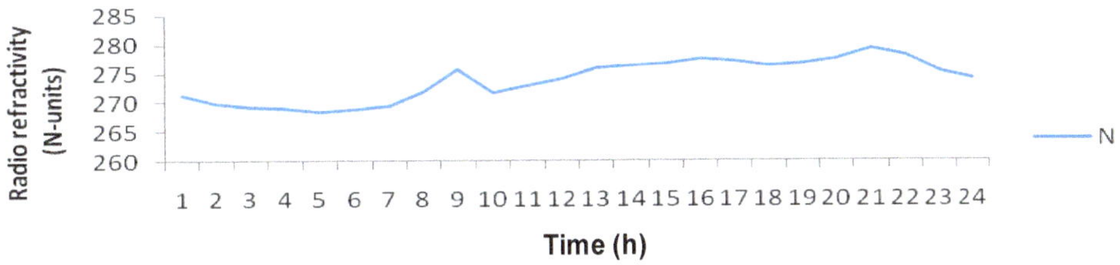

Figure 2. Variations of radio refractivity with time for quiet day during dry season.

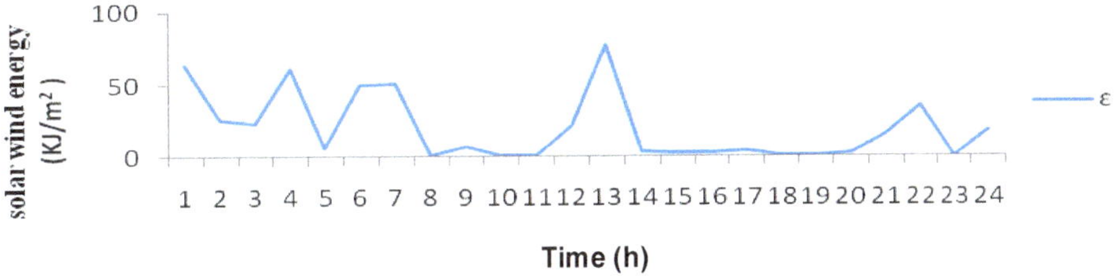

Figure 3. Variations of solar wind energy with time for disturbed day during dry season.

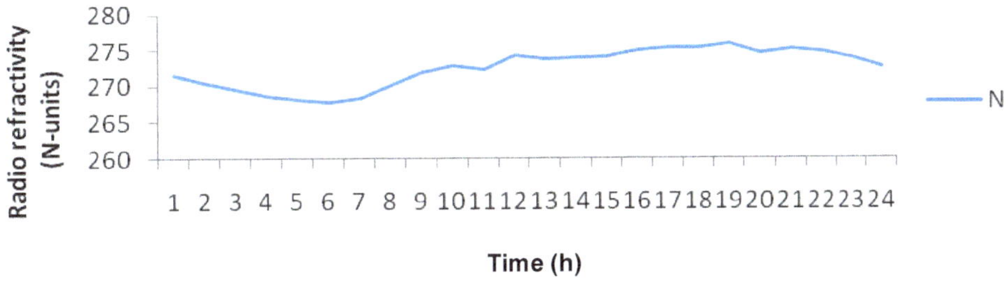

Figure 4. Variations of radio refractivity with time for disturbed day during dry season.

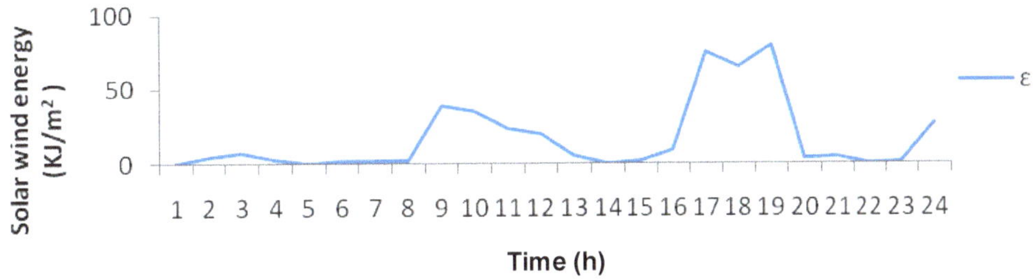

Figure 5. Variations of solar wind energy with time for quiet day during rainy season.

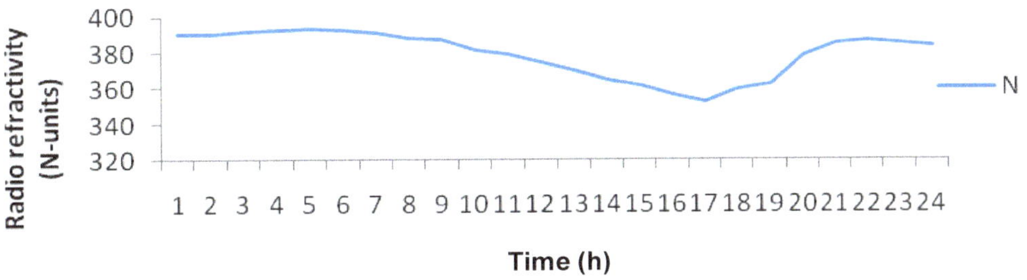

Figure 6. Variations of radio refractivity with time for quiet day during rainy season.

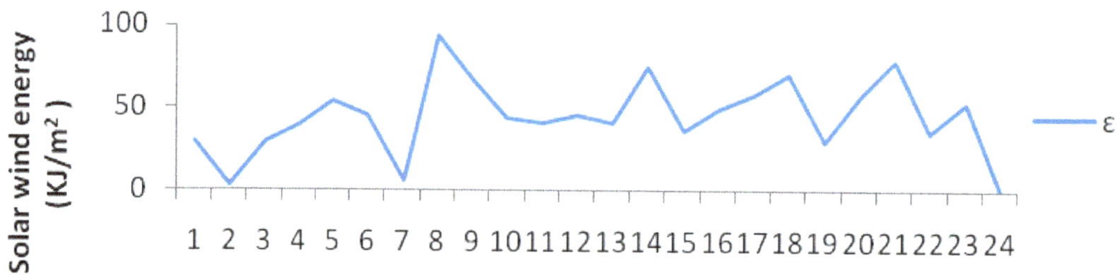

Figure 7. Variations of solar wind energy with time for disturbed day during rainy season.

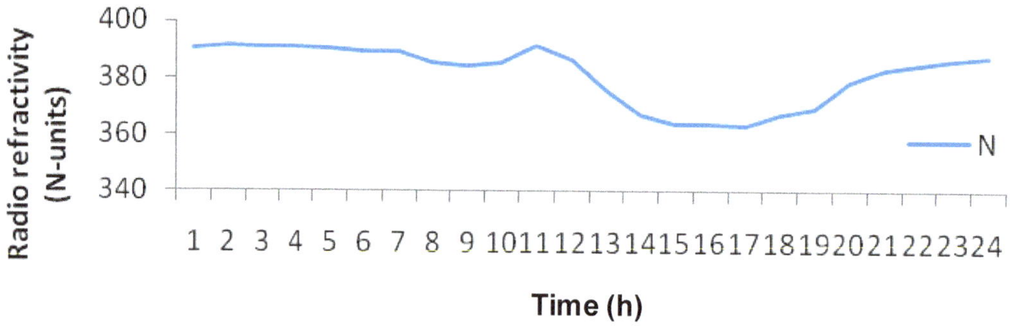

Figure 8. Variations of radio refractivity with time for disturbed day during rainy season.

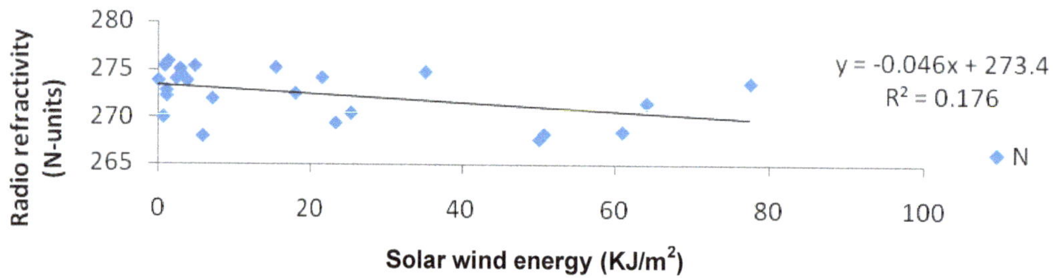

Figure 9. Correlation of solar wind energy with radio refractivity for quiet day during dry season.

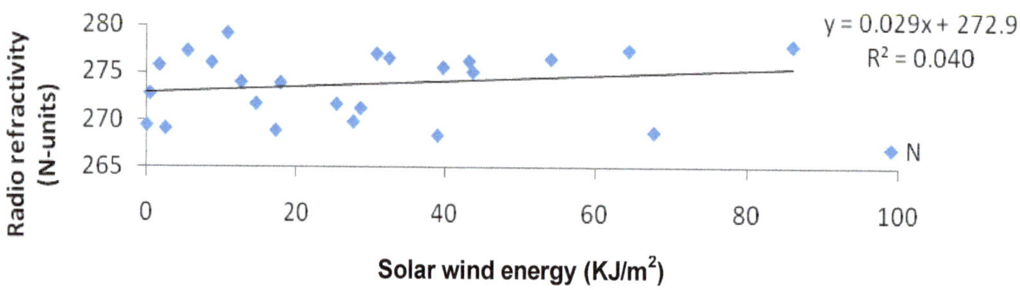

Figure 10. Correlation of solar wind energy with radio refractivity for disturbed day during dry season.

the sun during solar wind magnetosphere coupling activities. This process results in the ejection of high solar wind (stream of electrons and protons) which is the characteristics of southward continental drift observed

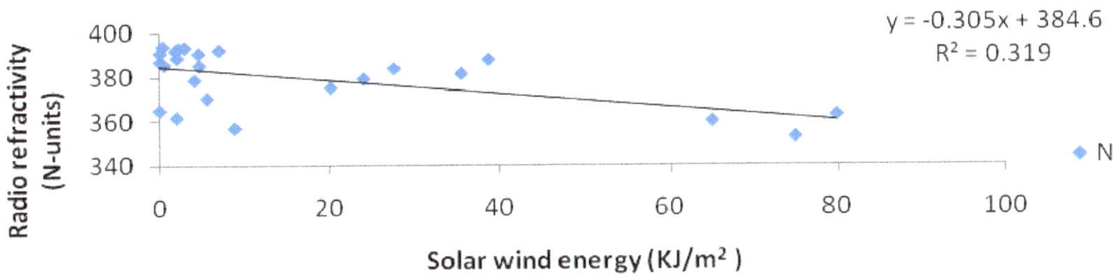

Figure 11. Correlation of solar wind energy with radio refractivity for quiet day during rainy season.

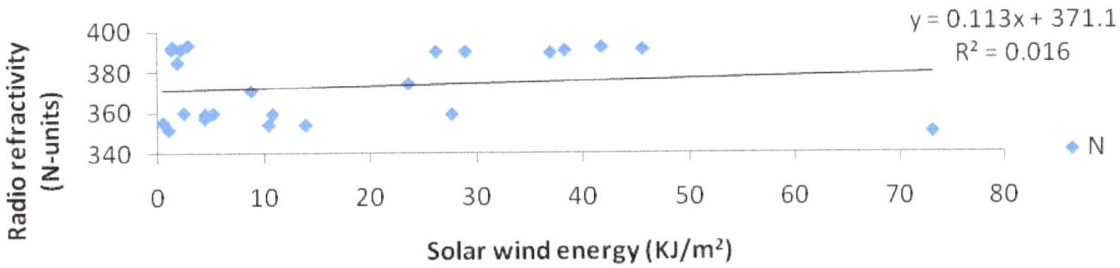

Figure 12. Correlation of solar wind energy with radio refractivity for disturbed day during rainy season.

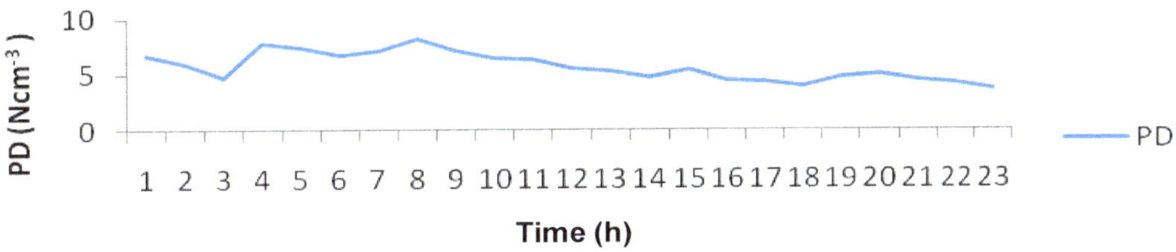

Figure 13. Variation of PD with time for quiet day during dry season.

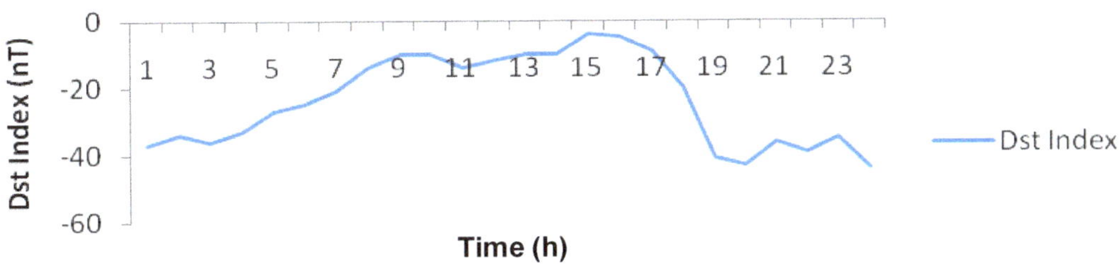

Figure 14. Variation of Dst index with time for quiet day during dry season.

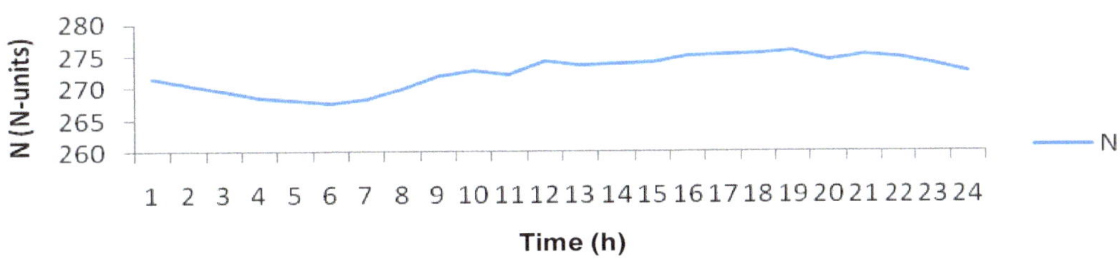

Figure 15. Variation of radio refractivity with time for quiet day during dry season.

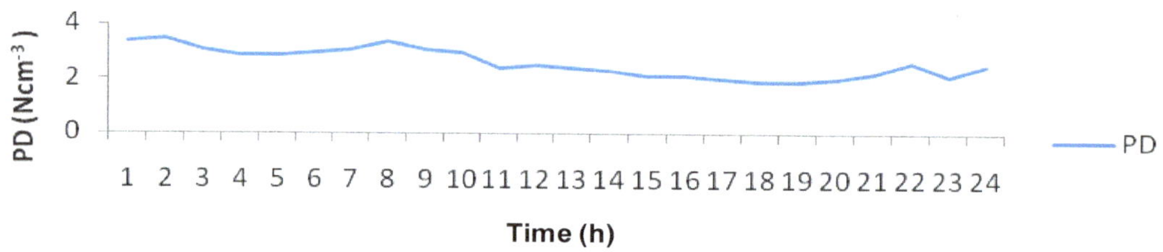

Figure 16. Variation of PD with time for disturbed day during dry season.

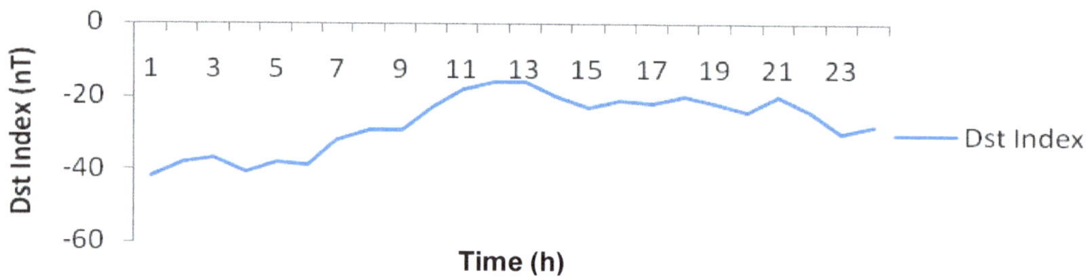

Figure 17. Variation of Dst index with time for disturbed day during dry season.

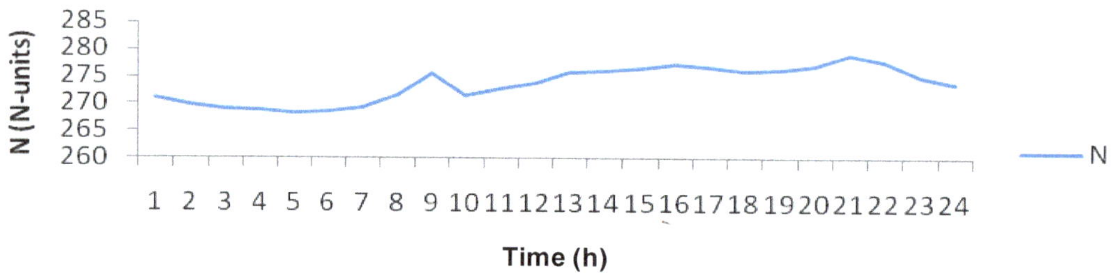

Figure 18. Variation of radio refractivity with time for disturbed day during dry season.

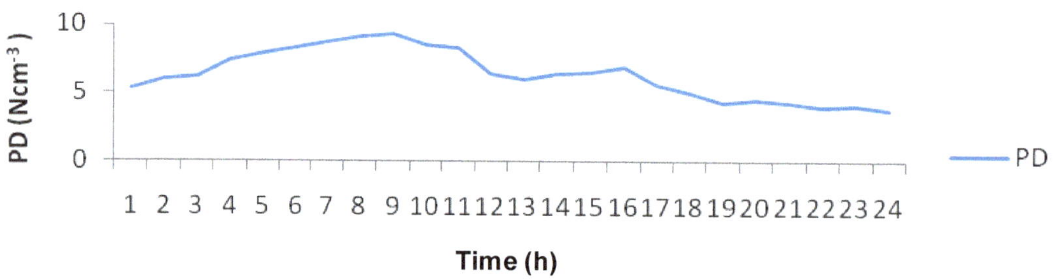

Figure 19. Variation of PD with time for quiet day during rainy season.

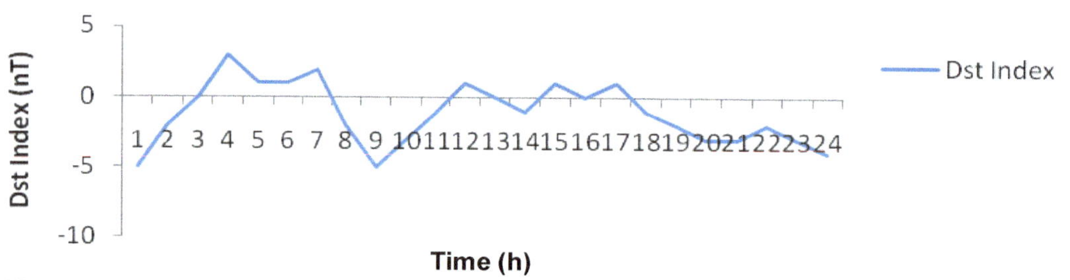

Figure 20. Variation of Dst index with time for quiet day during rainy season.

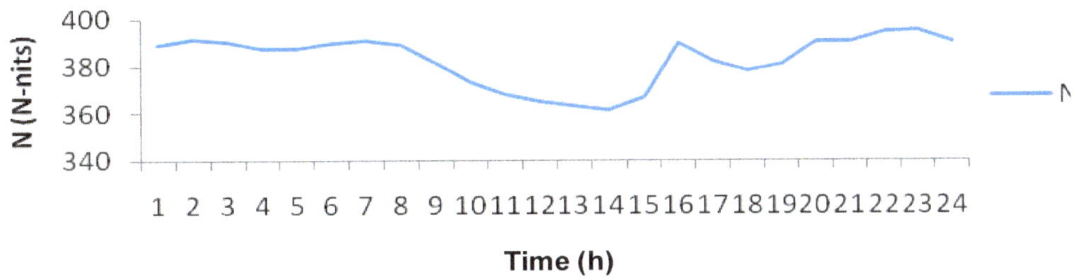

Figure 21. Variation of radio refractivity with time for quiet day during rainy season.

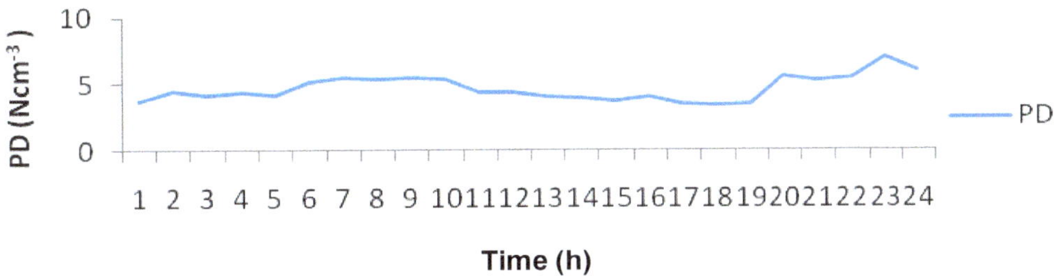

Figure 22. Variation of PD with time for disturbed day during rainy season.

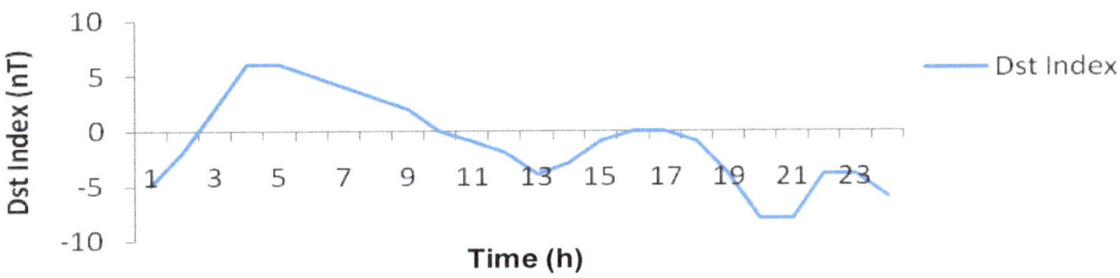

Figure 23. Variation of Dst index with time for disturbed day during rainy season.

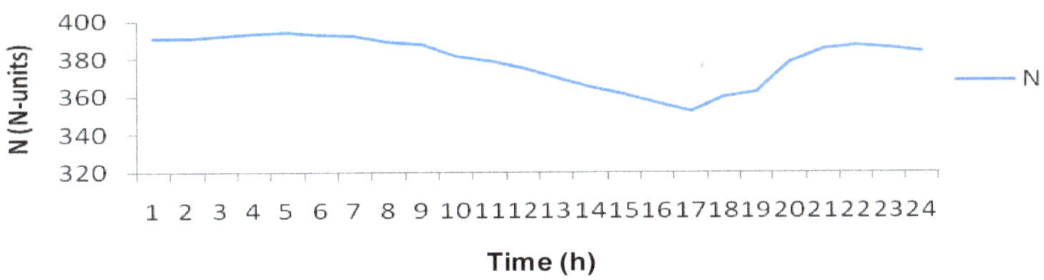

Figure 24. Variation of radio refractivity with time for disturbed day during rainy season.

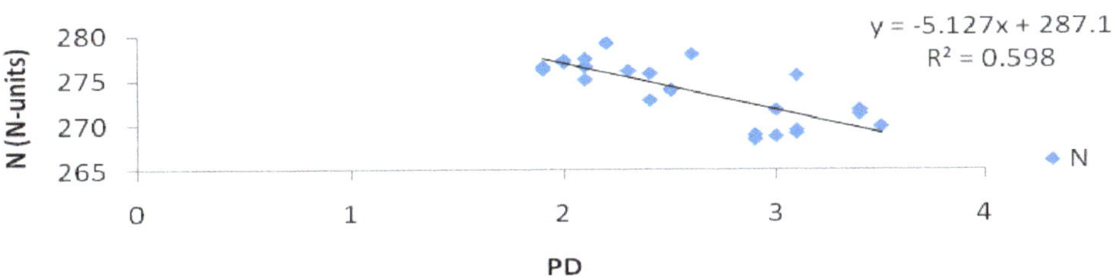

Figure 25. Correlation of PD with radio refractivity for quiet day during dry season.

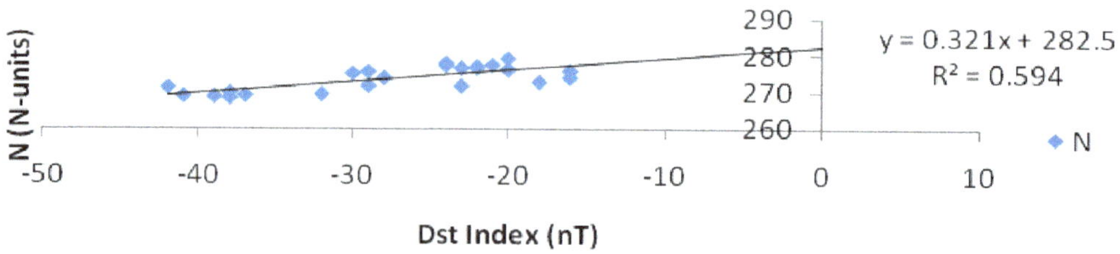

Figure 26. Correlation of Dst Index with radio refractivity for quiet day during dry season.

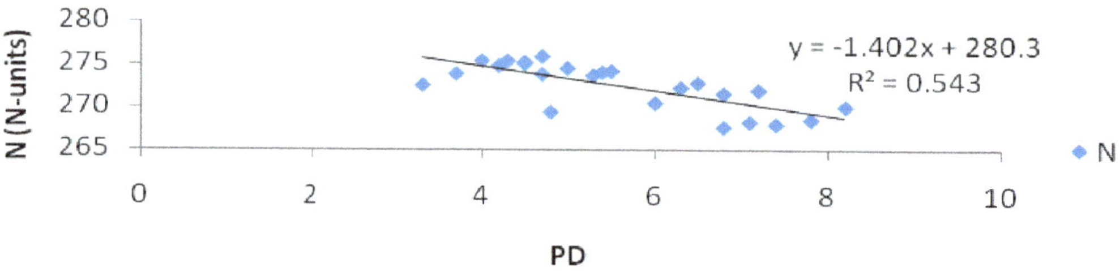

Figure 27. Correlation of PD with radio refractivity for disturbed day during dry season.

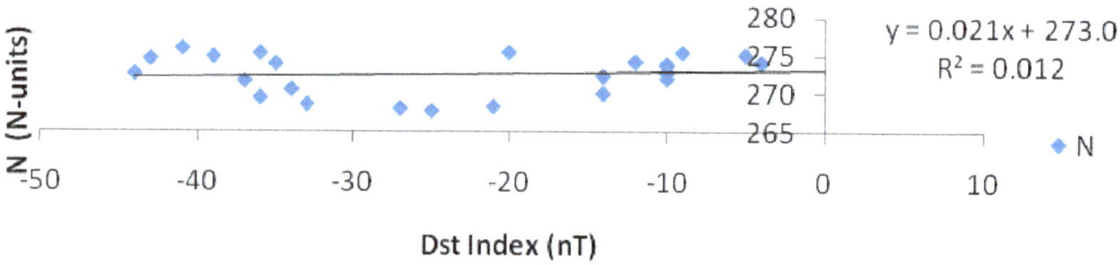

Figure 28. Correlation of Dst index with radio refractivity for disturbed day during dry season.

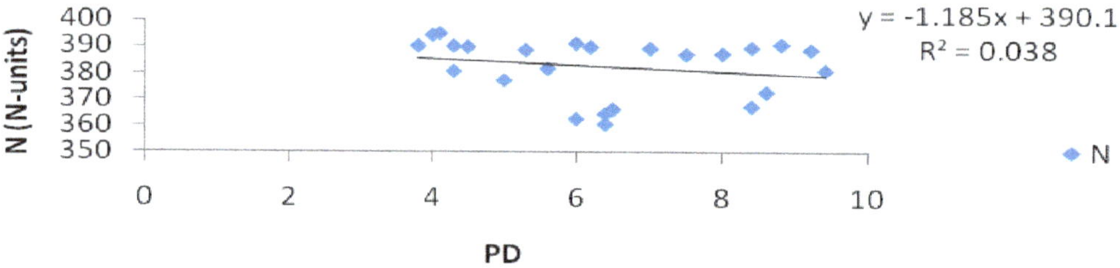

Figure 29. Correlation of PD with radio refractivity for quiet day during rainy season.

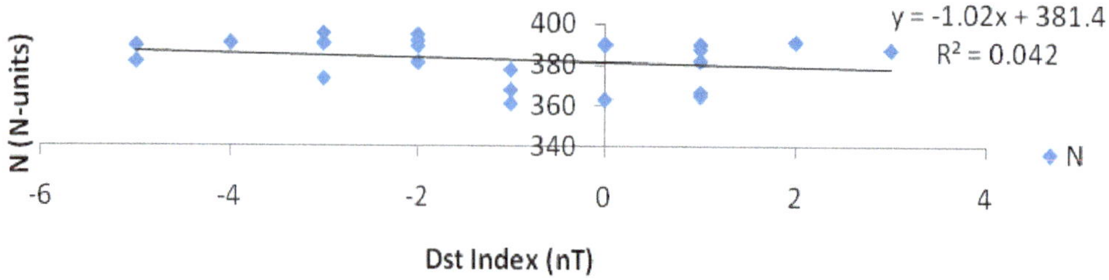

Figure 30. Correlation of PD with radio refractivity for quiet day during rainy season.

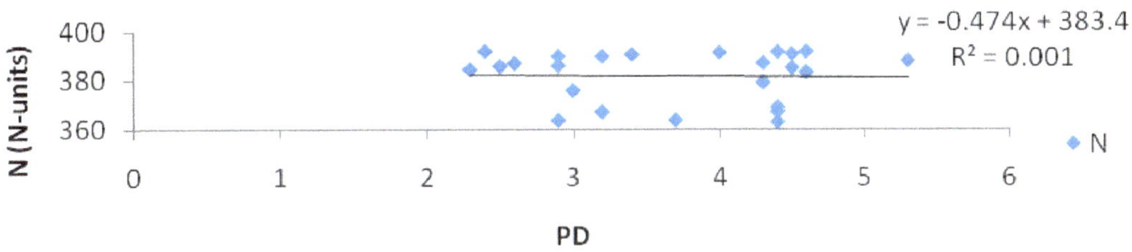

Figure 31. Correlation of PD with radio refractivity for disturbed day during rainy season.

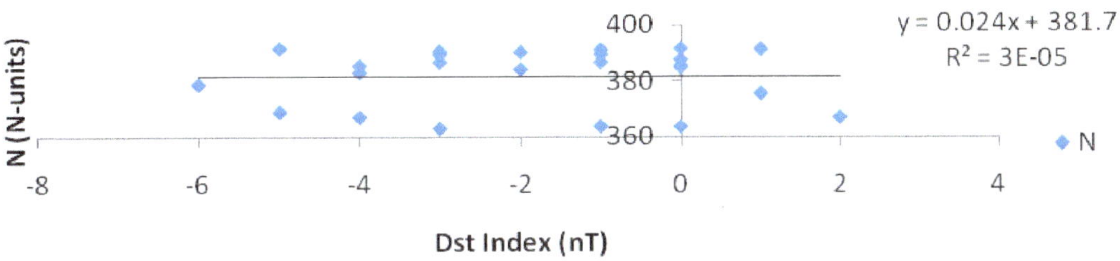

Figure 32. Correlation of Dst index with radio refractivity for disturbed day during rainy season.

during dry season having energies ranging from 10 to 100 kev. The maximum peaks observed in Figures 1 to 8 also signify the period known as "disturbed period"; which is the period when solar wind magnetic field and that of the earth magnetosphere reconnects resulting in the outflow of plasma aurora zones with a high energy into the magnetosphere. During the time of these events, the energy transported from solar wind into the magnetosphere creates geomagnetic variation and drives different types of pulsations such as changes in temperature, pressure, humidity in the atmosphere. Since radio refractivity is a function of temperature, it also changes as the solar wind changes during disturbed period.

As can be observed in Figures 5 to 8 during rainy season, there is a minimum decrease in solar wind activities (solar minimum) as indicated in the trend line of the figures, owing to the weather condition of the atmosphere during the period. This period which is noted by slow solar activity in the atmosphere is dominated by cool and moisture weather condition due to rainy season which leads to decrease in temperature during solar wind magnetosphere coupling within the period. This equally leads to decreased radio refractivity within the period as well for both quiet and disturbed days.

In Figures 9 to 12, the correlation analysis of solar wind energy with radio refractivity was done to estimate the extent at which radio refractivity depend on solar wind activities for both quiet and disturbed days during dry and rainy season. In Figures 9 and 10, the square of correlation coefficients of solar wind energy, with radio refractivity for quiet and disturbed days during dry season are, respectively 0.176 and, 0.0401. Also, in Figures 11 and 12, the square correlation coefficients of solar wind energy with radio refractivity for quiet and disturbed days during rainy season are 0.3198 and 0.0.16, respectively. These results clearly review that the there is weak correlation between solar wind energy and radio refractivity during solar wind-magnetosphere interactions with radio refractivity. This thus shows that solar wind is not the sole cause of fluctuations in radio refractivity, but the fluctuations should be attributed to other magnetospheric interplanetary parameters or local effects.

However, the slope of the trend lines shows that solar wind changes mostly with positive slope in rainy season within most of the times. This may be caused by advection processes, cooling of the earth's surface through radiation and compression of air masses during this period of the year.

Figures 13 to 24 are the hourly variations of interplanetary parameters and radio refractivity with time for both quiet and disturbed days during dry and rainy season. From time variations in the amplitude of interplanetary parameters and radio refractivity periodicity plots, it is observed that almost all the interplanetary parameters follow the same sporadically change in amplitude of their movement as the radio refractivity for both quiet and disturbed days during dry and rainy seasons. But in some cases, while radio refractivity changes with time in positive direction, some of the interplanetary parameters change in negative direction for both quiet and disturbed days during dry and rainy season. These concurrent disparities on the nature of the trend line within different intervals in the movement of interplanetary parameters, as well as radio refractivity, show that certain significant fluctuations on interplanetary parameters are marked by the absence of similar fluctuations on radio refractivity within that period.

The significant effect of the interplanetary parameters on radio refractivity has also been found when the square of the correlation coefficients values were considered. Figures 25 to 32 depict the correlations of interplanetary parameters with radio refractivity for both quiet and disturbed days during dry and rainy season for the year 2008. It is noticed from the figures that the square of the correlation coefficient is quite high for the cases of PD and Dst Index with their square of the correlation coefficients (R^2 = 0.598) and (R^2 = 0.5949) for quiet days during dry season (Figures 25 and 26). Similarly, for disturbed days during dry season, it was observed that only PD exhibits high correlation with square of the correlation coefficient (R^2 = 0.543). From these results, it can be clearly stated that the change in radio refractivity is sorely caused by the change in PD and Dst Index for both quiet and disturbed days during dry season. In contrast, during quiet and disturbed days for rainy season, it was observed that all the interplanetary parameters exhibit very weak correlation with radio refractivity (Figures 27 to 32). Therefore, the change in radio refractivity is not dependent on the change in interplanetary parameters for quiet and disturbed day during rainy season.

Conclusion

This work has clearly shown that radio refractivity has corresponding variations with solar wind energy as well as other magnetospheric interplanetary parameters with time. However, the trend lines of variation of solar wind energy with time, exhibit positive gradient within the period of the research. This may be caused by advection processes, cooling of the earth's surface through radiation and compression of air masses during this period of the year. However, from the result of the correlations, it was observed that solar wind energy exhibits weak correlation with radio refractivity, while magnetospheric parameters exhibit strong correlation with radio refractivity within all the periods of investigation. Having these facts in mind, one can say that, the random and systematic fluctuations, observed during radio wave propagation, emanating from the deviation of radio refractive index of the radio wave in the atmosphere, is due to the influence of magnetospheric parameters interaction with radio wave.

ACKNOWLEDGEMENT

We are grateful to the Centre for Basic Space Science (CBSS), University of Nigeria, Nsukka and OMINIWEB.com for the availability of the hourly means of the megnetospheric interplanetary parameters and parameters.

REFERENCES

Baker DN, Pulkkinen TI, Angelopoulos V, Baumjohann, W, McPherron RL (1996). Neutral line model of substorms: pastand present views, J. Geophys. Res. 101:12975-13010.

Dungey JW (1963). The structure of the exosphere or adventures in velocity space, in Geophysics; The Earth's Environment, Gordon and Breach, New York. pp. 505-550.

Fuselier SA, Klumpar DM, Shelley EG (1991). Ion reflection and transmission during reconnection at the Earth's subsolar magnetopause.

Fuselier SA, Shelley EG, Peterson WK, Lennartsson OW, Collin HL, Drake JF, Ghielmetti AG, Balsiger H, Burch JL, Johnstone A, Rosenbauer H, Steinberg JT (1997). Bifurcated cusp ion signatures: evidence for re-reconnection. Geophys. Res. Lett. 24:1471-4.

Okoro ON, Agbo GA, Ekpe JE, Obiekezie TN (2012). Solar Wind-Magnetosphere Coupling Effect on Radio Refractivity in 2008 for Minna. J. Nat. Sci. Res. 2:76.

Phan TD, Paschmann G, Baumjohann W, Sckopke N, Luhr H (1994). The Magnetosheath region adjacent to the dayside magnetopause: AMPTE/IRM observations. J. Geophys. Res. 99:121.

Rickett BJ (1990). Radio propagation through the turbulent interstellar plasma, Ann. Rev. Astron. Astrophys. 28:561-605.

Russell CT (2000). The solar wind interaction with the Earth's magnetosphere: A tutorial. IEEE Trans. Plasma Sci. 28:1818.

Russell CT, Gang L, Luhmann JG (1998). Lessons from the Ring Current Injection during the September 24-25, 1998 Storm. Geophys. Res. Lett. submitted 2000.

Tsurutani BT, Meng CI (1972). "Interplanetary magnetic field variations and substorm activity." J. Geophys. Res. 77:2964.

Willoughby AA, Aro TO, Owolabi IE (2002). Seasonal Variations of radio refractivity gradients in Nigeria. J. Atmos. Solar-Terrestrial Phys. 64:417-425.

Assessment of arsenic levels in Guguletu and Langa rivers in Cape Town, South Africa

Akinsoji O.[1], Fatoki O. S.[1], Ximba B. J.[1], Opeolu B. O.[2] and Olatunji O. S.[1]

[1]Department of Chemistry, Cape Peninsula University of Technology, Bellville, South Africa.
[2]Department of Environmental and Occupational Studies, Cape Peninsula University of Technology, Cape Town, South Africa.

This study investigates the baseline environmental levels of arsenic (As) in two rivers; one of which traverses through Guguletu (R_1) and the other running by the peripheral of Langa (R_2) residential areas in Cape Town. The rivers were monitored for As in upstream, midstream and downstream stations for twelve months, during consecutive summer and winter seasons between 2011 and 2012. The collected water samples were treated with HCl and KI, and heated in a water bath at 80°C for 30 min, and As was measured using hydride generation atomic absorption spectrometer (HG-AAS). The data obtained were subjected to descriptive statistics, while the Duncan Multiple Range tests was used to compare the mean concentrations of As in the different sampling stations, and between the two rivers. The mean concentration levels of arsenic ranged R_1, 0.98±0.34 to 1.23±0.43 (0.62 – 2.03) µg/L and 0.43±0.02 to 0.44±0.03 (0.08 – 0.68) µg/ℓ. The highest concentration of As (2.03 µg/ℓ) was observed in R_1 river water during summer, and the least concentration 0.08 µg/ℓ in R_2 water during winter. There was significant difference ($P \leq 0.05$) in the mean concentrations of As in R_1 and R_2 rivers water, while the seasonal mean concentrations during winter and summer, was not significantly different ($P \geq 0.05$). The concentrations of As detected in both rivers were lower than the recommended guideline value for freshwater.

Key words: Arsenic, concentration levels, seasonal, Guguletu and Langa rivers, Cape Town.

INTRODUCTION

Arsenic (As) occurs naturally in the environment; however the concentrations of arsenic can be elevated as a result of anthropogenic release, such as from mining and processing activities of variety of mineral ores such as copper, gold, nickel, lead, zinc etc, and during smelting metallurgical operations. Other anthropogenic sources of arsenic are from the use of arsenic containing insecticides and herbicides, combustion of wood treated with arsenic based preservative, seepages from hazardous waste sites, cotton and wool processing operations, and from coal combustion power generation plants, when arsenic contaminated coals are fired (Mondal et al., 2006).

Arsenic is mobilized and or remobilized by natural weathering reactions, biological activity, geochemical reactions, volcanic emissions and anthropogenic redistribution (Mohan and Pittman, 2007). Smedley et al. (2007); Smedley and Kinniburgh (2002) reported high arsenic levels in groundwater derived from zones of gold (Au) mineralization in the lower Proterozoic (Birimian) volcano-sedimentary rocks, where Au occurred along with quartz in its vein structures, and in altered sulphide minerals (pyrite, chalcopyrite, arsenopyrite). Soil erosion and leaching contribute between 6.12 × 1010 and 2.38 × 1011 g/year of arsenic in dissolved and suspended forms, respectively into the rivers and oceans (Mackenzie et al., 1979). Water bodies and drinking water sources may contain low concentration of anthropogenic or naturally occurring arsenic; however the level depends on the nature and source type. The increasing level of arsenic in water systems is of great concern all over the world due to human exposure via use of such water and its attendant effect on human and environment health.

The availability of arsenic in the environment arises from its persistence and ability to accumulate in different environmental compartment. The build-up of arsenic in the environment may result in toxic and deleterious consequences ranging from acute to chronic effects (IPCS, 2001; NRC, 2001; Some et al., 2012). The World Health Organization established that arsenic concentration greater than 10 µg/L in water, has adverse effect on human health (WHO, 2001). Studies revealed that skin disorder linked to the presence of arsenic in drinking water has been documented in many part of the world with the frequency of occurrence of the skin lesions associated with arsenic concentration in tube-well and groundwater (Charlet and Polya, 2006; Some et al., 2012). Elevated levels of arsenic in human blood causes skin disorders such as melanosis, keratosis and more rare skin tumour in contaminated populations (IPCS, 2001; Hashim et al., 2013). Other reported adverse effects of human exposure to arsenic include skin lesions, cardiovascular diseases, liver and kidney dysfunction cancer of the bladder, lungs, skin, kidney, nasal passages, liver, and prostate etc (NRC, 2001; Brinkel et al., 2009; Hashim et al., 2013).

The affinity of arsenic and selenium with gold, copper, nickel, zinc, lead, cobalt, silver and other ores mined in South Africa suggests that trace amount of arsenic may be widespread. Arsenic and selenium, unlike most other toxic metals and metalloids are fairly easily dissolved through a range of pH conditions and redox potential (Eh). This may lead to significant natural dissolution when high concentrations present in rocks and mineral formations are in contact with water. The use of As treated wood, or charcoal produced from the charring of As treated wood for domestic and commercial cooking, as well as other purposes such as braying, roasting, grilling and smoking in many homes and settlements in Cape Town and many part of South Africa, forms a major source of anthropogenic release and exposure of As. Information concerning the prevalence and distribution of arsenic in urban and rural rivers in Cape Town and its environs is scarce and scanty. There is need for the evaluation, characterization and prioritization of arsenic in the environment, in order to acquire basic data and critical information that may be required for developing policies, and making informed decision on environmental protection, management of arsenic contamination, as well as developing rational approach to mitigate the release of arsenic.

In this study, a systematic assessment on the sources, pathways and levels of arsenic in metropolitan streams and rivers traversing the Cape Town environment was investigated, in order to define the prevalence in Cape Town environment, and generate baseline data useful for monitoring purposes.

MATERIALS AND METHODS

Study area

Cape Town, the administrative seat of the Government of the Western Cape is located on the Western coast of South Africa geo-referenced on the Southern latitude S 33o and Eastern longitude E 18o. Outside the Cape Town city centre are a number of Peri-urban communities, settlements, and camps, many of which are populated beyond carriage capacity of the available residential. This meant the possibility of unsustainability use of available resources with high potential of negative consequences on the environment. The study covered stretch of major metropolitan rivers traversing through alternates of farmland and high density residential areas at Guguletu, (R_1) and offset areas of Langa (R_2) along Vanguard Road (Figure 1), which delivers urban sewerage into the central waste water treatment plant at Athlone, Cape Town environs.

The river flowing through Langa has its origin at the Southern area between Eppings Industrial Layout 1 and 2, traversing by the outskirt of the community. Industrial waste water released from the industries in Eppings 1 and 2 contributes significantly to the flow volume of the stream, and flow by the Athlone waste water treatment facility, releasing its content into Swart River which in-turn discharge it content in Atlantic Ocean in the North of the Peninsula. The river traversing Guguletu on the other hand, has it sourced from industrial waste water receiving pond at Belhar, traversing through the community, and with high human activities taking place around it. The Guguletu flow channels into Lotus River, which in-turn release it content into the Atlantic Ocean by the South of the Peninsula. Due to the high density residential through which both rivers flows, large quantity of domestic waste water, surface runoff and solid waste are added to the flowing waters.

At each of the rivers (Figure 1), water samples were collected at three different points marked as upstream, midstream and downstream. The geo-reference GPS co-ordinates for the Langa stream are; upstream (LS1) S 33° 56.698' E 18° 31.453', midstream (LS2) S 33° 56.994' E 18° 31.453 and downstream (LS3) S 33° 56.994' E 18° 30.500'. The GPS coordinates for Guguletu are upstream (GS1) S 33° 59.176′, E 18° 34.263′, midstream (GS2) S 33° 59.252′, E 18° 34.212′, downstream (GS3) S 33° 59.576′, and E 18° 34.055′.

Sample collection

Water samples were collected in pre-cleaned polyethylene containers. The containers were cleaned by soaking in a solution of 10% nitric acid and rinsed with deionized water. Samples were collected in triplicate from each sampling station along the rivers. Determination of physical and chemical properties of the water samples pH, conductivity, salinity and total dissolve solid (TDS).

The pH and conductivity of the water samples were determined in-situ using pre-calibrated Instrument (Hanna HI 4522), while water turbidity were measured with a pre-calibrated Turbidimeter 2100P. The salinity and total dissolve solid (TDS) of the water samples were determined according to methods prescribed by American Public Health Association (APHA, 1998). Recovery studies was

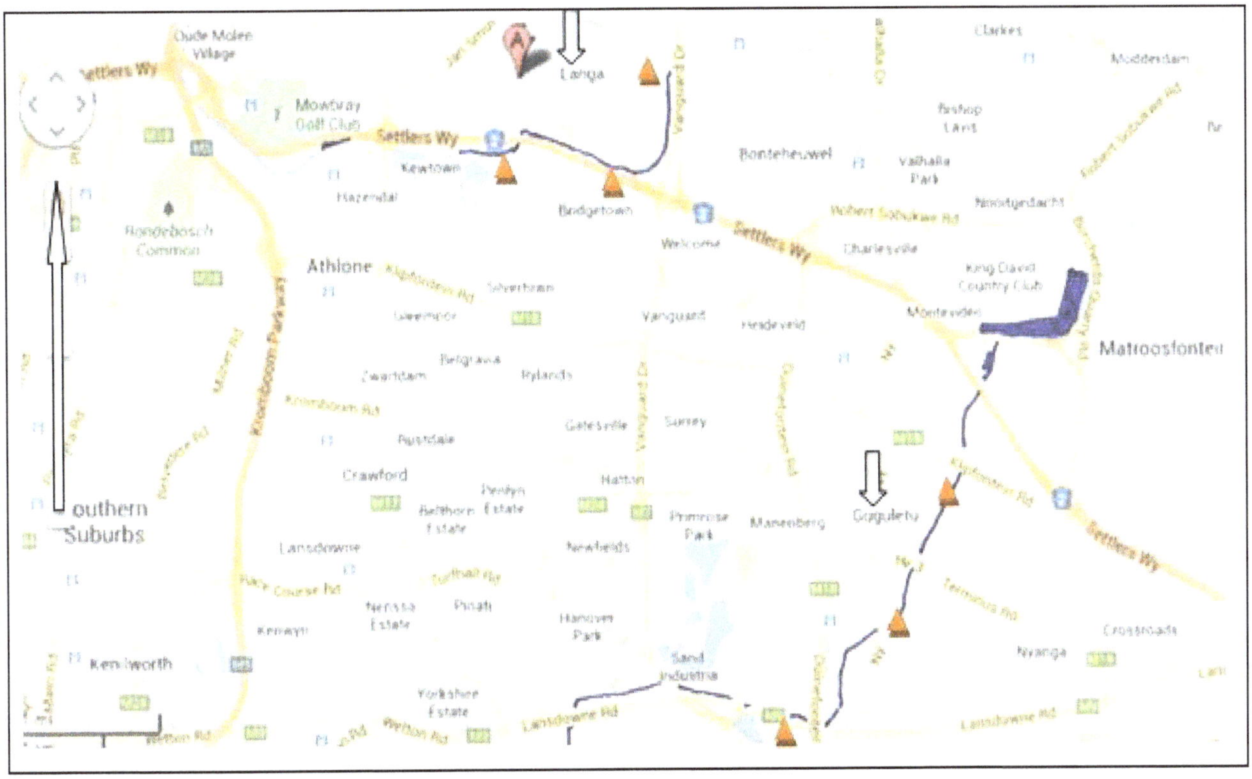

Figure 1. Map of Cape Town showing Guguletu and Langa study areas (Source: adapted from Google map). Legend: ▲ - Sampling Station; ⇩ - Study location.

also conducted on replicates (n=5) of spiked water samples and deionized water (n=5) to validate the efficiency of recovery of the method.

Sample preparation and estimation

Water samples were treated by the addition of 1 ml of 5 M HCl and 1 ml of 20% KI to a 10 mℓ water sample in test tubes and heated in a water bath at 80°C for 30 min (Wahed et al., 2006). The treated water were filtered into 50 ml volumetric flasks and made up to mark with deionized water. Procedural blank was prepared using the same method (Wahed et al., 2006), and aspirated to correct for background absorption. The concentrations of As were determined in the water samples using a Shimadzu atomic absorption spectrophotometer (AA–6300), coupled with a Hydride Vapour generator HVG-1 (HG- AAS) in stream of argon gas, at 0.32±0.02 MPA and a flow rate of 70 ml/min.

Recovery studies

Water samples and deionized water were spiked with different concentrations of As in 5- replicates. The spiked water samples and deionized water were subjected to sample preparation procedures, followed by the estimation of As in each of them.

RESULTS

The analysis of replicates spiked water samples and deionized water (n=5) showed good recovery between 100.3 to 104.4% and relative standard deviation of 3.21 to 12.42%. The percentage recoveries and relative standard deviation for arsenic (As) were 100.3 and 3.21% for spiked water; 104.4 and 12.42% for deionized water. The percentage recovery of As falls within the generally acceptable recovery 100±20%.

The mean concentration level of As in the water samples collected upstream (GS1, LS1), midstream (GS2, LS2) and downstream (GS3, LS3) at Guguletu (R_1) and Langa (R_2) were ranged 0.98±0.34 – 1.23±0.43 (0.62 – 2.03) µg/ℓ and 0.43±0.0.15 - 0.44±0.03 (0.08 – 0.68) µg/ℓ, respectively (Table 1). The highest concentration of As, 2.03 µg/L was observed in water collected from R_1 during summer, while the least concentration 0.08 µg/ℓ was measured in water collected from R_2 during winter.

The pH of R_1 water ranged between 6.7 upstream during winter and 8.4 midstream, during summer, while it ranged between 6.0 midstream during winter and 8.8 midstream during summer in R_2 river water. However, the mean pH of water measured at the different sampling stations along the rivers at R_1 and R_2 were 7.53±0.43 to 7.77±0.66 and 7.47±0.42 to 8.07±0.53, respectively (Table 1). The highest conductivity 1.67 mS/cm was measured in upstream water at R_1 during winter, with the least conductivity 0.49 mS/cm in downstream water at R_2 during winter as well. The mean conductivity of

Table 1. Result of selected physico-chemical properties and mean concentration (ug/l) of arsenic in water samples from the different sampling station along river course.

Sampling station	pH	Conductivity (mS/cm)	Turbidity (NTU)	Total dissolve solid (mg/L)	As (µg/L)
Guguletu					
Upstream	7.58±0.60	1.23±0.33	23.74±28.93	595.35±169.02	1.15±0.39
Midstream	7.58±0.12	1.06±0.50	44.33±33.99	530.27±248.86	0.87±0.18
Downstream	7.63±0.06	0.71±0.16	14.00±0.00	306.37±17.77	0.69±0.05
Overall mean	7.59±0.49	1.13±0.40	27.23±29.42	548.78±197.42	1.04±0.36
Summer	7.77±0.66	1.37±0.17	5.93±5.57	682.83±88.37	1.23±0.43
Winter	7.53±0.43	1.06±0.42	33.32±30.70	510.48±204.41	0.98±0.34
Mean	7.59±0.49	1.13±0.39	27.23±29.42	548.78±197.42	1.04±0.37
Langa					
Upstream	7.78±0.04	0.82±0.34	19.75±0.42	409.07±169.98	0.54±0.08
Midstream	7.59±0.62	0.95±0.26	17.29±14.30	471.98±132.45	0.47±0.10
Downstream	7.38±0.12	0.75±0.18	13.99±5.97	376.87±92.41	0.25±0.13
Overall mean	7.59±0.50	0.88±0.27	17.12±11.38	440.37±135.63	0.43±0.14
Summer	8.07±0.53	0.97±0.07	6.29±3.78	482.87±35.60	0.44±0.03
Winter	7.47±0.42	0.86±0.30	19.83±11.04	429.75±149.43	0.43±0.15
Mean	7.59±0.50	0.88±0.27	17.12±11.37	440.37±135.63	0.43±0.14

the water collected from the rivers at the different sampling stations ranged 1.06±0.42 mS/cm during winter to 1.37±0.17 mS/cm (Table 1) during summer at R_1, while that at R_2 ranged from 0.86±0.30 mS/cm to 0.97±0.07 mS/cm during winter and summer seasons, respectively.

The mean turbidity of the water samples were 5.93±5.57 - 33.32±30.70 NTU and 6.29±3.78 - 19.83±11.04 NTU in R_1 and R_2, during summer and winter seasons, respectively (Table 1). The lowest and highest water turbidity value of 0.37 NTU and 89.2 NTU were detected upstream at R_1 during summer and winter, respectively. The mean total dissolved solids (TDS) in water samples collected at the different sampling stations at R_1 and R_2 ranged 510.48±204.41 to 682.83±88.37 ppm and 429.75±149.43 to 482.87±35.60 ppm, respectively (Table 1). The lowest and highest TDS, 277.4 and 836.6 ppm were measured midstream at R_2 and upstream at R_1, respectively during winter. Generally, TDS detected in R_2 stream water ranged 277.4 - 779.5 ppm midstream, while TDS ranged 288.7 - 836.9 ppm in R_1 river.

DISCUSSION

The concentration of As varied between sampling stations in the upstream, midstream and downstream water courses; 0.62 - 2.03 µg/ℓ in the river traversing Guguletu (R_1) and 0.08 - 0.68 µg/ℓ in river running the peripherals of Langa (R_2). The variation in concentration levels of As depends on the physico-chemical properties of water, exposure of the rivers to either natural or anthropogenic source inputs and the chemical form of As in the rivers. In water, the concentration of As depends on the prevalence of AsIII oxide (As_2O_3), however the predominating form of As is a function of the water pH and redox conditions. For instance AsV oxide (HAs_2O_2-) is prevalent in water at circum - neutral pHs. In oxidizing or acidic water environment, As_2O_3

tends to form hydrated oxide (or arsenious acid), thereby decreasing the free and measurable As (As_2O_3) concentration. The concentration of As will therefore increase until at high pH, or strongly reducing water environment when ortho-arsenite (AsO33-) and meta-arsenite AsO_2- (that is, arsenites which are readily oxidized to arsenates) are formed from the reduction of As_2O_3 (Welch et al., 2000; Smedley and Kinniburgh, 2002; Charlet and Polya, 2006). There were significant differences (P≤0.05) in the levels of As detected in water samples collected at the different sampling stations in R_1 and R_2. This may be as a result of differences in volume and quantity of municipal and industrial runoff discharged into the main stream course, from adjoining settlements and industrial layouts at different intersections. According to EPA (1998), municipal waste water systems and septic tank in unsewered areas can contributes significant quantities of metals especially arsenic, mercury, chromium, lead, iron, manganese and biodegradable organic carbon to streams and rivers into which they empty their content. Exposure of surface water (especially those with sources from rocks/springs) or groundwater to geological sources rocks containing As minerals, and the nature of the hydrogeology of the water along the stream flow course as well as environmental factors can also exacerbate As levels. There may also be geological or anthropogenic hot spots, where higher levels of arsenic may occur along the river course (UNICEF, 2008). Human population and the food chain can inadvertently be exposed by drinking or using As-contaminated water (NRDC, 2000).

The concentrations of As detected in R_1 and R_2 waters were however below the 10 µg/L (0.010 ppm) maximum contaminant guideline level (MCL) set by World Health Organization (2001), however chronic exposure to lower levels of As is linked to adverse health effects in human populations. For instance, waters from tubes wells in some countries such as Bangladesh, Cambodia, Vietnam, Argentina, Chile, USA, Canada, Japan etc were reported to contain As at concentration levels that poses unacceptable risks (Hutton, 1987; Hashim et al., 2013). Although, As levels in the river systems were not detected at concentrations above the MCL, a maximum contaminant level goal (MCLG) of 0 ppm in water is recommended in order to avoid these adverse consequences (EPA, 1989). The concentration of As in R_1 water were higher compared to R_2 water, and this may be attributed to the proximal of R_1 river flow course to the Braai Centre at Guguletu, from where ash produced from open commercial meat roasting are washed into the main stream drainage, as well as other human activities such as the direct loading of combustion ash and other wastes into river.

The availability and prevalence of metals and many toxic substances in water systems, depends on the exposure of the water body to metals/toxic substances sources, the level they are exposed to, and the physico-chemical properties (pH, dissolved solids (TDS), turbidity conditions) of the water. The concentration of As varied with the pH of the river water at the different sampling stations (6.7 - 8.4 and 6.0 - 8.8 at R_1 and R_2, respectively), with higher As concentration observed as pH increased. Welch et al. (2000) reported that the concentration of As_2O_3 increases with increase in pH, and appeared to be constant at near neutral pH. The pH detected in the different sampling stations along the course of each of the rivers traversing Guguletu and Langa were not significantly different (P≥0.05). The observed conductivities of the waters in the different sampling stations in R_1 and R_2 at Guguletu and Langa were also not significantly different (P≥0.05). The ANZECC (2000) guideline (recommends that water pH should be within the range of 6.5 - 8.0 in healthy freshwater fit for sustenance of aquatic life and human health, while the South African Department of Water Affairs and Forestry and WHO water guideline standards (DWF, 1996) suggests a pH of range 6.0 – 9.0 for water used for domestic purposes.

Although, no definite relationship was established between pH and TDS (510.48±204.41 to 682.83±88.37 mg/l and 429.75±149.43 to 482.87±35.60 mg/l, in R_1 and R_2, respectively), results showed that the concentration of TDS increases with decrease in water pH. The conductivity of stream water (1.06±0.42 to 1.37±0.17 mS/cm and 0.86±0.30 to 0.97±0.07 mS/cm in R_1 and R_2, respectively) also increases with increase in TDS. The concentration of As also increases with increase in conductivity of the river water (with correlation coefficient ɣ=0.73 and ɣ=0.49 at R_1 and R_2, respectively). The conductivity of the water collected from all the sampling stations in R_1 and R_2 at Guguletu and Langa were higher than the 0.5 mS/cm, WHO recommended guideline limit for conductivity in water (Agbalagba et al., 2011).

Aside from insoluble substrates and particulates which may remain suspended in water, the turbidity of water is a function of the dynamics of water saturation and its capacity for dissolved solids. Guguletu river water showed higher turbidity (5.93±5.57 to 33.32±30.70 NTU) than R_2 at Langa (6.29±3.78 to 19.83±11.04 NTU). There is evidence of good correlation between water turbidity and TDS with correlation coefficient ɣ=0.68 and ɣ=0.54 for Guguletu and Langa river waters, respectively. This implies that turbid waters have high propensity for soluble metals than highly turbid water. On the other hand, it was reported that highly turbid waters tend to undergo self-purification by depositing excess particulate load in siltation and sedimentation process or as tidal deposits in flowing waters.

The turbidity of the water upstream, midstream and downstream were not significantly different (P≥0.05), although midstream turbidity was slightly higher probably because of human activities taking place about midstream. This study result is consistent with the findings of Amraoui et al. (2003) on turbidity dynamics in

Karstic systems of Ribaa and Bittit springs in the Middle Atlas (Morocco), Rowe et al. (2003), and Engstrom-Ost and Candolin (2006). Apart from a few samples (7.35%) where the turbidity of the water were less than 5 NTU during summer, all samples had turbidity values higher than the acceptable level, an indication of poor freshwater quality. The turbidity of the water in the two rivers were however not significantly different ($P \geq 0.05$).

The concentrations of arsenic were noted to be higher during summer than in winter. The low concentration of arsenic observed during winter may due to dilution, although this (dilution) may not totally account for the seasonal variation due to the likelihood of other arsenic reactions. The physico-chemical properties of the streams also varied with season. The pH, conductivity and TDS of the streams were higher during summer (dry season) than in during winter (wet season) while water turbidity was lower during summer (dry season). This is consistent with the findings of Ezekiel et al. (2011) on seasonal effect of pH condition of water measured in Sombreiro River. The conductivity, TDS, and salinity measured in water samples collected from Guguletu were also significantly ($P \leq 0.05$) higher than in water collected from Langa during both seasons.

The changing physico-chemical properties, such as TDS, conductivity, pH and redox potential, as well as flow volume and seasonal variations may probably explain the slight variation in the detected level As between different sampling stations on the rivers. The form and characteristics of occurrence in the water also have an effect on the available concentration levels. Thus, constant and consistent monitoring of water systems is desirable.

Conclusion

The concentration of As in the river waters were not appreciably high and within background levels. There were slight concentration levels variations in spatial distribution of arsenic in both river waters. Arsenic concentration levels in water samples were lower than the guideline minimum standard limit set by WHO, hence are below compliance level. The observed levels may serve as baseline data needed for environmental prioritization of arsenic prevalence in surface waters, since the observed concentration levels are not at a critical health thresholds limit. There is also need to develop national benchmark level for environmental monitoring against the buildup of As in streams and rivers in order to forestall exposure and adverse health effects in human populations.

REFERENCES

Agbalagba OE, Agbalagba OH, Ononugbo CP, Alao AA (2011). Investigation into the physico-chemical properties and hydrochemical processes of groundwater from commercial boreholes In Yenagoa, Bayelsa State, Nigeria. Afr. J. Environ. Sci. Technol. 5(7):473-481.

Amraoui F, Razack M, Bouchaou L (2003). Turbidity dynamics in Karstic systems: Example of Ribaa and Bittit springs in the Middle Atlas (Morocco). Hydrolog. Sci. J. 48(6):971-984.

ANZECC (2000). National Water Quality Management strategy:An introduction to the Australian and New Zealand guidelines for fresh and marine water quality. National water quality management strategy; No. 4a.

APHA (1998). Standard methods for the Examination of Water and Wastewater. 1268p 20th Ed. American Public Health, Association (APHA), American Water Works Association (AWWA), Water Pollution Control Federation (WPCF), Washington DC, USA.

Brinkel J, Khan MMH, Kraemer A (2009). A systematic review of arsenic exposure and its social and mental health effects with special reference to Bangladesh. Int. J. Environ. Res. Publ. Health 6:1609-1619.

Charlet L, Polya DA (2006). Arsenic in shallow, reducing ground waters in Southern Asia:an environmental health disaster. Elements 2:91-96.

DWF (Department of Water Affairs and Forestry) (1996). South African Water Quality Guidelines (second edition). Volume 1: Domestic Use.

Engstrom-Ost J, Candolin U (2006). Human-induced water turbidity alters selection on sexual displays in sticklebacks. Behavioral Ecology. Adv. Access Publ. pp. 393-398.

EPA (1989). Risk assessment guidance for superfund, Vol. 1: Human health evaluation manual Part A., interim final. Office of Emergency and Remedial Response, US Environmental Protection Agency, Washinghton DC, EPA/540/1-89/002.

EPA (1998). U.S. Environmental Protection Agency:Health Effects Test Guidelines, OPPTS 870.3250, 90-Day Dermal Toxicity.

Ezekiel EN, Hart AI, Abowei JFN (2011). The physical and chemical condition of Sombreiro River, Niger Delta, Nigeria. Res. J. Environ. Earth Sci. 3(4):327-340.

Hashim JH, Radzi RSM, Aljunid SM, Nur AM, Ismail A, Baguma D, Sthiannopkao S, Phan K, Wong MH, Sao V, Yasin MSM (2013). Hair arsenic levels and prevalence of arsenicosis in three Cambodian provinces. Sci. Tot. Environ. Article in press.

Hutton M (1987). Human health concerns of lead, mercury, cadmium and arsenic: In lead mercury, cadmium and arsenic in the environment, edited by TC Hutchinson and K Meema, Scope published by John Wiley and Sons Ltd.

IPCS (International Programme on Chemical Safety) (2001). Environmental Health Criteria 224: Arsenic and arsenic compounds. Geneva: World Health Organization.

Mackenzie ET, Lamtzy RJ, Petorson V (1979). Global trace metals cycles and predictions. J. Int. Assoc. Math. Geol. 6:99-142.

Mohan D, Pittman CU Jr. (2007). Arsenic removal from water/wastewater using adsorbents - A critical review. J. Hazard. Mat. 142:1-53.

Mondal P, Majumder CB, Mohanty B (2006). Laboratory based approaches for arsenic remediation from contaminated water:recent developments. J. Hazard. Mat. B. 137:464-479.

NRC (National Research Council) (2001). Arsenic in drinking water 2001 update. Washington DC; National Academy Press.

Rowe M, Essig D, Jessup B (2003). Guide to selection of sediment targets for use in Idaho TMDLs USEPA.

Smedley PL, Kinniburgh DG (2002). A Review of the source, behaviour and distribution of arsenic in natural waters. Appl. Geochem. 17:517-568.

Smedley PL, Knudsen J, Maiga D (2007). Arsenic in groundwater from mineralized proterozoic basement rocks of Burkina Faso. Appl. Geochem. 22:1074-1092.

Some IT, Sakira AK, Ouedraogo M, Ouedraogo TZ, Traore A, Sondo B, Guissou PI (2012). Arsenic levels in tube-wells water, food residents' urine and the prevalence of skin lesions in Yatenga province, Burkina Faso. Interdiscipl. Toxicol. 5(1):38-41.

UNICEF (2008). Arsenic Primer Guidance for UNICEF Country Offices on the Investigation and Mitigation of Arsenic Contamination. Water, Environment and Sanitation Section Programme Division UNICEF New York.

Wahed MA, Chowdhury D, Nermell B, Khan SI, Ilias M, Rahman M, Persson LA, Vahter M (2006). A modified routine analysis of arsenic content in drinking-water in Bangladesh by hydride generation-atomic absorption spectrophotometry. J. Health Popul. Nutr. 24(1):36-41.

Welch AH, Westjohn DB, Helsel DR, Wanty RB (2000). Arsenic in ground water of the United States: Occurrence and Geochemistry. Ground water 38(4):589-604.

World Health Organisation (2001). Draft: Abernathy C. and edited by Ann Morgan. Exposure and Health Effects. Office of Water, Office of Science and Technology, Health and Ecological Criteria Division, United States Environmental Protection Agency, Washington, DC, USA. Accessed; www.who.int/water_sanitation_health/dwq/arsenicun3.pdf on 10/09/2012.

4

Investigation of carbon monoxide concentration from anthropogenic sources in Lagos, Nigeria

Akinyemi M. L. and Usikalu M. R.

Department of Physics, Covenant University, P. M. B. 1023, Ota, Ogun State, Nigeria.

The work investigated the carbon monoxide (CO) concentration from anthropogenic sources in eight, different Local Council Development Authority (LCDA) in Lagos State Nigeria. The CO concentration obtained as a result of movement of cars, trucks, motorcycles, working of generators, commercial and residential activities etc., was measured using the digital carbon monoxide detector (model: DC01001) manufactured by General Tools and Instrument, New York. It was found that trucks emitted highest concentration of CO (289.64 ppm) followed by generator operated on diesel (116.23 ppm) while cooking with firewood emitted the lowest concentration of CO (5.75 ppm). Carbon monoxide concentration from residential and commercial areas, are moderate and considered safe for those in the area. On the other hand CO concentration from trucks is very high based on the US air quality index (AQI) and therefore considered unhealthy for the people. It is obvious from this study that CO concentration is generally associated with the use of fossil fuel which suggest that the less we rely on fossil fuel, the more healthy the environment.

Key words: Carbon monoxide, concentration, anthropogenic, fossil fuel, environment.

INTRODUCTION

There has been growing concern about environmental pollution problem all over the world. The people well-being is solely the mirror of a healthy environment. Air pollution can be defined as the release of substances into the air, some of which can cause problems for humans, plants, and animals. These include dust, gases, smog, acid rain, the greenhouse effect, and depletion in the ozone layer. Each of these problems has serious implications for our health and well-being as well as for the whole environment. The air pollution emanating from the anthropogenic sources has been reported to have damaging effects on the health of the exposed populace (Erhabor et al., 1992; Wanner, 1990) and in cases of extreme events, it can cause increased death rate among the exposed group.

The sources of air pollution can be grouped into 2: sources and man-made (anthropogenic) sources. Natural processes that affect air quality include volcanoes, which produce sulfur, chlorine, and ash particulates. Wildfires produce smoke and carbon monoxide. Cattle and other animals emit methane as part of their digestive process. Many forms of air pollution are man-made. Industrial plants, power plants and vehicles with internal combustion engines produce nitrogen oxides, VOCs, carbon monoxide, sulfur dioxide and particulates. In most mega cities, vehicles are the main source of these pollutants. Stoves, incinerators, and farmers burning their crop waste produce carbon monoxide, carbon dioxide, as well as particulates in rural areas. Other man-made sources include aerosol sprays and leaking refrigerators as well as fumes from paint, varnish, and other solvents (USEPA, 2009). Pollutants can cause respiratory disease, acute intoxication and adverse reactions in sensitive people. Symptoms such as breathing difficulties, eye irritation, skin rashes and intestinal upsets appear immediately after exposure. Other chronic effects occur only after long exposures. These are often difficult to predict, and are caused by the slow build-up of chemicals in the body or the gradual accumulation that damages the human tissues. One of the major pollutants that humans are exposed to is Carbon monoxide (CO). It is a colorless, odorless, non-irritant and tasteless poisonous gas. It is slightly lighter than air. It is highly toxic to humans and animals at high concentrations. The gas consists of one carbon atom and one oxygen atom connected by a triple bond which consists of two covalent bonds as well as one dative covalent bond. Carbon monoxide is formed when there is not enough oxygen to produce carbon dioxide (CO_2), such as when operating a stove or an internal combustion engine in an enclosed space. In the presence of oxygen, carbon monoxide burns with a blue flame, producing carbon dioxide. Carbon monoxide is produced by the incomplete combustion of solid, liquid or gaseous fuels. Appliances fueled with gas, oil, kerosene, wood or coal used in boilers, engines, oil burners, gas fires, water heaters, solid fuel appliances and open fires may produce CO. All fuel burning appliances have the potential to produce CO in varying concentrations. Biologically, Carbon monoxide poisoning is the most common type of fatal air poisoning in many countries (Walker and Hay, 1999). It combines with hemoglobin to produce carboxyhemoglobin (COHb) which is ineffective for delivering oxygen to bodily tissues termed anoxemia (Fairburn et al., 1992; Omaye, 2002). It is easily absorbed through the lungs (Ernst and Zibrak, 1998). Inhaling even relatively small amounts of the gas by humans can lead to hypoxic injury, neurological damage, and even death.

Different people and populations may have a different carbon monoxide tolerance level (Thom et al., 2000). On average, exposures at 100 ppm or greater is dangerous to human health (Chichkova and Prockop, 2007). Carbon monoxide exposure may lead to a significantly shorter life span due to heart damage (Henry et al., 2006). The carbon monoxide tolerance level for any person is altered by several factors, including activity level, rate of ventilation, a pre-existing cerebral or cardiovascular disease, cardiac output, anemia, sickle cell disease and other hematological disorders (Lipman, 2006). Symptoms such as delirium and hallucinations have led people suffering poisoning to think they have seen ghosts or to believe their house is haunted (Donnay, 2004). Carbon monoxide poisoning in pregnant women may cause severe adverse fetal effects.

Based on the different contamination due to emission from both natural and man-made sources, it is therefore important to have knowledge of the air quality in a particular environment. Air quality is an indication of the healthfulness of the air based on the quantity of polluting gases and particulates (liquid droplets or tiny solid particles suspended in air) it contains. Air is considered safe when it contains no harmful chemicals and only low levels of other chemicals that become harmful in higher concentrations to humans, animals, plants, or the ecosystems (Encarta, 2009). The aim of this work is to investigate the CO concentration in the selected LCDA in Lagos State.

MATERIALS AND METHODS

The CO concentration was detected and measured with the aid of a digital carbon monoxide detector (model: DC01001) manufactured by general tools and Instrument, New York. The equipment is designed for CO detection in the range of 0 and 999 ppm, with 1 ppm resolution and accuracy of ±20% at 0 to 100 ppm; ±15% at 100 to 500 ppm; (at 20 ± 5°C, 50 ± 20% RH). Eight (8) different local council development authority (LCDA) in Lagos, Nigeria were selected for the study. The choice of the location was based on the accessibility, activities peculiar to the location and congestion in the areas. Lagos was chosen because of its status as the economic capital of Nigeria that houses a lot of central business districts. The investigated areas are Ikeja, Festac, Ikoyi, Gbagada, Apapa, Kosofe, Surelere and Isolo which can be grouped into industrial area, commercial and residential areas according to the activities predominant in such vicinity. 5 different measurements were taken for 2 h in each location in 1 LCDA and average values were taken. The CO concentration from cars, trucks, motor cycles, petrol generators, diesel generators and household kitchens and firewood cooking areas were considered in each location the same numbers of point source were considered.

RESULTS AND DISCUSSION

Figure 1 displays the mean values of the CO concentrations emitted as a result of car movements in each LCDA over a period of 2 h. The highest CO concentration was obtained in Kosofe (76.7 ppm) which can be traceable to the fact that there is high vehicular movement in this area since this is where Ojota is located. The lowest CO concentration was obtained in Isolo (32.5 ppm) this may be associated with low vehicular movement in the area during this period. According to the US AQI limit, it is only Isolo that has good healthy concern as related to CO concentration while all other locations considered can be said to have moderate healthy concern to CO concentration. Figure 2 presents the mean values of CO concentrations from motorcycles and generators that are operated with petrol and diesel. The results showed that generators operated with diesel emit more carbon monoxide than those operated with petrol. The highest CO concentration (117.36 ppm) in this group was obtained from generator operated with diesel. It was also observed that though the locations are different there is no significant difference between the CO emitted from the same set of equipment considered elsewhere. There is also no significant difference between the CO values obtained in petrol generators and motorcycles in all the locations considered. This may be because the 2

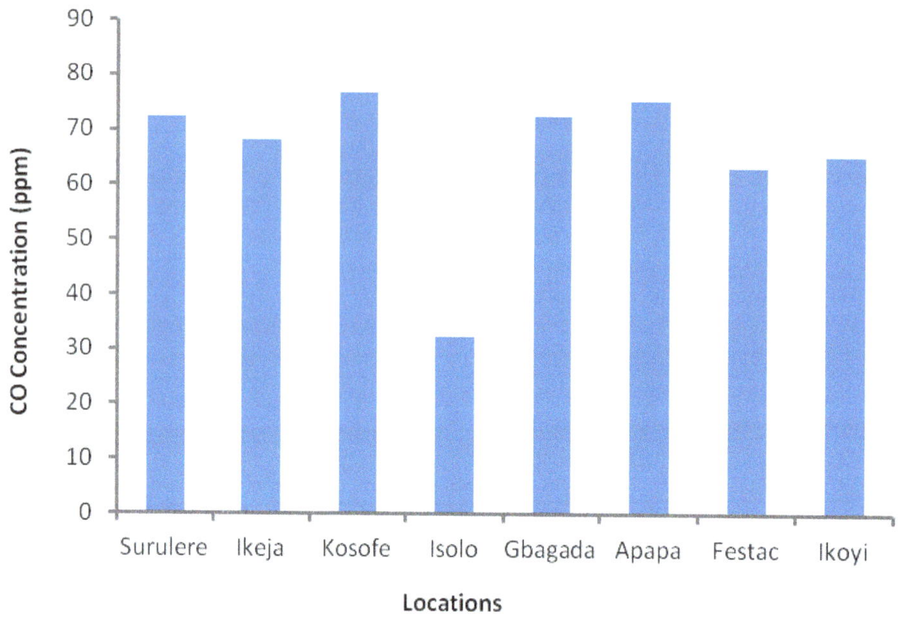

Figure 1. Mean Values of CO concentration from cars.

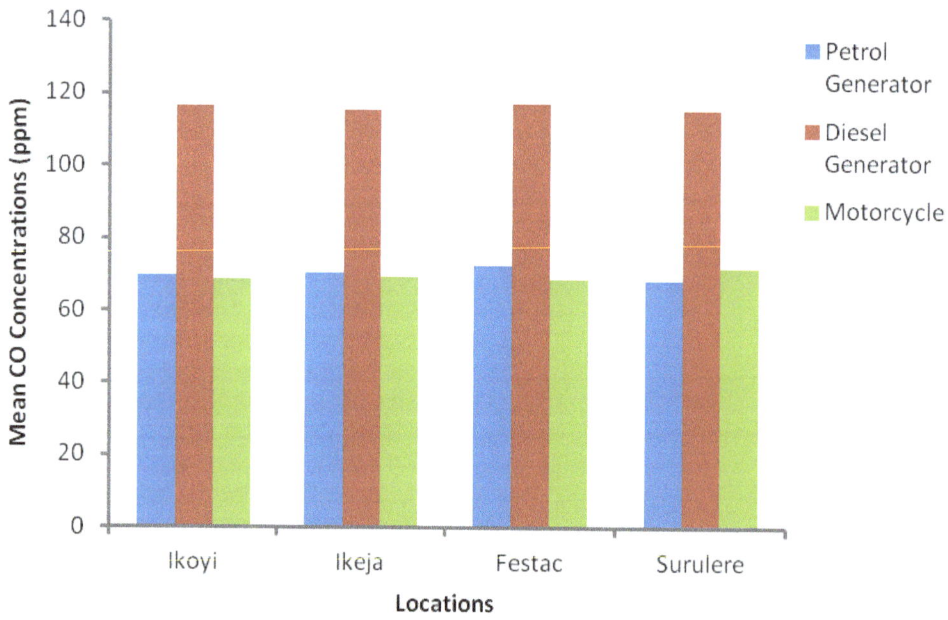

Figure 2. Mean CO Concentrations from Generators (Petrol and Diesel) and Motorcycles

equipments used petrol. Motorcycle and generator operated with petrol can be categorized as moderate healthy concern with CO concentration while those with diesel can be said to be unhealthy for sensitive people based on the US AQI limit.

Figure 3 is a bar chart representing the mean values of carbon monoxide concentration from different sources in the environment. It is observed that carbon monoxide concentration from commercial areas, residential areas and firewood cooking are relatively small compared to equipment operating on fossil fuels (that is, petrol generators, diesel generators, cars, motorcycles and trucks). This study revealed that trucks produce the highest carbon monoxide (289.64 ppm) in comparison to the other sources. The reason may be due to the type of combustion taking place in their engine as well as the fuel used (diesel). It is also observed that all equipment operated on diesel (generator and truck) emit more CO

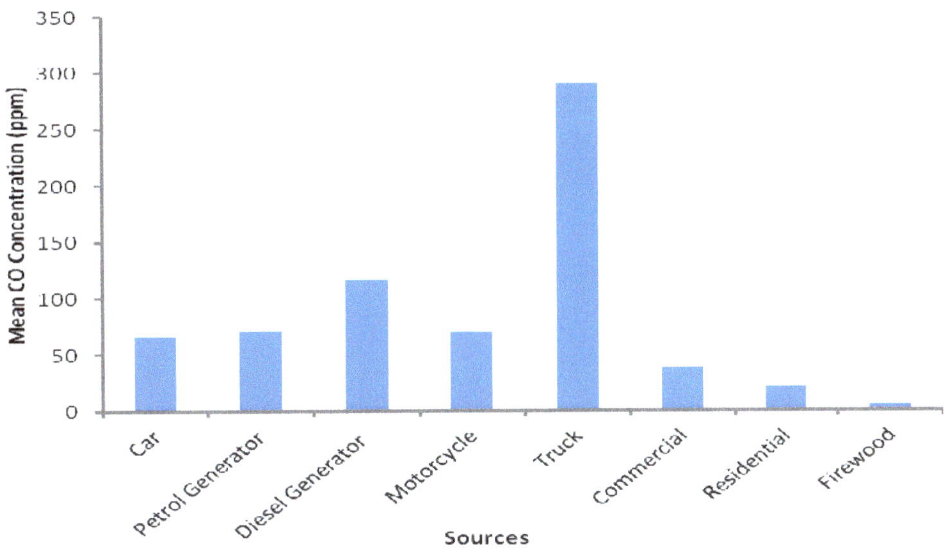

Figure 3. Variation of mean CO concentrations from all sources.

than those operated on petrol which therefore suggest that diesel combustion engine is a higher emitter of carbon monoxide than petrol. The CO concentration from the commercial area (38.17 ppm) is moderate and is within the specified healthy limit, though high when compared with that from domestic sources. This is due to many commercial activities (markets, grinding pepper, smoking, driving, wood burning, welding etc) taking place at that particular time. The source of CO from the residential area is majorly the kitchen using cooker, fridge and some household appliances. It was observed that the CO concentration from this group is within the healthy limit as it is low compared to concentration from commercial area. The CO concentration was also observed to be activity dependent, that is, it depends on what is going on in the area and does not vary with location but with what is released into the atmosphere from the different activities going on in the designated location. The area where firewood is used has the lowest CO concentration (5.75 ppm) than any other sources considered. This is because there is availability of abundant air (oxygen) for complete combustion with firewood thereby eliminating the production of carbon monoxide; a product of incomplete combustion. Chichkova and Prockop (2007) suggested that exposures to CO concentration at 100 ppm or greater are very dangerous to human health. Therefore, concentration of carbon monoxide from diesel engine and truck may be considered unhealthy to human beings and should be avoided.

CONCLUSION AND RECOMMENDATION

It has been established from this study that concentration of carbon monoxide can be obtained from anthropogenic sources. It was revealed that the equipment operated on diesel has highest concentration of carbon monoxide, followed by those operated on petrol. The major source of CO concentration observed from sources considered was from trucks and generators. It is obvious from this study that CO concentration is generally associated with the use of fossil fuel which suggests that the less we rely on fossil fuel, the more healthy the environment. The study also revealed that the more the activities in an area the higher the CO concentration from such location. Based on the results of this study, we are recommending that Government should make policy that restrict the time of movement (whether night or early in the morning) for trucks due to their high level of carbon monoxide concentration.

There is need for the government to encourage the exploitation of other sources of energy that are renewable in place of fossil fuel.

There should be frequent seminars and conferences on carbon monoxide for proper awareness of risks associated with it. It is good if each house (powered by generator) to have carbon detectors that will trigger an alarm when the CO safe limit is exceeded and the generator house should also be well ventilated.

REFERENCES

Chichkova RI, Prockop LD (2007). Carbon monoxide intoxication: an updated review. J. Neurol. Sci. 262 (1-2):122-130.

Donnay A (2004). A True Tale of A Truly Haunted House. Ghostvillage.com. Retrieved 2008-12-16.

Encarta (2009). Encyclopedia.com. <http://www.encyclopedia.com>. Assessed 23 Mar. 2011.

Encyclopedia.com. <http://www.encyclopedia.com>. Assessed 23 Mar. 2011.

Erhabor GE, Oluwole AF, Ojo JO (1992). Pulmonary effects of occupational exposure to cement dust in workers at Portland cement factory in Nigeria. Nig. Med. Practitioner. 24(3):34-37.

Ernst A, Zibrak JD (1998). Carbon monoxide poisoning. New Eng. J. Med. 339 (22):1603-1608.
Fairburn F, Tikuisis P, Kane DM, McLellan TM, Buick SM (1992). Rate of formation of carboxyhemoglobin in exercising humans exposed to carbon monoxide. J. Appl. Physiol. 72(4):1311-1319.
Henry CR, Satran D, Lindgren B, Adkinson C, Nicholson CI and Henry TD (2006).Myocardial Injury and Long-term Mortality Following Moderate to Severe Carbon Monoxide Poisoning. JAMA. 295 (4):398-402. DOI:10.1001/jama.295.4.398.
Lipman GS (2006). Carbon monoxide toxicity at high altitude. Wilderness Environ. Med. 17(2):144-145.
Omaye ST (2002). Metabolic modulation of carbon monoxide toxicity. Toxicology 180 (2):139-50. DOI:10.1016/S0300-483X(02)00387-6.
Thom SR, Raub JA, Mathieu-Nolf M, Hampson NB (2000). Carbon monoxide poisoning-a public health perspective. Toxicology 145(1):1-14. DOI:10.1016/S0300-483X(99)00217-6. PMID 10771127.US AQI http://www.thegoodairlady.com/about_air_quality_index_000284.html
USEPA (United States Environmental Protection Agency). (2009). Data from the National Concentrations Inventory, Version 2.0. Accessed 2009. http://www.epa.gov/ttn/chief/eiinformation.html.
Walker E, Hay A (1999). Carbon monoxide poisoning. BMJ.319:1082-1083.
Wanner HU (1990). Effects of atmospheric pollution on human health. J. Aerosol Sci. 21(1):5389-5396.
Zibrak JD, Ernst A (1998). Carbon monoxide poisoning. The New England J. Med. 339 (22):1603-1608.

Modelling of solar radiation for West Africa: The Nigerian option

Cecily Nwokocha[1], Raymond Kasei[2] and Urias Goll[3]

[1]Department of Physics, Alvan Ikoku Federal College of Education, P. M. B. 1033 Owerri, Imo State, Nigeria.
[2]Centre for Development Research, University of Bonn, Germany.
[3]Department of Environmental Studies, University of Liberia, Liberia.

The United Nations Framework Convention on Climate Change (UNFCCC) reports indicate that those who are least responsible for climate change are also the most vulnerable to its projected impacts. In no place is this more evident than in Sub Saharan Africa, where greenhouse gas (GHG) emissions are negligible from a global scale. In Africa, energy demands could be the major factor that may lead to the increase of its emissions in the very near future. Forests are being used up for domestic energy supply. Oil produced energy increases carbon foot prints and hydropower is unreliable due to uncertainties in rainfall patterns. By 2004, the energy consumption mix of West Africa was dominated by oil (58%), followed by natural gas (38%) and hydroelectric (8%) with coal and other energy forms not part of the mix. The Sayigh's universal equation is presented in this work, for estimating the global solar radiation analysing data from 1972 to 2004 in Nigeria using Umudike, as a case study. The global solar radiation within the region was noted to range from 1.99 to 6.75 kWh indicating that the method could be used in producing signatures of global solar radiation in Nigeria when actual measurements are not available.

Key words: Solar radiation, modelling, West Africa, Nigeria.

INTRODUCTION

Countries in West Africa face many of the same challenges as the rest of the world due to growing fuelling energy demand. Besides few exceptions such as Nigeria, the countries in the region do not have or have developed sufficient domestic resources so they have become dependent on imports. State companies which dominate the energy sector lack commercial incentives and tend to be inefficient; prices of fuels such as kerosene, diesel and electricity are heavily subsidized encouraging inefficient use of energy, deterring additional investment in energy supply and often resulting in shortages and rationing (Taimur et al., 2007).

Notwithstanding, combustion of these fuels for power supply unleashes intolerable amounts of carbon (iv) oxide as well as other harmful gases to the environment thus contributing in turning the earth's atmosphere to a greenhouse with the harmful effect of producing global warming. The environmental implications of gas emission from these sources including gas flaring are part of issues that informed the Copenhagen Climate Change Summit at Denmark in December 2009. The summit was aimed at creating global awareness of climate change as well as establishing treaties that can be supported by global businesses.

This paper aims at recommending West Africa and Nigeria in particular to renewable energy, having estimated levels of global solar radiation at Umudike (latitude 5.29°N, longitude 7.33°E) Eastern Nigeria, using the Angstrom universal method from the period 1972 to 2004. Already, many countries like the United States of America, China, India and Brazil etc, are employing this initiative by the United Nations to start-up the clean development mechanism project, such as the use of solar power generation; biofuel generation from waste and investment in carbon trading amongst other things, geared at making the environment free from harmful emissions. With advancement in science and technology, the initiation to use solar energy for power generation is already in use towards meeting electricity needs. There are indications that the desire to diversify sources of electricity generation may become visible, following plans by the country's first solar electricity generation company to commence operation before the end of the year (Chineke, 2002; Igbal, 1983). The purpose of the work is to use different models for estimating global solar radiations in Nigeria for locations where no measured data is available.

MATERIALS AND METHODS

Study area

Nigeria is located in the tropics between 3°E to 14°E of longitude and 4°N to 14°N of latitude. She is endowed with an annual average daily sunshine of 6.25 h ranging between 3.5 h at the coastal areas and 9.0 h at the far northern boundary. The minimum and maximum hours of sunshine amount to 0.1 and 9.9 h respectively. Similarly, it has an annual average daily solar radiation of about 5.25 kWh/m^2 per day varying between 3.5 kWh/m^2 per day at the coastal areas and 7.0 kWh/m^2 per day at the Northern boundary. The minimum and maximum temperatures are 9.7 and 41.5°C respectively (Akinbami, 2001).

Acquisition of data

A monthly average temperature data obtained from the National Root Crop Research Institute (NRCRI) Umudike, for the period of 30 years ranging from 1972 to 2004 was used for this research work. The three models of estimating solar radiation is presented in this work but only the Sayigh's method was used to plot the signatures because of its accuracy, hence the three models were compared by combining all the models. In spite of the importance of global solar radiation data, few meteorological stations, especially in developing countries such as Nigeria measure these data accurately and continuously. The situation can be solved using equation methods that estimate global solar radiation from available meteorological parameters such as sunshine duration, daily temperature, relative humidity e.t.c (Akpabio and Etuk, 2003; Akpabio et al., 2005; Chineke, 2008). The correlation model developed can be used in estimating global solar radiation in locations of similar latitude, altitude and climatology. Empirical modelling is an essential and economical tool for the estimation of global solar radiation. There are several methods available for estimating global solar radiation (Chiemeka and Chineke, 2009; Chineke, 2008); however, for this work the following were considered.

Sayigh's equation

Sayigh proposed his universal formula for the estimation of total radiation intensity (Sayigh, 1977; El-Sahaam and Sayigh, 1979; Oduro-Afriyie, 1997), which relates the global solar radiation Rg, to mean maximum temperature Tmax, mean relative humidity RH, and the ratio of sunshine hours to the length of the day S1, expressed as:

$$Rg = NK \exp\{\varphi(S1 - RH/15 - 1/Tmax)\} \tag{1}$$

Where

$$N = 1.7 - 0.458\varphi \tag{2}$$

and

$$K = 1.163\ (a\ Z + bij\ Cos\ \varphi) \tag{3}$$

is a constant ($KWh/m^2/day$)

$$a = 0.2/(1 + 0.1\varphi - - - - - - \tag{4}$$

φ, the latitude in degrees; Z is the length of day in hours; the monthly average is taken in the middle of the month; K, N and a, are factors which can be computed using the equation above. bi j is a humidity factor, i=1,2,3 where 1 is for RH < 65%, 2 is for RH > 70%, and 3 is for 65% $\leq$ RH $\leq$ 70%. On the other hand, j = 1, 2, 3... 12 which refers to the months of the year.

Hargreave's equation

Chineke and Jagtap (1995) compared 3 models and got the best fit with the modified temperature based Hargreave's method (Hargreaves and Samani, 1982). The temperature-based model is most appropriate when data on sunshine hours is lacking like in the case for Nigeria. Chineke et al. (1999) proposed the equation for estimating solar radiation and presented it in the form:

$$Rs = 0.16RaTd_{1/2} \tag{5}$$

Where R_S = solar radiation, T_d = daily temperature difference (maximum minus minimum) values, Ra = extraterrestrial solar radiation (generated by a computer routine and requiring the locations grid parameter).

Angstrom equation:

This is one of the most popular models. It is a regression equation (Angstrom, 1924; 1929 and Black et al., 1954), which relates global solar radiation H, to the duration of sunshine S, being the parameter mostly readily measured at meteorological stations.
The model is thus expressed:

$$H + Ho = a + bs/Z \tag{6}$$

H = total horizontal solar radiation
Ho = extraterrestrial solar radiation

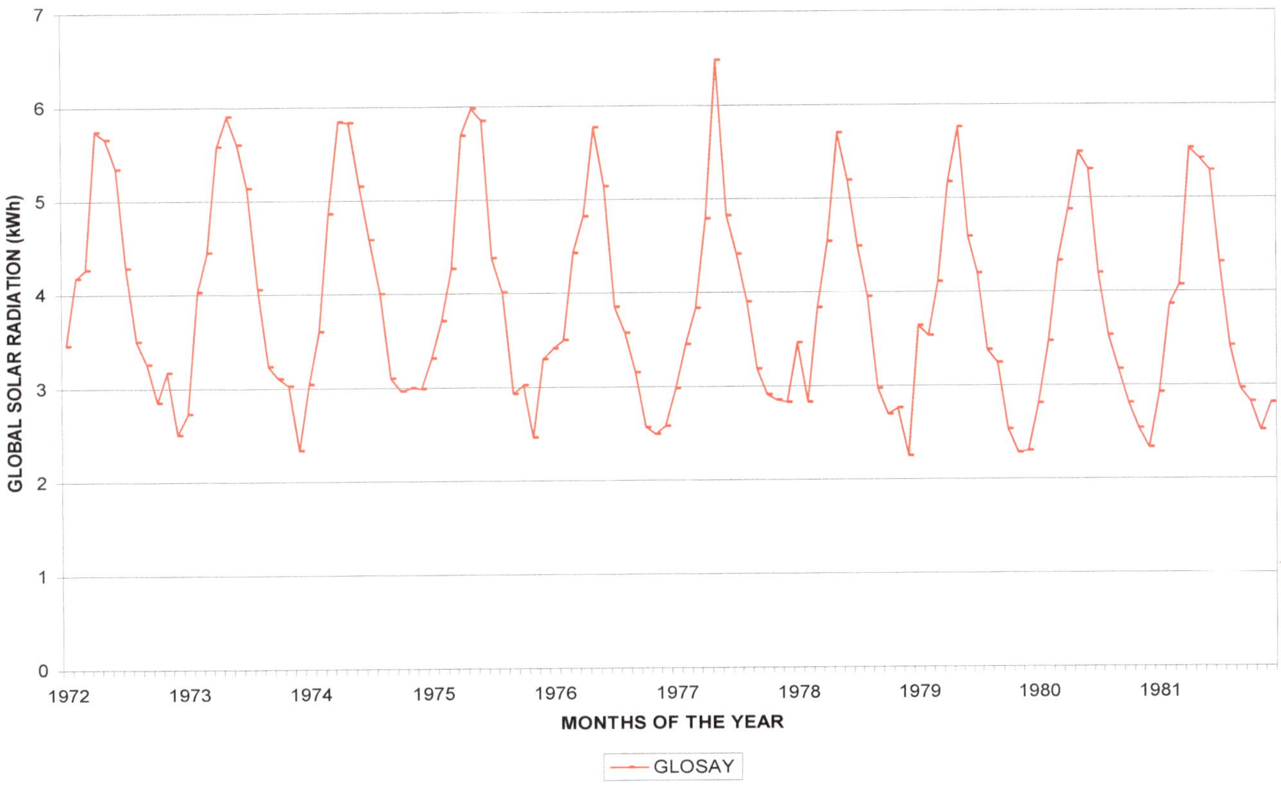

Figure 1. Global Solar Radiation for the first decade between 1972 to 1981.

S = Average sunshine hours per month
Z = length of the day in hours, the monthly average is taken in the middle of the month a and b are constants.

From the daily averages of the temperature and sunshine hour data, the global solar radiation was estimated for the region (Angstrom, 1924, 1929; Oduro-Afriyie, 1997; Chineke et al., 1999; Hargreaves and Samani, 1982; Taimur et al., 2007; Chineke, 2008; Chiemeka and Chineke, 2009). This would highlight how much the clearness index for the region could be established (Akpabio et al., 2005; Babatunde and Aro, 1995). For this analysis, the period of the study was grouped into four decades, ranging from 1972 to1981 for the first decade, 1982 to 1991 for the second decade, 1992 to 2001 for the third decade and the forth decade being the years from 2002 to 2004.

RESULT AND DISCUSSION

In Figures 1 to 3, the signatures of the estimates were shown creating a trend line for modelling and subsequently, the expected levels of solar radiation within the region. Within the first decade, the region recorded a maximum of 6.48 kWh in 1977 and a minimum of 2.25 kWh in 1978. However, the records of the estimated maximum levels of solar radiation that reached the region for the decade were 5.74 kWh in 1972; 5.9 kWh in 1973; 5.84 kWh in 1974; 5.98 kWh in 1975; 5.77 kWhin 1976; 5.7 kWh in 1978; 5.76 kWh in 1979; 5.49 kWh in 1980 and 5.53 kWh in 1981; as shown in Figure 1. From Figure 2, the trend line of the estimates of the second decade was similar to the first although with variation in the estimated maximum solar radiation, which was 6.23 kWh in 1988 and a minimum of 1.99 kWh in 1985. However, the records yet reflect the potentials of harnessing this source of renewable energy in areas of photovoltaic in rural electrification (Chineke, 2008; Chineke et al., 2007), evapotranspiration and agronomy (Okoro et al., 2008), to mention but a few.

The third decade ranging from 1992 to 2001 had its signature similar to those previously discussed. From Figure 3, the estimate showed the highest level in the maximum solar radiation within the period of study, which was 6.75 kWh in 1998, with similar high records of 6.06; 5.87; 6.05 and 5.98 kWh in the years 1995; 1996; 1997 and 2001 respectively. This could not be far from the truth as this period recorded the highest awareness on global warming and climate change. The gaps in the data were mainly due to unavailability of measurements of the variables for the months within the period.

Figure 4 showed that the levels of solar radiation were also high with maximums of 5.92, 6.09 and 6.1 kWh in the years 2002, 2003 and 2004 respectively; having minimums of 2.58, 2.38 and 2.5 kWh in the years 2002, 2003 and 2004 respectively. These earlier figures

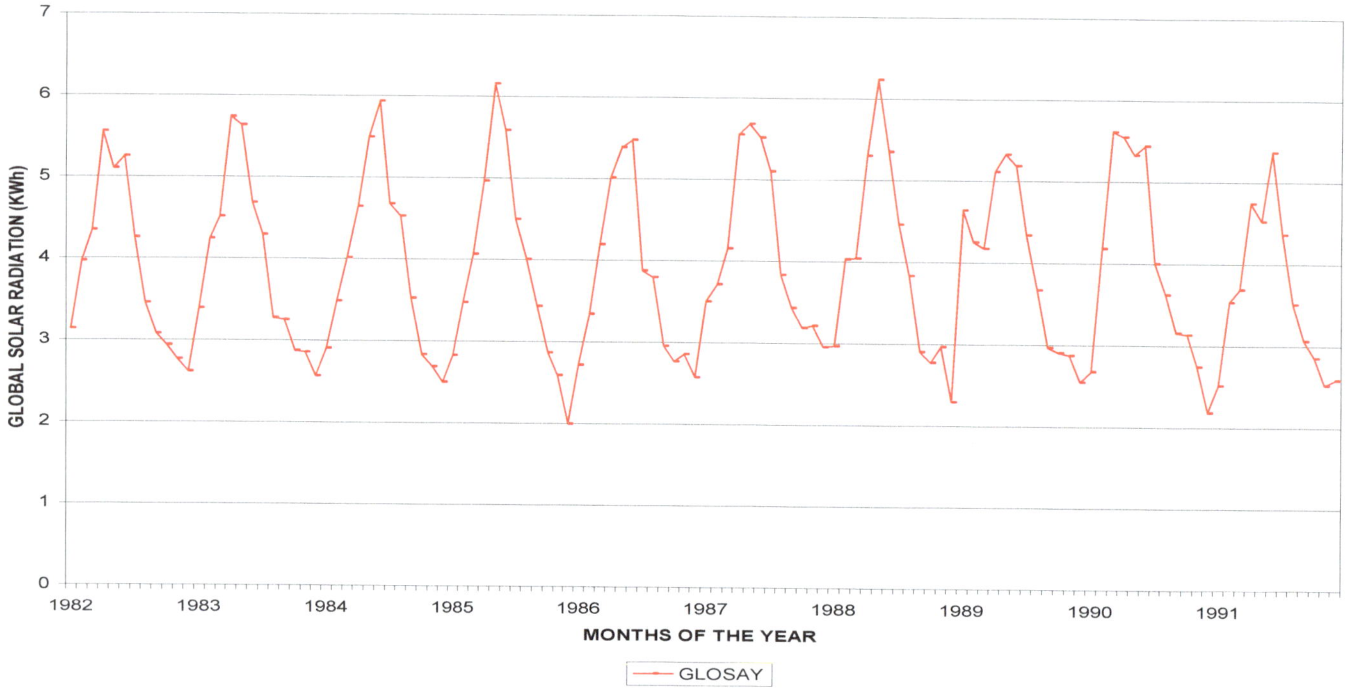

Figure 2. Global solar radiation for the second decade between 1982 to 1991.

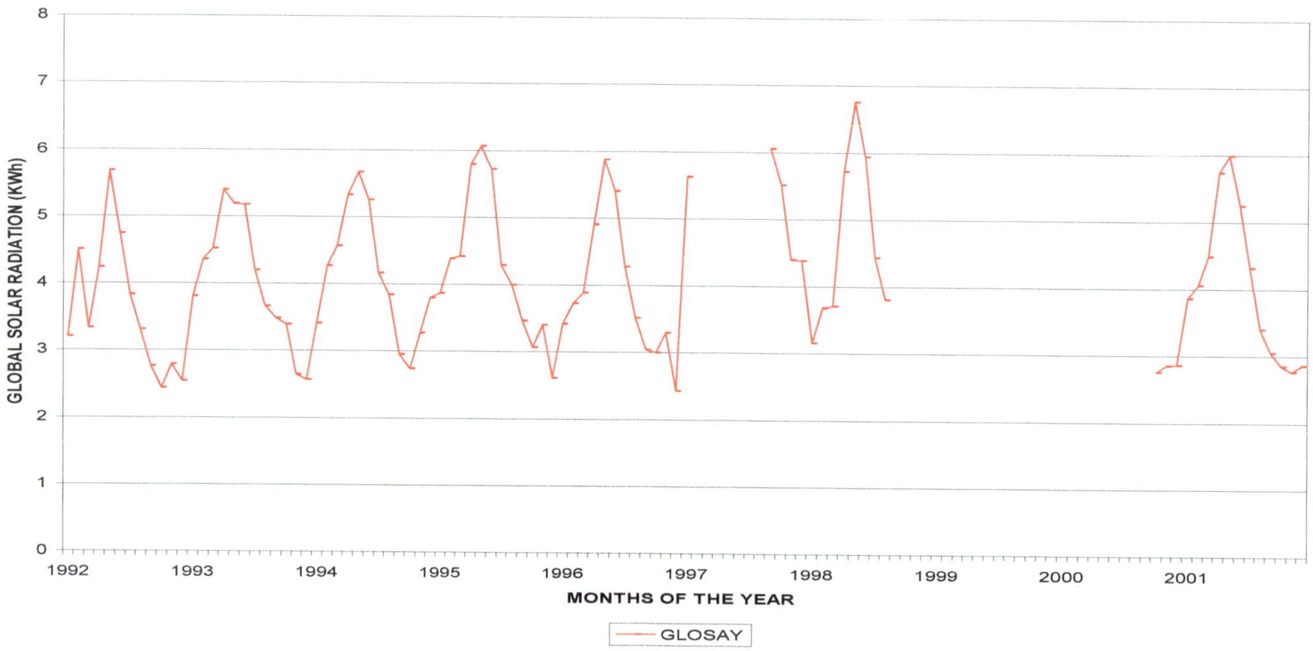

Figure 3. Global solar radiation for the third decade between 1992 to 2001.

revealed that the power need of Nigeria, most especially those within the South-eastern region, could be addressed if the potentials of this renewable energy is adequately harnessed. However, comparing the results from the Sayigh universal formula with those obtained using Angstrom and Hargreaves estimation methods yet over the region as shown in Figure 5; it could be seen that the trend line was similar hence showing an

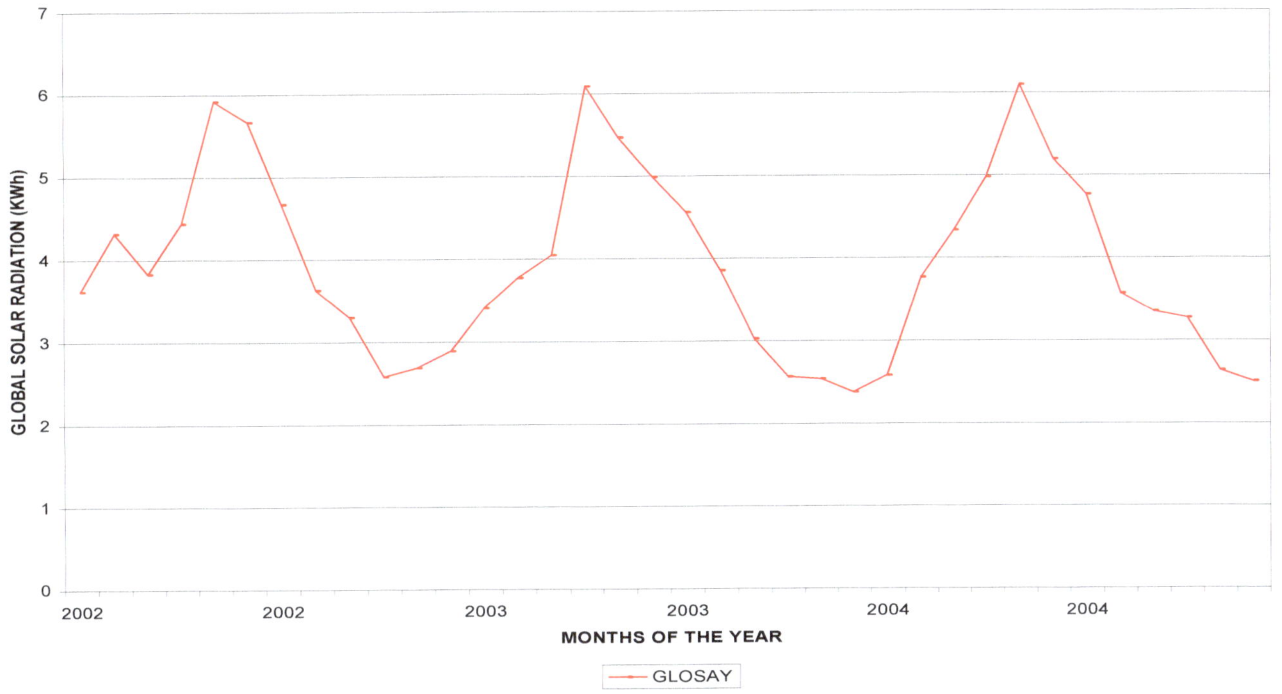

Figure 4. Global solar radiation for the last three years in the fourth decade between 2002 to 2004.

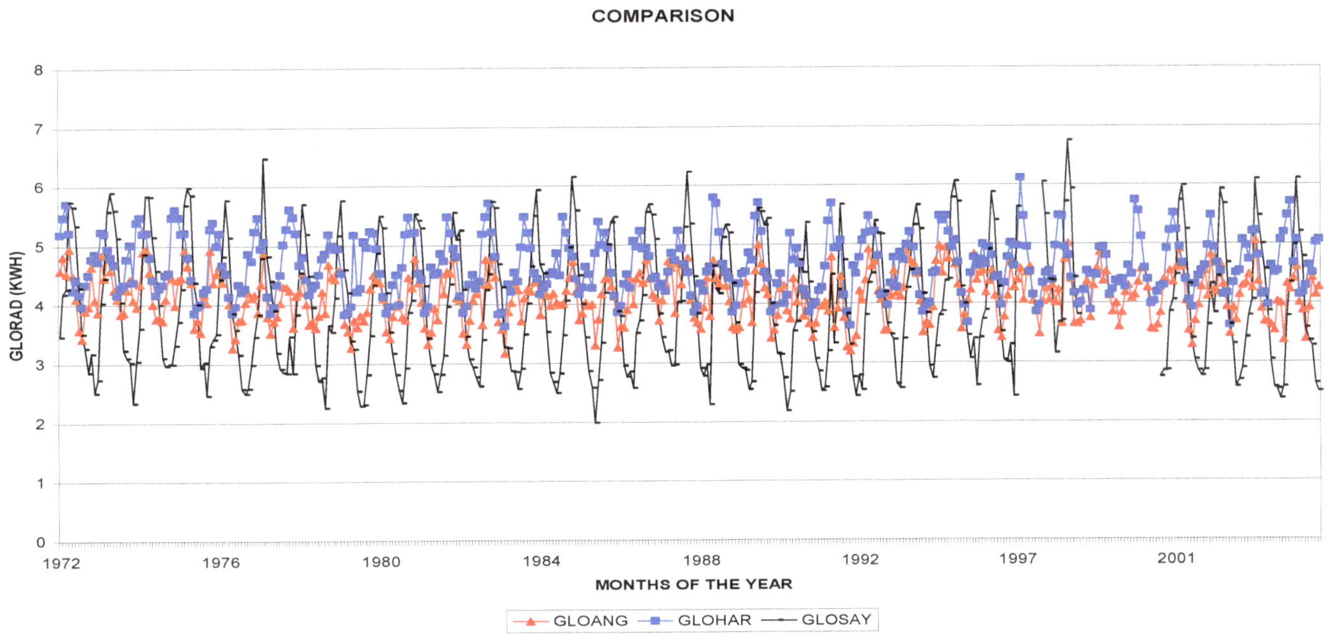

Figure 5. Comparison of levels of Global Solar Radiation at Umudike with the three estimation methods from 1972-2004.

agreement within each estimation method.

Conclusion

From the analyses, it is seen that the levels of global solar radiation recorded over the region were high having least values over the harmattan months as also reported in Chineke et al. (2007). These values are adequate in citing Photovoltaic systems (PV) as there is enough energy to power them all year round (Chiemeka and

Chineke, 2009). Designing and constructing solar cookers, incubators, ovens and preserving equipments that would be affordable but efficient for those living in the rural areas within the study area, becomes necessary as majority of the people living in these area are poor. This will in turn, enhance their living standard (Chineke and Igwiro, 2008), which is one of the objectives of the Millennium Developmental Goals (MDG) of the government.

Photovoltaic systems are now the lowest cost option for satisfying many of the electrical energy needs of areas not served by distributed electricity, particularly in developing countries, located in the tropics, where the amount of sunshine is generally high and rural household electricity consumption is comparatively low. Efforts should therefore be made to improve the number of meteorological stations available so that observations may be obtained from these stations to cover a wider scope.

REFERENCES

Akinbami JFK (2001). Renewable Energy Resources and Technologies in Nigeria: Present Situation, Future Prospects and Policy Framework. Mitigation and Adaptation Strategies for Global Change. 6:155-188. Kluwer Academic Publishers. Netherlands.

Akpabio LE, Etuk SE (2003). Relationship Between Global Solar Radiation And Sunshine Duration For Onne, Nigeria. Turk J. Physics, 27:161-167.

Akpabio LE, Udo SO, Etuk SE (2005). Modeling Global Solar Radiation For A Tropical Location: Onne, Nigeria. Turk J. Physics, 29:63-68.

Angstrom A (1924). Solar and terrestrial radiation. Q. J. R. Meteorol. Soc. 50:121-125.

Angstrom A (1929). Recording Solar Radiation. Medd. Statues meterol. Hyd. Anstalt. 3:4-10.

Black JR, Bonython CW, Precott J (1954). Solar Radiation And The Duration Of Sunshine. Quart J. Roy. METEOR. 80:231-235.

Babatunde EB, Aro TO (1995). Relationship Between "Clearness Index" And "Cloudiness Index" At A Tropical Station (Ilorin, Nigeria), Renewable Energy, 6(7):801-805.

Chiemeka IU, Chineke TC (2009). Evaluating the global solar energy potential at Uturu, Nigeria. Int. J. Phys. Sci. 4(3):115-119. Available Online at http://www.academic journals.org/IJPS.

Chineke TC (2002). A Robust Method Of Evaluating The Solar Energy Potential Of A Data Sparse Site, The Physical Scientist, 1:59-69.

Chineke TC (2008). Equations for estimating global solar radiation in data sparse regions. Renewable Energy 33(4):827-831.

Chineke TC, Aina JI, Jagtap SS (1999). Solar Radiation Database for Nigeria, Discovery & Innovation, 11(3/4):207-210.

Chineke TC, Igwiro EC (2008). Urban and rural electrification: enhancing the energy sector in Nigeria using photovoltaic technology, UNESCO/ANSTI Afr. J. Sci. Technol. 9(1):102-108.

Chineke TC,Jagtap SS (1995). A Comparison of Three Empirical Models For Estimating Solar Radiation In Nigeria. Afr. J. Sci. Technol. Series A, 11:45-48.

Chineke TC, Okoro UK, Igwiro CE (2007). The Imperatives Of Solar Photovoltaic As Viable Options for Rural Electrification in Nigeria. Int. J. Nat. Appl. Sci. 3(2):193-199.

El-Sahaam MA, Sayigh AA M (1979). Estimation of Diffuse Solar Radiation In The Arabian Pem Proc. Symp. On Radiation in The Atmosphere. Ed. J. A. Book Sc. Press, pp. 607-610.

Hargreaves GH, Samani ZA (1982). Estimating Potential Evapotranspiration. J. Irrig. Drainage Engr. 108(IR3):225-230.

Igbal M (1983). An introduction to solar radiation. Academic Press, New York.

Oduro- Afriyie K (1997). Performance of Sayigh's universal formula for the estimation of global solar radiation in Ghana. Renewable Energy, 10(1):91-106.

Okoro UK, Chineke TC, Nwofor Ok, Ibe A (2008). Evapotranspiration Levels At Owerri In The Niger Delta Nigeria. Int. J. Nat. Appl. Sci. 4(2):168-173.

Sayigh AAM (1977). Estimation of total radiation intensity a universal formula in 4th course of solar energy convention, Vol. II Mancin and Quercia I. F. (Eds), ICTP, Trieste, Italy.

Taimur B, Amine M, David C, Joseph N (2007). IMF Working Paper No. 07/71, March 2007.

Experimental study of air-water turbulent flow structures on stepped spillways

Mohammad Reza Beheshti[1] , Amir Khosrojerdi[1] and Seyed Mahmud Borghei[2]

[1]Department of Water Science and Engineering, Science and Research Branch, Islamic Azad University, Tehran, Iran.
[2]Department of Civil Engineering, Sharif University of Technology, Tehran, Iran.

Stepped chute flows are characterized by intense turbulence and strong flow aeration, but most studies did not investigate the turbulence characteristics. In this study, highly turbulent air-water flows skimming down a large-size stepped chute were systematically investigated. An experimental study of detailed air-water flow properties measurements were introduced in different types of flow regimes on a stepped chute (θ = 21.8°, h = 0.04 m, l = 0.10 m) model to investigate the location and the flow depth at inception point of air entrainment and velocity profiles distributions. Detailed velocity and turbulence intensity measurements in flow direction were performed by use of a phase detection conductivity probe which was designed, developed and calibrated by writers. The results showed that the turbulence characteristics vary in different regions. The study showed further that the turbulence intensity increases rapidly close to the step bottom at the viscous sub layer and maximized for 0.4 ≤ y/dc ≤ 0.5 at intermediate region then decreases gradually in the upper region. It is hypothesized that the high turbulence levels in the intermediate region were caused by the continuous deformations and modification of the air-water interfacial structure.

Key words: Air-water flows, conductivity probe, inception point, stepped chute, turbulence intensity.

INTRODUCTION

The interactions between flowing waters and atmosphere may lead to strong air-water mixing and complex multiphase flow situations. The evaluation of the air-water flow properties is particularly important on stepped spillways over RCC dams experiencing moderate unit discharges, because large quantities of air entrain upstream of the spillway toe.

Numerous studies of stepped spillway flows were conducted with different approaches over the past forty years. Studies of spillways with particular focus on stepped chutes include Sorensen (1985), Peyras et al. (1991), Christodoulou (1993), Chamani and Rajaratnam (1999), Chinnarasri and Wongwises (2006), Relvas and Pinheiro (2011), and others. Only a limited number of published papers refer to highly turbulent flows associated with strong free-surface aeration (Chanson and Toombes, 2002; Yasuda and Chanson, 2003).

In civil engineering applications, the flow velocity exceeds typically 5 to 10 m/s, the flow Reynolds number ranges from 1 E+7 to over 1 E+9 in large dam spillways, and free-surface aeration is nearly always observed. Such high-velocity highly-aerated flows cannot be studied analytically or numerically because of the large number of relevant equations and parameters. Current knowledge

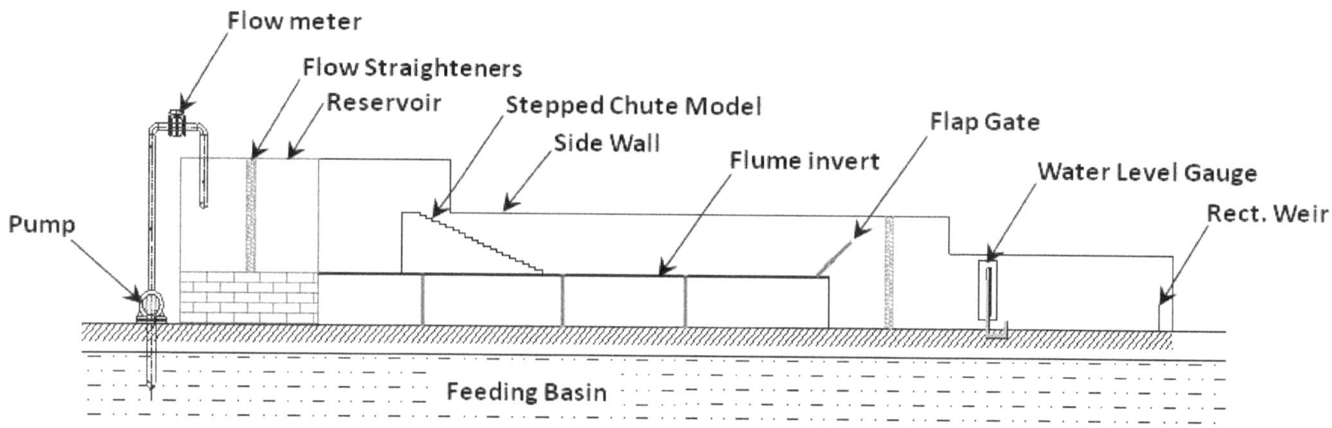

Figure 1. The experimental set-up of the stepped spillway (Hydraulic laboratory of WRI, Tehran, Iran).

relies upon physical modeling and experimental measurements. Accurate measurement systems for air-water flow measurements are limited to intrusive phase-detection probes, hot-film probes, and LDA/PDA systems.

According to the literature review, the measurement of the flow velocity concerns only the longitudinal component, parallel to the mean slope. Double fiber-optical and double-tip conductivity probe are at the moment the more efficient intrusive instrumentation to measure the mixture flow velocity. In fact, the Prandtl tube or back-flushing Pitot tube is limited to low aerated flow (that is, for air concentration less than 0.7).

In the present study, new experiments were conducted in a large stepped channel with a 21.8° slope and it is shown that the advanced developed dual-tip conductivity probe may provide new information on the air-water turbulent characteristics on stepped chutes. Detailed air-water flow properties were recorded systematically for several flow rates including turbulence levels, void fractions and velocity distributions. The results provide a new understanding of the complicated air-water structure on stepped spillways.

MATERIALS AND METHODS

Experimental setup

The experiments were conducted in a rectangular Perspex-walled flume in the Hydraulics laboratory at the water research Institute of Iran Power Ministry. The experimental channel was 5 m long and 0.90 m wide. Water flow was supplied from a large feeding basin under the laboratory leading to a concrete reservoir upstream of the flume.

The test section consisted of a broad crested weir (0.83 m wide, 0.45 m long, with upstream sharped corner) followed by twenty identical steps (h = 0.04 m, l = 0.10 m) made of Plexi-Glass plates. The stepped chute was 0.83 m wide with Perspex sidewalls followed by a horizontal steel-invert flume ending in a dissipation pit (Figure 1). The physical model is part of a closed water cycle.
The chute bottom angle, equal to the pseudo-bottom angle, was φ = 21.8° (1V : 2.5H).

The measurement reach located at 3.5 from the entrance channel section, was 3 m long. A pump controlled with an adjustable frequency AC motor drive delivered the flow rate, enabling an accurate discharge adjustment in a closed-circuit system.

At the downstream end of the flume, the tail gate controls the tail water depth. For all measurements, the gate was fully opened.

Instrumentation and measurement techniques

In present study, the measured parameters are included: clear-water flow depths, clear-water and air-water flow velocities and flow rate.

Clear-water flow depths were measured using a rail mounted vertical pointer gauge with 1 mm accuracy. Clear-water velocities were measured with a Pitot tube (3.3 mm external diameter, 20 mm distance between tip and lateral holes). The flow rate measurements were performed using a rectangular sharp-crested weir located in the outlet channel and an electromagnetic flow meter positioned in the inlet pipe, used for confirmation of the weir discharge measurements. Detailed information about air-water flow properties like the air-water flow velocity, the air concentration and bubble sizes and frequencies was gained using a double-tip conductivity probe (Ø = 0. 7 mm) which was developed and calibrated at Science and Research Branch of Islamic Azad University by authors.

The conductivity probe is a phase-detection intrusive probe designed to pierce the bubbles. The phase-detection relies on the difference in electrical resistance between air and water. The probe mounted on a support frame parallel to the chute slope and consists of two identical tips made of stainless steel surgical needle. The tips are aligned in the flow direction and the distance between the tips is 4 mm (Figure 2). The sampling frequency of the developed probe is 50 kHz with adjustable sampling time. The two-phase flow properties measurements were taken from the tip of the steps to as close to the water surface as possible. All measurements were performed on the channel centerline (z = 0) at step edges and adjacent step edges.

Experimental flow conditions

In the stepped chute, the basic flow regimes were inspected in a series of visual observation with discharges ranging from 0.012 m^3/s to 0.035 m^3/s (Table 1). The model development was based on the Froude similitude. The facility was designed to operate with flow conditions ranging from nappe to skimming flow regimes. The

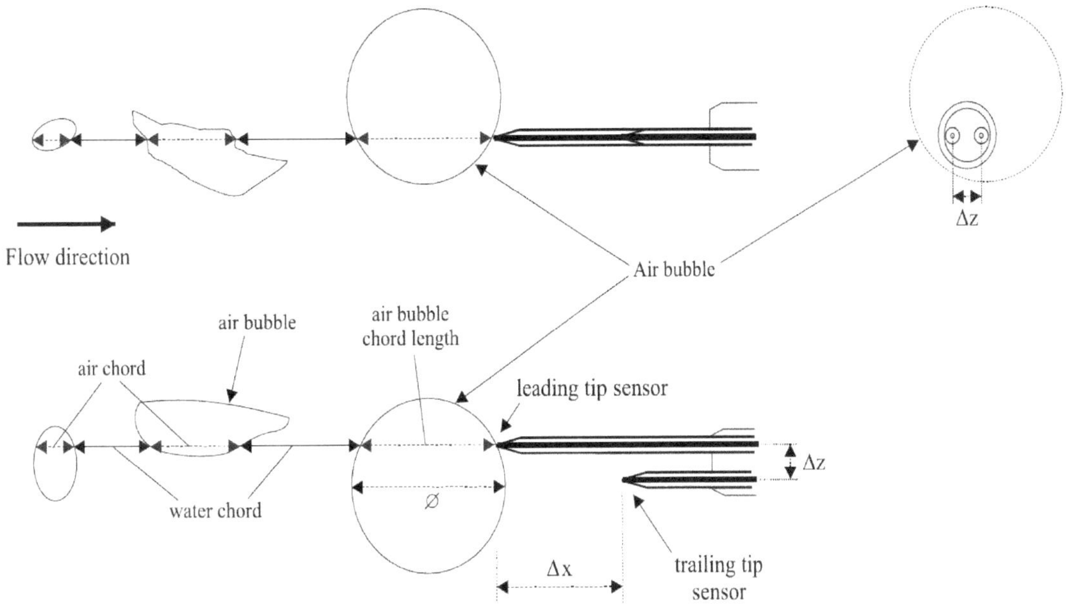

Figure 2. Sketch of the double-tip conductivity probe and its operation.

Table 1. Summary of experimental flow conditions (Present study).

Q_w, m³/s (1)	q_w m³/s/m (2)	d_c/h (3)	Re (4)	We (5)	Flow regime (6)	Location of inception of air entrainment (7)	Remarks (8)
0.012	0.0147	0.701	5.8×10^4	102.519	Nappe Flow	Step edge 4	W = 0.83 m, h = 0.04 m, l = 0.10m
0.015	0.0184	0.813	7.2×10^4	109.975	Transition Flow	Step edge 4	W = 0.83 m, h= 0.04 m, l = 0.10m
0.020	0.0241	0.981	9.6×10^4	119.380	Transition Flow	Step edge 6	W = 0.83 m, h= 0.04 m, l = 0.10m
0.025	0.0301	1.131	1.2×10^5	119.788	Skimming Flow	Step edge 7	W = 0.83 m, h= 0.04 m, l = 0.10m
0.030	0.0361	1.282	1.4×10^5	204.789	Skimming Flow	Step edge 7	W = 0.83 m, h= 0.04 m, l = 0.10m
0.035	0.0422	1.415	1.7×10^5	314.580	Skimming Flow	Step edge 8	W = 0.83 m, h= 0.04 m, l = 0.10m

flow conditions corresponded to flow Reynolds numbers ranging from 5.8×10^4 to 1.7×10^5. The Reynolds number defined in terms of the hydraulic diameter as:

$$R_e = V \times D_H / \upsilon_w \quad (1)$$

Where D_H is the hydraulic diameter, V and υ_w are the interfacial velocity and the kinematic viscosity of the water, respectively.
The measured flow rates, and other experimental parameters of the different conditions, are shown in Table 1.

Data processing

Air-water flow properties were recorded for nappe (that is, dc/h < 0.701), transition (0.701 < dc/h < 0.1.131) and skimming flows (that is, dc/h > 1.131). In rectangular channels, the critical depth is estimated by $d_c = \sqrt[3]{q_w^2 / g}$ where q_w is the water discharge per unit width and g is the gravity acceleration.

During each test, photographic observations recorded the inception point of air entrainment. The inception point was noted as the location where the flow changed from a glassy, smooth appearance to a more frothy appearance at the free surface. The inception point location was described quantitatively by its horizontal distance from the spillway crest and with regards to step location within the spillway chute. On stepped spillways, the position of the start of air entrainment is a function of the discharge, spillway roughness, and step geometry.

Velocity measurement is based upon the successive detection of air-water interfaces by two tips. This technique assumes that the probe tips are aligned along a streamline and that bubbles and droplets are little affected by the leading tip. When a proportion of bubbles strike the leading tip and trailing tip sequentially, the resulting signals from the two tips will be similar with a slight time delay. This time delay is the average time for a bubble to travel the distance between the two tips. If this distance is known, the time can be converted into a mean air-water interface velocity.
The time-averaged air-water velocity equals:

$$V = \frac{\Delta X}{T} \quad (2)$$

Where Δx is the longitudinal distance between probe sensors (Figure 2) and T is the air-water travel time between the two probe sensors. Air-water flow characteristics data were collected at 50 KHz frequency with a sampling duration of 90 s. The data analyses were done in Excel spread sheets and Grapher V.9.1.536 software.

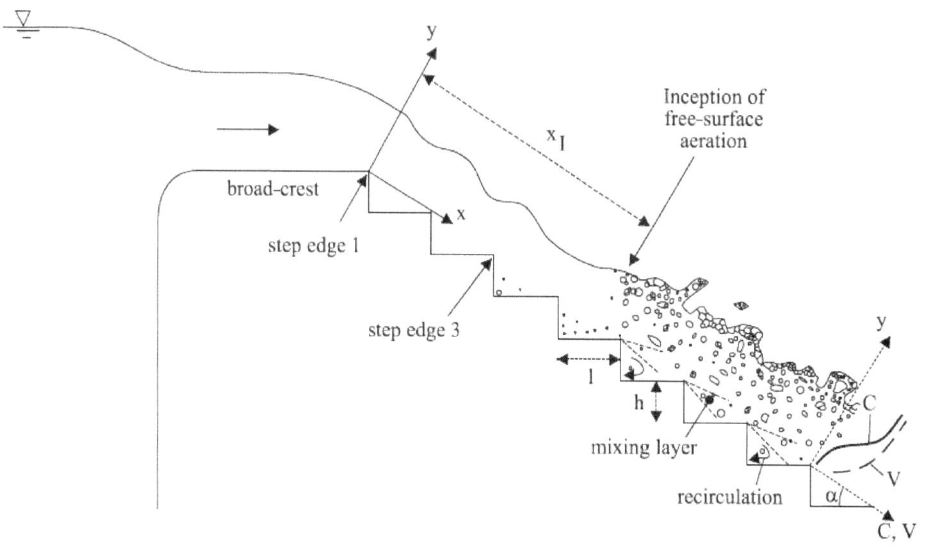

Figure 3. Definition sketch of flow over stepped chute.

In analysis, the distribution of the time averaged velocity, air concentration, fluctuating velocities and turbulence intensities as well as their decay along the downstream direction were considered.

EXPERIMENTAL RESULTS AND DISCUSSION

Detailed measurements of air-water flow properties were conducted for a number of dimensionless flow rates (d_c/h).

The maximum water heads above the spillway crest are between 86.25 and 93.95 cm (measured from the toe of the spillway as the datum) and corresponding unit width discharges between 0.0147 and 0.042 m^3/s/m. The approach flow conditions corresponding to Froude numbers between 0.643 and 1.007.

A right hand coordinate system was adopted at measuring sections in the flume with x-axis as the streamwise distance along the channel bottom, y-axis as the distance measured normal to the invert (or channel bed) and z-axis as the transverse distance from the channel centerline.

The dimensionless distributions of interfacial velocity and turbulent intensity showed some characteristic features.

Basic flow observations

Flow regimes

For a given chute geometry, the flow pattern above a stepped channel may be nappe flow, transition flow or skimming flow. The type of flow regime is a function of the discharge and step geometry (Chanson, 1994b; Chanson and Cummings, 1996). For low discharges, the water flow drops from step to step on a stepped chute. That is the nappe flow regime which was studied by Chamani and Rajaratnam (1994), Chanson (1994a), Toombes (2002), El-kamash et al. (2005), and Toombes and Chanson (2008).

For some intermediate discharge, the transition flow regime is observed which is characterized by a chaotic behavior and strong splashing and droplet projections downstream of the inception point of free surface aeration (Chanson, 2001b, Chanson and Toombes 2004). For this regime, the flow can appear as nappe flow (presence of air caity) for some steps and as skimming (onset of vortices in a filled step) for the rest. For larger flows, the waters skim above the pseudo-bottom formed by the step edges.

The skimming flow is characterized by strong cavity recirculation with three-dimensional vortical patterns (Rajaratnam, 1990; Chanson 1994b, 1995a; Chamani and Rajaratnam, 1999). The flow regime changes in the stepped chute was observed and analyzed in this study. The conditions linked to the flow configuration changes have been quantitatively studied. The parameters which have been considered to quantify these changes are the ones related to flow critical depth (dc) and to the spillway geometry, which are the non-dimensional ratios d_c/h and h/l where dc is the critical depth and h and l are the step height and length respectively. A sketch of stepped chute flow and other geometrical parameters are shown in Figure 3.

A nappe flow regime without hydraulic jump was observed for small flow rates ($d_c/h < 0.813$) and for large flow rates ($d_c/h > 1.131$), the water skims over the pseudo-bottom. At moderate discharge rates ($0.813 < d_c/h < 1.131$), a transition flow pattern was observed. The non-dimensional characteristic depth (d_c/h) values corresponding to the different flow regime observations

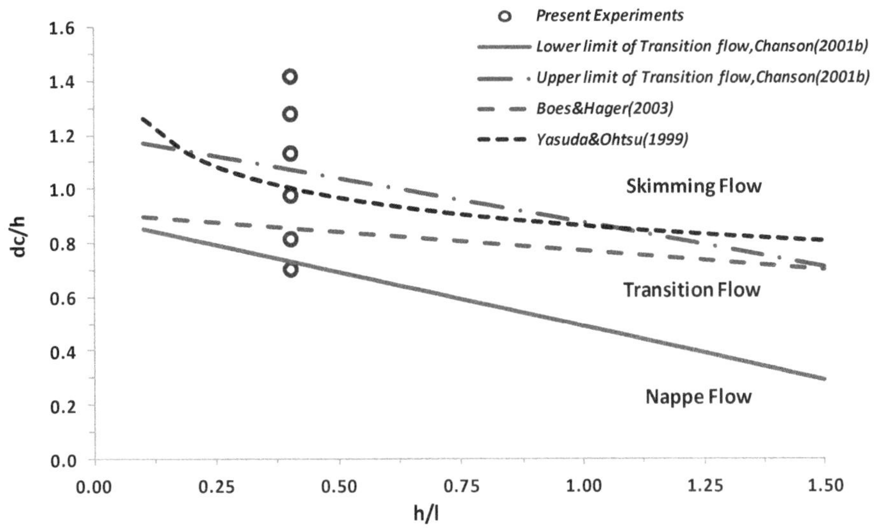

Figure 4. Flow regimes estimation on the stepped chute and comparison with the empirical relationships.

are shown in Figure 4 and compared with some researchers' observations.

The observation results for flow regime prediction were in agreement with onset/transition functions in the literature (Chanson, 1995a, 2001b).

Characterizing the inception point of air entrainment

At the upstream end of the stepped spillway, a turbulent boundary layer is generated by bottom friction and it develops in the flow direction. When the outer edge of the developing boundary layer reaches the vicinity of the free-surface, air entrainment is initiated.

This location is called the inception point of free-surface aeration. Downstream, the turbulence next to the free-surface is large enough to initiate substantial flow aeration. In the present study, the location of the inception point of free-surface aeration was recorded in all flow regimes. At the inception point, the air is entrained into the flow owing to a large degree of turbulence.

Determination of the inception point location along the spillway is important to estimate the unaerated zone, which is prone to the cavitation damage. Cavitation is a serious problem resulting from hydrodynamic pressures, on the step surfaces or at the step edges, which could fall below the vapor pressure (Boes and Hager, 2003).

Cavitation might cause severe damage to the steps and can be mitigated by a significant aeration along the spillway. Prevention of cavitation potential damage requires a mean air concentration of at least 5 to 8% (Peterka, 1953). For all the current models experiments, the location of the inception point was defined by visual observation and pointer gauge measurements of the water surface where the air was entrained into the flow for the whole spillway width. The non-dimensional parameter x_I/k_s, is plotted as a function of the roughness Froude number (Fr*) in Fig. 5 for current experimental data. The parameter k_s is defined as the step roughness height and equals to hcosθ. Fr* is a Froude number defined in terms of the step roughness:

$$F_r{*} = q_w \Big/ \sqrt{(h\cos\theta)^3 g \sin\theta} \qquad (3)$$

The available literature data are also shown in Figure 5 and compared with the current data. Figure 5 shows there are good agreement of the results with Boes and Hager (2003) and a fair agreement with others. For the present study, the results were best correlated by the following equation to locate the inception point of air entrainment on the 21.8° stepped chute for 0.701<dc/h<1.415:

$$\frac{X_I}{K} = 8.0833 F_*^{0.7723} \qquad (4)$$

Interfacial velocity and turbulent intensity distributions

In analogy with the studies of uniform rough flow for steep gravel beds in mountain streams, a few authors (Chamani and Rajaratnam, 1999) suggested that above the pseudo-bottom, the homogeneous air-water flow agrees with the rough boundary layer theory. However, because of the high relative roughness of the steps, the 3D pattern of the macro-turbulence structures as well as

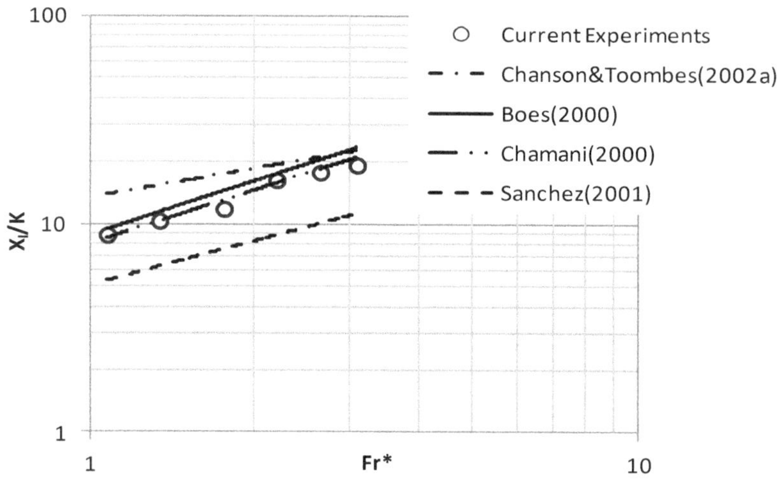

Figure 5. Dimensionless location of the inception point of free-surface aeration.

Table 2. Comparison of the power law coefficients for velocity distribution.

References	D	N	Remarks
Chanson (2001b)	1	5~6	Skimming flow
Chamani and Rajaratnam(1999)	1	6.3	$\theta = 30°$
Boes(2000)	1	6.3	$0.04 \le Y_{90} \le 0.5$
Boes and Hager (2003)	1.05	4.3	$0.04 \le Y_{90} \le 0.8$, $26° \le \theta \le 55°$
Yasuda and Chanson (2003)	1	9	$0.05 \le Y_{90} \le 1$, $\theta = 15.6°$
Current Study	1.012	5.21	$0.03 \le Y_{90} \le 0.92$, $\theta = 21.8°$

the drag around steps, the classical logarithmic equation as proposed by Karman-Prandtl is not fully applicable.

$$\frac{v(x)}{\sqrt{\tau_o / \rho_w}} = \frac{2.30}{\kappa} \log\left(\frac{x}{\kappa}\right) + C1 \quad (5)$$

In which k is the von Karman turbulence constant determined experimentally between 0.40 and 0.42,
C1 an integration constant, k the equivalent roughness size and τ_0 the boundary shear stress.

Nevertheless, most researchers proposed that above the pseudo-bottom the local mixture velocity depends on the maximum velocity and that the velocity distribution follows a power law given by:

$$U_{90} = D Y_{90}^{1/N} \quad (6)$$

Where $U_{90} = \frac{u}{u_{90}}$ and $Y_{90} = \frac{y}{y_{90}}$ are the dimensionless mixture flow velocity and depth respectively. u_{90} and y_{90} are the velocity and flow depth corresponding to 90% air concentration. D and N are coefficients obtained from experiments data. This result has been validated for model and prototype data of air-water open channel flows by several other researchers, including Chanson (1989), Wood (1991) and Chanson and Cummings (1996). However, depending on the tested discharges or the portion of flow layer considered in the empirical fitting, several power laws have been proposed with different coefficients and some of them were listed in Table 2.

In the present study, the velocity power law exponent was 1/5.21 in average (that is, N = 5.21), although it varied between adjacent step edges. Such fluctuations were believed to be caused by some interference between adjacent shear layers and cavity flows. In current study, the velocity distributions at each step edge compared favorably with the power-law function for $y/y_{90} < 1$ and with a uniform profile for $y/y_{90} > 1$:

$$\left(\frac{u}{u_{90}}\right) = D\left(\frac{y}{y_{90}}\right)^{1/N}, \quad 0 \le \frac{y}{y_{90}} \le 1 \quad (7)$$

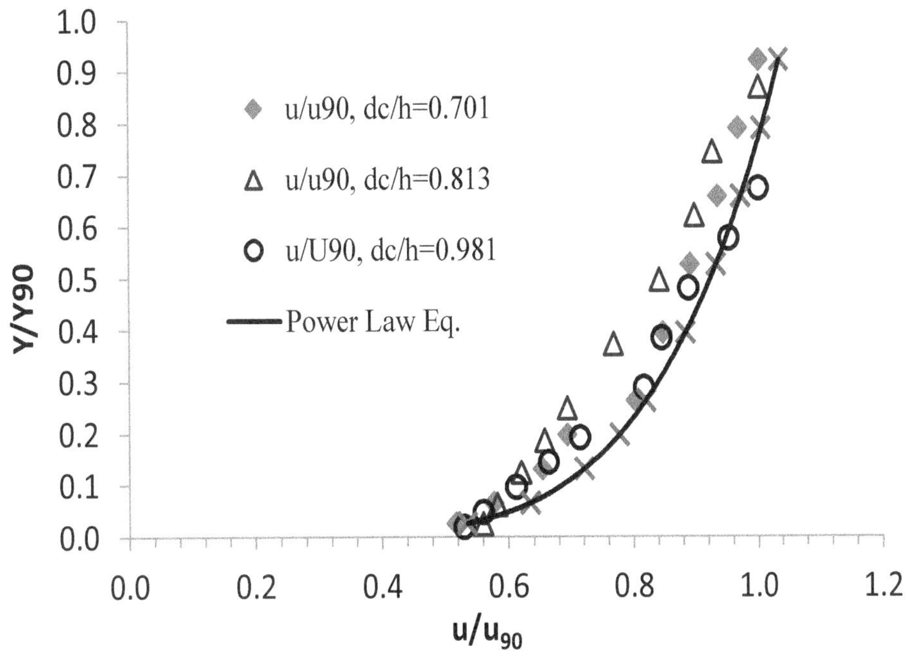

Figure 6. Dimensionless distributions of normalized interfacial velocity u/umax in different flow regimes at step edge 15.

$$\left(\frac{u}{u_{90}}\right)=1 \; , \quad 1 \le \frac{y}{y_{90}} \le 2.5 \tag{8}$$

Equation (7) is compared with experimental data in Figure 6 for several dimensionless discharges dc/h. The data does not include the profiles closest to the sidewall, which are likely affected by sidewall friction. In nappe and transition flows, the velocity measurements show quasi-uniform velocity profiles at step edges (Figure 6).

The result is relatively close to that of Chanson and Toombes (2002). It has no theoretical justification and it is only a rough correlation. The velocity distribution results in skimming flows are basically identical to measured velocity distributions in self-aerated flows on smooth-invert chutes (Cain, 1978), although the rate of energy dissipation is much greater on a stepped cascade and flow resistance is dominated by form drag.

Due to turbulent eddies; the local instantaneous flow velocity (u) within the boundary layer will fluctuate in a more or less random fashion. However, the velocity record for turbulent flow includes both a mean and a turbulent component.

The decomposition of the flow velocity is as follow:

$$u=\bar{u}+u' \quad \rightarrow \quad u'=u-\bar{u} \tag{9}$$

This is commonly called a Reynolds' decomposition, where:

$\bar{u}=$ Time-averaged velocity which equals to $\lim_{T\to\infty}\frac{1}{T}\int_0^T u(t)dt$

$u'=$ Velocity fluctuation in the stream-wise direction also called as characteristic turbulent velocity.

T= sampling time

Two methods are used to describe the turbulence of the flow; single point observations and correlations between multiple points. The more common method uses measurements of velocity fluctuations at one point in the flow. So, an accurate measure of the turbulent intensity can be obtained.

Herein, the turbulent intensity is defined in terms of the statistical quantities which describe the random fluctuations. A convenient measure of the fluctuating component of the flow velocity is the root-mean-square value as below:

$$u'_{rms}=\sqrt{\overline{u'^2}}=\sqrt{\frac{\sum_{i=1}^{n}\left(u_i-\bar{u}\right)^2}{n}} \tag{10}$$

Where n is the number of samples.

The normalized turbulence intensity in flows over a spillway dam is subject to the influence of flow patterns of the boundary layers as well as boundary conditions of different sections (e.g. the dam crest, downstream slope, surface roughness). Turbulence Intensity is a dimensionless measure of the turbulent fluctuations of

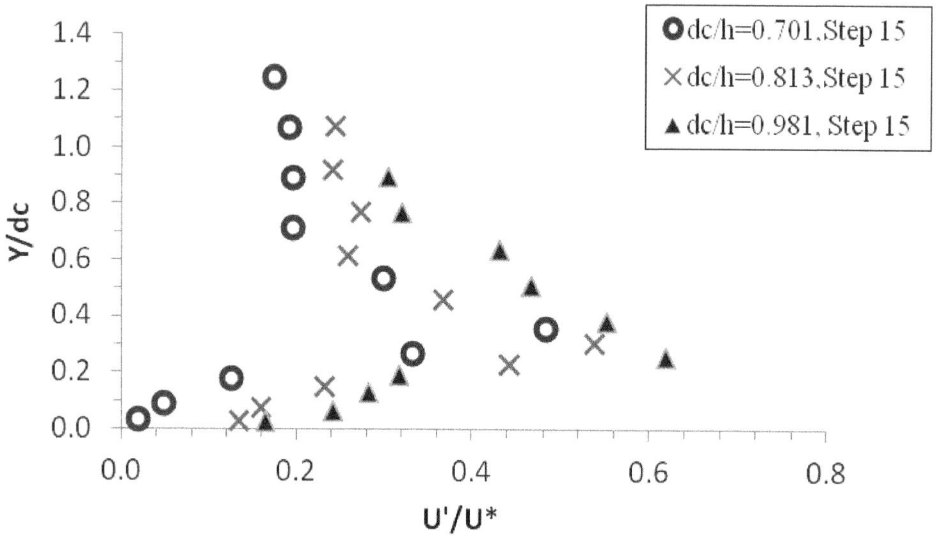

Figure 7. Vertical distribution of turbulence intensity in air-water flow over steps.

interfacial velocity which characterizing turbulence and defined as the root-mean-square value of the fluctuations in the flow direction divided by the shear velocity.

$$T_U = \frac{u'_{rms}}{u^*} \quad (11)$$

Where the u* is the shear or friction velocity and can be determined from the boundary shear stress distribution based upon the friction slope S_f, if the flow is gradually varied.

The shear velocity have a great impact on the distribution of the mean velocity and turbulence structure in the boundary layer and is the most fundamental velocity scale with which to normalize mean velocity and turbulence.

$$u^* = \sqrt{\tau_o / \rho_w} \quad (12)$$

$$\tau_0 = g\rho_w S_f \int_{y=0}^{y=y_{90}} (1-C)dy \quad (13)$$

Where the frictions slope $S_f = -\frac{\partial H}{\partial x}$ is the slope of the total head line.

Figure 7 shows the normalized turbulent intensities (Tu) distributions at different verticals and relative discharges (d_c/h). The data were obtained at step edges and at the same dimensionless distance from the inception point.

It can be seen from Figure 6 that the turbulence characteristics vary in different regions. Herein, the turbulence intensity increases rapidly close to the step bottom at the viscous sub layer and maximized for 0.4≤ y/dc ≤ 0.5 at intermediate region then decreases gradually in the upper region. A similar trend was observed in previous studies (Chanson and Toombes, 2002; Gonzalez, 2005; Chanson and Carosi, 2007a). It is hypothesized that the high turbulence levels in this intermediate region were caused by the continuous deformations and modification of the air-water interfacial structure.

Conclusions

Flows cascading down a stepped spillway with a moderate slope are characterized by some strong aeration and high turbulence of the flow. In this study, a set of air-water flow experiments were preformed in high velocity flows on a relatively large stepped channel based on Froude similitude.

The two phase flow measurements were preformed with a conductivity double-tip probe downstream of the inception point of aeration. The results included air water flow properties such as air concentration, flow velocity, turbulence, and bubble count rate. Based on measured air-water velocities, the flow shear stress, friction velocity and turbulence intensities were estimated accurately. The air-water flow properties presented some basic characteristics that were qualitatively and quantitatively in agreement with previous studies. High turbulence levels were recorded downstream of the aeration inception point. The distributions of turbulence intensity had a similar shape as clear-water flow data of Ohtsu and Yasuda down a stepped chute.

The measurements showed that immediately downstream of the inception point of air entrainment, the flow is rapidly varied and the maximum turbulence intensities occurred in the intermediate region between the spray and bubbly flow regions (that is, 0.3 < C < 0.7).

Although the findings were obtained for a moderate slopes (θ = 21.8°), it is believed that the outcomes are valid for a wider range of chute geometry and flow conditions typical of embankment chutes.

ACKNOWLEDGEMENT

The funding for these and related studies was provided by the Guilan Regional Water Company. Also, the authors acknowledge the helpful discussion with Prof. Chanson and Mr. Reza Rushan (Head of Hydraulics structures Department of the Water Research Institute, Power Ministry, Tehran, Iran).

Nomenclature: C, Void fraction defined as the volume of air per unit volume; **DH,** hydraulic diameter (m); **d,** characteristic depth (m); **dc**, critical flow depth (m); **Fr*,** Froude number defined in terms of the roughness height; **H,** total head (m); **Hmax,** maximum head available (m) using the spillway toe as datum; **H,** height of steps (m) (measured vertically); **ks,** step dimension (m) measured normal to the flow direction; **l,** length of steps (m) (measured horizontally); **Qw,** water discharge (m^3/s); **qw,** water discharge per unit width (m^2/s); **Re,** Reynolds number defined in terms of the depth-averaged flow velocity and **Sf,** hydraulic diameter friction slope; **T,** sampling time(s); **Tu,** turbulence intensity; **Uw,** equivalent clear-water flow velocity (m/s); **u',** root mean square of longitudinal component of turbulent velocity (m/s); $\bar{u}$, time-averaged velocity(m/s); **u*,** shear or friction velocity(m/s); **W,** stepped spillway width (m); **we,** Weber number; **x,** longitudinal/streamwise distance (m); **xl,** distance (m) from the start of growth of boundary layer to the inception point of air entrainment; **y,** distance (m) from the pseudo-bottom (formed by the step edges) measured perpendicular to the flow direction; y_{90}, characteristic depth (m) where the air concentration is 90%; τ_0, boundary shear stress (Pa).

Subscripts: c, Critical flow conditions; **f,** friction; **I,** characteristics of the inception point; **I,** a discrete point; **max,** maximum value in a cross-section; **rms,** root mean square; **w,** water flow; **90,** flow conditions where C = 0.90.

REFERENCES

Boes M (2000). Role of natural and immune IgM antibodies in immune responses. Mol. Immunol. 37:1141–1149.

Boes RM, Hager WH (2003). Hydraulic Design of Stepped Spillways. J. Hydr. Eng. pp. 671-679.

Chamani MR, Rajaratnam N (1994). Jet Flow on Stepped Spillways. J. Hydr. Eng. A.S.C.E. 120(2):254-259.

Chamani MR, Rajaratnam N (1999). Characteristics of Skimming Flow over Stepped Spillways., J. Hydr. Eng. ASCE. 125(4):361-368.

Chanson H (1989). Study of Air Entrainment and Aeration Devices. J. Hydr. Res. IAHR. 27(3):301-319.

Chanson H (1994a) "Hydraulic Design of Stepped Cascades, Channels, Weirs and Spillways, Pergamon, Oxford, UK, Jan. P.292 (ISBN 0-08-041918-6).

Chanson H (1994b). Hydraulics of Skimming Flows over Stepped Channels and Spillways, J. Hydr. Res. 32:445-460

Chanson H (1995a). Hydraulic Design of Stepped Cascades, Channels, weirs and Spillways, Pergamon, Oxford, UK.

Chanson H (2001b). A Transition Flow Regime on Stepped Spillways? The Facts. In Proceedings of the 29th IAHR Biennial Congress, Beijing, China, Theme D. Vol.1. Edited by G. LI.Tsinghuna University Press, Beijing, pp.490-498.

Chanson H, Cummings PD (1996). Air-Water Interface Area in Supercritical Flows down Small-Slope Chutes. Research Report No. CE151, Dept of Civil Engineering, University of Queensland, Australia, P. 60.

Chanson H, Toombes L (2002). Air-Water Flows down Stepped Chutes: Turbulence and Flown Structure Observations. Int. J. Multiphase Flow. 28(11):1737-1761.

Chanson H, Toombes L (2004). Hydraulics of Stepped Chutes: the transition flow. J. Hydr. Res. 42:43-54.

Chanson H, Carosi G (2007a). Advanced post-processing and correlation analyses in high-velocity air–water flows. Springer, J. Environ. Fluid Mechanics, 7:495–508.

Chinnarasri C, Wongwises S (2006). Flow patterns and energy dissipation over various stepped chutes, J. Irrig. Drain. Eng. 132(1):70–76.

Christodoulou GC (1993). Energy Dissipation on Stepped Spillways, J. Hydr. Eng. A.S.C.E. 19(5):644-650.

El-Kamash MK, Loewen MR, Rajaratnam N (2005). An Experimental Investigation of jet flow on a stepped chute, J. Hydr. Res. 43:31-43.

Gonzalez CA (2005). An Experimental Study of Free-Surface Aeration on Embankment Stepped Chutes, Ph.D. Thesis, Department of Civil Engineering, The University of Queensland, Brisbane, Australia, P.240.

Peterka AJ (1953). The effect of entrained air on cavitation pitting, Joint Meeting Paper, IAHR/ASCE, Minneapolis, Minnesota, Aug.

Peyras L, Royet P, Degoutte G (1991). Ecoulement et Dissipation sur les Déversoirs en Gradins de Gabion (Flow and Energy Dissipation of Energy on Gabion Weirs), La Houille Blanche, 1:37-47.

Rajaratnam N (1990). Skimming Flow in Stepped Spillways. J. Hydr. Eng. 116(4):587-591.

Relvas AT, Pinheiro AN (2011). Stepped chutes lined with wedge-shaped concrete blocks: Hydrodynamic pressures on blocks and stability analysis, Can. J. Civ. Eng. 38(3):338–349.

Sorensen RM (1985). Stepped Spillway Hydraulic Model Investigation, J. Hydr. Eng. ASCE. 11(12):1461-1472.

Toombes L (2002). Experimental Study of Air-Water Flow Properties on Low-gradient Stepped Cascades. Ph.D. thesis, Dept of Civil Engineering, University of Queensland, Brisbane, Australia.

Toombes L, Chanson H (2008). Flow patterns in nappe flow regime down low gradient stepped chutes, J. Hydr. Res. 46(1):4–14.

Wood IR (1991). Air Entrainment in Free-Surface Flows, IAHR Hydraulic Structures Design Manual, No. 4, Hydraulic Design Considerations, Balkema Publ., Rotterdam, The Netherlands, P.149.

Yasuda Y, Chanson H (2003). Micro -and Macro-Scopic study of two-phase flow on a stepped chute, Proc. XXX IAHR Congress, Thessaloniki, Greece, 695-703.

Macrophyte waste stabilization ponds: An option for municipal wastewater treatment

Mumtaz Shah[1]* and Hashim Nisar Hashmi[2]

[1]Department of Civil and Environmental Engineering, University of Engineering and Technology (UET) Taxila, Pakistan.
[2]Department of Civil Engineering, University of Engineering and Technology (UET) Taxila, Pakistan.

The objective of this research is to evaluate the performance of macrophyte waste stabilization pond system for municipal wastewater collected from Taxila (Pakistan). A model of macrophyte waste stabilization pond system was operated for six trials with each trials comprising different detention times that is 3, 5, 7 and 10 days, respectively. For the treatment, locally available macrophyte (water hyacinth) was used. To evaluate the performance of macrophyte waste stabilization pond, BOD_5, TDS, TSS, COD, faecal coliform and Nitrogen for the effluent from pond model were measured at each detention time of every trial after ensuring steady state conditions. The influent values of same parameters have been measured at the start of each trial as the wastewater sample was collected from the municipal sewer. The average reduction of effluent value of each parameter varies from 30 to 48%, that is BOD_5 (48%), TDS (31%), TSS (30%), COD (42%), faecal coliform (45%), and Nitrogen (40%). The mechanisms of pollutant removal in this system include both aerobic and anaerobic microbiological conversions, sorption, sedimentation, volatilization and chemical transformations. There is currently no macrophyte wastewater waste stabilization pond system under study/practice in many of the developing countries; therefore, for the promotion of macrophyte based waste stabilization system detailed pilot scale studies should be conducted for municipal sewage under local environmental conditions to explore the treatment efficiency of the system. Also, it is necessary that macrophyte system be used as treatment facility for wastewater because their performance is comparable to conventional wastewater treatment plants and since the system has very low operation and maintenance costs.

Key words: BOD_5, *Eicchornia crassipes*, macrophytes, municipal sewage, water hyacinth, wastewater treatment.

INTRODUCTION

Pakistan is located in a severe water-scarce zone, which increases the scope of the wastewater treatment and its reuse. Treatment plant has been constructed and being operated in different cities of Pakistan; however, due to financial constraints, non availability of skilled operators and electricity shortages most of them are in a non-functional condition (Aziz, 2005). Treatment plants have been installed in almost every large city of Pakistan, like Karachi (Two 20 MGD wastewater treatment plants employing trickling filters) and Islamabad (a 5 MGD capacity activated sludge plant), although most of them are non functional as far as performance is concerned. The main reasons of their failure are electricity shortages and lack of skilled operators.

Oxidation ponds have been found to be an effective means of low cost treatment of domestic and industrial wastewater, especially for those countries lying in tropical and subtropical regions of the world (Kharkar et al., 1972). The oxidation pond treatment technology has been recommended by a number of workers (McGarry and Pescod, 1970; Mara, 1975; Oswald, 1975; Middlebrooks et al., 1981). The technology has an edge over the mechanical systems as being technologically very simple, economically very cost effective and a higher degree of pathogens removal is achieved. The conventional wastewater treatment trickling filter plants are very efficient in removing suspended solids and organic matter (more than 85%) and usually very poor in removing nitrogen, phosphorous, heavy metals, non-biodegradable organics, bacteria and viruses. Oxidation ponds have been employed for the treatment of wastewater for over 3000

*Corresponding author. E-mail: mumtazshah@gmail.com.

Table 1. Summary of wastewater treatment plant in Pakistan.

Characteristics	Treatment plants			
Treatment technology	Trickling filter - Karachi	Trickling filter - Karachi	Aerated lagoon - Karachi	Activated sludge -Islamabad
Year of construction	1963	1965	1985	1962
Location	TPI Sher Shah	TPII Mahmoodabad	North Karachi	Islamabad
Capacity	20 MGD	20 MGD	5 MGD	5 MGD
Type	Two stage high rate	Two stage high rate	—	—
Design and Make	Dorr Oliver, Holland	Dorr Oliver, Holland	Karachi Development Authority	—
Primary clarifier	4 (138 ft dia, 12 ft) deep	4 (138 ft dia, 12 ft) deep	2 (35 ft × 20 ft)	2 (18 ft dia, 9 ft deep)
Trickling filter	8 (136 ft dia, 7 ft deep)	8 (136ft dia, 12 ft deep)	—	—
Secondary clarifier	4 (128 ft dia, 12 ft deep)	4 (138 ft dia, 12 ft deep)	2 (63 ft × 30 ft)	2 (120 ft dia, 12 ft deep)
Aerator	—	—	16 (Oxygen requirement = 980 kg/day)	—
Aeration unit/ Aeration tank	—	—	4 (Semi corrusel 6.5 ft deep, 1.25 MGD)	2 (111 x 30 × 18 ft)
Capital cost million Rs.	30.5	30.2	15.85	8.0
Operation and maintenance cost per year (Million Rs.)	0.084	0.084	0.168	0.275
Present status	Non operative	Non operative	Non operative	Non operative

WASA & NESPAK (2002).

years. Similarly, by employing the macrophyte in waste stabilization ponds their efficiency can be further enhanced. Macrophytes are common features of an aquatic ecosystem. Various types of macrophytes are generally observed in an aquatic ecosystem such as Floating plants (*Eichhornia crassipes* (Water Hyacinth), Spirodela (Duckweed), *Salvinia molesta* (Salvinia) and Emergent plants (*Schoenoplectus validus* (Great Bulrush), *Juncus ingens* (Giant Rush), Phragmites (Common Reed), *Typha spp.* (Cumbungi or Cattail). For the developing countries that have limited resources for the construction and operation of conventional treatment plants, waste stabilization pond with macrophyte is the most economical solution. Additional benefits are that they serve as wildlife habitat and recreation purposes. The cost effective method of treatment makes it a potentially economical and suitable option for bio- and phyto-remediation.

In Pakistan, the wastewater treatment plants installed in major cities are mostly non operative due to the non availability of the skilled workforce as well as due to electricity crisis. The details of some of the plants are given in Table 1.

Currently in Pakistan, waste stabilization ponds with macrophyte have not yet been recognized as a treatment option for the wastewater treatment. Failure for the acceptance is due to the lack of information, research work and government policy to prove the effectiveness of the system for wastewater treatment. Pakistan has wide area under natural wetlands in coastal areas of Sindh and Baluchistan province in form of mangrove forests that can be successfully used for treatment and polishing. In the proposed study the municipal wastewater of Taxila (Punjab, Pakistan) is taken for treatment from a bench scale laboratory model of a macrophyte waste stabilization pond. Since water hyacinth is the only locally available macrophyte therefore, the same has been chosen for the study.

MATERIALS AND METHODS

All the primary treated effluent is sent to the inlet zone of the model whereby passing through the wastewater reaches the main tank where it gets treatment with macrophyte (water hyacinth) under different detention times (3, 5, 7 and 10 days). After adding water hyacinth to the pond model different parameters of wastewater was monitored for both the raw sewage as well as for the wastewater treated by water hyacinth in the model. Both the raw sewage as well as the treated effluent has been tested for a set of parameters (BOD_5, TDS, TSS, COD, faecal coliform and Nitrogen). All tests performed on the wastewater were carried out according to the procedures laid down in "Standard Methods for the Examination of Water and Wastewater" (20th edition, 1998). Monthly average air temperature during the whole period of sampling varied between 25 and 29.4°C with an average of 26.9°C. The initial values of the selected parameters for raw sewage are given in Table 2.

Description of the treatment system

The work reported here was carried out during the dry season from April 2011 until June 2011. All effluent collected from the municipal sewer was first sent to a primary settling tank and then after passing through it is distributed in to a lagoon having the macrophyte (water hyacinth). The effluent has been retained for 3, 5, 7 and 10 days in lagoon during each trail, and total of six trials has been performed. After completion of each detention time sample was collected from lagoon and tested for effluent quality. The plants in lagoon were harvested two to three times per week. At each detention, the following parameters were measured during daytime: BOD_5, chemical oxygen demand (COD), total dissolved solids (TDS), total suspended solids (TSS), nitrogen, faecal coliform and pH. All analyses were done according to Standard Methods for the Examination of Water and Wastewater (1995). Monthly average air temperature during the whole period of sampling varied between 25 and 29.4°C with an average of 26.9°C. The plants maintained a surface covering of 80 to 100%. The plants were quite fragile; they cannot resist too much handling during collection and transportation. The construction detail of the lagoon is given below in the Table 3. The Schematic Diagram of the system is shown in

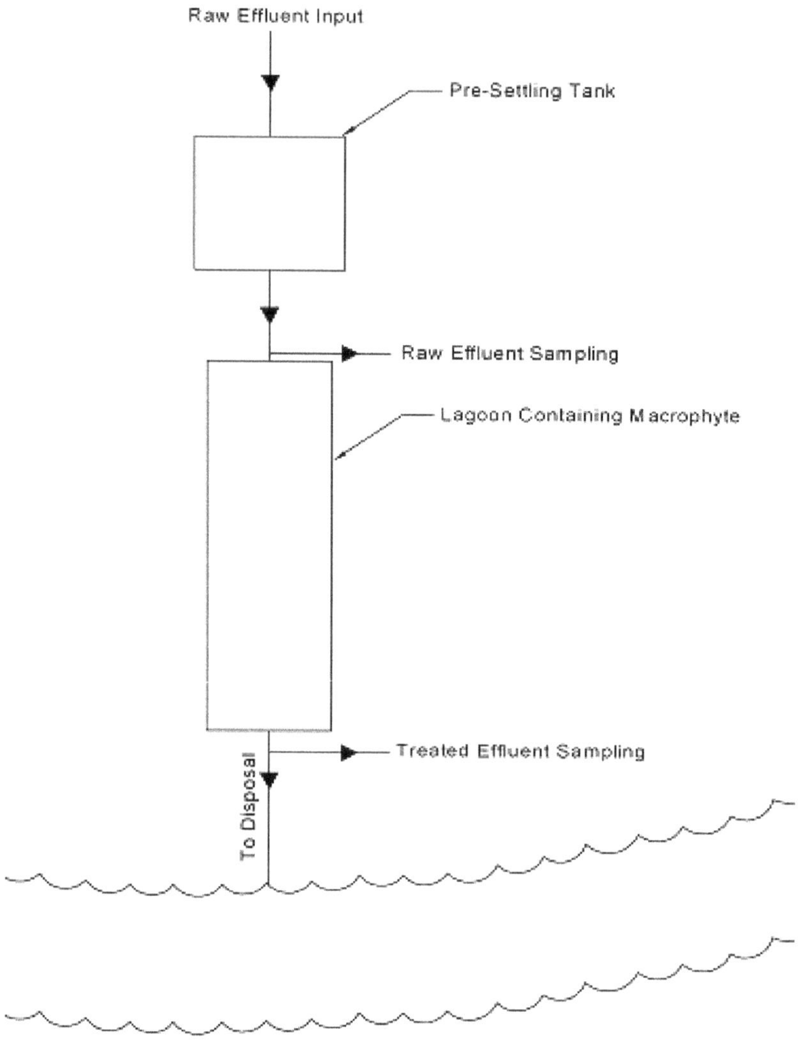

Figure 1. Schematic diagram of the treatment system.

Table 2. Raw sewage values.

S/N	Parameter	Unit	Value	Standard deviation
1	BOD_5	mg/L	132	23
2	COD	mg/L	236.33	41.29
3	Nitrogen	mg/L	2.65	1.0
4	TDS	mg/L	723.33	70.62
5	TSS	mg/L	98	7.18
6	Faecal coliform	No./100 ml	1361.67	320.53

Figure 1.

RESULTS AND DISCUSSION

Average pollutant concentrations of the treated effluent samples collected at different detention times during six trials is given in Table 4. Effluent concentrations of all the collected samples are used to compare the percentage degree of treatment by macrophytes under different detention times.

The results demonstrate the removal efficiencies of the waste stabilization pond system expressed as percentage of input concentration. The results clearly indicates average reduction of 30% in TDS, 31% in TSS,

Table 3. Model characteristics.

S/N	Characteristics	Unit	Value
1	Depth	ft	4
2	Length	ft	18
3	Width	ft	6
4	Cross section	--	Trapezoidal
5	Retention time	days	Variable

Table 4. Results of treated effluent.

Trials	Detention (days)	Concentration							Average reduction
		pH	TDS (mg/L)	TSS (mg/L)	Nitrogen (mg/L)	BOD_5 (mg/L)	COD (mg/L)	Faecal coliform (No./100 ml)	
1	3	8.4	810	84	1.06	106.9	180	904.7	TDS = 5% TSS = 59% Nitrogen = 33% BOD_5 = 41% COD = 63% F-Coli = 45%
	5	8.2	800	70	0.93	97.8	166	839.3	
	7	7.7	790	54	0.80	82.3	130	654	
	10	7.5	780	42	0.71	76	84	599.5	
2	3	8.4	694	77	2.80	124	227	946.2	TDS = 30% TSS = 31% Nitrogen = 32% BOD_5 = 48% COD = 40% F-Coli = 45%
	5	8.1	624	71	2.40	112	203	877.8	
	7	7.6	601	65	2.11	91	189	684	
	10	7.3	546	62	1.91	79	162	627	
3	3	8	486	82	2.3	78	143	1195.2	TDS = 30% TSS = 31% Nitrogen =41% BOD_5 = 47% COD = 40% F-Coli = 45%
	5	7.9	436.8	75	2.2	70	128	1108.8	
	7	7.7	420	68	2.2	57	119	864	
	10	7.5	382.2	66	2	50	102	792	
4	3	8	340	95	2.85	101	185	1494	TDS = 30% TSS = 31% Nitrogen = 40% BOD_5 = 48% COD = 40% F-Coli = 45%
	5	7.9	305.76	87	2.50	91	165	1386	
	7	7.7	294	79	2.11	74	154	1080	
	10	7.5	267.54	76	1.92	64	132	990	
5	3	7	238	84	1.07	110	202	863	TDS = 30% TSS = 31% Nitrogen =40% BOD_5 = 48% COD = 40% F-Coli = 45%
	5	6.9	214.032	77	0.94	99	180	801	
	7	6.9	206	71	0.79	81	168	624	
	10	6.5	187.278	68	0.72	70	144	572	
6	3	7	163	79	2.49	131	245	1361	TDS = 31% TSS = 34% Nitrogen = 40% BOD_5 = 48% COD = 38% F-Coli = 44%
	5	6.9	149.8224	71	2.18	118	219	1262	
	7	6.9	142	66	1.85	96	204	1014	
	10	6.5	129.2218	61	1.68	83	179	929.6	

40% in Nitrogen, 48% in BOD_5, 42% in COD and 45% in faecal coliform. The results show higher removal efficiency for the Nitrogen, BOD_5 and faecal coliform; however, removal of TDS and TSS is not much as expected.

The results shows after three days of detention there is not much reduction in pollutant values but as the detention time increase the reduction of pollutant in the sewage also increase. During the entire study water hyacinth shows a good growth rate; however, their growth is affected due to the decrease in temperature, that is, during winter. Therefore the system shows excellent performance during the summer season and so it can be used for the treatment of municipal wastewater of developing countries like Pakistan having tropical environment.

Conclusions

1) In many developing countries municipal sewerage systems carry mixed wastewater rich in inorganic and organic toxic pollutants which may inhibit microbial processes, and hence reduce the macrophyte based treatment system efficiency. Characteristics of the wastewater to be treated must be considered in the design.
2) Appropriate choice of species adapted to tropical environments is of great significance. In the tropics where growth rates are high, the frequency and hence the cost of harvesting has to be considered. Economic utilization of excess biomass and frequent harvesting costs should be well assessed before choosing such a plant.
3) Currently in Pakistan almost all installed mechanized wastewater treatment systems are not working properly due to operation and maintenance problems associated with these systems. Therefore there is dire need to search for an alternative wastewater treatment system like macrophyte waste stabilization pond system using the local macrophyte species for wastewater treatment.
4) The main reasons for opting for the macrophyte based treatment system includes commercial value of biomass. The sludge from the macrophyte waste stabilization pond can be used as fertilizer after some primary precautions and the sludge can be used for producing energy by anaerobic digestion.
5) The macrophyte wastewater stabilization pond system should be promoted for the municipal wastewater treatment of developing countries as it reduces the major wastewater pollutant to a large extent. Additionally, this system of treatment is also self sustainable which is a main reason for opting for this system than other mechanized systems. This will help in protecting the water sources from the pollution as well.

REFERENCES

Aziz JA (2005). Management of source and drinking-water quality in Pakistan. Eastern Med. Health J. 11:5-6.

Kharkar CB, Venkatesam TL, Tiwari AR (1972). Review of performance of the stabilization ponds at Bhilai. In: Proceedings of the symposium of low cost waste treatment, India.

Mara DD (1975). Proposed design for oxidation ponds in hot climates. Jour Inc. New York.

McGarry MG, Pescod MB (1970). Stabilization pond design criteria for tropical Asia. In: 2nd International Symposium for waste treatment lagoons, Kansas City.

Middlebrooks J, Middlebrooks CH, Reynods JH, Watters GZ, Reed SC, George DB (1981). Wastewater stabilization lagoon designing performance and upgrading. Macmillan, N. Y.

Oswald WJ (1975). Experience with new pond design in California. Presented before the conference "Ponds as wastewater treatment alternative. University of Texas. Austin.

WASA & NESPAK (2002). Lahore Master Plan. Lahore Development Authority (LDA). Pakistan.

Characteristics of total solar radiation in an urban tropical environment

Adeniyi M. O.*, Nymphas E. F. and E. O. Oladiran

Department of Physics, University of Ibadan, Ibadan, Nigeria.

The maximum total solar radiation in Ibadan (7.38° N, 3.93° E), Nigeria occurred around 13 and 14 h local time (LT) and it varied from an average value of 584.3±10.9 Wm^{-2} in the wet season (April to October) to 642.5±10.8 Wm^{-2} in the dry season (November to March) throughout the period of 1997 to 2001. Maximum total solar radiation had two peaks, one in March/April/May with value 750.3±41 Wm^{-2} and the other in October/November with value 633.1±63.5 Wm^{-2}. The minimum was in July/August with the value 404.8±54 Wm^{-2}. Diurnal total solar radiation had a similar variation with mean day total solar radiation; the latter had its maxima and minimum values to be 564.7±19, 488±36 and 308.4±40 respectively. The annual trend of solar radiation was positive throughout the period considered, which revealed global brightening. The values of total solar radiation observed in Ibadan compared well with those obtained at Ife (7.55° N, 4.56° E) a nearby tropical station. A time lag of about 3 h between the peak of the diurnal temperature and solar radiation was observed. During the month of July or August as the case may be, there was little or no time lag between the time of maximum total solar radiation and maximum diurnal temperature. Significant intra-annual solar radiation cycles (at 0.05 significant level) of 360 to 370 days (1 per year); 180 to 200 days (2 per year), 90 to 121 days (3 per year), 45 to 70days (5 per year); 20 to 35 days (10 per year), 4 to 16 days (20 per year) were obtained for all the years (1997 to 2001).

Key words: Total solar radiation, lag, intra-annual solar radiation cycles, global brightening.

INTRODUCTION

Total solar radiation, which is global radiation on a horizontal surface, is the sum of incident diffuse radiation and the direct normal radiation projected onto the horizontal surface. If the surface under study is tilted with respect to the horizontal, the total solar radiation is the sum of incident diffuse radiation, the direct normal radiation projected onto the tilted surface and ground reflected radiation that is incident on the tilted surface. The combination of diffused and direct solar energy incident on a horizontal plane at the Earth's surface is referred to as global solar energy and all three quantities (specifically their rate or irradiance) are linked mathematically by the following expression:

$$I_g = I_d + I_b \cos z \qquad (1)$$

where I_g = global solar radiation on a horizontal surface, I_d = diffuse solar radiation, I_b = direct solar radiation on a surface perpendicular to the direct beam, z = Sun's zenith angle.

All the atmospheric, earth, underground and ocean processes depend mainly on solar radiation. Solar radiation is used as passive sensors in remote sensing. It has also been harnessed and used in many forms like (i) direct thermal application: for heating and cooling of buildings and water, distillation, for drying and processing of food stuffs, in irrigation and generation of power (ii) solar electrical application: conversion to electricity. (iii) Fuels from Biomass: the heat from the sun serves as catalyst in decomposition or any chemical reactions that involve heat/light, for example, photosynthesis and in the development of plants for medical use (Monteith, 1975).

There was global dimming before 1985, but global radiation has started increasing again; that is, global

*Corresponding author. E-mail: mojisolaadeniyi@yahoo.com.

brightening has set in from that time on (Zurich, 2009). Variations in solar radiation have impact on weather and climate (Zurich, 2009). Solar eclipse of 29 March, 2006 was reported to influence solar radiance, temperature and wind speed (Founda et al., 2007; Gerasopolus et al., 2007; Nymphas et al., 2009).

Some works have been done on solar radiation all over the globe of which only few are in the tropics. These include Kane (1977) 's work on interannual variability of global radiation at Wageningen. Jegede (1995) worked on seasonal characteristics of global, diffuse and long wave radiation at Ile-Ife, Nigeria. El Bakry (1994) worked on net radiation over the Aswam high dam lake. Jegede (1997a) worked on diurnal variations of net radiation in a tropical station-Osu, Nigeria. Jegede (1997b) also presented the daily averages of net radiation measured at Osu in 1995. Results of works done in urban tropical area are scarce. Jegede et al. (2004) did micro-meteorological experiment where net radiation and global radiation were measured but for short period.

Ibadan is an urban tropical city having a population of about 5 million with a land mass of 20 × 20 km^2. It has been shown that urbanization leads to a very high spatial concentration of energy use, as urban population increases by 1%, the growth in consumption of energy is found to be 2.2% (Jones and Williams, 1992). As urbanization is increasing in Ibadan so also the energy consumption is increasing. Nigeria's urban growth rate is 3 to 5 times greater than rural growth rate and this figure doubles almost every decade (nigeriafirst.com). Ibadan is one of the 15 cities, which the Federal Government of Nigeria has designed the urban management strategy and facility for; these are provided in these cities (Kwanashie, 2003). Ibadan is also one of the twelve cities chosen in each twelve countries of the world, selected as demonstration models in preparation for the World Summit, Habitat II in Istanbul, Turkey in June 1996. The project was tagged "Sustainable Ibadan Project" (SIP) under the World Bank, United Nations Development Programme (UNDP) and the United Nations centre for Human Settlements (UNCHS) (Habitat) funded the Urban Management Program (UMP) for the whole world (Ogunlami, 2004).

The fact that Ibadan is a recognized growing urban city made it clear that it needs a lot of energy for everyday use. Because of the erratic nature of power supply and economic considerations, the energy need of the people has not been met. Interest in research in climate change (global warming) necessitates analysis of solar radiation data, also optimal sitting of solar energy systems requires knowledge of the solar radiation characteristics at any given location. Information on inter-annual solar cycles is available but not intra-annual cycles (Mordvinov, 2003). The data sets used in this investigation happen to be the available long period data in this area. This necessitates this investigation. This investigation provides solar radiation variation that can represent other urban tropical stations. It is therefore necessary to investigate the characteristics of total solar radiation, which is expected to be abundant in a tropical urban area like Ibadan, to verify the availability of the solar energy round the year; for some years. This will provide a basis for harnessing it.

METHODOLOGY

Diurnal variation of total solar radiation and temperature for the individual months from 1997-2001 at Ibadan were found and compared. Standard deviations were also obtained for the wet season and dry season maximum total solar radiation. The two maxima and one minimum values for the total solar radiation were also found for all the years considered. Annual trend of mean day total solar radiation was also found.

Daily averaged total solar radiation for the period 1997 to 2001 was subjected to harmonic analysis to detect the Amplitudes and cycles existing intra-annually. The Amplitudes and intra-annual cycles for the various years were compared.

Data acquisition

Total solar radiation and temperature data used in this investigation are hourly and daily averaged values for the years 1997 to 2001. The data collection station is located at International Institute of Tropical Agriculture (IITA) Ibadan Observatory where HMP 35C temperature, relative-humidity probe with accuracy of 0.2°C was used to monitor temperature and L1200X Pyranometer with accuracy of ±3% was used to measure global radiation. The diurnal data for the following periods were not available: July 1997, June 1999 and April 2001. Observation time was recorded as the mean local time in Nigeria. The height of the grass on the measurement site was not allowed to exceed 0.45 m.

Harmonic analysis

Harmonic analysis is used to reproduce atmospheric phenomena on different temporal and geographical scales (Coe and Stern, 1982; Kirkyla and Hameed, 1989; Sansigolo, 1993). Harmonic analysis makes use of Fourier series to determine the amplitude and the period of a certain harmonic in a set of data. A Fourier series is an expansion of a periodic function X_t in terms of an infinite sum of sines and cosines based on the orthogonality relationship between sine and cosine as shown in Equation (2).

$$X_t = \bar{X} + \sum_{i=1}^{N/2} a_i \cos\frac{2i\pi}{P}t + b_i \sin\frac{2i\pi}{P}t \qquad (2)$$

i takes values from 1 to $N/2$ and not to N because the remaining $N/2$ values is a conjugate harmonic of the first $N/2$ harmonics.

$$X_t = \bar{X} + \sum_{i=1}^{N/2} A_i \sin(\frac{2i\pi}{P}t + \phi_i) \qquad (3)$$

where $A_i = \sqrt{a_i^2 + b_i^2}$,

$$\phi_i = \tan^{-1}(\frac{-a_i}{b_i})$$

$$a_i = \frac{2}{N}\sum_{t=1}^{N}\left[X_t - \overline{X}\right]\cos\frac{2i\pi}{P}t$$

$$b_i = \frac{2}{N}\sum_{t=1}^{N}\left[X_t - \overline{X}\right]\sin\frac{2i\pi}{P}t$$

Here, $\overline{X}$ is the arithmetic mean, A_i is the amplitude of the harmonic, ϕ_i is the phase angle of the corresponding harmonic, N is the number of observations, P is the period of observation and X_t is the value of total solar radiation at time *t*. In this case N and P are equal to 365 in a normal year or 366 in a leap year. Comparison of the sizes of the amplitudes reveals the particular harmonics dominating the curve, which is represented by a series of trigonometric functions. The phase angle represents the day of the year when the maximum or minimum of a given harmonic occurs.

Determination of significant Harmonics

When the sizes of the amplitudes of all the harmonics are compared, and the standard deviation is found; the significant harmonics at 95% confidence level are those that are with amplitudes greater than 2 standard deviations (σ) (Kane, 1997).

RESULTS

The monthly diurnal variations of total solar radiation and the mean temperature for the period under consideration were found to be similar for each year. Figure 1 illustrates the diurnal variation of these parameters.

During the month of July or August as the case may be there was little or no time lag between the time of maximum total solar radiation and maximum diurnal temperature.

Figures 3 (a) and (b) revealed the similarity between the annual variation of maximum diurnal total solar radiation and mean diurnal total solar radiation, with greater magnitude of the maximum total solar radiation as expected.

There was a positive annual trend of total solar radiation during the considered period (Figure 4).

DISCUSSION

Diurnal variation of temperature and total solar radiation

Total solar radiation and mean temperature were observed to have the same diurnal variation with one peak period daily. Total solar radiation and temperature had low values in the morning and evening, and reached their peaks in the afternoon. There was an average time lag of about 3 h between the time of maximum temperature and maximum total solar radiation. In Ife (7.55° N, 4.56° E), another tropical station in Nigeria, afternoon peak of temperature lags behind that of global radiation as found here in Ibadan (Jegede et al., 2004). In Osu (7.43° N, 4.58° N) peak time of temperature lags behind that of net radiation by about 2 h (Jegede, 1997a). This time lag is expected due to the delay in heating the atmospheric column (which starts at the ground) by the incoming solar radiation (Garrat, 1992). This is explained thus: the atmosphere itself does not receive enough insolation that will raise its temperature to the maximum, the atmosphere is heated by the earth; the 3 h time lag amounts to the space of time between receiving of insolation from the sun by the earth and releasing some of the energy into the atmosphere to heat it and raise its temperature to the maximum. There was no time lag between the maximum total solar radiation and maximum temperature in August 1998. This was also noticed in July 1999 when the time of maximum total solar radiation and temperature was at 14 h local time. In July 1997, July 2000 and July 2001 there was reduced time lag (Figure 2). During these periods the maximum total solar radiation was small compared with all the other months. This may be due to the presence of clouds in the atmosphere since this is part of the rainy season. The insolation received before the maximum was able to raise the temperature to almost the maximum for that month and the maximum global radiation had little or no impact on the maximum temperature since it was not that higher than the incoming solar radiation a few hours before it.

The wet season is characterized by lower daytime total solar radiation of an average value of about 584.3±10.9 Wm^{-2}, while the dry season value is about 642.5±10.8 Wm^{-2} throughout the years1997 to 2001. The wet season value in a typical year is 592.7±107.1 Wm^{-2} and the dry season value is 644±76.5 Wm^{-2}. Annual fluctuations are very small (of the order of 10) Fluctuations of the hourly averages for the dry months are lower (σ about 76.5 Wm^{-2}) than that for wet months (σ about 107.1). The cloud and water vapour that are predominant during the wet season vary both spatially and temporally. These are responsible for the low intensity of the incoming solar radiation and the high standard deviation during these months (Iqbal, 1983; Kyle, 1991). During the dry months, there is no cloud cover to attenuate the incoming solar radiation and so the intensity of total solar radiation is high and the standard deviation is low. These values compared well with those measured at Chian Maig and Bangkok (13° 50`N, 100° 29`E) and Thailand by Exell (1976). Variation of diurnal total solar radiation obtained is comparable with those obtained in Ife a nearby tropical station. The maximum values are between 600 and 800 Wm^{-2} in Ife towards the end of February (Jegede et al., 2004).

Comparison of annual and diurnal characteristics of total solar radiation

Figure 3 (a) and (b) revealed the similarity between the

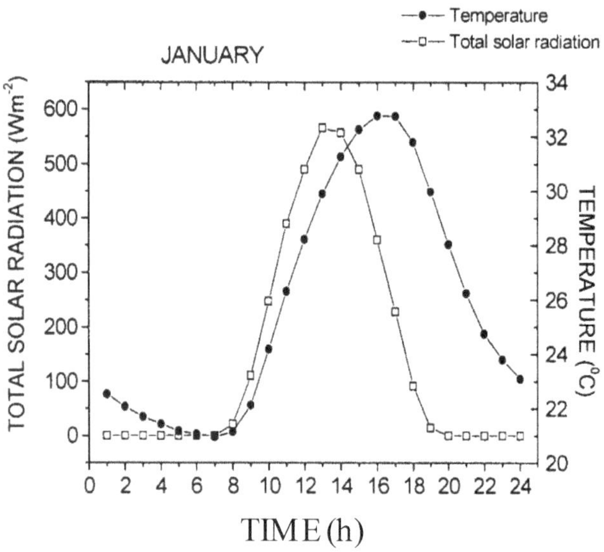
JANUARY
Temperature
Total solar radiation
TOTAL SOLAR RADIATION (Wm-2)
TEMPERATURE (0C)
TIME (h)

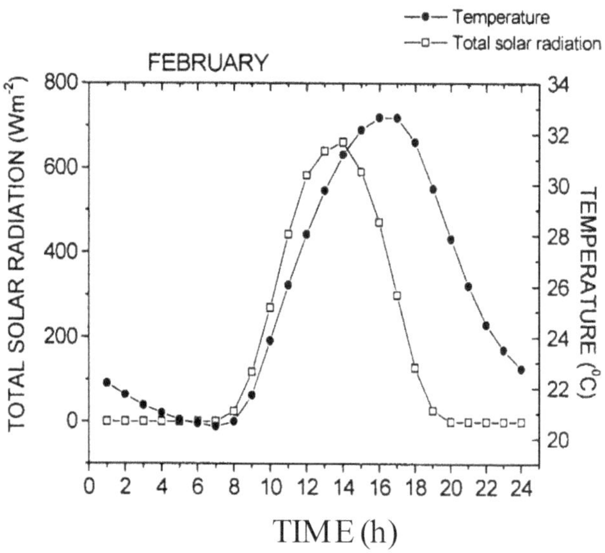
FEBRUARY
Temperature
Total solar radiation
TOTAL SOLAR RADIATION (Wm-2)
TEMPERATURE (0C)
TIME (h)

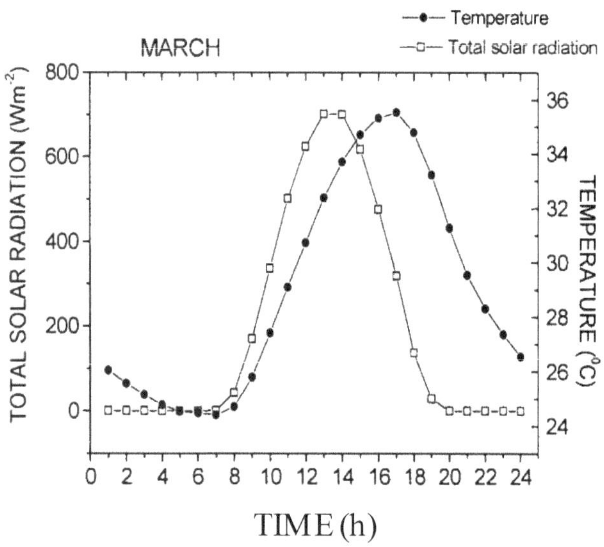
MARCH
Temperature
Total solar radiation
TOTAL SOLAR RADIATION (Wm-2)
TEMPERATURE (0C)
TIME (h)

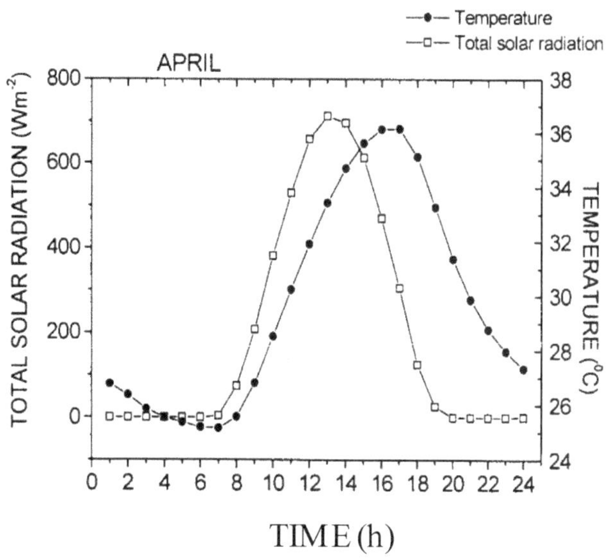
APRIL
Temperature
Total solar radiation
TOTAL SOLAR RADIATION (Wm-2)
TEMPERATURE (0C)
TIME (h)

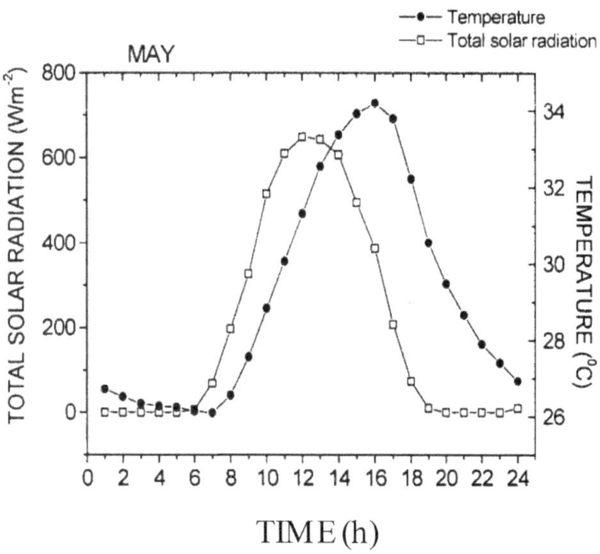
MAY
Temperature
Total solar radiation
TOTAL SOLAR RADIATION (Wm-2)
TEMPERATURE (0C)
TIME (h)

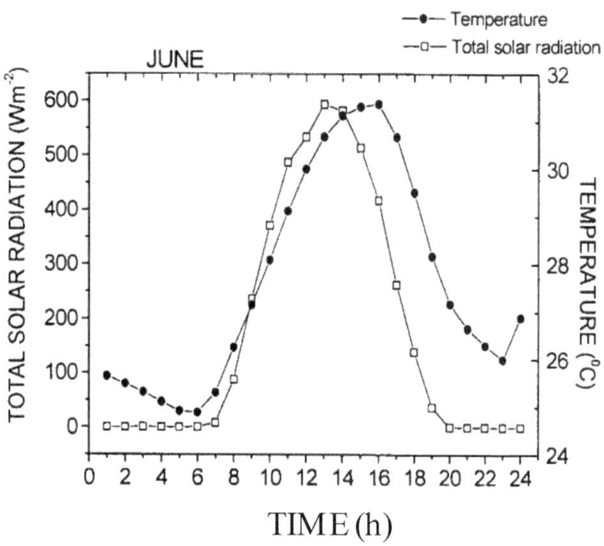
JUNE
Temperature
Total solar radiation
TOTAL SOLAR RADIATION (Wm-2)
TEMPERATURE (0C)
TIME (h)

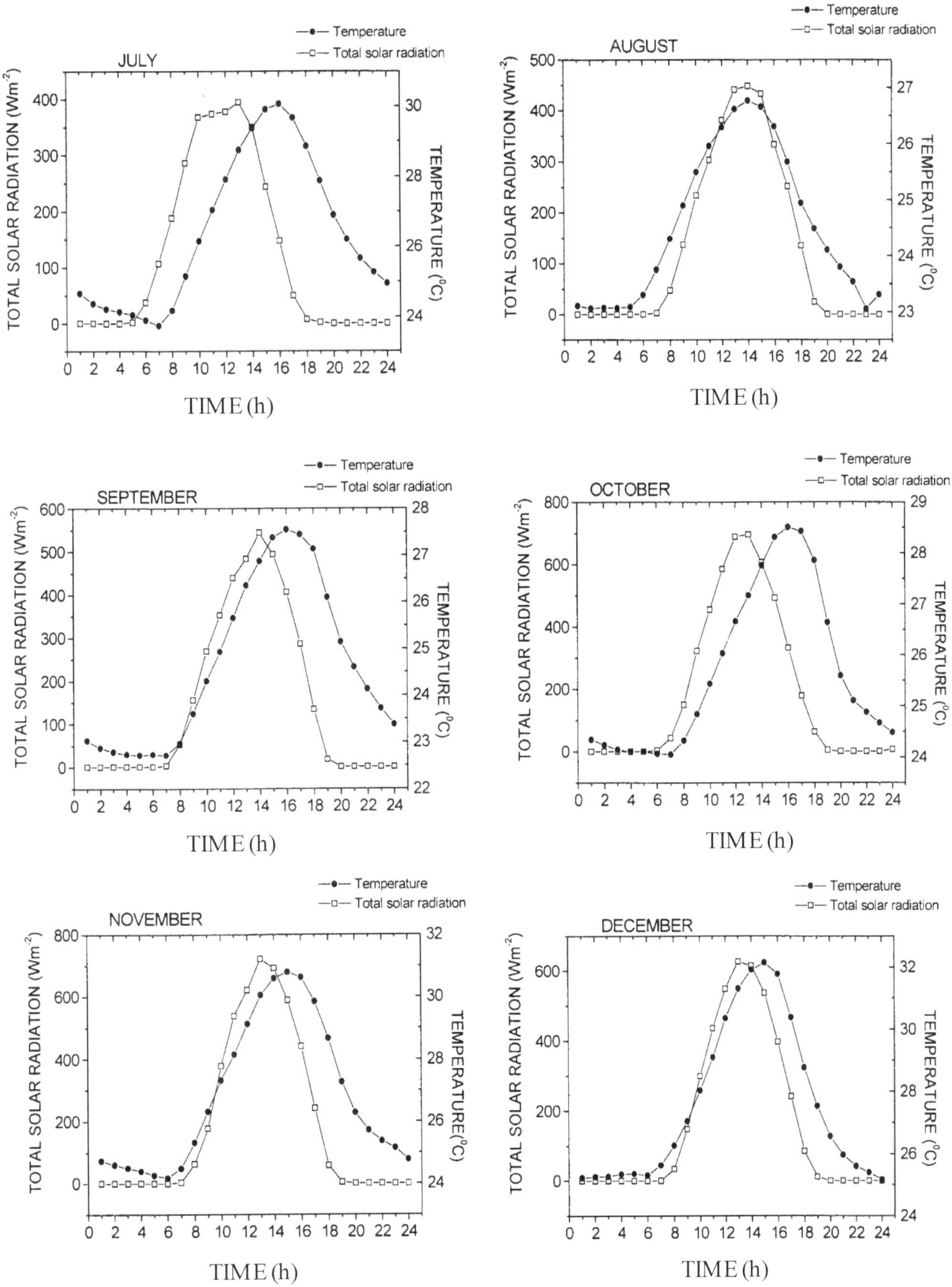

Figure 1. Diurnal variations of total solar radiation and temperature at Ibadan in 1998.

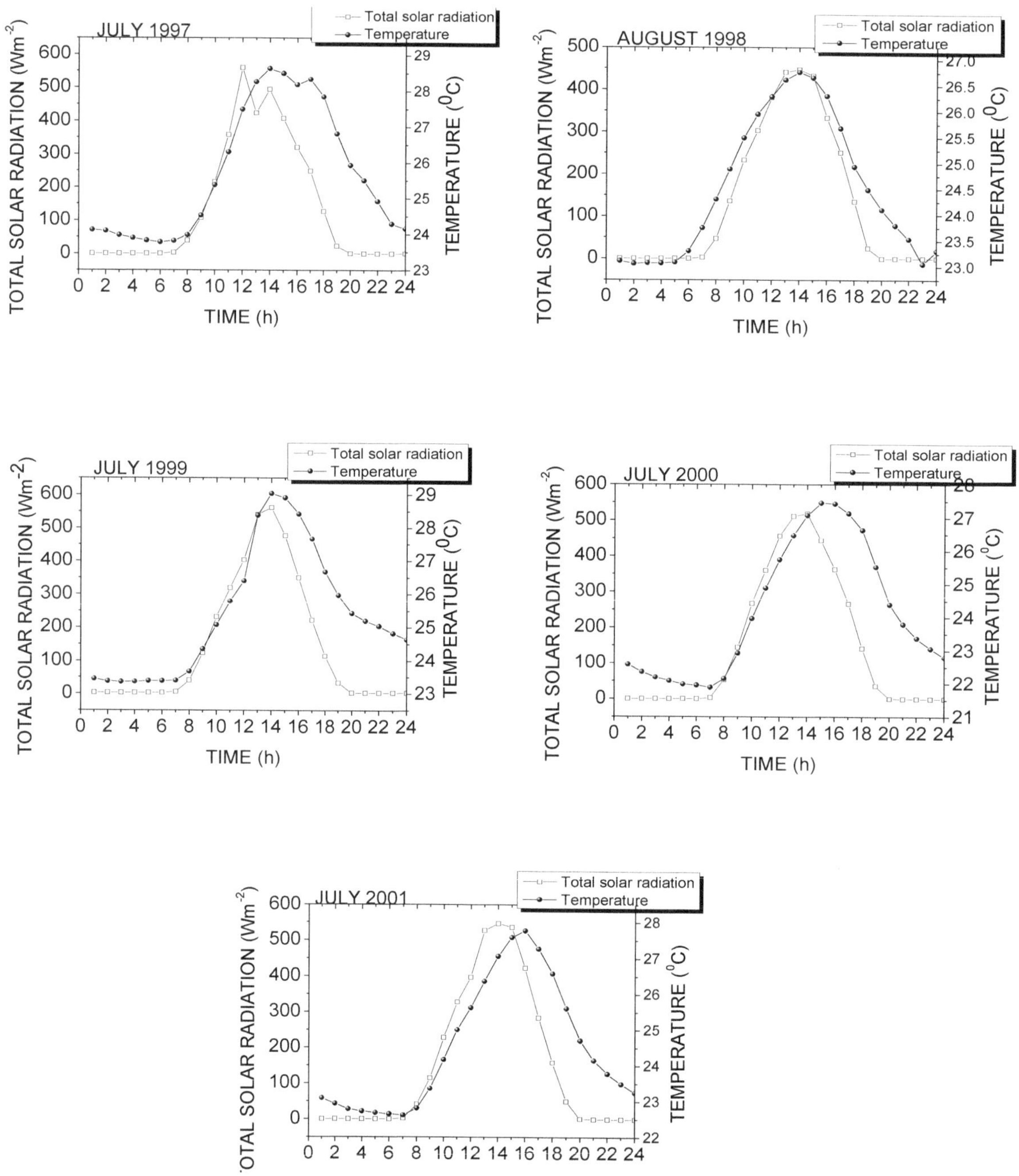

Figure 2. Time lag between the peak of total solar radiation and temperature in July/August at Ibadan from 1997 to 2001.

annual variation of maximum diurnal total solar radiation and mean diurnal total solar radiation, with greater magnitude of the maximum total solar radiation as expected. There are two peaks and one minimum in a year for both maximum and mean day total solar radiation. There is one peak in March/April/May with value 750.3±41 Wm^{-2} and the other in October/November with value 633.1±63.5 Wm^{-2} (for maximum). The minimum is in July/August with the value 404.8±54 Wm^{-2} (for maximum). Otunla et al. (2008) reported a similar variation in monthly temperature and total solar irradiance. The mean day total solar radiation values are 564.7±19, 488±36 and 308.4±40 respectively for the peak and minimum periods. Annual trend of total solar radiation was found to be positive with slope 3.9, this is similar to a report by Zurich (2009).

The following total solar radiation cycles are significant throughout the years 1997 to 2001:360 to 370 days (1 per year); 180 to 200 days (2 per year), 90 to 121 days (3 per year), 45 to 70 days (5 per year); 20 to 35 days (10

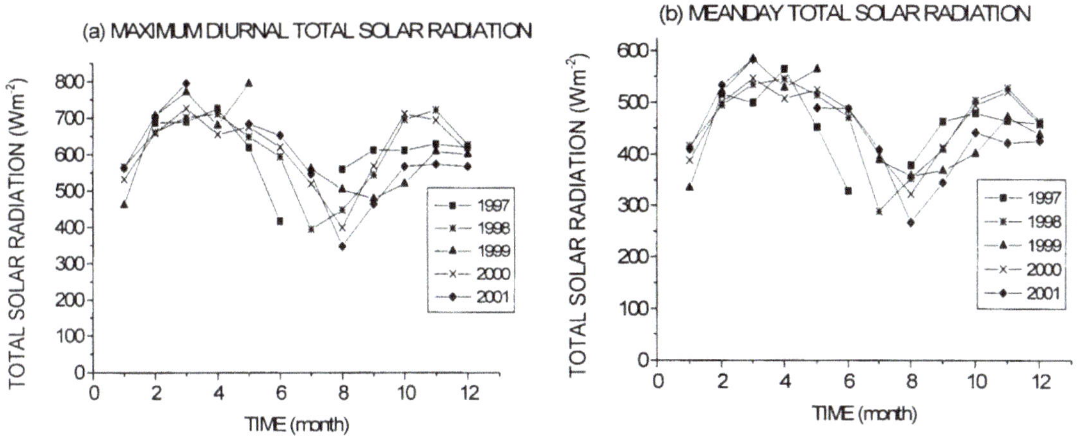

Figure 3. Annual variation of (a) maximum total solar radiation and (b) mean day total solar radiation in Ibadan (7.38°N, 3.93°E).

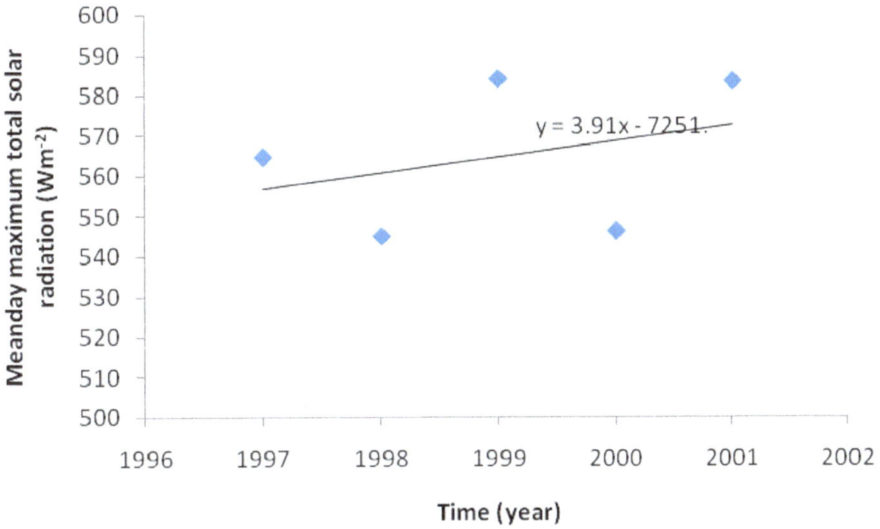

Figure 4. Annual trend of total solar radiation from 1997 to 2001.

per year), 4 to 16 days (20 per year) are obtained for all the years. Intra annual cycles of solar radiation has not been reported but the variation of solar radiation reveals the first and second cycles of 1 and 2 per year (Otunla et al., 2008). Information on solar cycles exist but there is none on intra-annual cycles (Mordvinov, 2003). There is no difference in the cycles obtained for the year 1997 to 2001. Generally in the tropics the annual and seasonal range of solar radiation and temperature are very low compared with the higher latitudes and also less than the diurnal ranges. The intra-annual solar radiation cycles show the possible variations of solar radiation during the year. The one cycle per year revealed that any variation occurring at a particular time in the year will re-occur after twelve months, the two cycles per year showed that the particular occurrence will be repeated after six months, this is in agreement with the two seasons we have in a year in this area. The other variations are revealed in the remaining cycles. The knowledge of the intra-annual cycles of total solar radiation and their amplitudes is useful for industrial and agricultural planning. Appropriate industrial and agricultural policies can be formulated based on the cycles. Capital investment needs can be formulated, insurance coverage requirements can be determined, seasonal variation in labour needs can be estimated. Whenever solar radiation is needed for any purpose (plant growth, rearing of animals, electricity and so on) depending on the quantity needed, plans could be made to repeat the processes once, twice, thrice, five times, ten times and twenty times in a year with the assurance of maximum yield. If the cycles are not known, this can only be done on risk.

Conclusion

Diurnal, seasonal and annual variations of total solar radiation and mean temperature are presented. The dry months have larger values (642.5±10.8 Wm^{-2}) of total solar radiation than wet months (584.3±10.9 Wm^{-2}). The variations are attributed to clouds, dust haze and harmattan. There are two peaks and one minimum in a year for both maximum and mean day total solar radiation. There is one peak in March/April/May with value 750.3±41 Wm^{-2} (for maximum) and the other in October/November with value 633.1±63.5 Wm^{-2} (for maximum). The minimum is in July/August with the value 404.8±54 Wm^{-2} (for maximum). There was positive trend of total solar radiation throughout the period of consideration. The mean day values are 564.7±19, 488±36 and 308.4±40 respectively for the peak and minimum period of total solar radiation.

The times of maximum total solar radiation and maximum mean temperature diurnal variation show a lag of about 3 h in almost all the months except for July/August in most of the years.

Intra-annual cycles and amplitudes of all available harmonics of averaged daily total solar radiation for the years 1997 to 2001 were found. Significant cycles of 360 to 370 days (1 per year); 180 to 200 days (2 per year), 90 to 121 days (3 per year), 45 to 70 days (5 per year); 20 to 35 days (10 per year), 4 to 16 days (20 per year) are obtained for all the years. The cycles are useful for the formulation of capital investment needs, determination of insurance coverage requirements and estimation of seasonal variation in labour needs.

ACKNOWLEDGEMENT

We acknowledge the International Institute of Tropical Agriculture (IITA) Ibadan for the provision of the data used in this investigation.

REFERENCES

Coe R, Stern RD (1982). Fitting models to daily rainfall data. J. Appl. Meteorol. 15:1024-1031.

El Bakry MM (1994). Net radiation over the Aswam high dam lake. Thoer. Appl. Climatol. 49:129-133.

Exell RHB (1976). Fluctuations of solar radiation in Thailand. J. Solar Energy Sci. Technol. 18(6):349-354.

Founda D, Melas D, Lykoudis S, Lisarsidis I, Gerasopoulos E, Kouvarakis G, Petrakis M, Zerefos C (2007). The effect of the total solar eclipse of 29 March 2006 on meteorological variables in Greece. Atmos. Chem. Phys. Discuss. 7:10631-10667.

Garratt JR (1992). The atmospheric boundary layer, Cambridge University Press p.316.

Gerasopoulos E, Zerefos CS, Tsagouri I, Founda D, Amidiris V, Bais AF, Belehaki A, Christou N, Economou G, Kanakidou M, Karamanos A, Petrakis M, Zanis P (2007). The total solar eclipse of March 2006: overview. Atmos. Chem. Phys. Discuss. 7:17663-17704.

Iqbal M (1983). An introduction to solar radiation, Ontario Academic Press p. 390.

Jegede OO (1995). Seasonal charateristics of global diffuse and longwave radiation at Ile-Ife Nigeria (unpublished).

Jegede OO (1997b). Daily averages of net radiation measured at Osu, Nigeria in 1995. Int. J. Climatol. 17:1357-1367.

Jegede OO (1997a). Diurnal variations of net radiation at a tropical station- Osu; Nigeria. Thoer. Appl. Climatol. 58:161-168.

Jegede OO, Mauder M, Okogbue EC, Foken T, Balogun EE, Adedokun JA, Oladiran EO, Omotosho JA, Balogun AA, Oladosu AR, Sunmonu LA, Ayoola MA, Aregbesola TO, Ogolo EO, Nymphas EF, Adeniyi MO, Olatona GI, Ladipo KO, Ohamobi SI, Gbobaniyi EO, Akinlade GO (2004). The Nigerian Micrometeorological Experiment (NIMEX-1): An overview. Ife J. Sci. 6(2):191-202.

Jones BG, Williams AK (1992). Population growth, urbanization, disaster risk and vulnerability in metropolitan areas: a conceptual framework, in Kreimer, Akira, Mohan, Munasinghe. 1992. Environmental management and urban vulnerability, World Bank discussion paper No. 168.

Kane RP (1997). Interannual variability of global radiation at Wageningen. Int. J. Climatol. 17:1487-1493.

Kirkyla KI, Hameed S (1989). Harmonic analysis of the seasonal cycle in precipitation over the United States: a comparison between observations and a general circulation model. J. Climate 2:1463-1475.

Kwanashie E (2003). Housing and urban development. Available at http://www.nigeriafirst.org/article-267.shml Feb 22, 2003, 10:39

Kyle TG (1991). Atmospheric transmission, emission and scattering. Oxford, Pergamon Press p.288.

Montheith JL (1975). Principles of Environmental Physics, Contemporary Biology. Edited by: Barrington EJW, Author FRS and Willis J. Published by; Edward Arnold (Publishers) Limited 25 Hill Street London, Wix8 LL. Reprinted with corrections.

Mordvinov AV (2003). Secular total solar irradiance trend during solar cycles. Geophys. Res. Lett. 30(5):199.

Nigeriafirst (2003). The new challenge in urban planning. Available at http://nigeriafirst.org/printer_1351.shtm. Oct 10, 2003, 16:55.

Nymphas EF, Adeniyi MO, Oladiran EO (2009). Micrometeorological measurement in Nigeria during the total solar eclipse of 29 March, 2006. J. Solar–Terrestrial Phys. 71:1245-1253.

Ogunlami JA (2004). The use of remote sensing/GIS in planning of sustainable land management in Nigeria. RECTAS experience. Available at: http:144.16.93.203/energy/HC270799/LM/SUSLUP/Thema5/658/658.pdf.

Otunla TA, Oladiran EO, Adeniyi MO (2008). Variability and probabilistic extremes of some climatic elements over Ibadan. Online J. Earth Sci. 2(4):124-129.

Sansigolo CA (1993). A bivariate stochastic model for daily rainfall at Sao Paulo city, Brazil; precipitation variability and climate change in Sevruk B and Lapin M (eds), Proceedings of symposium on Precipitation and Evaporation, Bratislava, 2: 20-24 September.

Zurich ETH (2009). Role of solar radiation in climate change. Science Daily. Retrieved January 20, 2010.

Novel method of iron removal from underground dug well waters in communities around Okada town, Ovia North-East L.G.A. Edo State

Chike Luke Orjiekwe, Kehinde Oguniran and Solola Saheed Abiodun

Department of Chemical Sciences, Igbinedion University, Okada, Edo State Nigeria.

Some indices of pollution were studied in water samples from 25 dug wells randomly selected from Okada town and it's environ. Physicochemical parameters measured include temperature, pH, colour, turbidity, total dissolved solids (TDS), toxic shock syndrome (TSS), total hardness, Ca, Mg, Fe, Mn, nitrate and phosphate etc. The results of the study showed that dug wells from this area contain Fe and Mn concentrations in nuisance quantity, far in excess of WHO recommended limits, with obvious consequence for drinking and domestic uses. The use of sodium hydroperoxide, a powerful oxidant for the removal of the Fe and Mn from the dug well water samples is described. This novel method is preferable to most conventional methods because it is cost effective, environmentally friendly, very fast and easy to apply.

Key words: Iron removal, dug well water, sodium hydroperoxide.

INTRODUCTION

Ground water in form of springs, dug wells or borehole are major sources of drinking and domestic water usage throughout the world (Masters, 1998; Sharmer et al., 2005). It is considered better than surface water because it is free from pollutants and other harmful pathogens (Sharmer et al., 2005). However, in some locations, ground water contains dissolved ions such as ferric and ferrous ions, either from geological formation or from iron pump components.

Iron can be present in water either as ferrous iron, ferric iron, organic iron or iron bacteria (Machmeier, 1971). Based on taste and nuisance considerations, the World Health Organization (WHO) recommends that the iron concentration in drinking water should be less than 0.3mg/l (WHO, 1996). Iron concentration in excess of 0.3mg/l is undesirable in drinking water as it causes several aesthetic and operational problems including bad taste, discolouration, staining and deposition in distribution systems (Sharmer et al., 2005).

Several methods namely: oxidation-precipitation, filtration, lime softening, ion exchange e.t.c have been employed for iron removal from underground waters (Wong, 1984; Michalakos et al., 1997 and Twort et al., 2000). The oxidation-precipitation-filtration method is the preferred method in most developing and developed countries because it is more economical and less complicated (Sogaard et al., 2000). In oxidation-precipitation-filtration method, two basic physicochemical mechanisms are involved (Rott, 1985), namely, oxidation floc formation and absorption-oxidation mechanisms (Sharmer et al., 2005). Both mechanisms occur simultaneously in most conventional iron removal plants. However, under certain conditions, the dominant mechanism depends on the water quality and process conditions applied.

In this research work, we report the physicochemical properties of dug well waters in Okada town and its environs with a view to determine the extent of

Figure 1. Edo state senatorial districts.

pollution. Also included is the novel method of iron removal from the dug well waters by treatment with sodium hydroperoxide.

MATERIALS AND METHODS

Apparatus

The equipments used were Celsius thermometer (0-100°C), pH meter scale, Gooch funnel, oven, filter papers and flasks, conductivity cell, Hach colorimeter, Atomic Absorption spectrophotometer, 200 litre pressure tank, Electric motor mixer, water pump, vacuum pump, Calcite filter and storage tank.

Reagents

The reagents used for this study were purchased from Aldrich Chemical Company. They are all of analytical grades and were used directly without further purification. They include hydrogen peroxide, H_2O_2, sodium carbonate, Na_2CO_3 and distilled water.

Sampling area

25 underground dug well water were randomly selected and sampled within five communities in Ovia North-East Local Government area of Edo State. These are labeled DW_1 to DW_{25}. The map of the sampled area is shown in Figures 1 and 2.

Physicochemical properties

The measurement of temperature, pH, conductivity, TDS, TSS, total hardness, colour, turbidity, nitrate, phosphate, dissolved oxygen, calcium, magnesium, iron and manganese ion in the dug well water samples were carried out according to previously reported standard methods (Ademoroti,1996 and Chike et al., 2006).

Treatment of water samples with sodium hydroperoxide

About 180 L of dug well water was taken into the pressure tank with the aid of a water pump. 50% H_2O_2 solution and 2 M Na_2CO_3 solutions was prepared from the stock samples provided. 30 ml of 50% H_2O_2 and 5 ml of 2M Na_2CO_3 were mixed together in a beaker

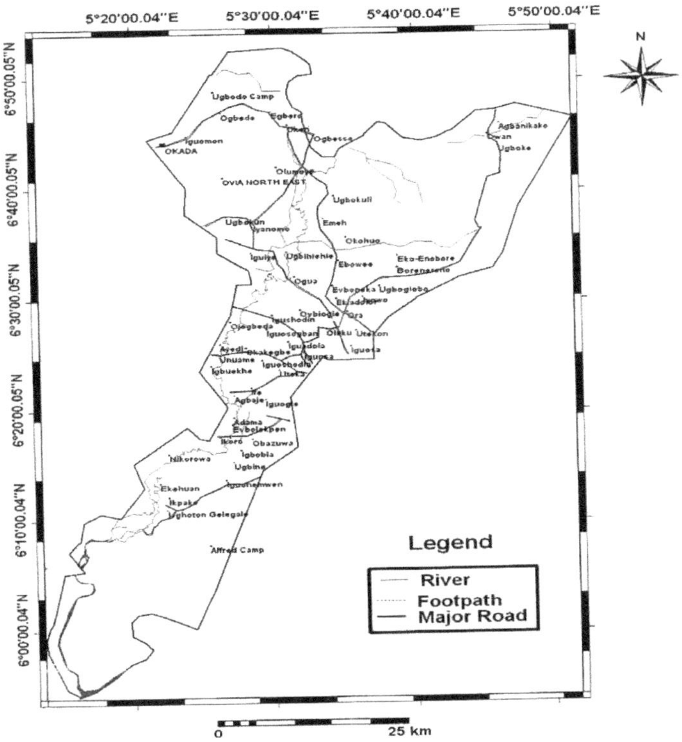

Figure 2. Map of Ovia North-Esat Local Government Area.

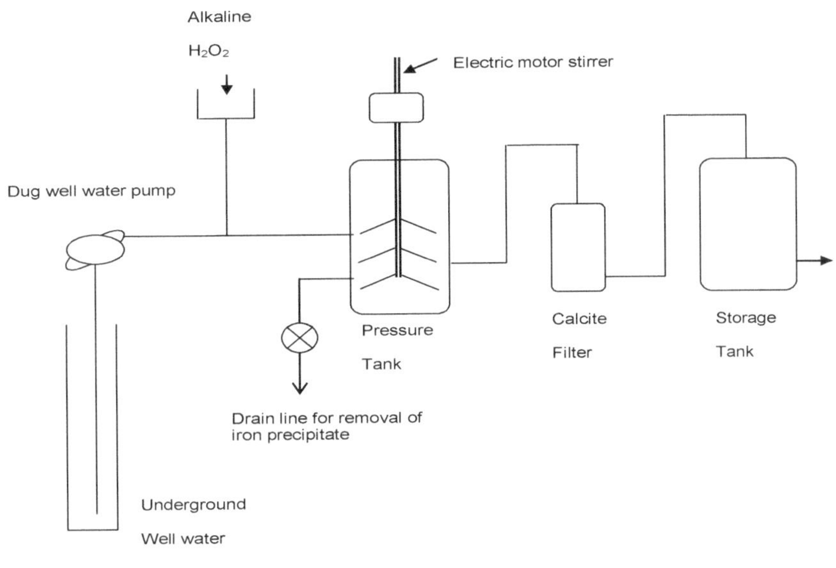

Figure 3. Iron removal by treatment with sodium hydroperoxide.

and allowed to stand for about 3 min to generate sodium hydroperoxide, NaOOH which is the oxidant. The NaOOH mixture was then poured into the tank and stirred with the aid of electric motor mixer for about 1 min. There was immediate precipitation of ferric hydroxide, Fe $(OH)_3$. This was consequently allowed to settle for about 30 min, filtered and stored in the storage tank before distribution, (Figure 3).

RESULTS AND DISCUSSION

The results of the physiochemical characteristics of dug well water samples from Okada town and other surrounding community is presented in Table 1. The temperature of the dug well water samples ranged from

Table 1. Results of physicochemical properties of dug well water samples from Okada Town and its environs.

Community	Dug well Water sample	Temp. (°C)	pH	Λm (ms/cm)	TDS (mg/L)	TSS (mg/L)	Total hardness (mg/L)	Ca (mg/L)	Mg (mg/L)	Fe (mg/L)	Mn (mg/L)	Colour (Pt Co)	Turbidity (fau)	Nitrate (mg/L)	Phosphate (mg/L)	(DO) (mg/L)
Iguomon	DW_1	28.19	7.06	88	44	56	36	28	8	5.8	1.8	334	93	0.15	0.09	5.20
	DW_2	28.10	7.10	90	50	54	36	27	8	6.0	2.0	324	88	0.13	0.08	5.40
	DW_3	28.00	6.90	94	55	55	40	26	9	6.2	1.6	320	86	0.20	0.09	5.50
	DW_4	28.10	7.10	90	50	56	40	27	9	5.8	1.8	324	84	0.14	0.10	5.30
	DW_5	28.10	7.06	90	45	57	36	28	8	6.0	2.0	340	95	0.16	0.10	5.30
Ogbese	DW_6	28.05	6.90	45	23	13	18	14	4	4.0	1.4	12	16	0.18	0.06	4.76
	DW_7	27.90	7.00	44	22	12	16	12	5	4.2	1.2	13	15	0.17	0.07	4.90
	DW_8	27.80	7.10	40	22	11	17	13	6	4.4	1.6	13	17	0.17	0.06	4.90
	DW_9	27.90	7.00	44	24	14	17	12	5	4.2	1.2	13	16	0.18	0.07	4.70
	DW_{10}	28.10	6.90	46	23	13	18	14	4	4.0	1.4	12	15	0.20	0.08	4.80
Okada	DW_{11}	28.28	7.35	198	99	10	70	50	20	8.0	2.6	22	16	0.09	0.06	5.19
	DW_{12}	27.90	7.20	182	90	14	68	49	21	7.8	2.4	21	15	0.08	0.05	5.20
	DW_{13}	28.00	7.45	170	85	12	66	48	22	8.4	2.8	24	15	0.11	0.07	5.30
	DW_{14}	27.90	7.30	183	90	14	67	51	20	8.4	2.4	21	16	0.10	0.06	5.10
	DW_{15}	28.30	7.20	190	100	12	72	49	21	7.8	2.6	23	18	0.09	0.05	5.20
Uhen	DW_{16}	28.17	6.89	96	48	119	42	31	11	6.0	2.0	495	184	0.12	0.07	5.21
	DW_{17}	28.00	7.00	88	45	124	40	29	12	6.2	1.8	490	180	0.11	0.06	5.30
	DW_{18}	27.90	7.10	84	40	130	39	30	13	6.4	2.2	500	190	0.14	0.08	5.40
	DW_{19}	28.00	7.00	86	46	123	40	32	12	6.0	2.0	490	182	0.12	0.07	5.10
	DW_{20}	28.20	6.90	94	50	120	41	29	12	6.2	1.8	492	188	0.13	0.07	5.30
Utese	DW_{21}	28.19	7.07	46	23	8	18	12	6	4.0	1.4	7	14	0.12	0.06	5.20
	DW_{22}	28.10	7.20	55	22	7	19	13	7	3.8	1.2	7	13	0.13	0.04	5.20
	DW_{23}	28.00	6.90	45	20	9	20	14	6	4.4	1.4	8	15	0.14	0.05	5.30
	DW_{24}	28.10	7.20	56	22	7	18	12	7	4.2	1.2	7	14	0.11	0.05	5.10
	DW_{25}	28.20	7.10	60	24	10	18	13	8	4.0	1.6	7	14	0.12	0.06	5.30
WHO		24.50-8.80	6.50 - 8.00	< 10	< 6	< 2	< 100	< 75	< 25	0.3	0.05	< 500	< 0.1	< 0.3	< 0.10	> 100

Table 2. Iron and manganese content of the dug well water samples before and after treatment With NaOOH.

Communities	Dug well water sample	Fe and Mn content before treatment with NaOOH		Fe and Mn Content after treatment with NaOOH	
		Fe (mg/L)	Mn (mg/L)	Fe (mgIL)	Mn (mg/L)
	DW_1	5.8	1.8	0.15	0.02
	DW_2	6.0	2.0	0.18	0.03
	DW_3	6.2	1.6	0.20	0.02
Iguomon	DW_4	5.8	1.8	0.20	0.03
	DW_5	6.0	2.0	0.15	0.03
	DW_6	4.0	1.4	0.10	0.02
	DW_7	4.2	1.2	0.15	0.02
	DW_8	4.4	1.6	0.15	0.03
Ogbese	DW_9	4.2	1.2	0.10	0.02
	DW_{10}	4.0	1.4	0.15	0.03
	DW_{11}	8.0	2.6	0.20	0.03
	DW_{12}	7.8	2.4	0.15	0.02
	DW_{13}	8.4	2.8	0.20	0.03
Okada	DW_{14}	8.4	2.4	0.15	0.02
	DW_{15}	7.8	2.6	0.20	0.03
	DW_{16}	6.0	2.0	0.15	0.02
	DW_{17}	6.2	1.8	0.10	0.02
	DW_{18}	6.4	2.2	0.10	0.03
Uhen	DW_{19}	6.0	2.0	0.09	0.03
	DW_{20}	6.2	1.8	0.09	0.04
	DW_{21}	4.0	1.4	0.10	0.03
	DW_{22}	3.8	1.2	0.08	0.03
	DW_{23}	4.4	1.4	0.10	0.03
Utese	DW_{24}	4.2	1.2	0.10	0.02
	DW_{25}	4.0	1.6	0.09	0.02
WHO		0.3	0.05	0.3	0.05

27.8 - 28.3 °C which is within the acceptable range recommended by World Health Organization (WHO). The maximum allowable limit for pH by WHO for safe drinking water is 6.50 - 8.00. The pH of all the dug well water samples also fell within this range. The conductivity of the water samples is a measure of the dissolved ions in the water.

Dug well water samples from Okada have the highest conductivity values, which corroborate its highest values for Ca, Mg, Fe and Mn ions. Dug well water samples from Uhen community has the highest average TSS value of about 123 mg/L which surpassed the WHO guideline value of 2.0mg/L for TSS. This corroborates very well with the highest average values of 495 PtCo and 185mg/L measured for colour and turbidity respectively. All the dug well water samples from the five communities contain Fe and Mn exceedingly in nuisance quantity far above the WHO allowable limit. Dug well water samples from Okada community have the highest values. The results were expected considering the fact that the soil from these areas is mostly laterite, with high iron content. Therefore the iron must have got into the underground water through leaching and seepage from the soil. The high content of Fe and Mn ions has obvious deleterious consequences on the use of the dug well waters for both domestic and drinking purposes. Such consequences include bad taste, staining and browning of clothes, food, containers and skin.

In this regard, a pretreatment procedure is advocated for the dug well water to bring the concentration of Fe and Mn to a tolerable level. Table 2 shows the content of Fe and Mn of the dug well water samples after treatment with sodium hydroperoxide and NaOOH. The dug well water was clear when drawn but gradually turned brown due to precipitation of ferric hydroxide and $Fe(OH)_3$ caused by atmospheric oxidation of the dissolved ferrous iron. The result indicates that the iron is present mostly as dissolved ferrous bicarbonate and $Fe(HCO_3)_2$. After treatment, the iron and manganese concentrations in the dug well water samples are considerably reduced to below 0.2 and 0.04 mg/L respectively. These values are tolerable concentration level for these ions since they fall within the WHO allowable limits.

The Na_2CO_3 solution was used to generate the sodium hydroperoxide, which is the powerful oxidant (Equation 1). The sodium hydroperoxide decomposes insitu to generate oxygen atom, [O] which then oxidizes the iron (II) to insoluble iron (III) (Equations 2 and 3).

The following mechanistic steps have been suggested for the reaction:

$$Na_2CO_3 + 2H_2O_2 \longrightarrow 2NaOOH + CO_2 + H_2O \quad (1)$$

$$2NaOOH \longrightarrow 2NaOH + 2[O] \quad (2)$$

$$Fe^{2+} + 3/2[O] + 3/2H_2O \longrightarrow Fe(OH)_3 \quad (3)$$

The overall equation of the reaction is therefore given by equation below:

$$2Fe(HCO_3)_2 + ½O_2 + H_2O \rightarrow 2Fe(OH)_3\downarrow + 4CO_2\uparrow \quad (4)$$

The ebullition of hydrogen peroxide aids the coagulation of the suspended ferric hydroxide particles. Apart from iron removal, this method is particularly better than most conventional methods for the following reasons:

(1) The hydrogen peroxide, H_2O_2 is a very important powerful germicide and is capable of disinfecting the water.

(2) The reaction is very fast and cost effective since reaction is completed in a few minutes and just about 35ml of the NaOOH mixture is required per 180 L of raw water (cf. 1Kg of Calcium hypochlorite per 100 Kg of raw water)

(3) It is environmentally friendly when compared with other conventional methods such as chlorination or ozonation e.t.c.

(4) It does not require external coagulant.

Conclusion

Raw water from dug wells from Okada town and neighbouring communities have very high iron concentration because of the high iron content of the laterite soil. Treatment with alkaline hydrogen peroxide is one surest way of removing dissolved iron from underground dug well or borehole waters. The method is preferable than most conventional methods because it is fast, cost effective, environmental friendly and does not require external coagulant.

ACKNOWLEDGEMENT

The Authors acknowledge the financial assistance of Igbinedion University management and the Marine Research Institute of Oceanography, Lagos, Nigeria for the use of their Analytical Laboratory.

REFERENCES

Ademoroti CMA (1996). Standard Methods for Water and Effluent Analysis. Foludex Press Ltd., Ibadan, Nigeria, pp. 20-30.

Machmeier RE (1971). Iron in Drinking Water. University of Minnesota Agricultural Extension Services. pp. 2–5.

Masters GM (1998). Introduction to environmental Engineering. 2ND Edition, Upper Saddle River, NJ: Prentice Hall pp. 11–12

Michalakos GD, Nieva JM, Vaynes DV, Lybertos G (1997). Removal of iron from potable water using trickling filter. J. Water Res. 31(5):991-995

Chike OL, Oshin D, Orjiekwe IU (2006). Water quality index assessment of Evebonede and Ovia Rivers in Ovia North-East LGA of Edo State, Nigeria. Int. J. Chem. 16(4):223 - 228

Rott U (1985). Physical, chemical and biological aspects of the removal of iron and manganese underground. J. Water Supply. 3(2):143-150

Sharmer SK, Petrusevski B, Schippers JC (2005). Advance ground water treatment: iron, manganese, fluoride and boron removal. J. Water Supply: Res. Technol. 54(4):239–244.

Sogaard EG, Modenwaldt R, Abraham-Peskir JV (2000). Conditions and rates of biotic and abiotic iron precipitation in selected fresh water plants and microscopic analysis of precipitate morphology. J. Water Res. 34(10):2675-2682.

Twort AC, Ratnayaka DD, Brandt MJ (2000). Standards for physical and chemical quality of water supply. 5TH Ed. Arnold London. pp. 4-46.

WHO (1996). Guidelines for Drinking Water Quality. Health Criteria and other supporting information, 2ND Ed., WHO Geneva.

Wong JM (1984). Chlorination-filtration for iron and manganese removal. J. Am. Water Works Assoc. 76(1):76 –79.

Application of regression and multiple correlation analysis to morning hours solar radiation in Lapai

Agbo G. A.[1] , Alfa B.[2], Ibeh G. F.[1] and Adamu I. S.[2]

[1]Department of Industrial Physics, Ebonyi State University, Abakaliki, Nigeria.
[2]Department of Physics, Ibrahim Badamasi Babangida University, Lapai, Nigeria.

Solar radiation in Lapai within the morning hours has been estimated by correlating meteorological parameters. This was achieved by applying the first and second order regression and multiple correlation analysis method. The correlation coefficient based on the first and second order equations in temperature and relative humidity were 0.828, 0.692, 0.860 and 0.622 while the correlation coefficient based on the multiple correlations between solar radiations, temperature and relative humidity were 0.351. Validity tests were carried our using mean bias error (MBE), root mean square error (RMSE) and mean percentage error (MPE). The tests show that the errors were minimal in the first order equations in temperature and relative humidity. Comparison of the measured and predicted values of solar radiation based on the relative humidity and average temperature first order equations show a close agreement, and suggests the best equations to be used in estimating solar radiation in Lapai and its similar climatic condition.

Key words: Solar radiation, multiple regression, correlation coefficient, meteorological parameter-temperature, relative humidity.

INTRODUCTION

Solar radiation is the energy from the sun reaching the earth and is important in understanding the land atmosphere energy exchange (Liang et al., 2012). It is the driving force of both the physical and biological cycles on the earth. Consequently, it then becomes pertinent to have very good knowledge of current and past records of solar radiation at a location so as to aid in the estimation of the performance of any solar energy system. Pyrheliometer and pyranometer are instruments used readily to obtain the diffuse component of solar radiation and the global solar radiation, respectively. Meteorological stations have been used mostly for this purpose. However, there are a few of such stations across the globe and worse still in the developing countries. In order to develop solar radiation data, researchers had extrapolated values from one location for application in a different location. Hence, solar radiation prediction from estimation models has been widely utilized globally to generate solar radiation database for various location of the world. There are numerous statistical techniques of estimating solar radiation and each method is based on different principles. The first empirical correlation using sunshine hours for estimation of solar radiation was proposed by Angstrom (Ahmad and Ulfat, 2004). This model was later modified by Prescott and Page (Page, 1961; Ahmad and Ulfat, 2004; Maghrabi, 2009) in which the model is given as:

$$\frac{H}{H_o} = a + b\left(\frac{S}{S_o}\right),$$

Table 1. Climate parameters.

Parameter	Unit
Average temperature (T_{av})	°C
Relative humidity (R)	%
Solar radiation (H)	W/m^2

Where H is the monthly average daily global radiation on horizontal surface, H_o is the monthly average daily extra-terrestrial radiation, S is the length of the day, S_o is the maximum possible sunshine duration, and *a* and *b* are constants.

Since this development, there have been other models such as Rietveld, Bahel, Glover, Hay and Grag models (Mahdi et al., 1992). Akpabio et al. (2005) developed the quadratic form of Angstrom-Prescott model and used it to estimate the global solar radiation at Onne, Nigeria (latitude 4° 46′ N, Longitude 7° 10′ E). Multi-linear polynomial form of the Angstrom-Prescott model was employed by Agbo et al. (2010) to estimate global solar radiation at Minna, Nigeria. Various other models employing other meteorological parameters with one or more variables with solar radiation have also been used. The estimation of the solar radiation at Uturu in Nigeria was carried out using an equation relating solar radiation and temperature (Chiemeka, 2008). Agbo (2012) had also estimated 'global' solar radiation at Onitsha using regression

analysis and artificial neural network models with the attendant parameters of temperatures and relative humidity. Lapai is in Niger State of Nigeria and is located at latitude 9.05° N and longitude 6.05° E. The main stay of the economy in Lapai is agriculture and solar radiation is an important parameter for agriculture. However, the availability of such vital data is not readily available. Due to the present economic hardship in Nigeria, there is every tendency for having problems of maintaining and purchasing the instrument used in measuring the meteorological parameters in the meteorological stations.

The solution to this problem has motivated one into the need for an empirical modeling for estimation of solar radiation that will be affordable and easily maintained as it uses mathematical, statistical correlation and regression analysis for computation. The first time solar radiation was being measured in Lapai is in 2010 and even as that the measurements were not consistent due to instrument and power failure. Sometimes, when one wants to get the value of instant solar radiation to test the functioning of solar devices or to have idea or past records at any particular time so as to forecast crop yield for the next season, the instruments for measuring the solar radiation are not found or faulty. It is by this reason that this paper is written. Thus, the aim of this work is therefore to develop a relation between solar radiation with temperature or relative humidity or both for morning hours in Lapai. The morning hour records of 5 min intervals are the only record retrieved and made available by the recorder.

MATERIALS AND METHODS

The data (maximum and minimum temperatures, relative humidity and solar radiation) used in this study were obtained from the weather station located at Geography Department of Ibrahim Badamasi Babangida University, Lapai (IBBUL), Niger State, Nigeria. The data were collected at 5-min interval in the year 2010 and cover a time period of 9.30 to 10.30 am as there were inconsistency and missing data at early hours of the morning. The data were averaged over days that have meaningful records. The minimum and maximum temperatures were averaged to obtain the average temperature at each time interval. Table 1 shows the climatic parameters obtained with their units. In regression analysis, the first and second order regressions of normal equations were employed to estimate the morning hours global solar radiation as, thus, following the treatment in Agbo (2012); we have:

$$a + bx = y \tag{1}$$

$$a + bx + cx^2 = y \tag{2}$$

Where *a* and *b* are constants which will be determined. *y* is the same as solar radiation (H) and it is the dependent variable. *x* is the independent variable which can be replaced by any of the meteorological parameters such as average temperature and relative humidity.

To carry out the regression analysis of the first order, both sides of Equation 1 is multiplied by one and summed on both sides to yield Equation 3. Equation 1 also is multiplied by *x* and summed on both sides to yield Equation 4:

$$aN + b\Sigma x = \Sigma y \tag{3}$$

$$aN\Sigma x + b\Sigma x^2 = \Sigma xy \tag{4}$$

Applying the variables T_{av} and R for average temperatures and relative humidity respectively, as the independent variable in Equation 1 for first order regression we obtain:

$$a_1N + b_1T_{av} = H_1 \tag{5}$$

$$a_2N + b_2R = H_2 \tag{6}$$

To obtain the second order regression equation, Equation 2 is multiplied by 1, *x* and x^2 successively and summed to obtain the following equations:

$$a_1N + b_2\Sigma x + c_3\Sigma x^2 = \Sigma y \tag{7}$$

$$a_1N\Sigma x + b_2\Sigma x^2 + c_3\Sigma x^3 = \Sigma xy \tag{8}$$

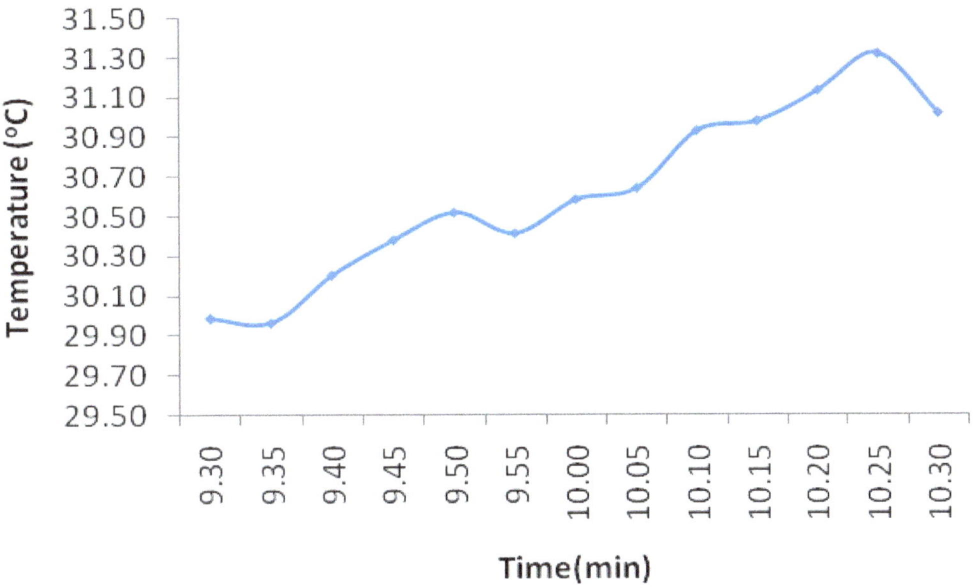

Figure 1. Variation of temperature with time in Lapia.

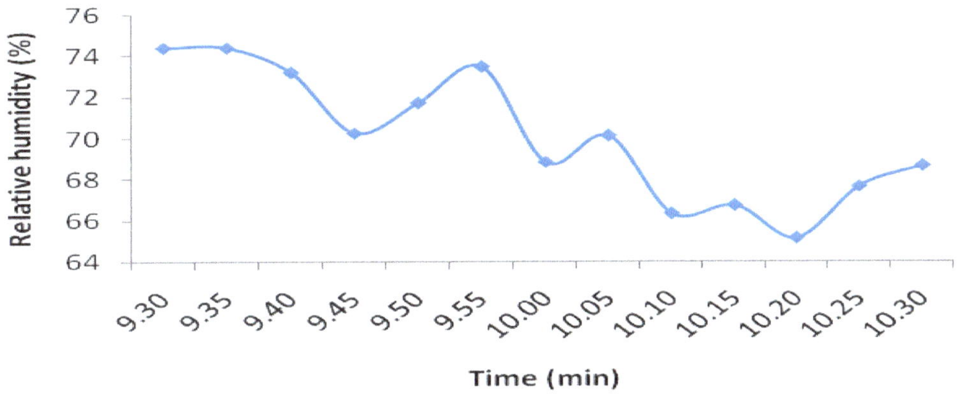

Figure 2. Relative humidity variation during the morning hours in Lapia.

$$a_1 N\Sigma x^2 + b_2\Sigma x^3 + c_3\Sigma x^4 = \Sigma x^2 y \quad (9)$$

Applying these equations by using our independent variables such as average temperature and relative humidity, we obtain:

$$a_3 + b_3 T_{av} + C_3 T_{av}^2 = H_3 \quad (10)$$

$$a_4 + b_4 R + C_4 R^2 = H_4 \quad (11)$$

RESULTS AND DISCUSSION

Figures 1 and 2 are the plots of the measured temperature and relative humidity as against time in Lapai. From the temperature plot (Figure 1), it shows that there is a continuous increase in temperature at early morning hours of the day while the relative humidity plot (Figure 2) shows a downward trend from early hours to before noon time hours. The trend variation as observed in temperature and relative humidity suggests that solar insolation at early morning hours in Lapai causes changes in the weather parameters. Figures 3 and 4 show the scatter plot of solar radiation against average temperature, and solar radiation against relative humidity, respectively. Despite the large amount of scattered points, definite correlations between them do exist. The large amount of scattered data is partly attributed to error in estimation of the solar radiation at 5-min intervals.

The first and second order regression equations obtained in this work using the average temperature and relative humidity are as follows:

$$H_1 = -2715.96 + 105.72 T_{av} \quad (12)$$

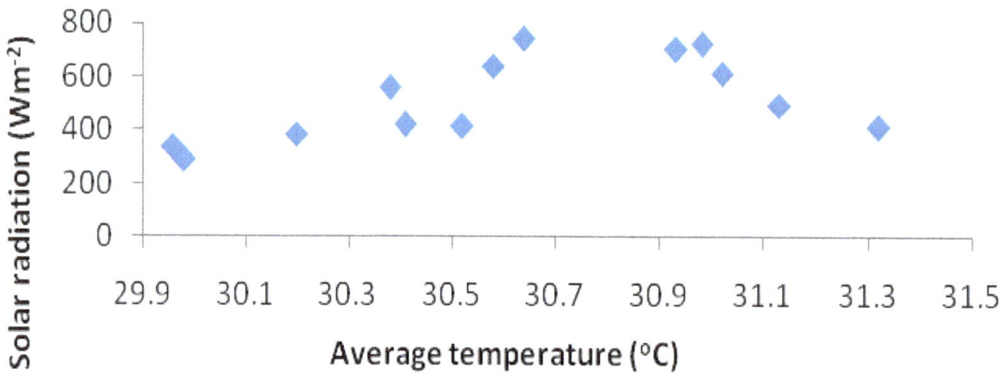

Figure 3. Scatter plot of solar radiation against average temperature.

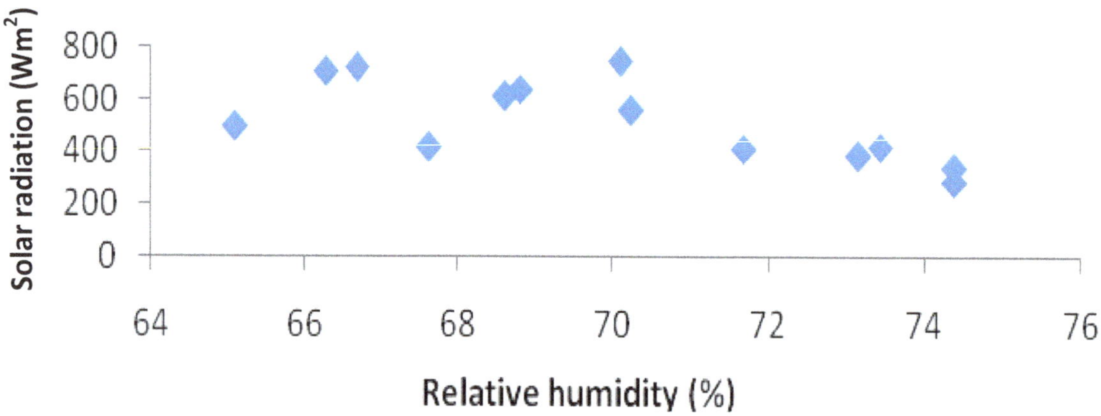

Figure 4. Scatter plot of solar radiation against relative humidity.

$$H_2 = 2883.77 - 33.094R \quad (13)$$

$$H_3 = -5011.04 + 175.08T_{av} + 0.81T_{av}^2 \quad (14)$$

$$H_4 = 698.60 - 1.70R - 0.012R^2 \quad (15)$$

The multiple regression equation of solar radiation with average temperature and relative humidity were obtained as:

$$H_5 = -20.89R + 64.77T_{av} \quad (16)$$

The correlation based on the average temperature alone is obtained as 0.828 and relative humidity alone is 0.860. The estimated results from each of the average temperature and relative humidity show a closer agreement to the measured data than that predicted from the correlation based on the average temperature and relative humidity taken together. A comparison of the result of Equation 12, 13, 14, 15 and 16 indicates that the correlation of Equation 13 based on the relative humidity alone and Equation 12 based on average temperature alone should be employed for computing solar radiation in the morning hours in Lapai. The deviations between the measured and estimated values of the solar radiation have been evaluated. This is in an effort to test the validity of the equations for the solar radiation. As seen from Table 2, the percentage errors are very small and the equations may be taken as valid equations for estimating solar radiation in the morning hours in Lapai. The validity of the equations were also tested using mean bias error (MBE), root mean square error (RMSE) and correlation coefficient (CC) which gave good results as seen in Table 2.

Figure 5 shows the relationship between the measured solar radiation and the estimated solar radiation using each of the equations as aforementioned.

Conclusion

In this study, an attempt has been made at estimating solar radiation based on meteorological data at IBBUL.

Table 2. Summary of calculations.

Equations	CC	CD	MBE	RMSE	MPE
$H_1 = -2715.96 + 105.72T_{av}$	0.828	0.686	-0.189	0.682	0.003
$H_2 = 2883.77 - 33.094R$	0.860	0.740	0.004	0.016	-0.001
$H_3 = -5011.04 + 175.08T_{av} + 0.81T_{av}^2$	0.692	0.479	0.419	1.152	-0.006
$H_4 = 698.60 - 1.70R - 0.012R^2$	0.622	0.387	-0.217	0.782	0.003
$H_5 = -20.89R + 64.77T_{av}$	0.351	0.123	0.538	1.939	0.008

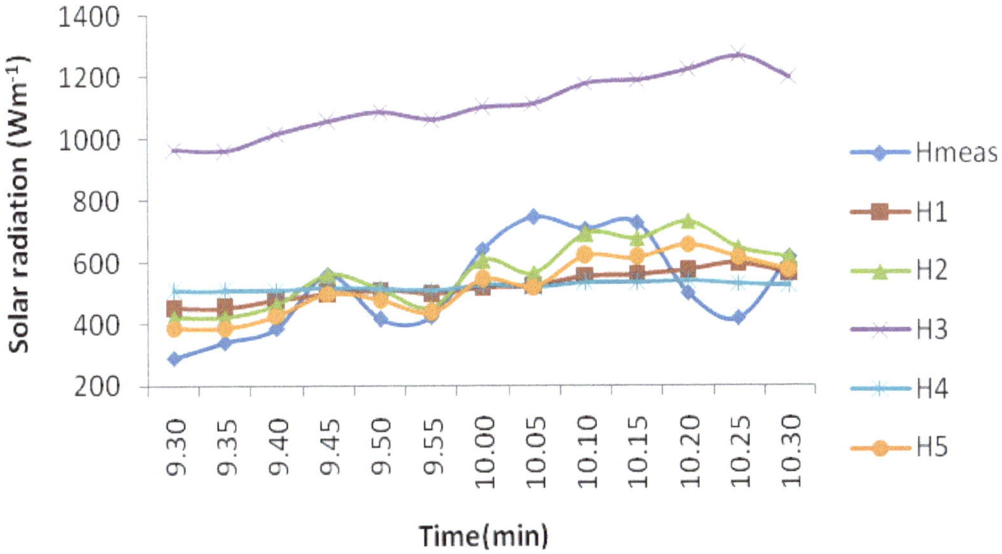

Figure 5. Measured and estimated values of solar radiation.

The results of the estimated values show a closer agreement to the measured data. The results were obtained using first and second order regression and multiple regression analysis which portrays a good correlation. The deviations between the measured and estimated values were minimal. The results from this analysis will greatly reduce the problem associated with waste of time and lack of efficient instruments. It will also be an easy way to present data and ensure that data never runs out. This will also allow user of the data to track information of solar radiation data in Lapai.

ACKNOWLEDGEMENT

The authors are grateful to the Geography Department of the IBBUL for making the data obtained from the automatic weather station available. The first author thanks the vice-Chancellor of IBBUL for granting him the honour of spending one-year sabbatical leave with the Physics Department of the University. Thanks to Mr. Samuel in the Geography Department of the University.

REFERENCES

Agbo GA, Baba A, Obiekezie TN (2010). Empirical Models for the correlation of monthly average global solar radiation with sunshine hour at Minna, Nig. J. B. Res. 1 (1):41-47.

Agbo GA (2012). Estimation of Global Solar Radiation at Onitsha with Regression analysis and artificial Neural Network Models. Res. J. Recent. Sci. 1 6):1-8.

Ahmad F, Ulfat I (2004). Empirical Models for the correlation of monthly average daily global solar radiation with hours of sunshine on a horizontal surface at Karachi, Pakistan. Turk. J. Phys. 28:301-307.

Akpabio LE, Udo SO, Etuk SE (2005). Modeling Global Solar Radiation for a Tropical Location: Onne, Nigeria. Turk. J. Phys. 29:63-68.

Chiemeka IU (2008). Estimation of solar radiation at Uturu, Nigeria. Inter. J. Phys. Sci. 3 (5):126-130.

Liang H, Zhang R, Liu J, Sun Z, Cheng X (2012). Estimation of Hourly Solar Radiation at the Surface under Cloudless Conditions on the Tibetan Plateau Using a Simple Radiation Model Adv.Atm.Sci. 29(4):675-689.

Maghrabi AN (2009). Parameterization of a simple model to estimate monthly global solar radiation bases on meteorological variables and evaluation of existing solar radiation models for Tabouk, Saudi Arabia. Energy conversion and management. 50:2754 –2760.

Mahdi NL, Baharn NS, Zaki FF (1992). Assessment of solar radiation models for the Gulf Arabian countries. Renew. Energy 2(1):65-71.

Page JK (1961). The estimation of monthly mean values of daily total shortwave radiation on vertical and inclined surfaces from sunshine records for latitudes 40°N – 40°S. In proceedings of UN conference on new sources of energy. pp. 378-390.

11

Dissolved and particulate trace elements' configuration: Case study from a shallow lake

Feza Karaer[1], Aslıhan Katip[1], Saadet İleri[1], Sonay Sarmaşik[1] and Nurcan Aydoğan[2]

[1]Department of Environmental Engineering, Görükle Campus, 16059, Uludag University, Faculty of Engineering-Architecture, Bursa, Turkey.
[2]Bursa Metropolitan Municipality-General Directorate of Water and Sewerage Administration (BUSKİ), Bursa Eastern Wastewater Treatment Plant, Küçük Balıklı, Osmangazi, Bursa,Turkey.

Trace element distribution was monitored for Uluabat lake water polluted with point and non-point sources in Bursa. An examination of Zn, Cr, Ni, Cu, Pb, As, B concentrations during the study period indicated that the particulate phase at the bottom of the water column were high in regard to dissolved phase. The source of trace elements in the deeper parts of the water column was derived from the higher concentrations of suspended solids present. Storm conditions and resuspension process in the lake was targeted as the source of trace elements' configuration in the dissolved and particulate phases. Significant correlations were found among temperature, pH, organic matter percentage, suspended solids, chlorophyll-a and particulate trace elements parameters. Factorial Anova Analysis was performed to help to understand the effects of parameters on adsorption of trace elements. Decrease in water level, increase of pH, organic matter percentage and suspended matter were appeared responsible for the increase of dissolved and particulate trace elements concentrations in summer season. Especially Cr, Zn, Pb and Cd concentrations were detected higher than the standards of EPA and WHO. The concentrations of trace elements in water column have to be monitored and the external loading with heavy metals coming from point and non-point sources has to be reduced for the conservation of ecological values.

Key words: Trace elements, dissolved phase, suspended particulate matter, Lake Uluabat.

INTRODUCTION

Certain heavy metals and trace elements in the form of dissolved and suspended particulate matter can introduce easily into the sediment layer (Stumm and Morgan, 1996; Sigg et al., 1987). Dissipation of trace elements from the water column is originated by the cycles of major elements in lake as photosynthetic production, $CaCO_3$ formation, iron and manganese oxides, aluminium and silicates (Stumm and Morgan, 1996). The pH interval in natural waters is between 7 to 8.5 which is the most acceptable range for adsorption of trace element ions to hydroxides and organic particulates. The heavy metal cations are completely released under circumstances of extreme acidic conditions (Shukla et al., 2002). Temperature, pH, redox potential, organic matter, ion exchange processes and microbiological activity affect some characteristics of trace elements such as mobility and bioaccumulation properties (Filgueiras et al., 2004). International standard values are used for determination of the evaluation of heavy metal's toxicity. Values of WHO and EPA guidelines can be given as an example for the standards (Anonymous, 2006a, b). The study site is shallow Lake Uluabat. It is subject to the "Convention on Wetlands of International Importance, especially as Waterfowl Habitat" (the RAMSAR Convention), which was approved in Ramsar, Iran in 1971. It was selected as the study area due to the international importance of this lake. The surface area of the lake is 161 km^2 at maximum water level and 138 km^2 at minimum water level and its average depth is 3 m (Dalkıran et al., 2006). Lake Uluabat has been contaminated by domestic and industrial waste waters and polluted runoff waters resulting from rainwater for many years. Particular problems of the Lake are eutrophication (Akdeniz et al., 2011) and heavy metal pollution. There are 67 residential areas, factories, workplaces, agricultural operations and mines in the basin of the Mustafakemalpaşa (MKP) creek which feeds the lake (Dalkıran et al., 2006). Furthermore, the MKP creek carries a significant amount of suspended solids, which is linked to a 12% reduction in the surface area and the volume of the lake, recorded between 1984, 1993 and 1998 (Aksoy and Özsoy, 2002). One of the most important centers in the MKP Basin, Tavşanlı has a sewage system without any treatment facility and the wastewater of Tunçbilek Thermal Power Plant and lignite plant are discharged to Orhaneli creek after being physically treated. The sewage water of MKP creek, wastewater of the leather organized industrial zone with physical treatment, wastewater of 27 dairy and butchery plants are discharged to the MKP creek without any treatment. In addition, there are 16 residential areas along the creek. The domestic and industrial wastewater of Akçalar is discharged to Akçalar creek, and then finally to Uluabat Lake (Dalkıran et al., 2006). Uluabat, Atabay and Karaoğlan pumping stations is that some of the neighbouring residential areas and factories discharge their wastewaters into the drainage canals of State Hydraulic Works (SHW). Reverse current occurs in the Kocasu creek draining the lake, because of flow rate is increased by rainfall during winter. The Kocasu creek formed by the merging of the Nilüfer and Simav creeks carry pollutants into the lake.

The objective of this study is to examine the concentration of selected trace elements in dissolved and particulate phase at 8 stations in Lake Uluabat annually between May 2008 and 2009. In addition, it is aimed to evaluate the effects of pH, temperature, organic matter percentage, suspended solids, chlorophyll-a and electrical conductivity parameters against trace element adsorption.

MATERIALS AND METHODS

Study area and sampling

The water and sediment samples were taken once in a month for annually, between May 2008 and 2009 in order to observe seasonal differences. The samples were taken on the same day from 8 stations, which were selected according to their distances from the contaminant sources in the lake and their hydrodynamic characteristics. The sampling stations presented different degrees of pollution around the lake. The first station is located in the outflow of Lake Uluabat. The second station is located near to the Atabay pumping station. Whereas the third station represented the north part of the lake, the fourth station is located in the entrance of the Mustafakemalpaşa creek. The fifth station represented the southwestern part of Island Mutlu which is located in nearby of Eskikaraağaç village. The sixth station is situated between Island Mutlu and Gölyazı village. The seventh station is presented at the near of Eskikaraağaç village. The new planned highway would pass close to these villages and finally the eighth station is located at the near of Akçalar village. The coordinates of sampling stations were determined with GPS and are shown in the map (Figure 1) and in Table 1. Determination of coordinates was based on European 1950 UTM coordinate system by the help of Magellan XL handheld GPS.

Samples were taken from all the sampling stations at two heights within the water column. The water samples at 0.5 m below the surface were taken by using the precleaned dark polyethylene bottles (Nguyen et al., 2005; An and Kampbell, 2003; Katip, 2010; İleri, 2010) and the others were taken from the lake bottom (just above the layer of the sediment) by using a precleaned standard sampling device (Hydro-Bios brand), which can take samples along the depth. Samples were transferred into dark-colored polyethylene (PE) bottles that had been washed with HNO_3 and deionized water. During sampling in situ, Temperature (T), pH and Electrical conductivity (EC) were measured using a HACH brand Sension 156 device (Burton and Pitt, 2002; Katip et al., 2012a, b). Sediment samples were taken from the top 5 cm of the sediment with an Ekman sediment sampler (Hydro-Bios brand) and transferred to the laboratory in plastic bags (Burton and Pitt, 2002). Arm-length gloves were used during sampling (Nguyen et al., 2005). A 1-liter water sample was taken along the water column in order to determine the chlorophyll-a concentrations, which is known to be linked to suspended solid concentration. In summer, as the algae increases, the SS concentrations also increase (Stumm and Morgan, 1996). The samples were transferred into the laboratory in dark-coloured PE bottles (Parsons and Strickland, 1963).

Sample handling and analysis

Suspended solids were determined gravimetrically by filtering 100 ml sample through a pre-weighed Milipore filter paper (pore size of 0.45 µm) (APHA, 1998). After the 1 L of sampled water was filtered through a GF/C filter paper, chlorophyll-a concentrations were measured by extracting the filtered water in 90% acetone and then reading the results on a spectrophotometer at 665, 645 and 630 nm wavelengths (Parsons and Strickland, 1963). Percentage of organic matter of sediment was measured by incineration method. pH and conductivity parameters of sediment were measured by 1 : 10 sediment-water mixture (Radojevic and Bashkin, 1999).

For trace element analyses, the following procedures were implemented. Water samples were filtered through a Milipore filter paper with preweighed 0.45 µm pore-size. The filter papers were then put into precleaned petri dishes and stored in a freezer. The filtered water samples were acidified with 0.2% (v/v) concentrated nitric acid and kept in glass bottles cleaned with detergent, water,

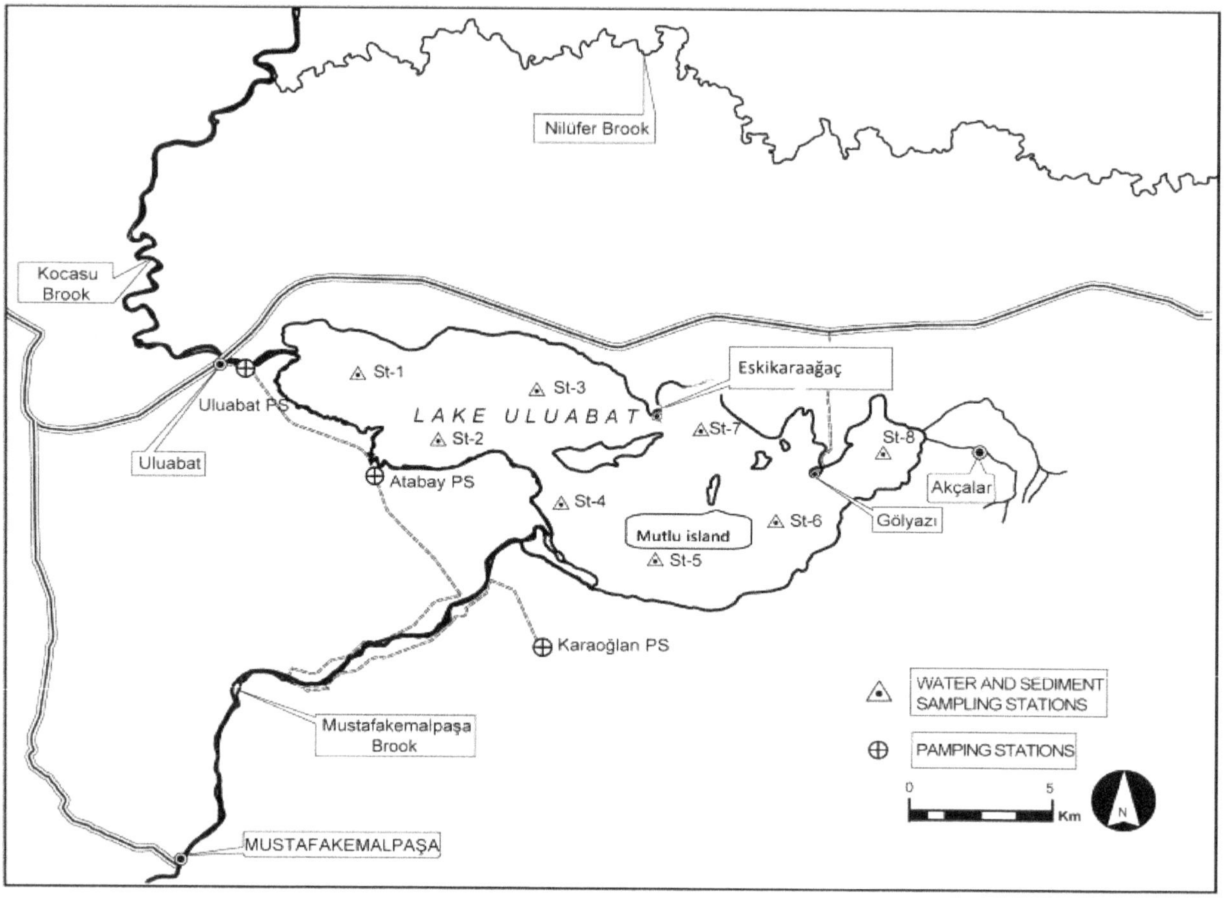

Figure 1. Sampling Stations in Lake Uluabat.

Table 1. The coordinates of sampling stations in Uluabat Lake.

Station No	X (East) (1950 UTM m)	Y (West) (1950 UTM m)
1	626865	4451240
2	629634	4448777
3	633217	4450699
4	633953	4446460
5	637299	4444284
6	641498	4445781
7	638853	4449214
8	645313	4448410

nitric acid and Mili-Q water respectively (An and Kampbell, 2003). The filter papers containing the suspended solids were air dried and reweighed again. They were digested with 4/1 (v/v) HNO_3/HCl mixture using a microwave device. After cooling, digestions were diluted to 30 ml with Mili-Q water. Samples were placed in Teflon cups and digestion operations were performed in a CEM brand Mars 5 model microwave device. Microwave decomposition operations were programmed as a 3-phase process. The device operated at 34.45 kPa for 1 min in the 1st stage, at 172.25 kPa for 5min in the 2nd stage and at 826.8 kPa for 60 min in the 3rd stage (Nguyen et. al., 2005). Trace elements were determined using the VISTA-MPX model of the VARIAN brand ICP-OES device. The operating conditions of the device, the wavelengths used in the detection of the elements and detection limits for analytical methods are shown in Tables 2 and 3 respectively (Anonymous, 2007a).

Three different groups were formed while preparing the standards in order to prevent elements from interfering with each other during detection. The grouping is as follows: 1st Group, Fe, Zn, Cr, Mn, Ni, Cu; 2nd Group, B, Cd, Pb; 3rd Group, As. The element As, in the 3rd group, was analyzed using a hybrid system. A standard stock solution of 1000 µg ml^{-1} (5% HNO_3) was used in the preparation of standards. As the concentrations of trace elements

Table 2. Operating conditions of the ICP-OES.

Parameter	
Power (kW)	1.2
Plasma flow (L min^{-1})	15
Auxiliary flow (L min^{-1})	1.5
Nebulizer flow (L min^{-1})	0.9
Replicate read time (s)	5
Instry stabilization delay (s)	30
Sample uptake rate (s)	40
Pump rate (rpm)	15
Rinse time (s)	10
Replicates	3
Display mode	Axial
Gas	Argon

Table 3. Wavelengths used in the detection of the elements.

Elements	Wavelength (nm)	Detection limit (µg L^{-1})
As	188.98	3
B	249.772	0.1
Cd	214.439	0.2
Cr	267.716	0.5
Cu	327.395	0.9
Fe	238.204	0.3
Mn	257.61	0.1
Ni	231. 604	0.7
Pb	220.353	1.5
Zn	213.857	0.2

dissolved and particulate in water are very different concentrations according to the each other, three different standard series (low, mid and high) were prepared at ppb level. The device was calibrated using a solution of 5 mg l^{-1} prepared from a Merck Mn solution of 500 mg L^{-1}. The ambient temperature was kept at 25°C to prevent possible expansion during the calibration (Anonymous, 2007a).

Statistical analysis of dissolved and particulate forms of trace elements in water was done. Correlation coefficients (r values) determine the correlations between parameters and p values show the significance levels. Factorial ANOVA Analysis was carried out by using General Linear Model method to be able to understand whether the differences of the trace element concentrations in the dissolved and particulate forms among stations, months and depth were significant. Study sample size is 192. All analyses were done in duplicate. The significance level was detected as p=0.05. All statistical calculations were done with Minitab 15.0 Program (Anonymous, 2007b).

RESULTS AND DISCUSSION

Dissolved trace elements

The concentrations of dissolved trace elements in the samples from Lake Uluabat varied between 0.0653 and 15.7298 mg L^{-1} (Fe), 0 and 0.1732 mg L^{-1} (Mn), 0.0011 and 1.3471 mg L^{-1} (Zn), 0 and 0.3508 mg L^{-1} (Ni), 0.0016 and 1.9597 (Cr), 0 and 0.3846 (Cd), 0 and 0.116 (Cu), 0.0012 and 0.0933 (As), 0.2253 and 5.5038 (B), 0 and 0.5028 mg L^{-1} (Pb), respectively. A comparison between the dissolved trace element concentrations found in Lake Uluabat and other fresh waters is presented in Table 4. It is seen that Fe, Zn, Cr, Ni, B and Pb concentrations in Lake Uluabat are higher than those of other water resources and other elements are within the similar ranges.

The degree of toxicity of dissolved metals according to their annual mean concentrations was evaluated with the help of international standards. Accordingly, As, Ni and B concentrations were found to be high according to standard values for potable water published by the World Health Organization (WHO) (Anonymous, 2006a) and low (except B) according to EPA guidelines (Anonymous, 2006b). Pb concentrations were found to be high according to WHO values and chronic effect values published by the EPA, and low according to the level of acute effect (Anonymous, 2006b). It was determined that mean Mn and Fe concentrations were below the standards. It was determined that Cd concentrations were higher than the standards of both WHO and EPA; Cu concentrations exceeded the chronic toxicity level and were very close to the acute level; Cr concentrations were higher than the chronic toxicity level; and Zn concentrations were higher than both the chronic and acute toxic limit values (Anonymous, 2006b). Table 5 shows the mean dissolved trace element concentrations in Lake Uluabat in comparison with international standard values.

Factorial ANOVA analysis of dissolved trace elements

Differences in the annual mean concentrations of dissolved trace element according to stations, depths and months; and differences in depths according to stations (interaction between station and depths) were examined using analysis of variance. The Factorial ANOVA results for dissolved trace elements are shown in Table 6. Analysis of variance showed significant differences between sampling stations in terms of Cr, Cd, Cu, Ni, B and Pb levels, and non-significant in differences between levels of As, Fe, Mn and Zn. The concentrations of Cr, Cu, Ni, B and Pb were highest at the 1st station, which is close to the outlet of the lake. The Kocasu creek which provides the outflow of the lake and reverse currents occasionally occur in it and Uluabat and Atabay Pumping Stations which are point sources of pollution are close to the 1st station. As the current in the stations 6, 3, 4 and 8 are faster than the other sampling locations, trace elements have a higher chance of being adsorbed to suspended solids and they are far from the pollution source. Trace elements concentrations were determined

Table 4. Dissolved trace element concentrations in Lake Uluabat and other fresh waters.

Elements	References	Mean ± SD
As (mg L^{-1})	Lake Texoma (An and Kampbell, 2003)	<0.033
	(Lake Uluabat)	0.023 ± 0.0179
B (mg L^{-1})	Lake Texoma (An and Kampbell, 2003)	0.225 ± 0.049
	(Lake Uluabat)	2.1668 ± 0.9964
Cd (mg L^{-1})	Lake Balaton (Nguyen et al., 2005)	0.000002
	River Vistula (Gue´guen and Dominik, 2003)	0.00056 ± 0.000018
	Lake Texoma (An and Kampbell, 2003)	0.020 ± 0.061
	(Lake Uluabat)	0.0116 ± 0.0427
Cr (mg L^{-1})	Lake Texoma (An and Kampbell, 2003)	0.004 ± 0.002
	River Vistula (Gue´guen and Dominik, 2003)	0.0013 ± 0.00005
	(Lake Uluabat)	0.0992 ± 0.2347
Cu (mg L^{-1})	Lake Balaton (Nguyen et al., 2005)	0.000475
	Lake Texoma (An and Kampbell, 2003)	0.024 - 0.020
	River Vistula (Gue´guen and Dominik, 2003)	0.03085 ± 0.0008
	(Lake Uluabat)	0.0115 ± 0.0142
Fe (mg L^{-1})	Lake Texoma (An and Kampbell, 2003)	0.119±0.093
	(Lake Uluabat)	0.5344 ± 1.3989
Mn (mg L^{-1})	Lake Texoma (An and Kampbell, 2003)	0.007 ± 0.018
	River Vistula (Gue´guen and Dominik, 2003)	0.35813 ± 0.00491
	(Lake Uluabat)	0.0116 ± 0.02
Ni (mg L^{-1})	Lake Texoma (An and Kampbell, 2003)	0.005 ± 0.003
	Lake Balaton (Nguyen et al., 2005)	0.0006
	(Lake Uluabat)	0.0304 ± 0.0448
Pb (mg L^{-1})	Lake Balaton (Nguyen et al., 2005)	0.00009
	Lake Texoma (An and Kampbell, 2003)	<0.015
	River Vistula (Gue´guen and Dominik, 2003)	0.00023 ± 0.00001
	(Lake Uluabat)	0.0355 ± 0.0567
Zn (mg L^{-1})	Lake Balaton (Nguyen et. al., 2005)	0.00085
	Lake Texoma (An and Kampbell, 2003)	0.059 ± 0.036
	River Vistula (Gue´guen and Dominik, 2003)	0.09225 ± 0.000966
	(Lake Uluabat)	0.1955 ± 0.2384

as minimum at these sampling stations. It was determined that while elements like Cu, Pb, Hg and Zn are affected by organic ligands, Cd formed a complex with inorganic ligands especially in muddy and summer conditions (Bradl, 2005). Highest Cd levels were found in the 2nd and 1st stations. The relation between the suspended solid substances and the trace elements was presented in Suspended particulate trace elements section.

The molar ratios of B:Fe:Zn:Mn:Cr:Ni:As:Cu:Cd:Pb are 200425:9160:2439:2238:1631:517:240:201:86:171 respectively and their order of size according to their concentrations was B>Fe>Zn>Cr>Pb>Ni>As>Cu>Mn>Cd, which were determined using their annual mean concentrations. Figure 2 shows the changes in the dissolved forms of Cr, Cd, Cu, Ni, B and Pb according to sampling stations. The concentration levels were observed to be low during

Table 5. Mean values of dissolved trace element concentrations in Lake Uluabat and international standard values (Anonymous, 2006 a, b).

Trace elements (mg L^{-1})	WHO	EPA, water quality criteria		Lake Uluabat
	Drinking water	CMC	CCC	
As	0,01 ^{P}	0.34 A,D,K	0.15 A,D,K	0.0182 ±0.0176
B	0,5 ^{T}			2.1668 ± 0.9964
Cd	0,003	0.002 D,E,K	0.00025 D,E,K	0.0097 ± 0.0371
Cr		0.57 D,E,K	0.074 D,E,K	0.0848 ± 0.2092
Cu	2	0.013	0.009 D,E,K	0.0128 ± 0.0141
Fe			1	0.5116 ± 1.2313
Mn	0,4 ^{C}			0.0123 ± 0.0217
Ni	0,02 ^{P}	0.47 D,E,K	0.052 D,E,K	0.0304 ± 0.0416
Pb	0,01	0.065 D,E	0.0025 D,E	0.0355 ± 0.0567
Zn		0.12 D,E,K	0.12 D,E,K	0.1595 ± 0.2169

CMC, Acute; CCC, Chronic; P, Provisional value. It is known to be hazardous but there is limited data about its effects on health; T, Provisional data; The measured values are below the refinable level; A, Recommended criterion derived for As (III). Adopted for total As because it shows similar effects with As (V); D, Recommended water quality criterion, based on the criteria for aquatic organisms by use of a conversion factor (CF); E, Standard values measured as a function of the hardness in the water column; K, Adapted from the criteria developed for aquatic organisms in 1995.

Table 6. Factorial ANOVA results of dissolved trace elements.

Elements	Stations	Depths	Months	Stations*Depths
As	0.411	0.123	0 *	0.959
B	0 *	184	0 *	0.613
Cd	0.002 *	0.545	0.010 *	0.983
Cr	0.008 *	0.763	0.057	0.6
Cu	0.007 *	0.081	0 *	0.632
Fe	0.767	0.621	0.394	0.376
Mn	0.330	0.600	0.516	0.651
Ni	0.004 *	0.175	0.003 *	0.987
Pb	0 *	0.202	0.001 *	0.384
Zn	0.439	0.513	0 *	0.563

* $p < 0.05$.

rainy season with respect to arid season. The monthly variations were statistically significant for As, Cd, Cu, Ni, Zn, B and Pb, and non-significant for Cr, Fe and Mn.

The trace elements with significant variations in monthly concentrations had high concentrations in summer and at the beginning of autumn in September, when there is no rain, and low concentrations in spring, in mid autumn and in winter, when concentrations were diluted by the increased rainfall. While the highest value of Cu was observed in summer, its lowest value was observed in October. This may be because of rainfall increase during this month, transporting more suspended solids to the lake by surface flows. Because if the lake is shallow compared to the winter months it mixes more easily, thereby increasing adsorption. In contrast to other metals, B levels were highest in November. It is thought that this was due to the increase in the flow rate and runoff in this month because of the pollution sources and rainfall. The monthly variations in the levels of dissolved As, Cd, Cu, Ni, Zn, B and Pb are presented in Figure 3.

Stronger winds cause vertical mixing of bottom sediment within the water column. This promotes to bonding between dissolved trace elements and particulates. The annual mean concentrations of suspended solids at the surface and the bottom of the water column were 37.15 and 60.70 mg L^{-1}, respectively. According to Table 6, concentrations showed non-significant differences depending on regions and significant differences depending on sampling stations. Factorial ANOVA analysis showed non-significant variation between the concentrations of dissolved trace elements at the surface and the bottom layers of the

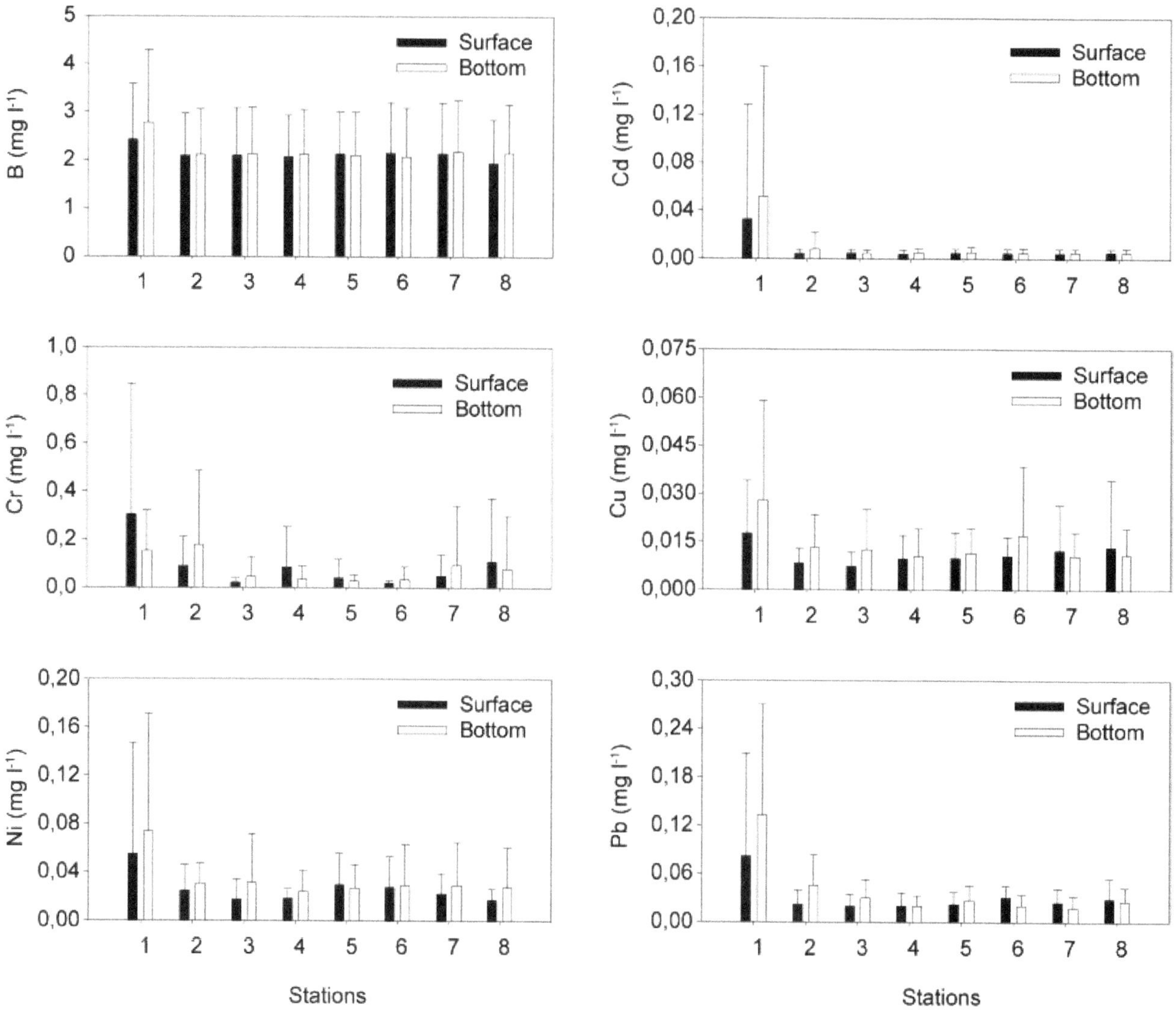

Figure 2. Variations in the concentrations of dissolved Cr, Cd, Cu, Ni, B and Pb according to sampling stations.

water column.

Suspended particulate trace elements

Factorial ANOVA analyses of suspended particulate trace elements

Table 7 shows annual means, standard deviations and concentration intervals of the particulate trace elements. In general concentration intervals are wide range. Adsorption-desorption mechanisms affected the particulate trace element concentrations within the suspended solids in the lake.

Factorial ANOVA was used to examine variations in the annual mean concentrations of particulate trace elements according to stations, surface and bottom distinction and season. Variations in Cr, Zn and B according to stations were found to be significant, and variations in the other trace elements were found to be non-significant. The maximum concentrations of Cr, Zn and B were observed at the 2nd, 3rd and 1st stations respectively. This may be because of those regions are exposed to a lot of wind and are also affected by the nearby SHW pumping stations, both of which cause water turbulence and increase the suspended solids concentration. The highest level of suspended solids (94.04 mg L^{-1}) was recorded at the 2nd station and the second-highest level of suspended solids (45 mg L^{-1}) was recorded at the 3rd station. Variations in the concentrations of particulate Cr, Zn and B according to sampling stations are shown in Figure 4. The statistical analysis showed that the variations in monthly means were significant for all of the elements. Therefore, the monthly variations in particulate trace elements are presented graphically in Figure 5. The annual mean of organic matter percentage in sediment

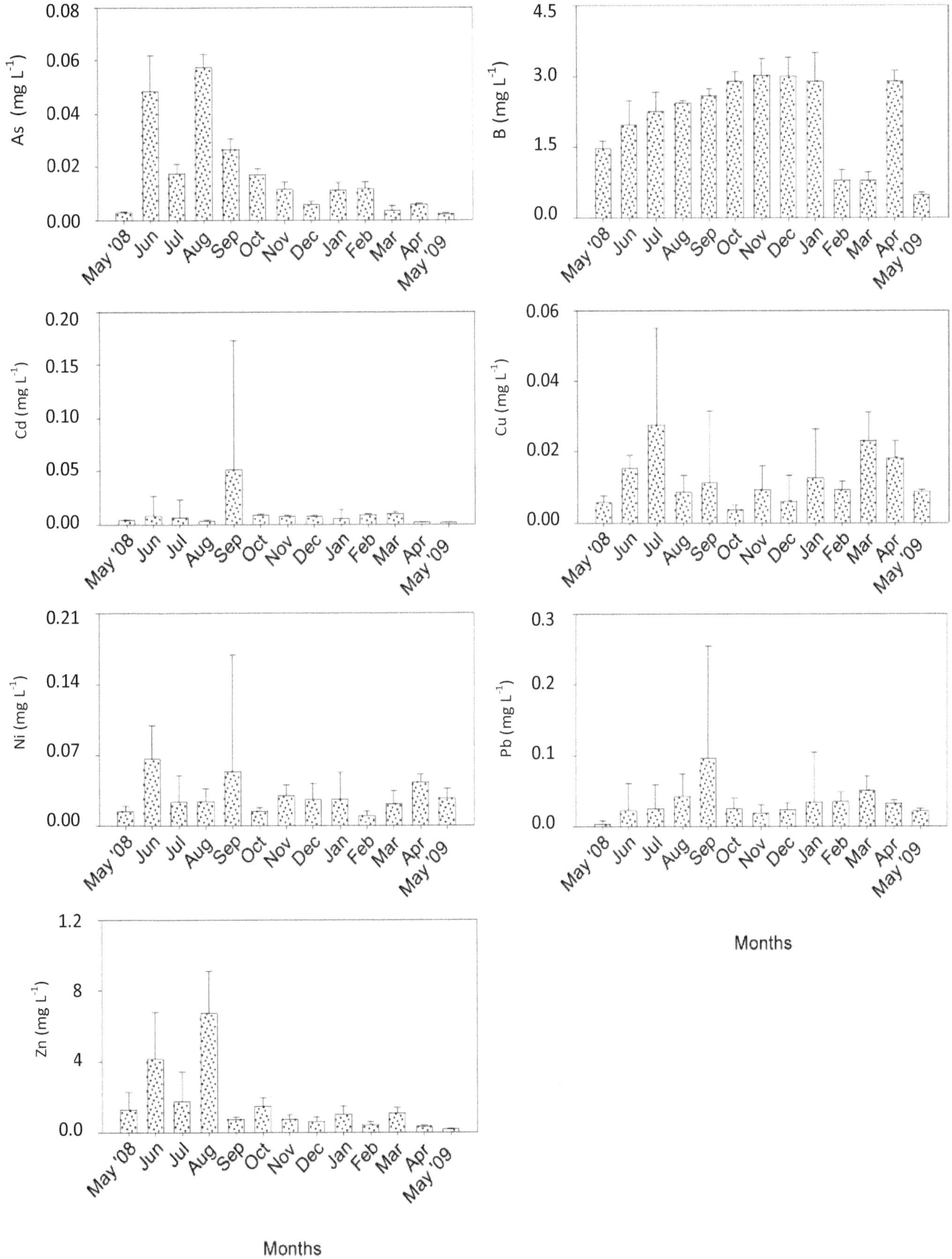

Figure 3. Monthly variations in dissolved As, Cd, Cu, Ni, Zn, B and Pb.

Table 7. Particulate trace element concentrations in Lake Uluabat and other fresh waters.

Elements	References	Min. - Max.	Mean ± SD
Cd (mg kg^{-1})	Changjiang and Hangzhou Bay (Che et. al., 2003)	0.04 - 8.39	0.565 ± 0.82
	Lake Balaton (Nguyen et al., 2005)	0.11 - 1.25	0.385
	River Danube (Yiğiterhan and Murray, 2008)		1.14
	Lake Uluabat	0 - 6.1877	1.1667 ± 0.5874
Cr (mg kg^{-1})	Lake Balaton (Nguyen et al., 2005)		19
	Lake Uluabat	0.0140 - 168.1396	13.4087 ± 20.2079
Cu (mg kg^{-1})	Changjiang and Hangzhou Bay (Che et al., 2003)	0.5 - 157.5	46.05 ± 25.6
	Lake Balaton (Nguyen et al., 2005)		16.5
	River Danube (Yiğiterhan and Murray, 2008)		115
	Lake Uluabat	0 - 49.9319	5.3701 ± 5.2744
Fe (mg kg^{-1})	Lake Balaton (Nguyen et al., 2005)	3700 - 27300	9950
	River Danube (Yiğiterhan and Murray, 2008)		36000
	Lake Uluabat	33.1149 - 26509.08	2039.2269 ± 2949.51
Mn (mg kg^{-1})	Lake Balaton (Nguyen et al., 2005)	199 – 12500	1300
	River Danube (Yiğiterhan and Murray, 2008)		1700
	Lake Uluabat	3.9596 - 565.8088	50.9093 ± 66.6561
Ni (mg kg^{-1})	Lake Balaton (Nguyen et al., 2005)		25
	Lake Uluabat	0.0161 - 123.9227	18.6455 ± 26.0972
Pb (mg kg^{-1})	Changjiang and Hangzhou Bay (Che et al., 2003)	2 – 175	32.5 ± 18.3
	Lake Balaton (Nguyen et al., 2005)		29.5
	River Danube (Yiğiterhan and Murray, 2008)		84
	Lake Uluabat	0 - 105.247	7.61223 ± 9.4288
Zn (mg kg^{-1})	Lake Balaton (Nguyen et al., 2005)	18-147	71.5
	River Danube (Yiğiterhan and Murray, 2008)		248
	Lake Uluabat	0.2246 - 162.4596	26.8742 ± 27.4318

was found to be 3.81% for the lake; this increased to 3.99% in summer and decreased to 3.55% in winter. The highest organic matter content for an individual sampling station was 4.2%, recorded at the 8th station. This station had the lowest values of Cr, Zn and B. Moreover, it was determined (Table 8) that the level of Zn is related to the percentage of organic substances in the sediment. When monthly variations of trace elements were examined, it was found that concentrations tended to increase in summer and in September while to decrease from the mid autumn to the end of spring in general.

This increase in summer was thought to occur as a result of elements transferred from the sediment to the water column, as a result of wind induced mixing (Bailey and Hamilton, 1997) and the decrease in dilution and the reduced water level. The transfer of elements from the sediment to the water showed a particular increase as a result of the increase in the solubility of metals as temperature increases, and elements bounded to the suspended solids again. Cd, Cu, Zn, As and B were found in high concentrations during rainy seasons. This may be resulted by the adsorption of those elements to solid substances which are increased with surface flows and then carried into the lake and by the amount of phytoplankton increases especially in spring (Straškraba et al., 1993).

The molar ratios of B:Fe:Mn:Zn:Ni:Cr:Cu:As:Cd:Pb, were37002:365145:9266:4110:3175:2606:845:127:103:367, respectively, which were determined by taking the annual mean of the particulate trace elements in the lake. Their order of size according to their concentrations was Fe>Mn>B>Zn>Ni>Cr>Pb>Cu>Cd>As. This differed from the size order of dissolved trace elements. A comparison of the concentrations of the particulate trace elements with the other fresh waters sources is presented in Table 7. According to this comparison, except for Cd, the concentrations of the other metals examined were at low levels. The reason is that Cd can form a complexation

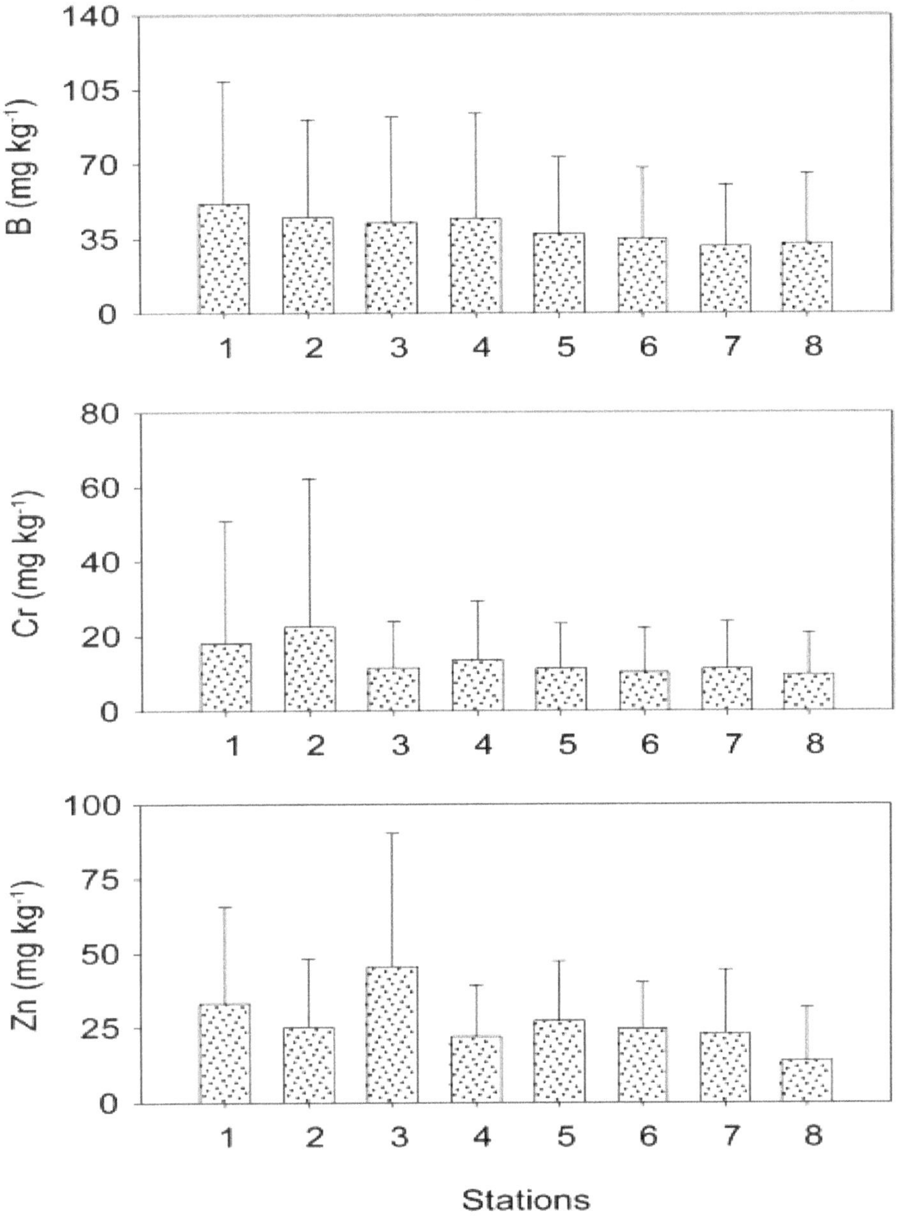

Figure 4. Variations in the concentrations of particulate Cr, Zn and B according to sampling stations.

with inorganic ligants in mudier and arable soils (Bradl, 2005). Therefore, Cd was accumulated in the mudier sediment layer of Uluabat lake. Concentration of particulate form of Cd was increased by flows in water and resuspension process. Some dissolved trace elements like Fe, Zn, Cr, Ni, B, Pb concentrations are higher than the water sources. Another reason is that heavy metals coming from point sources such as creeks and pump stations located around the lake are in dissolved form.

Factorial ANOVA analysis showed higher concentrations of particulate Cu, Ni, Fe and Mn within samples at the bottom of the water column compared to surface samples. The other elements showed non-significant variations between surface and bottom samples. This was resulted from the greater amount of suspended solids at the bottom of the water column and the relationship between chlorophyll-a and suspended solid concentrations. Due to these differences between surface and bottom concentrations, variation graphs of the particulate trace element concentrations were produced by calculating the surface-bottom means of each station. Figure 6 showed examples of concentration difference for particulate Cu and Fe at the surface and

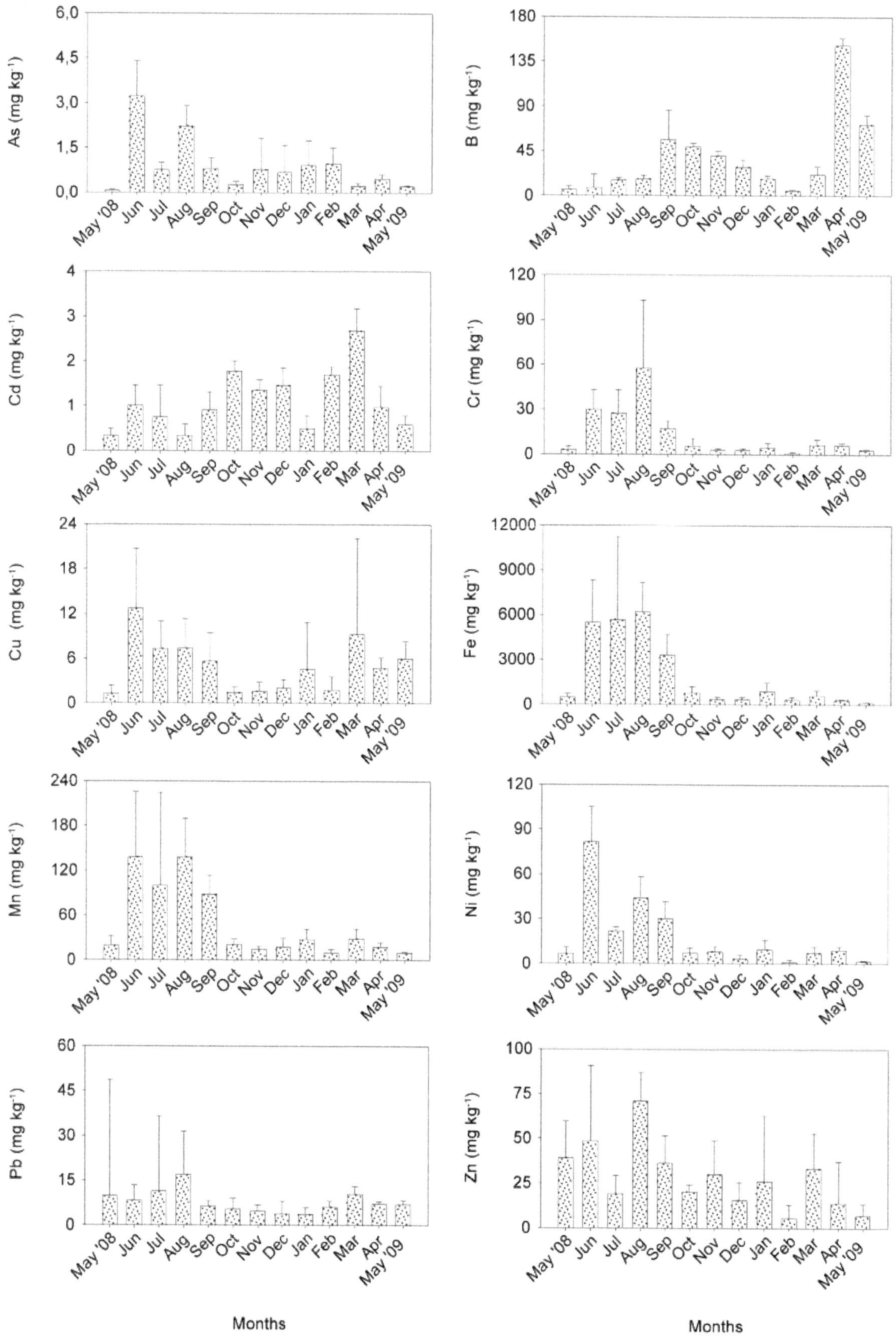

Figure 5. Monthly variations in particulate trace elements.

Table 8. The matrix of correlation coefficients (r-values) among particulate trace elements with water and sediment quality parameters.

	As	Cr	Cd	Cu	Ni	Fe	Mn	Zn	B	Pb	pH	T	EC	Chl-a	Organic Matter
Cr	0.414 *														
Cd	-0.21 *	-0.12													
Cu	0.203 *	0.288 *	0.012												
Ni	0.709 *	0.578 *	-0.196	0.428 *											
Fe	0.4 *	0.747*	-0.061	0.292 *	0.614 *										
Mn	0.445 *	0.731 *	-0.029	0.327 *	0.677 *	0.961 *									
Zn	0.359 *	0.436 *	-0.076	0.294*	0.447 *	0.363 *	0.37 *								
B	-0.33 *	-0.2	-0.136	-0.12	-0.23 *	-0.23 *	-0.226 *	-0.14							
Pb	0.272 *	0.318 *	-0.102	0.224 *	0.356 *	0.29 *	0.295 *	0.284 *							
pH	0.35 *	0.062	-0.039	-0.14	0.134	0.071	0.08	-0.07	-0.284 *	-0.142					
T	0.288 *	0.525 *	-0.3 *	0.385 *	0.589 *	0.524 *	0.505 *	0.287 *	0.083	0.382 *	-0.101				
EC	0.231 *	-0.01	-0.027	0.011	0.225 *	0.079	0.117	0.036	-0.096	-0.064	-0.099	0.099			
Chl-a	0.326 *	0.295 *	-0.112	0.278 *	0.509 *	0.398 *	0.445 *	0.047	-0.145	0.015	0.209 *	0.549 *	0.331 *		
Organic Matter	0.326 *	0.153	-0.067	0.192	0.361 *	0.158	0.191	0.362 *	-0.047	0.293	-0.014	0.335 *	0.303*	0.241 *	
SS	0.228 *	0.445 *	-0.033	0.21 *	0.614 *	0.519 *	0.607 *	0.237 *	-0.051	0.16	0.006	0.332 *	-0.008	0.445 *	0.116

*$p < 0.05$.

the bottom.

Relationship between the particulate trace elements and the water-sediment quality parameters

Relationship between the particulate trace elements and the Chlorophyll-a concentrations (Stumm and Morgan, 1996) were studied. The effects of temperature, conductivity and pH on trace elements are emphasised (Ikem et al., 2003). Dissipation of trace elements from the water column is realised with absorption to algae and precipitation. The decrease in pH and the increase in temperature affect the disssolution of metals (Fernández et al., 2000). Therefore, the relationships among organic matter percentage, temperature, pH and conductivity parameters in the sediment layer were analysed. Also, suspended solids, particulate trace elements and chlorophyll-a parameters in lake water were evaluated statistically. The results showed a significant relationship between all of the elements, except for Cd, which only showed a significant relationship with As. Particularly the correlation between Fe with Mn, Cr and Ni, and the relation of Mn with Cr and Ni, were found to be higher compared to the other correlations. This findings reflected that the movements of elements can be similar to each other. Significant relationships were found between temperature and all of the elements except B; between suspended solids and all the elements other than Cd, B and Pb; and between chlorophyll-a and all the elements other than Cd, Zn, B and Pb. The positive correlations were found between suspended solids and chlorophyll-a showing that elements can be adsorbed to solid substances. It was determined that the amount of organic matter percentage in the sediment was related with As, Ni, Zn and Pb; conductivity was related with As and Ni; and pH was related with As and B. It was thought that there may be transfer from the sediment to the water as a result of decomposition of organic matter and an anaerobic condition resulting from the increase in temperature, the percentage of organic matter and therefore the conductivity. The matrix of correlation coefficients (r - values) among particulate trace elements with water and sediment quality parameters are presented in Table 8 and the correlation graphs between particulate Cr, Ni, Fe Mn and suspended solids are presented in Figure 7.

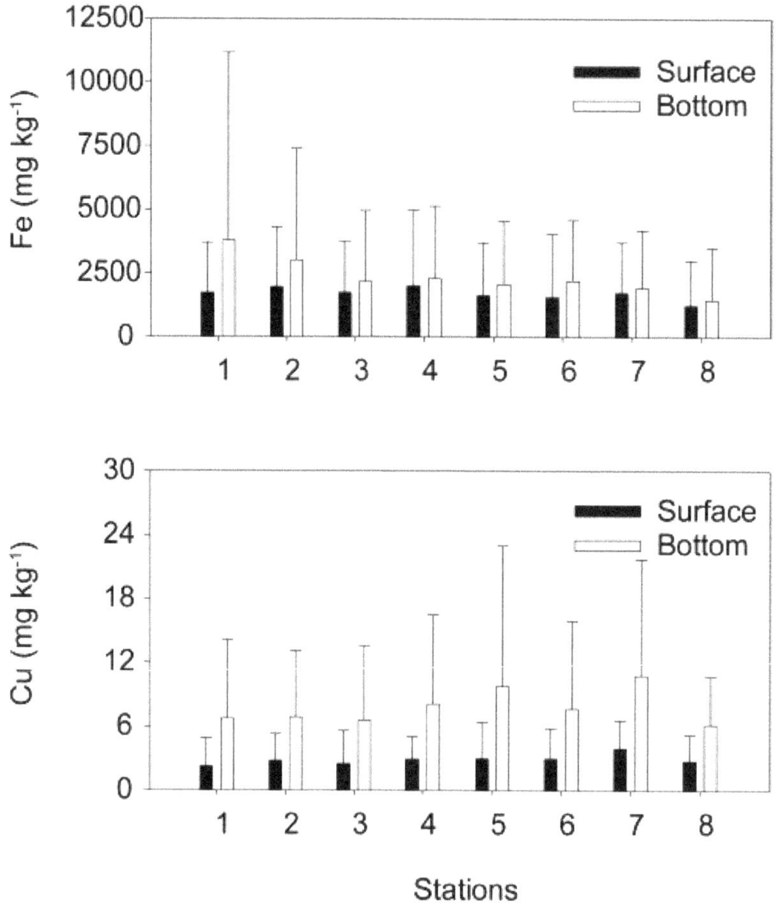

Figure 6. Surface and bottom concentrations of particulate Cu and Fe.

Conclusion

Dissolved and particulate trace elements concentrations were examined in lake water and sediment for 8 sampling stations representing the different regions of Lake Uluabat in Turkey. The dissolved and particulate trace elements concentrations were determined at the surface and the bottom layers within the water column. In general the highest concentrations of trace elements were observed in the particulate form. Lower organic content in water indicated that the suspended solids were transferred from the sediment. Concentrations of particulate Cr, Zn and B were detected lowest at the 8th sampling station where organic matter percentage in the sediment was highest.

The size of the trace elements according to the concentrations of the elements were B>Fe>Zn>Cr>Pb>Ni>As>Cu>Mn>Cd for the dissolved trace elements; and Fe>Mn>B>Zn>Ni>Cr>Pb>Cu>Cd>As according to particulate trace elements.

The differences between the surface and the bottom layers of the water column were non-significant in terms of dissolved trace elements, but significant in terms of some of the particulate trace elements (Fe, Mn, Cu and Ni). The concentrations of dissolved trace elements at the bottom of the water column were observed to be higher than those of surface samples. This difference was resulted from the higher concentrations of suspended solids present in the deeper parts of the water column.

Mostly, the dissolved and particulate trace elements were detected high in summer season. The main reason was the increase of concentrations of dissolved trace elements depending on solubility, decrease in water level and increase in air temperature. Furthermore, increase in photosynthetic production by implication of pH, concentration of $CaCO_3$ and adsorption capacity of iron and manganese oxides have caused this situation. In addition increasing wind velocity in summer season induced the mixing of sediment layer into the water column bringing about suspended trace elements. Significant correlations were found between suspended solids and particulate Fe, Mn, Zn, Cr, Cu, As, Ni elements; between Chl-a and particulate Fe, Mn, Cr, Cu, As, Ni elements. Same correlation was observed for particulate trace elements (except B) and temperature;

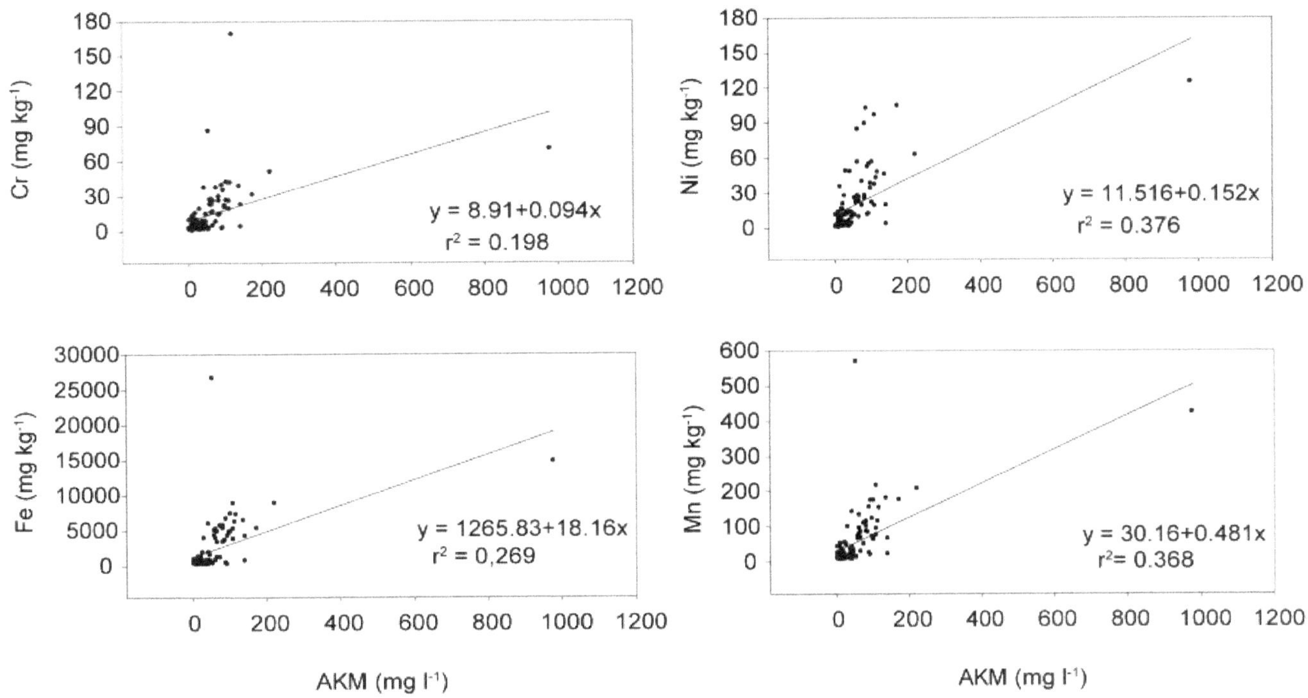

Figure 7. Correlation graphs between particulate Cr, Ni, Fe Mn and suspended solids.

pH and As, B and organic matter percentage with As, Ni, Zn elements. Annual means of dissolved trace elements were compared with WHO and EPA international standards. In lakes which have eutrophic status like Uluabat, trace elements absorbed to algae can pass to food web by release easily. Consequently, the lake will become more toxic when the probable decrease of pH. As a result, the long term monitoring and control of domestic and industrial wastewaters, restricted uses of agricultural pesticides are the important factors have to be taken into consideration in lake management. It is necessary to produce and apply new management strategies or plans in controlling point and non-point sources.

ACKNOWLEDGEMENTS

The project was funded by the Environment, Atmosphere, Earth and Marine Sciences group (ÇAYDAG- Project no: 107Y278) of the Scientific and Technical Research Council of Turkey (TÜBİTAK) and Scientific Research Foundation (Project no: M-2007/27) of Uludag University.

REFERENCES

Akdeniz S, Karaer F, Katip A, Aksoy E (2011). A GIS-based Method for Shallow Lake Eutrophication Assessment. J. Biol. Environ. Sci. 5(15):195-202.

Aksoy E, Özsoy G (2002). Investigation of multitemporal land use/cover and shoreline changes of the Uluabat Lake Ramsar site using RS and GIS, paper presented at the International Conference on Sustainable Land Use and Management, 13 October 2002, Turkey.

An Y, Kampbell DH (2003). Total, Dissolved, and Bioavailable Metals at Lake Texoma Marinas. Environ. Poll. 122:253-259.

Anonymous (2006a). *WHO,* A Compendium of Drinking-Water Quality Standards in The Eastern Mediterranean Region.

Anonymous (2006b). USEPA National Recommended Water Quality Criteria Correction Office of Water, EPA, 822-z-99-001, P. 25.

Anonymous (2007a). Terra/ Varian and Cem Brand ICP-MS/ICP-OES/AAS Microwave Digestion Seminar Notes, P. 81.

Anonymous (2007b). Minitab Inc. Meet Minitab, Release 15 for Windows, State College, PA.

APHA (1998). AWWA Standart Methods for the Examination of Water and Wastewater. American Public Heallth Association, 20th Edn. Washington, D. C.

Bailey MC, Hamilton DP (1997). Wind Induced Sediment Resuspension: a Lake-Wide Model. Ecol. Model. 99:217-228.

Bradl HB (2005). Heavy Metals in The Environment: Origin, Interaction and Remediation. Elsevier Academic Press. P. 269.

Burger J, Gochfeld M (2003). Heavy Metals in Commercial Fish in New Jersey. Environ. Res. 99:403-412.

Burton GA, Pitt RE (2002). Stormwater Effects Handbook, Lewıs Publıshers, CRC Press LLC. P. 929.

Che Y, He Q, Lin WQ (2003). The Distributions of Particulate Heavy Metals and Its Indication to The Transfer of Sediments in The Changjiang Estuary and Hangzhou Bay, China. Mar. Poll. Bull. 46:123-131.

Dalkıran N, Karacaoğlu D, Dere S, Sentürk E, Torunoğlu T (2006). Factors Affecting The Current Status of a Eutrophic Shallow Lake (Lake Uluabat, Turkey): Relationships Between Water Physical and Chemical Variables. Chem. Ecol. 22:279-298.

Fernández M, Cuesta S, Jıménez O, García MA, Hernández LM, Marina ML, González MJ (2000). Organochlorine and heavy metal residues in the water/sediment system of the Southeast Regional Park in Madrid, Spain. Chemosphere 41:801-812.

Filgueiras AV, Lavilla I, Bendicho C (2004). Evaluation of distribution, mobility and binding behaviour of heavy metals in surficial sediments of Louro River (Galicia, Spain) using chemometric analysis: A case

study. Sci. Total Environ. 330:115-129.
Gue´guen C, Dominik J (2003). Partitioning of Trace Metals Between Particulate, Colloidal and Truly Dissolved Fractions in a Polluted River: The Upper Vistula River (Poland). Appl. Geochem. 18:457-470.
İleri S (2010). The Assessment of physico-chemical parameters and geographic information system analysis of water and sediment quality of Lake Uluabat in terms of environment. MSc Thesis, Uludag University, Institute of Science, Bursa.
Ikem A, Egıebor NO, Nyavor K (2003). Trace Elements in Water, Fish and Sediment from Tuskegee Lake, Southeastern USA. Water, Air Soil Poll. 149:51-75.
Katip A (2010). Water quality monitoring of Lake Uluabat. PhD Thesis, Uludag University, Institute of Science, Bursa.
Katip A, Karaer F, Başkaya HS, İleri S, Sarmaşık S (2012a). Fraction distribution and risk assessment of heavy metals and trace elements in sediments of Lake Uluabat. Environ. Monit. Assess. 184:5399-5413.
Katip A, Karaer F, İleri S, Sarmaşık S, Aydoğan N, Zenginay S (2012b). Analysis and assessment of trace elements pollution in sediments of Lake Uluabat, Turkey. J. Environ. Biol. 33:961-968.
Nguyen HL, Leermakers M, Osán J, Török S, Baeyens W (2005). Heavy metals in LakeBalaton: Water Column, Suspended Matter, Sediment and Biota. Sci. Total Environ. 340:213-230.
Parsons RT, Strickland JD (1963). Discussion of Spectrofotometric Determination Of Marine Plant-Pigments With Revised Equations For As Certaining Chlorophyll and Corotenoids. J. Mar. Resour. 21:155-163.
Radojevic M, Bashkin, VN (1999). Practical Environmental Analysis., UK. Royal Soc. Chem. P. 466.
Shukla A, Zhang YH, Dubey P, Margrave JL, Shukla SS (2002). The role of sawdust in the removal of unwanted aterials from water. J. Hazard. Mater. B95:137-152.
Sigg L, Sturm M, Kistler D (1987). Vertical transport of heavy metals by settling particles in Lake Zurich. Limnol. Oceanogr. 32(l):112-130.
Straškraba M, Blažka P, Brandl Z, Hejzlar P, Komárková J, Kubečka J, Nesměrák I, Procházková L, Straškrabová V, Vyhnálek, V (1993). Framework for Investigation and Evaluation of Reservoir Water Quality in Czechosiovakia. Developments in Hydrobiology, Comp. Reservoir Limnol. Water Qual. Manage. 77:169-212.
Stumm W, Morgan J (1996). Aquatic Chemistry: An Introduction Emphasizing Chemical Equilibria in Natural Water, Wiley, New York, Third edn. pp. 655-666.
Yiğiterhan O, Murray JW (2008). Trace Metal Composition of Particulate Matter of The Danube River and Turkish Rivers Draining into The Black Sea. Mar. Chem. 111:63-76.

12

Meditator's non-contact effect on cucumbers

Osamu Takagi[1], Masamichi Sakamoto[2], Hideyuki Kokubo[1], Hideo Yoichi[1], Kimiko Kawano[1] and Mikio Yamamoto[1]

[1]Information and Research Center, International Research Institute (IRI), 1108-2 Sonno, Inage, Chiba 263-0051, Japan.
[2]Aquavision Academy, 1228-3 Tsubuura, Narita, Chiba 287-0236, Japan.

We clearly show the existence of an example of non-contact effect in which the "presence" of a meditator affects bio-samples without physical contact. This is the first report in the world to show this type of effect by scientific measurements. We used edible cucumber slices as bio-sensors and measured the concentrations of gas emitted from the slices by a technique developed by our group. The concentrations of gas emitted from cucumber slices were measured for a total of 672 sample petri dishes; each dish contained four cucumber slices so that a statistically meaningful comparison could be made. We found a statistically significant difference ($p=3.13\times10^{-10}$, t-test, two-tails) in the concentrations of emitted gas between the "presence" and the "absence" of the meditator. Our experimental results clearly indicated that there was a scientifically measurable effect on biological objects with which the meditator had no direct physical contact.

Key words: Non-contact effect, meditator, bio-sensor, cucumber, gas.

INTRODUCTION

It is a scientifically interesting theme whether or not humans can affect plant samples without coming into physical contact with them. However, no research has been conducted which scientifically measures this effect except some studies such as Grad (1976). Yamamoto et al. (2000, 2002) have conducted scientific research projects including the above theme at the National Institute of Radiological Sciences (Japan) since 1995. They have done measurements of bio-photons (Cohen and Popp, 1997; Inaba, 1997) (very low intensity visual range light emitted from bio-samples) for fundamental experiments to utilize them as bio-sensors since 1988 (Parkhomtchouk et al., 2000). Kokubo et al. (2010) have developed a bio-photon method, fluorescence method and gas method which utilize edible cucumber slices as bio-sensors. The gas method, which measures gas emitted from cucumber slices with gas detection tubes, enables measurement of a number of samples simultaneously at low cost. Therefore, we have used that method here.

The purpose of this paper is to verify by scientific measurements a case of the non-contact effect in which the presence of a meditator affects bio-samples without physical contact. A strict scientific comparison of the effect on bio-samples was conducted excluding all other factors between when a meditator is present ("presence") or absent ("absence"), where "absence" means the mediator was more than 5 km away from the place where the experiments were conducted. We experimentally investigated the effect on sliced cucumber samples by the gas method. The samples were slices of the fruit of *Cucumis sativus* 'white spine type'. A statistically very significant difference was obtained between the "presence" and "absence" with a sufficient number of experimental runs and the analysis of 672 measured data.

EXPERIMENTAL

Figures 1 and 2 show how cucumber samples were prepared. One set of samples uses four cucumbers A to D (Figure 1). Four pairs of sterilized plastic dishes (20 mm high and 90 mm diameter) were prepared (representing a standard set) (Figure 2). After the slices were placed in the dish, it was covered by a lid. We used 12 standard sets for one experimental run, representing 48 cucumbers per run; after use, dishes were discarded. As shown in Figure 3, EXP2 dish was stacked above EXP1 dish and both were placed above the meditator, while the remaining six dishes (EXP3 and EXP4 and CONT1 to CONT4) were placed at the Calibration Control Point, 8 m away from the meditator, where each pair was stacked one above the other. These dishes were placed 1.8 m above the floor in either location. Only one meditator (male, with 20 years of meditation experience using Hem-Sync® (Atwater, 1997)) participated in the experimental runs reported here, and the environment which he prefers for meditation was set up. In particular, a pyramid shaped structure was prepared. Written consent of the meditator to be involved in the meditation experiments was obtained, as well as approval from the ethics committee of the International Research Institute (IRI).

After a set of eight dishes (four pairs of dishes) were placed in their positions (Figure 3), the meditator meditated for 30 min. Immediately after the meditation, all the eight dishes were removed from their positions, their lids were taken off, and each dish was placed within a 2.2 L closed polypropylene container. After that, they were kept for approximately 36 h in a temperature-controlled room (at about 24°C) with no direct sunlight. Eight containers were stacked in two piles in such a way that the containers which have paired dishes were placed side by side. All containers were kept in the place 3.5 m away from pyramid before the gas measurement. Throughout this 36 h period, the cucumbers continued to emit gas. It is known that the concentrations reach the maximum level after about 12 h and remain constant afterward. Measurement of the gas contained in the closed container was done as follows. A gas sampling pump (GV-100: Gastech, Japan) to extract gas and gas detection tubes (ethyl acetate short-term gas-measuring detector tube 141L: Gastech, Japan) were used to measure the concentrations of gas generated from the cucumbers. The gas detection tubes can measure concentrations from 20 to 800 ppm. After 300 ml gas was extracted from the closed container using the gas sampling pump, the gas detection tube gauge was visually read. The accuracy of reading was on the order of 10 ppm.

One experimental run had four segments which were conducted continuously from early in the morning till sunset on one day to make the experiments as reliable as possible. The meditator was at least 5 km away from the pyramid during the first and the fourth segments. The first segment (before meditation, pre-med) was composed of three sets of experiments, in each of which eight dishes were placed in their positions for 30 min without the meditator inside the pyramid. The second segment (Meditation I, med-I) was composed of three experiments, in each of which the meditator meditated inside the pyramid for 30 min and all the other conditions were the same as in the first segment. There was about a 10 min break between the experiments. The third segment (Meditation II, med-II) was basically the same as the second. Between the second and the third, there was an hour lunch break. The fourth segment (after meditation, post-med) was basically the same as the first. We conducted seven experimental runs from April to October, the results of which were statistically analyzed and reported here.

The meditation was performed as follows. The meditator sat on a chair which was placed directly below the pyramid apex. During the 30 min meditation, he listened to Hemi-Sync®, which is an audio technology which helps a person to meditate for an extended period of time. The meditator wore a set of stereo headphones so that there was no sound leak to the environment. During the meditation, he imagined that he was breathing in energy from the earth as he breathed in and he was breathing in energy from heaven as he breathed out. He sometimes vocalized vowel sounds during meditation. He did not focus his attention onto the EXP1 and EXP2 samples which were placed right above him. Also, all the room lights were off and the blinds were closed so that the room was dimly lit even during the daytime.

The gas measurement results for the seven experimental runs were conducted on seven different days. The average values are in the range of 300 to 400 ppm for all four segments, before meditation, after meditation and Meditations I and II and the standard error is 21.1 to 25.5 ppm. Since the reading error is on the order of 10 ppm, the obtained data are very reliable.

MODE OF ANALYSIS

In order to see whether or not there is any influence from the meditator's presence on the cucumbers without having any direct physical contact, we introduce a new number which can be used to evaluate this. Such an effect can be buried within the noise of the data. The reason is that cucumbers as bio-sensors are very sensitive to many factors such as individual cucumber differences and environmental conditions including how samples are treated (temperature, humidity, electromagnetism, intensity of illumination, vibration and biorhythm). In order to minimize such variables we adopted a paired sample method (Figure 2) where EXP and CONT are paired and compared. A new number J is introduced to compare such a pair (Kokubo et al., 2010):

$$J = \ln (C_{EXP}/C_{CONT})$$

where C_{EXP} is the concentration of gas for EXP sample, while C_{CONT} is that for CONT sample. The J value eliminates the environmental changes and the effects of systematic deviations which may exist in the experimental environment, enabling us to highlight the difference between when the meditator is present ("presence") or absent ("absence"). When such factors as the individual differences and the experimental conditions for the EXP sample and the CONT sample are the same, C_{EXP} and C_{CONT} become the same, resulting in J=0.

We can calculate four J values for each set of experiments by using the gas measurement data. For each set of experiments, two pairs of samples EXP1 and CONT1 and EXP2 and CONT2 are called the main experiment samples and the each pair has a J value, which is termed J_E. At the same time, two pairs of samples EXP3 and CONT3 and EXP4 and CONT4 are called calibration experiment samples and the each pair has a J value with the average of these two J values being termed J_{CAL}.

Now in order to compare J_Es from different sets, we introduce a zero-point calibration method in which a new value $J_{E/CAL} = J_E - J_{CAL}$ is introduced. The reason behind using this calibration method is that the two pairs of samples EXP3 and CONT3 and EXP4 and CONT4 are placed in nearly equal conditions, which can be treated as the baseline in each set of experiments (Kokubo et al., 2010). In each segment, three sets of experiments are conducted with a total of six $J_{E/CAL}$s and three J_{CAL}s. Now we average the six $J_{E/CAL}$s to obtain $J_{E/CAL/AVE}$. This way, in each of the four segments (pre-med, med-I, med-II and post-med), $J_{E/CAL/AVE}$ is calculated. Furthermore, in order to compare when the meditator is absent and present, we average the $J_{E/CAL/AVE}$ for pre-med and that for post-med and

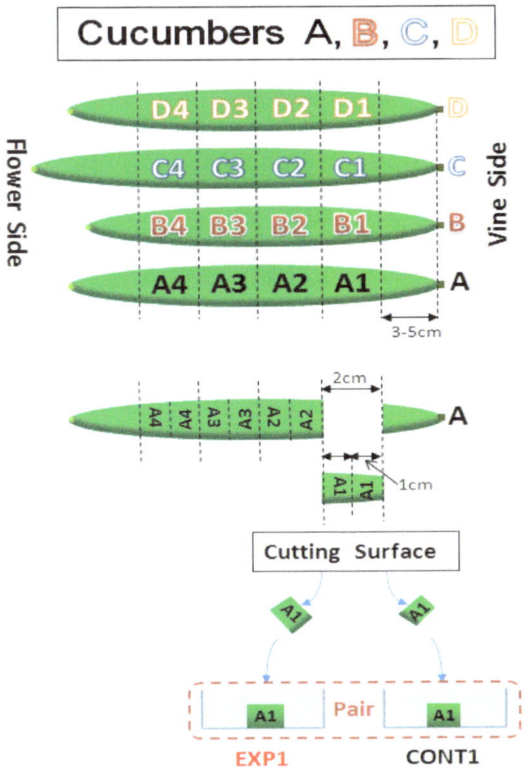

Figure 1. Preparation of cucumber slice samples.

subtract this value from each $J_{E/CAL}$ to obtain a new value J_C defined by the following equation:

$J_C = J_{E/CAL} - \{J_{E/CAL/AVE}\,(\text{pre-med}) + J_{E/CAL/AVE}\,(\text{post-med})\}/2.$

Each experimental run is composed of four segments and each segment has six J_Cs.

RESULTS AND DISCUSSION

Figure 4 (main) clearly indicates that the average values of J_C when meditator is absent ("absence" J_C=0.0) differ significantly from those for the case in which the meditator is present ("presence" J_C=-0.179). The J_C values are plotted with an error bar which corresponds to the 95% confidence interval for n=84. By analyzing the J_C values for these two cases, a p value of 3.13×10^{-10} (t-test, two tails) is obtained between them indicating a statistically significant difference. Figure 4 (inset) indicates that the average values of J_C when the meditator is absent (pre-med J_C=0.028 and post-med J_C=-0.028) differ from those when the meditator is present (med-I J_C=-0.173 and med-II J_C=-0.184). The horizontal axis corresponds to the flow of time during the day when the experiments took place. The J_C values are plotted with an error bar which corresponds to the 95% confidence interval for n=42. There is a lunch break between med-I and med-II. The data for med- I and med-II are nearly equal, while the data for pre-med and post-med are also nearly equal.

In order to explain the observed results, here we would like to review various known effects as causes. First, the effect of the electrostatic field created by the meditator can be excluded as a cause of the observed results, since the cucumber samples were in a Faraday cage as shown in Figure 3. Second, to consider the effect of temperature change by the presence of the human meditator, the temperature of the sample circumference was monitored during the experiments. No temperature rise due to the meditator's presence was detected, which indicates that the temperature effect of the meditator can also be excluded as a cause. It is thought that the effects of all other known factors are negligible. Therefore, the observed results may well be caused by a phenomenon which is not yet known scientifically but may well be caused by the presence of the meditator. Revealing the cause of this phenomenon would advance modern science greatly. The next theme should be conducting experiments with a different meditator.

Conclusion

We have investigated whether or not the presence of a meditator has a scientifically measurable influence on biological objects to which he has no direct physical contact. In this reports, cucumber slices were used as bio-sensors to detect such an influence. We developed experimental and analytical methods by which the "presence" and the "absence" of a meditator can be strictly compared excluding all other factors. The concentrations of gas emitted from cucumber slices were measured for a total of 672 samples. These experiments were conducted on seven different days to minimize any possible erroneous influence that could occur within a day. By statistically analyzing these data, we have demonstrated that there is a statistically significant difference ($p=3.13\times10^{-10}$, t-test, two tails) between the "presence" and the "absence" of a meditator in the concentrations of emitted gas. We have verified that this is a case of a non-contact effect in which the presence of a meditator affects bio-samples without physical contact. The observed results may well be due to a phenomenon which is not yet known scientifically but may well be caused by the presence of the meditator. Revealing the cause of this phenomenon would advance modern science greatly.

ACKNOWLEDGEMENT

This study was supported by the Masamichi Sakamoto Hypertech Project which is funded by the Aquavision Academy.

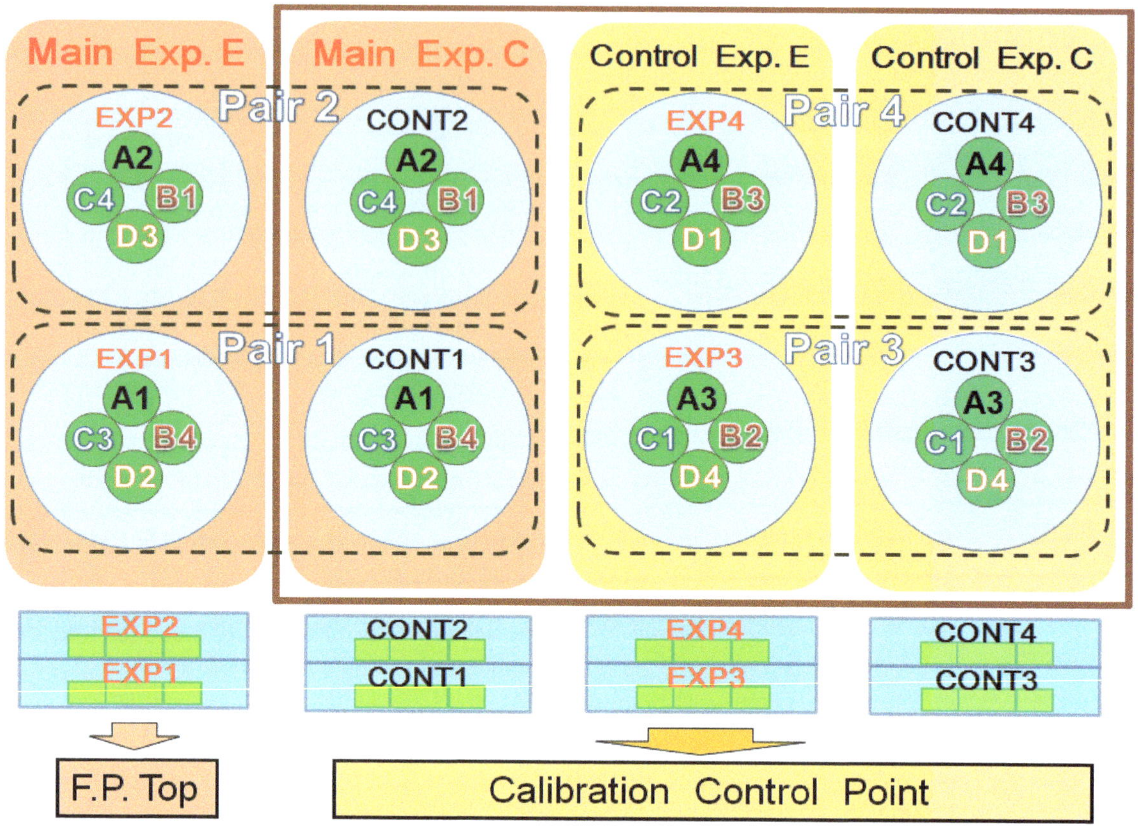

Figure 2. Preparation of cucumber slice samples on petri dishes.

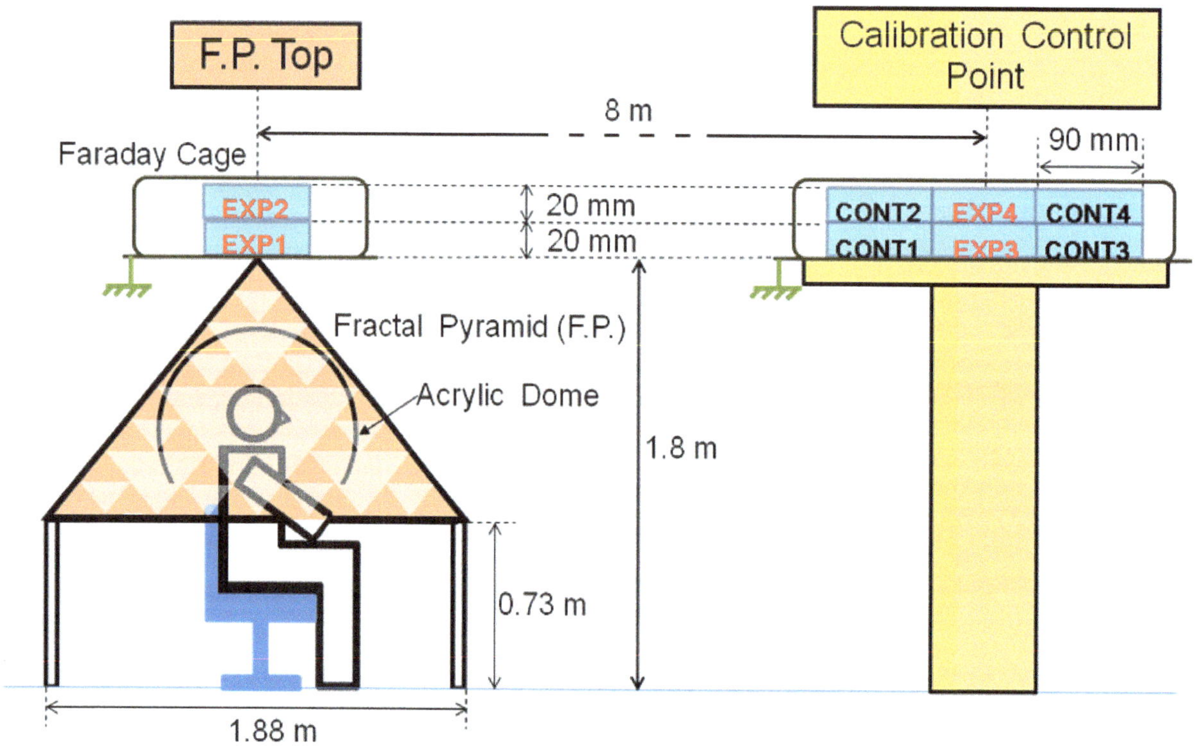

Figure 3. Position of cucumber slice samples relative to meditator position.

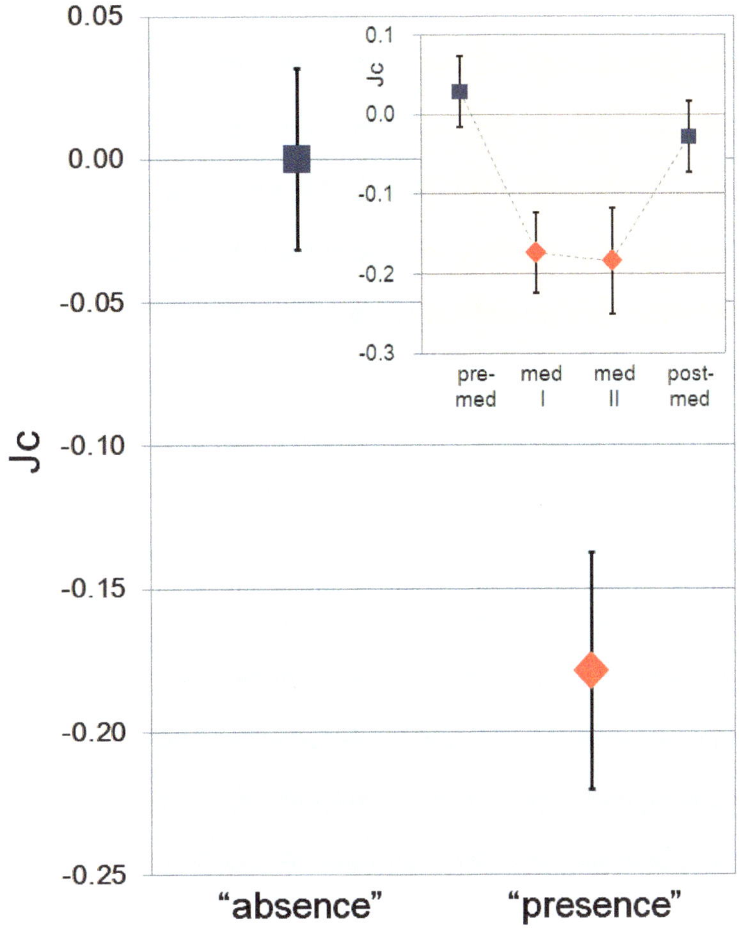

Figure 4. Average value of J_C.

REFERENCES

Atwater FH (1997). Accessing anomalous states of consciousness with a binaural beat technology. J. Sci. Explor. 11(3):263.

Cohen S, Popp FA (1997). Biophoton emission of the human body. J. Photochem. Photobiol. B: Biol. 40:187.

Grad B (1976).The biological effects of the "laying on of hands" on animals and plants: Implications for biology. in Parapsychology, G. R. Schmeidler, Eds. (Metuchen, NJ, Scarecrow Press) P. 76.

Inaba H (1997). Photonic sensing technology is opening new frontiers in biophotonics. Opt. Rev. 4(1A):1.

Kokubo H, Takagi O, Koyama S (2010). Application of a gas measurement method –Measurement of ki fields and non-contact healing . J. Intl. Soc. Life Info. Sci. 28(1):95.

Parkhomtchouk D, Hirasawa M, Yamamoto M (2000). Low noise system with optical cable probe for ultra-weak photon emission measurements from living matter. J. Intl. Soc. Life Info. Sci. 18(1):109.

Yamamoto M, Hirasawa M, Kokubo H, Tanaka M, Kawano K, Dmitri V, Zhang T, Fukuda N, Hirata T, Murakami S, Moroe T, Makino K, Tsuchiya K, Kokado T, Taniguchi J, Nakamura A, Kurano M, Furukawa A, Furukawa M, Matsumoto T, Sakaida H, Machi Y (2000). Study on analyzing methods of human body functions using various simultaneous measurements. J. Intl. Soc. Life Info. Sci. 18(1):61.

Yamamoto M, Kokubo H, Haraguchi S, Kokado T, Tanaka M, Zhang T, Chen W, Kotake J, Yoichi H, Kawano K, Fukuda N (2002). Are there any effects other than suggestion in external Qi. J. Intl. Soc. Life Info. Sci. 20(2):317.

13

Fourier transform infrared spectroscopic characterization of clay minerals from rocks of Lalibela churches, Ethiopia

Alemayehu Kiros[1]*, A. V. Gholap[1] and Giovanni E. Gigante[2]

[1]Department of Physics, Faculty of Science, Addis Ababa University, P. O. Box -1176, Addis Ababa, Ethiopia.
[2]La Sapienza University, Rome, Italy.

This study demonstrates the Fourier transform infrared (FTIR), Raman spectroscopy and scanning electron microscope - energy dispersive X-ray spectrometer (SEM-EDS) techniques which has been applied to characterize decay products and processes occurring at the surface of rock-hewn churches at the UNESCO's World Heritage site of Lalibela, Northern Ethiopia. Spectroscopic analyses comparing the observed frequencies with the available literature, minerals were identified. A special emphasis is devoted to the presence of whewellite/weddellite and kaolinite minerals, never pointed out before this study in the rock hewn churches of Lalibela, and their possible roles on rock decay. The identification of montmorillonite as a weathering product up to 20 mm from the rock/atmosphere interface beyond the level at which kaolinite, whewellite and gypsum are found, suggests that montmorillonite may penetrate even deeper into the rock substrate causing major rock decay. Results indicate that biological attack by green algae and lichens is currently responsible for severe stone surface, physical and chemical weathering leading to considerable weakening of the churches walls. A prompt and careful removal of the biological and other deposits with the non-destructive method (laser cleaning) treatment is therefore recommended.

Key words: Fourier transform infrared (FTIR), Raman, whewellite, kaolinite, rock hewn churches, mineral characterization.

INTRODUCTION

Lalibela is known throughout the world for its extraordinary complex of 11 monolithic churches (Figure 1), carved, some 800 years ago during the Empire of King Lalibela. The main cluster of 11 churches is located in the middle of the village since 1978; the churches and their surrounding area have been included in UNESCO's World Heritage List (Asrat and Ayallew, 2011). Water absorption represents the main contributing factor of damage for high exposed rock materials and structures, in terms of direct rainfall, soil infiltration, capillarity and diffuse humidity, for the alteration of basalt (Delmonaco et al., 2009). Another study (Asrat and Ayallew, 2011) reported material loss due to deep weathering activated by rain water penetration and the leakage of groundwater affects most of the Lalibela churches. This exposure has resulted in the severe degradation of the churches, most of which are now considered to be in a critical condition. A first restoration attempt of the Lalibela rock-hewn churches was carried out in 1920 (Delmonaco et al., 2009). Currently, most of the churches are protected by shelters to prevent the erosive effects of rainfall, especially on the roof rock cover, and temporary scaffolding to prevent collapse of the most exposed structures.

In many cases, the presence of biological colonization by lichens and other major plants is quite evident. For the study of biological aspects of weathering of rocks and minerals, lichen-encrusted rocks provide an ideal environment. The effects of lichens on their mineral substrates can be attributed to both physical and chemical processes. Lichens also have significant impact

*Corresponding author. E-mail: ftkiros@gmail.com.

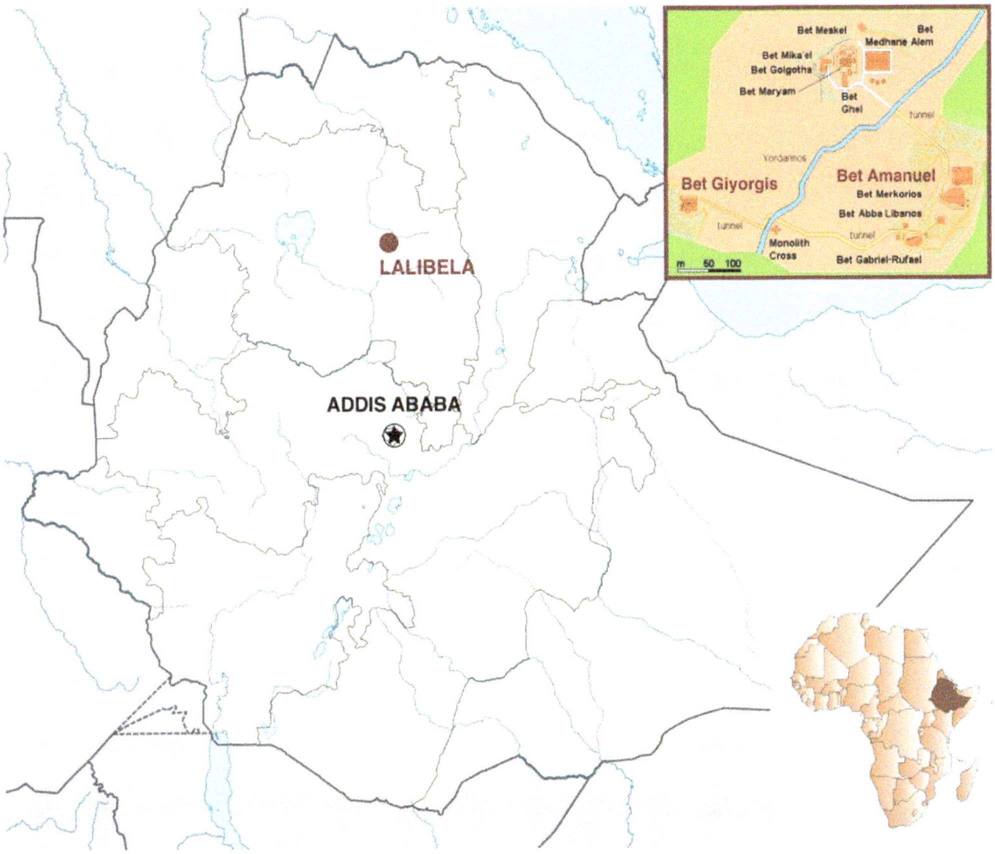

Figure 1. Lalibela location map. Inlet (top right) shows the geographical distribution of the rock-hewn churches at the Lalibela monumental site.

in the chemical weathering of rocks by the secretion of various organic acids, particularly oxalic acid, which can effectively dissolve minerals and chelate metallic cations (Adamou and Violante, 2000; Chen et al., 2000). As a result of the weathering induced by lichens, many rock-forming minerals exhibit extensive surface corrosion.

Raman spectroscopy, scanning electron microscope - energy dispersive X-ray spectrometer (SEM-EDS) and Fourier transform infrared (FTIR) techniques found many applications in the field of geology, archaeology and cultural heritage analysis due to its non-invasiveness. The infrared spectroscopic investigation is found to be an ideal tool for structure elucidation and for estimating quartz crystallinity of the natural samples (Saikia et al., 2008), and characterization of natural kaolinite (Saikia and Parthasarathy, 2010). FTIR and EDS are established tool for the ceramic body identification (Ravisankar et al., 2010), to study the ancient potteries (Ravisankar et al., 2010; Ravisankar et al., 2012). FTIR is also a reliable method for the characterization of the organic matter preserved in carbonates, and generally in all types of sedimentary rocks (Ravisankar et al., 2010; Schiavon, 2007). Raman spectroscopy and micro-Raman study provides a tool for examination of building stones, and is particularly useful for the investigation of surface degradation (Perez-Alonso et al., 2004; Potgieter-Vermaak et al., 2005).

Clay mineral characterization could be carried out employing spectroscopic methods for various purposes in the building stones (Schiavon, 2007; Davarcioglu and Ciftci, 2010). In the present work, we report and discuss the experimental results obtained through FTIR transmittance, Raman spectroscopy and EDS measurements performed on rock samples to report qualitative and quantitative data for the deleterious components of the rock hewn churches of Lalibela. Our previous X-ray diffraction (XRD) data on whole rock samples confirm the presence of peaks attributable to zeolites (in order of peak strength: analcime, heulandite, thomsonite, wairakite, pollucite, clinoptilite, scolecite), clinopyroxene, hematite, calcite, magnetite and minor smectites.

Climatic and geological setting

The Lalibela area has two rainy seasons from June to September and from March to April. The annual rainfall

Figure 2. Sampling sites. Bete Giyorgis with widespread lichenous cover and Bete Mikael Golgotha with widespread green algae and black crust cover.

varies from 500 to 1000 mm, which is believed to be the main direct/indirect cause of stone deterioration (Smith et al., 2008; McCabe et al., 2011). Lalibela area is characterized by mean annual temperature of 24.5°C and relative humidity is 52.9% (Asrat and Ayallew, 2011; Delmonaco et al., 2009).

Northern Ethiopia's geology is characterized by thick sequences of tholeiitic to transitional continental flood basalts overlain by minor rhyolitic-trachytic lavas and pyroclastic rocks of Oligocene to Miocene belonging to the Northern Ethiopia Plateau (Beccaluva et al., 2009; Natali et al., 2011; Asrat and Ayallew, 2011). Within the plateau, three magma types have been distinguished: two high- titanium (Ti) groups (HT1 and HT2), and one low-Ti group (LT). The central-eastern sector of the plateau in particular, that is, the one comprising the Lalibela area and facing the Afar triangle, is mainly characterized by a 1700 m-thick sequence of high TiO_2 (HT2 magma type) picrite/basalt lavas capped by about 300 m of rhyolites, linked to the magmatic activity at the Afar plume axial zone (Beccaluva et al., 2009). The HT2 volcanics (Pik et al., 1998) are represented by sub-alkaline olivine-clinopyroxene aphyric or glomerophyric basalts and picrites with phenocrysts of euhedral olivine (FO_{86}-FO_{81} range) and of Mg-Ti-Al rich augite. The ground mass contains olivine, clinopyroxene, Fe-Ti oxides and plagioclase (An85-An48 range) microlites. The Oligocene volcanism of the northwestern Ethiopian Plateau has been subdivided into three formations (Berhe et al., 1987); the Ashangi and Amba Aiba basaltic units, separated by an angular unconformity (Zanettin, 1992) from the upper ignimbritic Alaji unit. In the Lalibela monumental site, the Amba Aiba basalts are typically overlain by a thick horizon of reddish highly hydrothermally weathered basaltic scoria (commonly referred to as "tuffaceous material"). It is within this "softer" unit that the rock-hewn churches were carved.

MATERIALS AND METHODS

Samples were collected from internal and external walls of three Lalibela churches: Bete Michael Golgotha-external (BM-R; Figure 2), Bete Giyorgis-external (BG-E; Figure 2), Bete Gebriel-Rufael-internal (BR-I), Bete Giyorgis Mahilet (BGM-B, BGM-R; Figure 3) and tunnel from Bete Gabriel to Bete Merkorios (TG-M). For comparison and depth profiling analysis, we also performed the measurements on two rock samples from the surrounding: one from the tunnel in front of Bete Amanuel, bulk tuff weathered (Ts) and bulk tuff lateral cut (T4) and the other one collected on the north side of Bete Giyorgis, bulk basalt weathered (Bs) and bulk basalt lateral cut (B4) (Figure 3). The samples were submitted for analyses according to the specific requirements of the following analytical techniques: SEM-EDS, FTIR, µ-Raman and Raman spectroscopy.

In FTIR spectroscopy, samples are exposed to IR radiation and a fraction of the IR radiation is absorbed at wavelengths that are specific to the molecular excitation states of covalently bonded atoms. Each molecule has its own specific absorption spectrum, and the IR spectra are represented as a plot of transmittance across the IR band. A portion of rock samples was pulverized to a fine powder in an agate mortar and pestle. A small portion of the material weighing approximately 2 mg was then ground and mixed thoroughly with 40 mg of dried potassium bromide (KBr). The powder was distributed evenly within the die by first rotating the plunger, and the assembly was placed in a press. A load of 10 tones was applied for 3 to 4 min to form a transparent KBr disc. The die was disassembled and the disc carefully removed and placed in a sample holder. The Perkin Elmer Spectrum 65 Spectrum BX FTIR (at Addis Ababa University, Department of chemistry, Ethiopia) was used. The beam splitters used were KBr for the mid-IR region 4000

Figure 3. Bulk samples from rock hewn churches of Lalibela: basalt(surface) the arrow indicates the analyzed crystal using Raman, basalt(lateral cut) the arrow indicates the analyzed white zone using Raman and Bete Giyorgis Mahilet-Black (BGM-B), Bete Giyorgis Mahilet-Red (BGM-R) analyzed using FTIR.

to 400 cm^{-1} and the resolution was 4 cm^{-1}. The moving mirror velocity 0.6 cm/s and bi-directional Interferogram were used. For room-temperature measurements, we used Mercury Cadmium Telluride (MCT) detector for mid-IR spectra.

Micro-Raman characterization was performed using a Renishaw 2000 μ-Raman instrument equipped with a Peltier refrigerator-cooled charge-coupled device (CCD) camera in conjunction with a Leica optical microscope at CNR, Institute for the Study of Nanostructure Materials Research Area of Rome 1- Monte libretti - Italy. The laser was focused on the sample through a ×100 or ×50 objective lenses. The excitation light was the 514.5 nm line of an Ar^+ laser and a 785 nm line of a diode laser with laser power in the range of 10 to 60 μW. Care was taken to avoid sample damage and the typical spectrometer resolution was about 2 cm^{-1}.

Measured spectrum of calcite on the crystal and white area (Figure 3) of basalt rock is qualitatively determined by using Avantes AvaRaman-532 TEC Raman instrument equipped with a TE cooled AvaSpec-2048TEC-USB2 Spectrometer with 1200 lines/mm grating set of 535 to 752 nm, 25 μm slit, DCL-UV/VIS at (La Sapienza University, Department of Physics Rome). The excitation light was the 532 nm line of solid state laser with laser power of 50 mW, and FWHM 0.2 nm. The instrument scanned the spectra 10 times in integration time of 3 s.

SEM-EDS micro-morphological and chemical investigations were carried out by a Cambridge 360 SEM, equipped with a LaB_6 filament at CNR, Institute for the Study of Nanostructure Materials Research Area of Rome 1 - Monte libretti - Italy. Images with a higher spatial resolution were obtained by a high brilliance LEO 1530 field emission-SEM (FE-SEM) apparatus. Both the SEM and FE-SEM instruments were equipped with an EDS, INCA 250 and INCA 450, respectively, and a four sectors backscattered electron detector (BSD). SEM images were recorded both in the secondary electron image (SEI) and BSD image mode at an acceleration voltage at 20 kV. FE-SEM images were recorded both in SEI and BSD mode at different acceleration voltage ranging from 20 to 3 kV. Samples were coated with a thin layer of carbon or chromium in order to avoid charging effects. The carbon coating was deposited by using an Emitech sputter coater K550 unit, a K 250 carbon coating attachment and a carbon cord at a pressure of 1 × 10^{-2} mbar in order to produce a carbon film with a constant thickness of about 3.0 nm. The chromium coating was deposited by using a Bal-Tech SCD 500 equipped with turbo pumping for ultra clean preparations at a pressure of 5 × 10^{-3} mbar in order to produce a chromium film with a constant thickness of about 0.5 nm.

RESULTS AND DISCUSSION

FTIR results are summarized in Tables 1 and 2 where a series of analytical values wavenumber (in cm^{-1}) of the absorption bands of FTIR spectra of the powder samples are taken from the bulk tuff, bulk basalt and from internal and external walls of rock hewn churches of Lalibela. A typical example of FTIR spectra of surface crusts and in depth profile are shown in Figure 4. FTIR analysis comparing the observed frequencies with the available literature, secondary minerals such as calcite, kaolinite, whewellite, montmorillonite, gypsum, magnetite and hematite was identified.

FTIR spectrum obtained on the basalt samples are shown in Figure 4. The characteristic bands of calcite are

Table 1. Absorption frequency in the region of 4000 to 400 cm^{-1} of the bulk basalt and bulk tuff of Lalibela, together with minerals identification.

Sample	Kao	Mont	Cal	Gy	Hm	Mg	Ww	Wd
Bs	3694 3619 1031 444	3427 1642	2514 1799 1421 876 712	-	-	-	1320	-
B4	1032 438	3435 1628	2513 1799 1421 876 711	-	-	-	-	-
Ts	3677 1034	1637	-	3541 3406 1671 1622 1140 1116 670 603	540	582	1311	1355
T4	1043	3434 1638	-	-	539	578	-	-

Bs, Bulk basalt surface; B4, bulk basalt 20 mm deep; Ts, bulk tuff surface; T4, bulk tuff 20 mm deep; Kao, Kaolinite; Mont, Montmorillonite; Cal, calcite; Gy, gypsum; Hm, hematite; Mg, Magnetite; Ww, whewellite; Wd, Weddellite.

present, the sharp one at 876 cm^{-1}, is due to its asymmetric bending. Calcite FTIR spectrum shows (Table 1) peaks at 1424 to 1421, 876 and 712 to 711 cm^{-1} due to the C−O stretching of carbonate, and the weak peaks at 1799 and 2514 to 2512 cm^{-1}, which are combination and overtone bands (Gunasekaran and Anbalagan, 2008; Ravisankar et al., 2012; Ravisankar et al., 2010; Hans and Adler-Paul, 1962; Makreski and Jovanovski, 2003; Plav et al., 1999; Larson and Kerr, 1963). Hematite shows the characteristic peaks at 536 to 544 cm^{-1} (Ravisankar et al., 2010; Ravisankar et al., 2012), magnetite is related to 579 to 583 cm^{-1} peak.

The gypsum spectrum at v = 3547, and 3407 cm^{-1} (OH stretching), 1685, and 1622 cm^{-1} (O–H–O bending of the crystallization water) and at 670 cm^{-1} corresponding to the sulphate ions (Lane, 2007; Schiavon, 2007). The representative of gypsum at v = 670 cm^{-1} shows only on the surface of weathered tuff (bulk), in samples collected from BR-I, BG-E and TG-M.

Among the oxalates are the two calcium oxalates known as weddellite and whewellite. The secondary carbonyl stretching band, the metal-carboxylate stretch, v_s (COO-), is located at 1322 cm^{-1} for whewellite and at 1355 cm^{-1} for weddellite (Ruffolo et al., 2010; Hernanz etal., 2006; Kuzmanovski et al., 1999; Sofia et al., 2010; Manjula et al., 2012; Frost et al., 2003; Valarmathi et al., 2010). The characteristic feature of whewellite at v = 1322 cm^{-1} shows weak and weddellite at v = 1355 cm^{-1} is very weak.

The most characteristic spectrum of montmorillonite is the broad absorption band that varies from 3300 to 3500 cm^{-1}. Usually, this band centered at about 3400 cm^{-1} is due to stretching of the O-H molecule of water present in the interlayer region of montmorillonite (Madejova and Komadel, 2001). From Tables 1 and 2, the presence of peaks at 3435 to 3427 and 1642 to 1628 cm^{-1} in all samples indicates the presence of montmorillonite, as reported by many workers (Ravisankar et al., 2010; Saikia and Parthasarathy, 2010; Ravisankar et al., 2012). The characteristic spectrum of montmorillonite at v = 3435 cm^{-1} is only present in samples BG-E, BM-R and bulk samples (basalt, tuff).

Kaolinite shows a sequence characteristic IR absorption in the range of 3700 to 3600 cm^{-1}. The peak at v = 3694 cm^{-1} is due to OH stretching frequencies associated with the hydroxyl and the band observed at approximately 3620 cm^{-1} for kaolinite is as a result of an inner hydroxyl stretch (Madejova and Komadel, 2001;

Table 2. Absorption Frequency in the region of 4000 to 400 cm^{-1} of the samples collected from internal and external walls of rock hewn churches of Lalibela, together with minerals identification.

Sample	Kao	Mont	Gy	Hm	Mg	Ww	Wd
TG-M	1035 1016 448	1638 520	3410 674 603	-	579	-	-
BR-I	3676 1031	-	3538 3406 1622 1685 1115 671 602	-	579	1326	1354
BG-E	1032 444	3427 1632	3415 670	536	583	1324	-
BM-R	1035 445	3434 1639	-	-	580	1315	1356
BGM-B	3681 1039 444	1643	3533 3401	533	-	1322	-
BGM-R	1040 444	1639	3534 3408	540	-	1323	-

TG-M, Tunnel from Bete Gebriel to Bete Merkorios; BM-R, Bete Mikeal Golgota-Right; BG-E, Bete Giyorgis-External; BR-I, Bete Gebriel Rufael; BGM-B, Bete Giyorgis Mahilet-Black; BGM-R, Bete Giyorgis Mahilet-Red; Kao, Kaolinite; Mont, Montmorillonite; Cal, calcite; Gy, gypsum; Hm, hematite; Mg, Magnetite; Ww, whewellite; Wd, Weddellite.

Farmer, 1998). This inner hydroxyl group results from bonding between a proton and an oxygen that is also coordinated to Al^{3+} in an octahedral site. The IR absorption peaks appear at 3694, 3619, 1040 to 1043, 1031 to 1034, 1016 to 1023 and 438 to 444 cm^{-1} in the samples as reported (Saikia and Parthasarathy, 2010; Ravisankar et al., 2012; Davarcioglu and Ciftci, 2010; Ravisankar et al., 2010; Schiavon, 2007; Madejova and Komadel, 2001; Farmer, 1998), which indicates the presence of kaolinite (Tables 1 to 2). The characteristic feature of kaolinite appear only on the surface of the weathered basalt (Bs), very weak at $v = 3694$ cm^{-1}.

Raman spectroscopy revealed the bands of mineral at the surface of samples basalt, the crystal on the surface (Figure 3) and the white zone on the surface of basalt (lateral cut) (Figure 3), and the absorption intensity of the samples are compared with the reference samples, marble and quartz (Figure 5). The peak intensities of the samples are highly associated with the marble reference sample at peak intensities 713, 1077 and 1428 cm^{-1}. The intense band (v_1) of calcite spectrum corresponds to the symmetric stretching of CO_3 group at 1077 cm^{-1}. The v_2 (asymmetric deformation) vibration mode is not active in Raman. The values attributed to v_3 (asymmetric stretching) mode are 1428 cm^{-1} and v_4 (symmetric deformation) mode at 706 cm^{-1} (Gunasekaran and Anbalagan, 2008; Edwards et al., 2000). The band intensity of the crystal shows higher than the white area, whereas the FTIR analysis show higher in the sample B4 (lateral cut); it is perhaps during sample preparation, high concentration of calcite were taken from basalt (lateral cut) in the FTIR analysis.

Micro Raman on the tuff sample confirmed the extensive laterization experienced by the basaltic pyroclastic scoria deposit with the identification (beside peaks attributable to zeolitic (analcime) minerals) of well-defined peaks of hematite, lepidocrocite and combined hematite/lepidocrocite (Figure 6). The positions of the hematite spectral peaks are shown at 227, 293, 409, 499, and 609 cm^{-1} (Franquelo et al., 2009). According to Edwards et al. (2000), the Raman spectrum of lepidocrocite (γ-FeOOH) exhibits a broad band

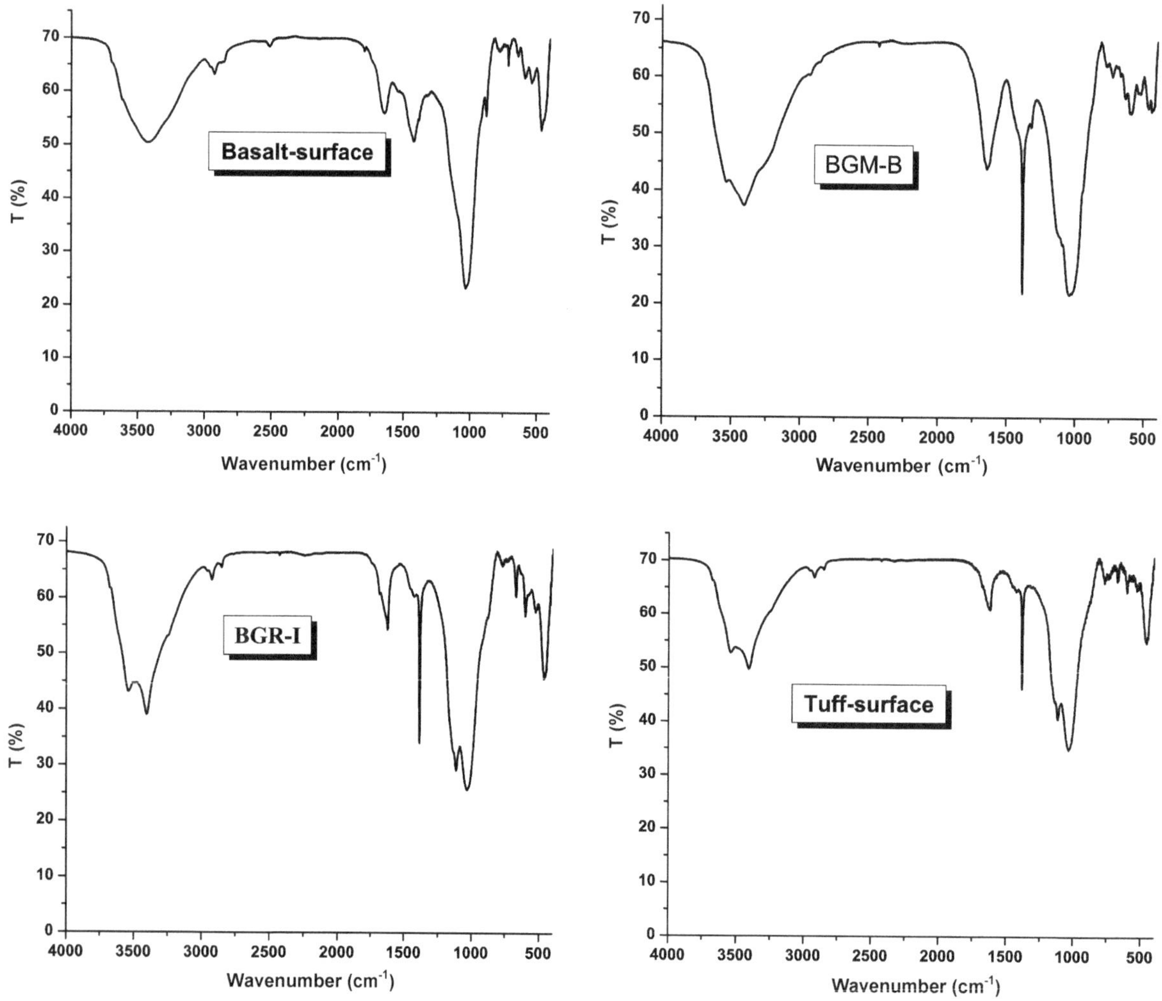

Figure 4. FTIR spectrum of the rock samples taken from the bulk basalt surface, bulk tuff surface, Bete Gebriel Rufael-internal (BGR-I) and Bete Giyorgis Mahilet-Black (BGM-B).

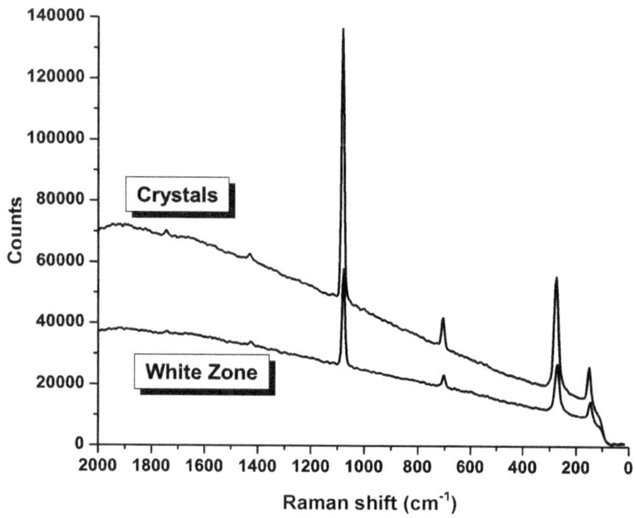

Figure 5. Raman spectrums of the rock samples taken from the crystal (Bs) and white zone (B4) bulk basalt.

about 1312 cm^{-1}.

The SEM-EDS major elements composition of the samples representing BG-M, tuff and basalt of the rock hewn churches of Lalibela are given in Table 3. All samples contain Carbon (C), Oxygen (O), Magnesium (Mg), Aluminium (Al), Silicon (Si), Potassium (K), Calcium(Ca), Titanium (Ti) and Iron (Fe). Samples BG-M and tuff also contain detectable levels of Na and P, while sample BG-M contain S. Coarse differences in composition are clear between the sample basalt and the others. Sample basalt appears to be particularly C and Ca rich, sample tuff contain high quantity of Na, sample BG-M contains more Fe and S than the other samples. The nature of these compositional differences can be investigated further through the individual element maps and the results of targeted area analysis of different surface chemical phases. The Mg value ranges between 1.98 to 3.06%, while Ca content varies between 5.11 to 14.2%, Si and Fe concentrations range from

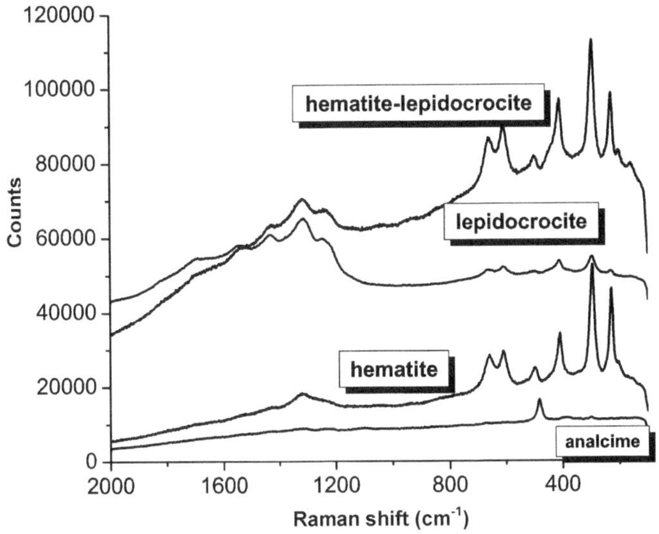

Figure 6. Raman spectrum of bulk tuff, collected from rock hewn churches of Lalibela.

Table 3. Elemental composition of rock samples of Rock hewn churches of Lalibela (Mean % weight).

Element	BGM	Basalt	Tuff
C	9.71	13.35	11.02
O	43.77	45.62	45.27
Na	1.70	-	2.79
Mg	3.06	1.98	2.23
Al	3.90	3.01	4.86
Si	12.33	10.99	13.43
P	0.78	-	0.78
S	1.17	-	-
K	2.50	3.45	0.56
Ca	6.33	14.2	5.11
Ti	2.65	1.79	2.1
Fe	15.32	6.21	11.84

BGM, Bete Giyorgis Mahilet.

10.99 to 13.43% and from 6.21 to 15.32%, respectively. From the EDS analysis, the abundant amount of Si, Al, Ca, Mg, Fe, K and Ti was found and it supports the vibrational spectroscopic findings; the presence of aluminosilicates (Si, O, Al, K, Na, Mg), Fe-rich oxide (Fe–O) and gypsum (Ca–S–O) (Ravisankar et al., 2010).

FTIR spectroscopy, Raman spectroscopy and EDS analysis result confirms that the Lalibela churches were widespread, abundant presence within the stone fabric of secondary minerals belonging to the zeolite, smectite (montmorillonite), kaolinite, whewellite/weddellite, hematite, lepidocrocite, calcite and gypsum has surely had important bearings on the stone susceptibility to weathering.

The presence of montmorillonite and analcite are reported by other workers (Delmonaco et al., 2009; Renzullia et al., 2011) as the sources of the progressive deterioration of the geomechanical properties of the rocks in Lalibela. In the present study, we bring attention to the presence of kaolinite, gypsum and whewellite/weddellite not identified before on the rock hewn churches of Lalibela, besides the presence of montmorillonite in depth of bulk samples.

The presence of lower amounts of olivine phenocrysts have been detected as compared with typical Ethiopian HT2 basalts and indeed no peaks unequivocally ascribable to olivine have been found in the FTIR analyses: this can be due to the highly hydrothermally altered nature of the olivine crystals that are often almost completely altered (laterized) to secondary Fe-rich oxide-hydroxides [hematite, magnetite and lepidocrocite (γ-FeOOH)]. The presence of Fe-rich oxide-hydroxides in the rock samples were evidenced by μ-Raman (Figure 6) and SEM-EDS (Table 3). In the most highly weathered sample BGM-B (Figure 3), the texture of the original basalt has been completely destroyed and the bulk of the rock is an intimate mixture of clay and hematite (Figure 7).

The identification of specific zeolite, in the Raman analyses should help to distinguish hydrothermal alteration in the Lalibela rocks. These minerals have a great ability to reversibly adsorb and desorb water molecules as a function of temperature and/or humidity. The potential role of the zeolite in the decay processes of the Lalibela architectural complex should be taken into account and considered as one of the main stone basic causes (Delmonaco et al., 2009).

Among the highly active in synergistically causing severe degradation in the rock hewn churches of Lalibela are: microbial biofilms which more than often act as precursors to the subsequent development of stone surface covering lichenous mats and are widespread in Lalibela, modify the capillary water uptake of porous stone material causing measurable alterations in the water-vapor diffusion properties of the lithic material. Furthermore, the water and moisture retention properties of the lichen patinas may enhance physical decay stresses associated with naturally occurring freeze-thaw cycles. In urban environments, biofilms have also been found to enhance the dry/wet deposition of atmospheric gaseous and particulate pollutants on outdoor stone facades (Schiavon, 2007) thus, playing a role in the bio-mediated precipitation of sulphate compounds such as gypsum and providing an additional route for salt weathering decay and growth of surface weathering black crusts. Field experience has suggested that in the Lalibela monumental site, wood burning and/or solid waste are the major issue in the area exhaust emissions-linked air pollution. The calcite identified by FTIR analysis in the basalt sample is also confirmed by the Raman, XRD and SEM-EDS analysis. The high peak of calcite may be due to the reaction of CO_2 with Ca oxides.

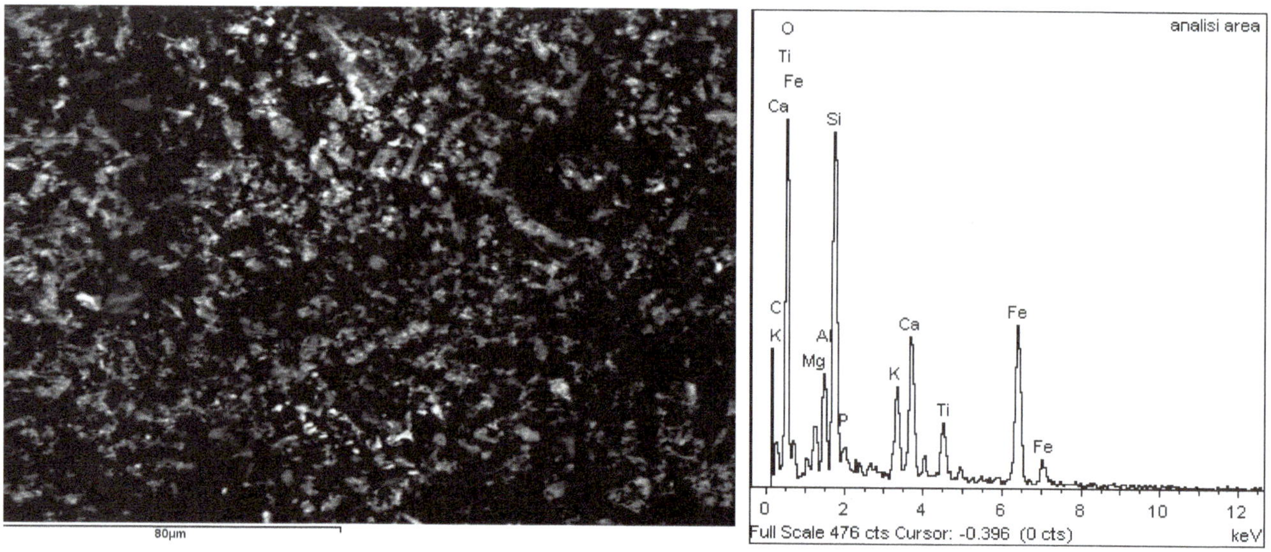

Figure 7. SEM photomicrographs with related EDS spectra, intimate mixture of clay (white) and hematite (black) in red weathered sample of the rock-hewn churches Bete Giyorgis Mahilet-Black (BGM-B).

This oxide is prevalent in silicate minerals.

$$Ca^{2+} + CO_2 + H_2O = CaCO_3 + 2H^+$$

In the FTIR and XRD analysis, Ca- plagioclase (anorthite) is not identified in basalt sample. Neutralization of H^+ will take place by reaction with the primary mineral phases in the basaltic host rock, as simplified in reaction below. Plagioclases are always replaced by kaolinite.

$$\text{(anorthite) } CaAl_2SiO_8 + 2H^+ + H_2O \rightarrow Ca^{2+} + Al_2Si_2O_5(OH)_4 \text{(kaolinite)}$$

Kaolinite is a non-swelling clay mineral; it does not have a permanent negative charge and therefore has no exchangeable cations present in the interlayer. The kaolinite found on samples of basalt stone could have the origin: deposition of kaolinite bearing airborne soil dust and/or surface wash on the monument surface and natural weathering in the original basalt stone, such as organic acids excretion by the mycobiont of lichens. Lichens enhance the rate of dissolution of plagioclase (Steven, 2005).

Biogeochemical weathering also plays a major role in the deterioration of the volcanic stone in the Lalibela monuments, acidolysis/dissolution and metal complexation (Adamou and Violante, 2000; Chen et al., 2000). The excretion by the mycobiont of lichens of low molecular weight organic carboxylic acids, such as oxalic, citric, gluconic, lactic acids, with combined chelating and acidic properties, and the production of slightly water soluble polyphenolic compounds called 'lichen acids', are processes known to lead to the formation of complexes with the metal cations(such as Mg^{++} and Fe^{++}) in the rock-forming minerals, eventually leading to their extraction by chelation to be used as nutrients and energy suppliers by green algae and lichen organisms (Adamou and Violante, 2000; Chen et al., 2000). In Lalibela, oxalate compounds (that is, whewellite/weddellite), have only rarely been detected by FTIR analysis.

The variation in intensity values along the weathering depth profile of the bulk samples for selected absorption bands, are chosen as representative of kaolinite (3694 cm^{-1}), gypsum (670 cm^{-1}), montmorillonite (3427 to 3435 cm^{-1}), and whewellite (1311 to 1328 cm^{-1}). The bands of kaolinite, gypsum and whewellite only appear on the surface of the bulk rocks (Tables 1 to 2). Montmorillonite band at 3427 to 3435 cm^{-1} shows a different trend in depth profile of FTIR analysis from the bulk tuff and basalt (that is, from surface down to unweathered tuff and basalt substrate with 5 mm difference). The identification of montmorillonite as a weathering product up to 20 mm from the rock/atmosphere interface beyond the level at which kaolinite, gypsum and whewellite are found suggests that montmorillonite may penetrate even deeper into the rock substrate causing major rock decay. The presence of swelling montmorillonite could result from surface weathering or hydrothermal alteration, minor dissolution-crystallization processes of soluble salts and the whole inexorable lateritization processes, typical of regions with dry and rainy seasons, are all consistent with the deterioration of the rocks of Lalibela.

Conclusion

The qualitative identification of minerals in the rock samples was carried out by FTIR, XRD, and Raman spectroscopy techniques. The spectroscopic analyses of rock samples from the rock hewn churches of Lalibela indicate the presence of secondary minerals zeolite-group, calcite, kaolinite, whewellite/weddellite, montmorillonite, gypsum, magnetite, lepidocrocite and hematite. Among the different secondary minerals, montmorillonite is present more or less in all samples, however, gypsum and calcite, were the most abundant minerals in samples tuff (Ts), BR-I and bulk basalt, respectively. The high presence of calcite in the sample basalt (bulk) is also confirmed by Raman spectroscopy techniques.

According to the Department of Authority for Research and Conservation of Cultural Heritage (ARCCH), Ethiopia, deposits like lichens and green algae were removed from the churches using mechanical method. The use of rotating abrasive discs or brushes is normally not advisable because they are both inefficient and damaging. Currently, most of the churches are protected by shelters to prevent the erosive effects of rainfall, especially on the roof rock cover, and temporary scaffolding to prevent collapse of the most exposed structures. Results indicate that biological attack by green algae and lichens is currently responsible for severe stone surface, physical and chemical weathering leading to considerable weakening of the churches walls. A prompt and careful removal of the biological and other deposits with the non-destructive method (laser cleaning) treatment is therefore recommended, together with understanding of the indicators of climatic and environmental change (Smith et al., 2008; McCabe et al., 2011).

REFERENCES

Adamou P, Violante P (2000). Weathering of rocks and neogenesis of minerals associated with lichen activity. Appl. Clay Sci. 16:229-256.

Asrat A, Ayallew Y (2011). Geological and geotechnical properties of the medieval rock hewn churches of Lalibela, Northern Ethiopia. Afr. J. Earth Sci. 59:61-73.

Beccaluva L, Bianchini G, Natali C, Siena F (2009). Continental flood basalts and mantle plumes: A case study of the Northern Ethiopian Plateau. J. Petrol. 50(7):1377.

Berhe SM, Desta B, Nicoletti M, Teferra M (1987). Geology, geochronology and geodynamic implications of the Cenozoic magmatic province in W and SE Ethiopia. J. Geol. Soc. London 144:213.

Chen J, Blume HP, Beyer L (2000).Weathering of rocks induced by lichen colonization: A review. Catena 39:121-146.

Davarcioglu B, Ciftci E (2010). The clay minerals observed in the building stones of Aksaray Guzelyurt area (Central Anatolia-Turkey) and their effects. Int. J. Phys. Sci. 5(11):1734-1743.

Delmonaco G, Margottini C, Spizzichino D (2009). Analysis of rock weathering and conservation strategies for rock-hewn churches of Lalibela (Ethiopia). In Protection of Historical Buildings, PROHITECH 09 ed. Mazzolani (ed). Taylor & Francis Group, London. pp. 137-142.

Edwards HGM, Newton EM, Russ J (2000). Raman spectroscopic analysis of pigments and substrata in prehistoric rock art. J. Mole. Struct. pp. 550-551, 245-256.

Farmer VC (1998). Differing effects of particle size and shape in the infrared and Raman spectra of kaolinite. Clay Miner. 33:601-604.

Franquelo ML, Duran A, Herrera LK, Jimenez MC, Perez-Rodriguez JL (2009). Comparison between micro-Raman and micro-FTIR spectroscopy techniques for the characterization of pigments from Southern Spain Cultural Heritage. J. Mole. Struct. pp. 924-926, 404-412.

Frost RL, Yang J, Ding Z (2003). Raman and FTIR spectroscopy of natural oxalates: Implications for the evidence of life on Mars. Chi. Sci. Bull. 48(17):1844-1852.

Gunasekaran S, Anbalagan G (2008). Spectroscopic study of phase transitions in natural calcite mineral. Spectrochim. Acta. 69:1246-1251.

Hans H, Adler-Paul FK (1962). Infrared study of Aragonite and Calcite. Am. Miner. P. 47.

Hernanz A, Gavira-Vallejo JM, Ruiz-Lopez JF (2006). Introduction to Raman microscopy of prehistoric rock paintings from the Sierra de las Cuerdas, Cuenca, Spain. J. Raman Spectrosc. 37:1054-1062.

Kuzmanovski I, Trpkovska M, Soptrajanov B, Stefov V (1999). Target-testing factor analysis of Fourier transform infrared spectra as a tool for the determination of the composition of human urinary calculi. Vibrat. Spectrosc. 19:249-253.

Lane MD (2007). Mid-infrared emission spectroscopy of sulfate and sulfate bearing minerals. Am. Miner. 92:1-18.

Larson HH, Kerr PF (1963).Infrared Spectra, Symmetry and Structure Relations of some Carbonate Minerals. Am. Miner. P. 43.

Madejova J, Komadel P (2001). Baseline Studies of the Clay Minerals Society Source Clays: Infrared Method. Clays Clay Miner. 49:410-432.

Makreski P, Jovanovski G (2003). Mineral from Macedonia, Distinction between some Rhombohedral Carbonates by FTIR Spectroscopy. Maced. Chem. Techno. Bull. 22(1):25-32.

Manjula K, Pazhanichamy K, Kumarans S, Evera T, Dalekeefe C and Rajendran K (2012). Growth Characterization of Calcium Oxalate Monohydrate Crystals Influenced Costus Igneus Aqueous Stem Extract. Int. J. Pharm. Pharm. Sci. P. 4.

McCabe S, Smith BJ, Adamson C, Mullan D, McAlister J (2011). The "Greening" of Natural Stone Buildings: Quartz Sandstone Performance as a Secondary Indicator of Climate Change in the British Isles. Atmos. Clim. Sci. 1:165-171.

Natali C, Beccaluva L, Bianchini G, Siena F (2011). Rhyolites associated to Ethiopian CFB: Clues for initial rifting at the Afar plume axis. Earth Plan. Sci. Lett. 312:59.

Perez-Alonso M, Castro K, Martine-Arkarazo I, Angulo M, Olazabal MA, Madariaga JM (2004). Analysis of bulk and inorganic degradation products of stones, mortars and wall paintings by portable Raman microprobe spectroscopy. Abstract, Anal. Bioanal. Chem. 379(1):42.

Pik R, Deniel C, Coulon C, Yirgu G, Hofmann C, Ayalew D (1998). The northwestern Ethiopian Plateau flood basalts: Classification and spatial distribution of magma types. J. Volcan. Geotherm. Res. 81:91.

Plav B, Kobe S, Orel B (1999). Identification of Crystallization forms of $CaCO_3$ with FTIR Spectroscopy. KZLTET 33(6):517.

Potgieter-Vermaak SS, Godoi RH, Grieken RV, Potqieter JH, Oujja M, Castillejo M (2005). Micro-structural characterization of black crust and laser cleaning of building stones by micro-Raman and SEM techniques. Abstract, Spectrochim. Acta. Mol. Biomol. Spectrosc. 61(11):2460.

Ravisankar R, Annamalai GR, Rajan K, Naseerutheen A, Senthilkumar G (2012). Mineral Analysis in Archaeological Pottery from Porunthal, Dindigaldist, Tamilnadu, India by FTIR Spectroscopic Technique. Int. J. Sci. Innov. Discov. 2(1):53-60.

Ravisankar R, Senthilkumar G, Kiruba S, Chandrasekaran A, Jebakumar PP (2010). Mineral analysis of coastal sediment samples of Tuna, Gujarat, India. India J. Sci. Technol. 3(7):858-862.

Renzullia A, Antonellia F, Margottinib C, Santia P, Fratini F (2011). What kind of volcanite the rock-hewn churches of the Lalibela UNESCO's world heritage site are made of J. Cult. Herit. 12:227-235.

Ruffolo SA, Russa MF, Barca D, Casoli, Comite, Nava G, Crisci GM, Francesco AM, Miriello D (2010). Mineralogical, petrographic and chemical analyses for the study of the canvas "Cristo alla Colonna "from Cosenza, Italy: A case study Period. Mine. Spec. pp. 71-79.

Saikia BJ, Parthasarathy G (2010). Fourier Transform Infrared Spectroscopic Characterization of Kaolinite from Assam and Meghalaya, Northeastern India. J. Mod. Phys. 1:206-210.

Saikia BJ, Parthasarathy G, Sarmahn C (2008). Fourier transform infrared spectroscopic estimation of crystallinity in SiO_2 based rocks. Sci. Bull. 31(5):775-779.

Schiavon N (2007). Kaolinisation of granite in an urban environment.

Environ. Geol. 52:399-407.
Smith BJ, Gomez-Heras M, McCabe S (2008). Understanding the decay of stone-built cultural heritage. Prog. Phys. Geog. 32:439-461.
Steven J (2005). Effect of environmental factors on the chemical weathering of plagioclase in Hawaiian basalt. Phys. Geog. 26(1):69-84.
Valarmathi D, Abraham L, Gunasekaran S (2010). Growth of calcium oxalate monohydrate crystal by gel method and its spectroscopic analysis. Indian J. Appl. Phys. 48:36-38.
Zanettin B (1992). Evolution of the Ethiopian Volcanic Province. Mem. Lincee Sci. Fis. Nat. 1:155.

Assessment of leachate effects to the drinking water supply units in the down slope regions of municipal solid waste (MSW) dumping sites in Lahore Pakistan

Khalid Mahmood[1], Syeda Adila Batool[1], Asim Daud Rana[1], Salman Tariq[1], Zulfiqar Ali[2,3] and Muhammad Nawaz Chaudhry[2]

[1]Department of Space Science, University of the Punjab, Lahore, Pakistan.
[2]College of Earth and Environmental Sciences, University of the Punjab, Lahore, Pakistan.
[3]Nano Science and Catalysis Division, National Centre for Physics, Q.A.U Campus, Islamabad, Pakistan.

The aim of this study was to evaluate seasonal as well as moderate temporal effects on deep groundwater quality by the dumpsites leachate in the Lahore Metropolitan with a population of about 8 million. Groundwater samples for this study ware drawn from groundwater supply units installed by local government for the provision of drinking water to the residents. The groundwater in down slope regions from municipal solid waste (MSW) dumping sites has been identified by using Inverted Watershed technique. Eleven water supply units have been spotted that fall within the identified leachate plumes. Parameters of pH, turbidity, electric conductivity, total dissolved solids (TDS) and *Escherichia coli* were determined on water samples from these identified units. The quality analysis indicates that the effect of the leachate is more prominent in the hot and dry pre-monsoon season due to the domination of base flow for recharge of the groundwater, whereas dilution observed in leachate contamination during and immediately after the monsoon rainy season is due to the recharge through seepage of the rain water which dominates the base flow. Variation in the correlation factor between electric conductivity and TDS has been used to estimate constituents of TDS contributed both by rainfall seepage and base flow. An increasing tendency, showing influence of the leachate on the deep water quality, has been observed while analyzing the correlation between distance of the tube wells from leachate sources and measurements of quality parameters.

Key words: Base flow, deep groundwater, down slope regions, landfill leachate, rainwater seepage.

INTRODUCTION

Almost 70% of the earth is covered with water but only 2.5% of this is fresh water (PCRWR, 2007; Limouzin and Maidment, 2009) is suitable for human use. Approximately 99% of fresh water is in aquifers (ESA, 2001; Gabriel and Khan, 2010) and being used as the major source of drinking water in Pakistan. Lashari et al. (2007) estimated that about 60 to 70% population of Pakistan depends directly or indirectly on groundwater for their livelihood. However, a number of locations in Punjab province of Pakistan have been reported to be contaminated by by-products of waste materials (World Bank, 2006). Therefore, qualitative and quantitative monitoring of these fresh water reservoirs is mandatory for human health.

Underlying aquifer is the main source of water for the city of Lahore like the rest of the Punjab province with a population of 80 million. The extraction of this groundwater can be categorized in three types depending on depths. The first category ranges in depth from 9 to 18 m depending upon the aquifer level. This serves the poor households around the dumpsites and cottage industry. The second category serves the agriculture; the water is drawn by relatively large water extraction tube wells with almost the same depth range as for the first category. The third type of extraction wells are installed by government agencies for the provision of drinking water to residents of the city. These tube wells are considered most sophisticated and deepest (150 to 180 m).

Provision of water supply to about 90% population of Lahore is the responsibility of Water and Sanitation Agency (WASA), which has installed over 450 tube wells and given over 531,336 connections in the city (Gabriel and Khan, 2006). Other agencies like Lahore Cantonment Board, Model Town Society etc. serve the remaining 10% population of Lahore. Gabriel and Khan (2006) reported that these tube wells are of varying capacity and are operated for an average duration of 16 to 18 h a day and the depth of these tube wells, varies from 150 to 180 m. Hydro-geologically, Lahore is a part of large interfluvial Bari Doab, bounded by River Ravi to the north-west and Rivers Satlej and Beas to the south-east. Its aquifer is un-confined with a thickness of about 400 m (Gabriel and Khan, 2006) and is composed predominantly of fine to medium grained sands and small lenses of silty clay.

Monitoring of underlying aquifer in the vicinity of a dumping site is mandatory for the protection of groundwater. Dumping sites provide an easy and the cheapest way for waste management all over the world (Mahini and Gholamafard, 2006; Zhang et al., 2011; Butt et al., 2008). However, without properly designed and maintained, microbial and chemical products from the dumping site may produce harmful effect to the environment (Butt and Oduyemi, 2003). Hazards associated with landfills require proper evaluation even after their construction, in order to estimate and eradicate the bad effects to environment and public health (Gorsevski et al., 2012). Contamination of the underlying aquifer by leachate from the dumping sites is the most highlighted environmental issue found in the reviewed literature (Butt and Oduyemi, 2003; Santos et al., 2006; Demitriou et al., 2008; Din et al., 2008; López et al., 2008; Singh et al., 2009; Li et al., 2012). For examples, groundwater contamination near the Ano Liosia landfill was reported in Attica region, Greece, which was unsuitable for irrigation (Feta et al., 1996). Another example is in Delaware New Castle where wells located downstream of the Llangollon landfill were found to be heavily contaminated and subsequently abandoned (Chian and DeWalle, 1976).

In 2009, the areas under the jurisdiction of Lahore city produced about 5000 tons/day of municipal solid waste (MSW), where 60% of the waste was collected and disposed of in the open dumpsites (Batool and Chohdary, 2009). Two major sites for municipal waste dumping in Lahore are located in Saggian and Mahmood Booti as shown in Figure 1. These dumping sites have neither a leachate collection system nor an impermeable liner system. The dumpsite leachate percolates through underlying sediments and contaminates the source of drinking water (Biswas et al., 2010; Li et al., 2012). Once the groundwater is contaminated, they are unable to be cleaned to a level that can meet international drinking water quality standards (Biswas et al., 2010). Horizontal spread of percolated landfill leachate is very limited (Cherry, 1990) but it moves down slope as a wide front of contamination (Lee and Lee, 1993). Municipal landfill's leachate is heavier than groundwater, therefore, after percolation through lithological cross section it will sinks to the depth where it is diluted to a level where its density becomes comparable to that of the uncontaminated groundwater (Cherry, 1991; Lee and Lee, 1993).

In the pre-monsoon season, lateral flow of groundwater is dominant while in the post-monsoon, vertical flow dominates the groundwater recharge. Lateral flow brings contamination to a location from its neighboring sources and the vertical flow brings contamination from the location itself. Most of the studies that analyzed the leachate hazards to the groundwater had used shallow (1 to 10 m below surface) water samples, where the effects are more pronounced (Jensen et al., 1998; Nyenje et al., 2013). This study is intended to delineate down slope groundwater regions of Mahmood Booti and Saggian dumping sites and the quality assessment of the drinking water supply tube wells that are located within the identified hazardous zones. The groundwater sampling depth ranges from 79 to 137 m from the tube wells installed by the government for the provision of drinking water. Timely detection of groundwater pollutants may lead to remedial measures for sustainability of the drinking water source. For this purpose some indirect methods have been used for the analysis of contamination caused by leachate to these depths.

MATERIALS AND METHODS

Methodology for this study has been divided into two steps. In the first step, down slope groundwater areas of each landfill have been identified and hence delineated. Afterward, in the second step tube wells falling within this area were assessed for the quality of drinking water.

Down slope area identification

WASA periodically measures the depth of groundwater on each of its tube wells installed within Lahore district. For this study, static water level (SWL) values for the months of April, July and October from 2009 to 2011 were used. Geo-referencing of the tube well sites was accomplished by a field survey identifying each tube well site using Garmin GPS map 76CSx with an accuracy of ±3 m. In order to develop water table, the point measurements had been interpolated after comparison of many techniques available for interpolation as shown in Table 1. This comparison had optimized Kriging (simple kriging) as the most suitable for the interpolation.

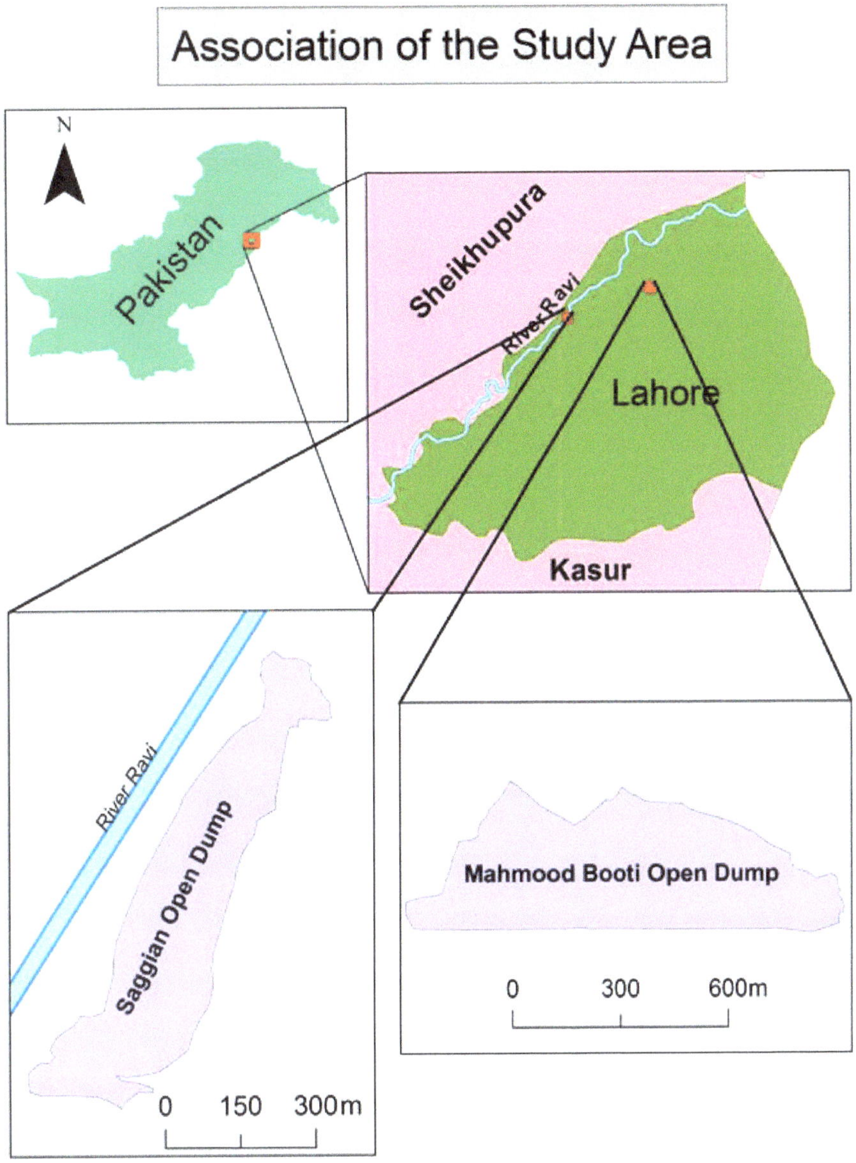

Figure 1. Dumping sites location in Lahore, Pakistan.

Table 1. Comparison of interpolation techniques.

Method	RMSE	Correlation	Mean
IDW	3.492	0.890536	0.1231
Spline	3.399	0.897549	0.1748
Kriging	3.37	0.898589	0.02551

For the down slope area identification some of the geographic information systems (GIS) software only facilitates in finding down slope trajectories for individual pixels; whereas, one of the objectives of the study was to delineate all the area that receives water after passing through the landfills, therefore an indirect approach has been applied for this purpose. The method devised in this study for the down slope area identification is an inverse watershed technique. In this technique, the water table surface is inverted to get its reflection surface which shows higher areas at low levels and vice versa. The down slope areas in the actual raster surface will appear to be up slope areas or in other words, watersheds, in the output surface. The process of inversion of water

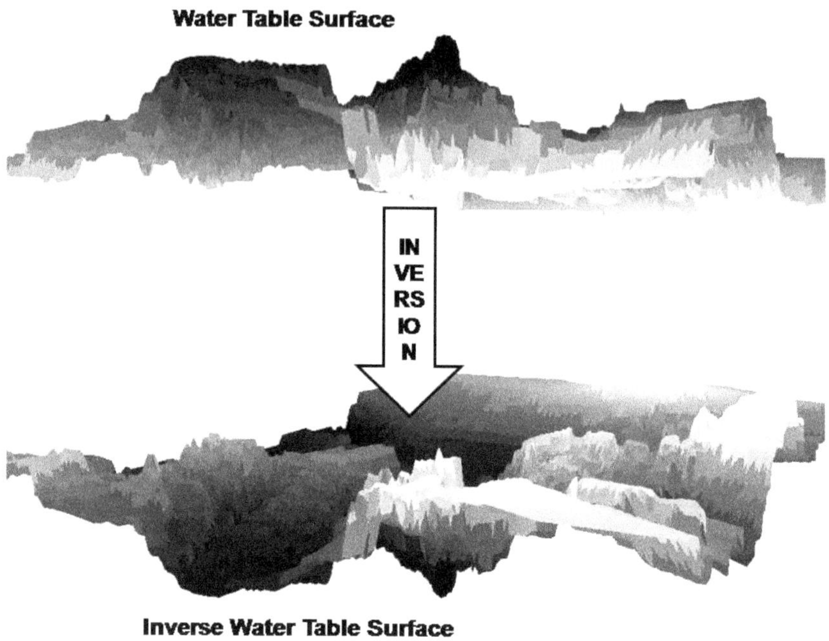

Figure 2. Inversion of groundwater surface.

table surface is shown in Figure 2.

Both landfills were used as target feature for delineating watersheds based on the inverted water table surface. This watershed gives the required down slope regions of the groundwater from landfills. To incorporate possible variations in the down slope area a time series of water table surfaces consisting of nine datasets spanning from April, 2009 to October, 2011 were used in this study. Groundwater surface changes with time as a consequence of rates and position of water extraction and recharge, the combined output of all down slope regions is shown in Figures 3 and 4 for Mahmood Booti and Saggian landfills, respectively. As a hypothesis, the estimated down slope areas from various input surfaces must overlap as they belong to the same region. Therefore, in the final down slope region only those areas which overlapped by at least two of the datasets were selected. The outputs of all the overlapping datasets are united for the demarcation of the down slope areas of both the dumping sites.

Quality assessment of risk prone tube wells

The final down slope area for each of the dumping sites identifies the possible hazardous region prone to leachate contamination. Tube wells installed by WASA, Lahore for drinking water supply were identified in this study. After identification of expected risk prone water supply units, an assessment was performed to check the water quality of these tube wells. The qualitative report of expected risk prone tube wells for pre-monsoon 2010 was taken from WASA, Lahore. The other data set for the water quality assessment was acquired especially for this study in post-monsoon 2012. The available datasets made it possible to provide both seasonal as well as moderate temporal changes in the drinking water quality. This time gap provided a fair chance to evaluate the impacts of leachate, having a slow lateral movement (Cherry, 1990), even at the farthest tube well identified. Water samples of the tube wells were collected in neat, airtight and sterilized polyethylene 500 ml bottles. Samples were immediately brought to laboratory for the analysis of selected water quality parameters.

In the pre-monsoon season, lateral flow of ground water is more dominant, whereas in the post-monsoon season, vertical flow dominates the groundwater recharge. Both seasons were taken into account to determine the influence of lateral as well as vertical flow of recharge to water quality. In order to highlight the influence of landfill leachate to water quality parameters a basic principle of likelihood has been used by virtue of which any effect caused by a source is inversely related to the distance (Nyenje et al., 2013). A correlation of each quality parameter with distance from dumping site was established. The distance from the dumping site had been taken, keeping in view, both the filter depth of tube wells and its ground distance from the dumping sites as shown in Figure 5.

RESULTS AND DISCUSSION

The delineated down slope regions for both municipal dumpsites show a general trend of groundwater flow from dumping sites towards the residential area of the city as shown in Figure 6. The extracted down slope regions using inverse watershed technique are verified by the cited literature, as explained in the following points:

1) It is a fact that almost all the extraction of groundwater for livelihood is made in central portions of the city leading to a depression zone as mentioned by a number of studies (Mahmood et al., 2013). So, in order to maintain surface level, water flows from outer boundaries, beneath municipal dumpsites, to inner regions of the city.

2) The River Ravi exists as one of the major recharge sources of Lahore's groundwater at northern and western edges of the city (Basharat and Rizvi, 2011; Mahmood et al., 2013). Since the leachate producing bodies are

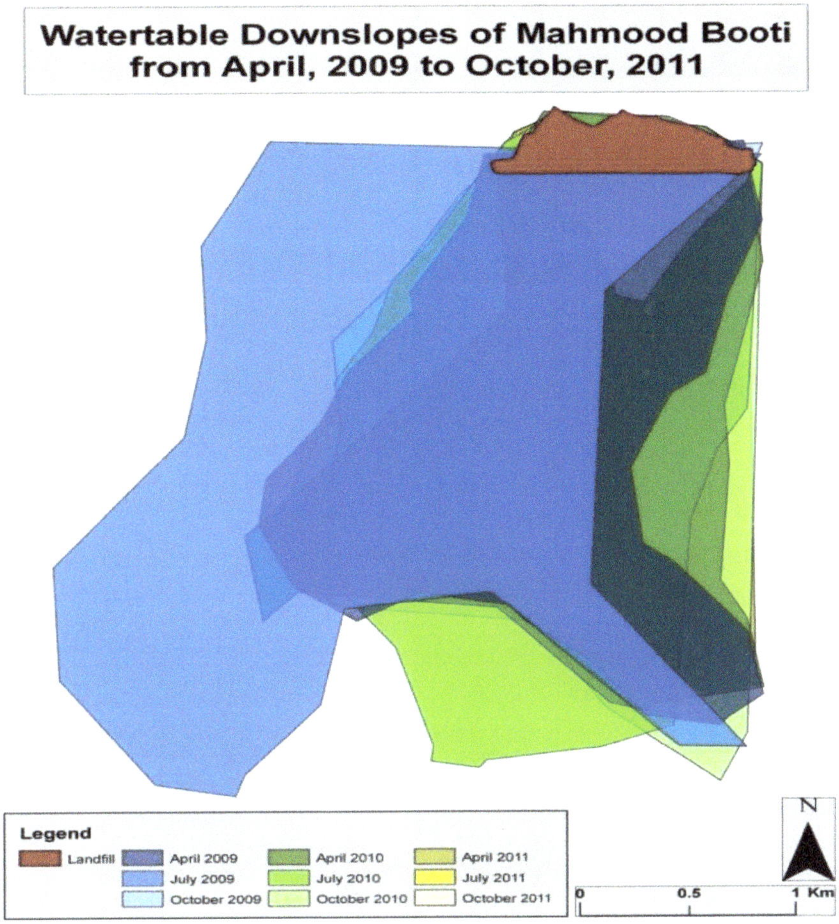

Figure 3. Time series of Water table down slopes of Mahmood Booti dumping site.

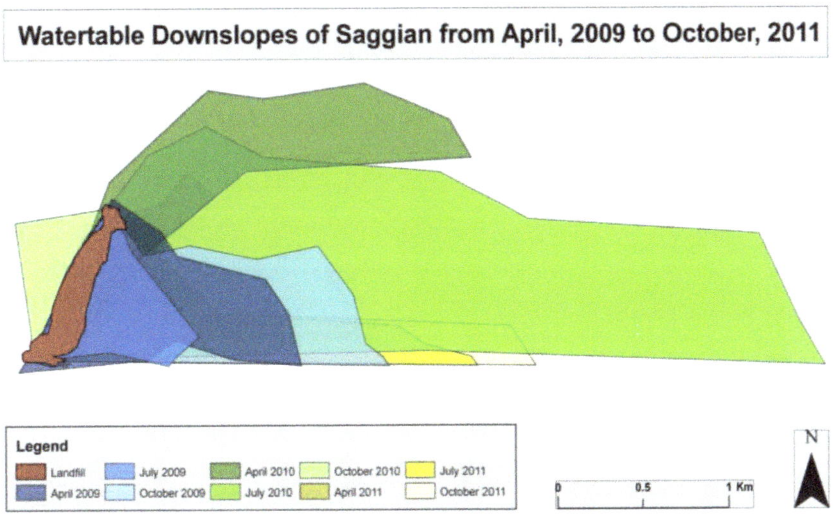

Figure 4. Time series of Water table down slopes of Saggian dumping site.

situated away from River Ravi to the residential part of the city, therefore, flow of the water for recharging the underlying aquifer is from dumpsites to central city. In this way, leachate contaminates that part of the groundwater

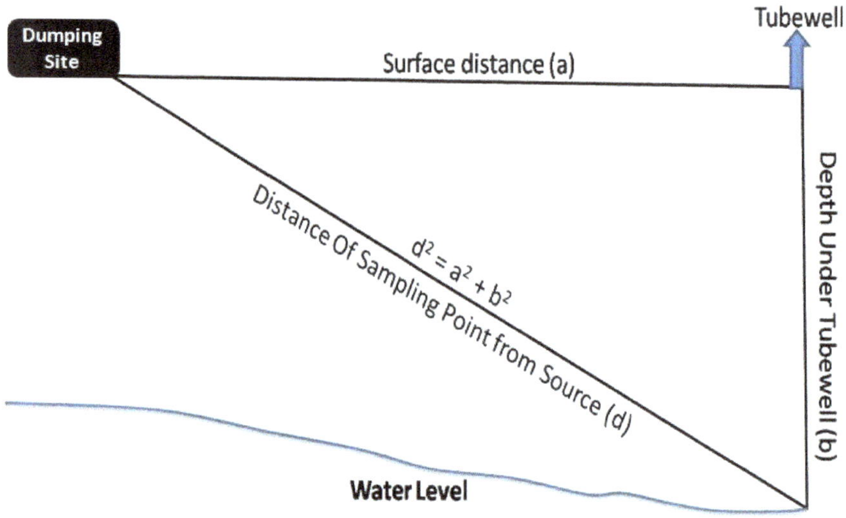

Figure 5. Distance from dumping site to tube well filter

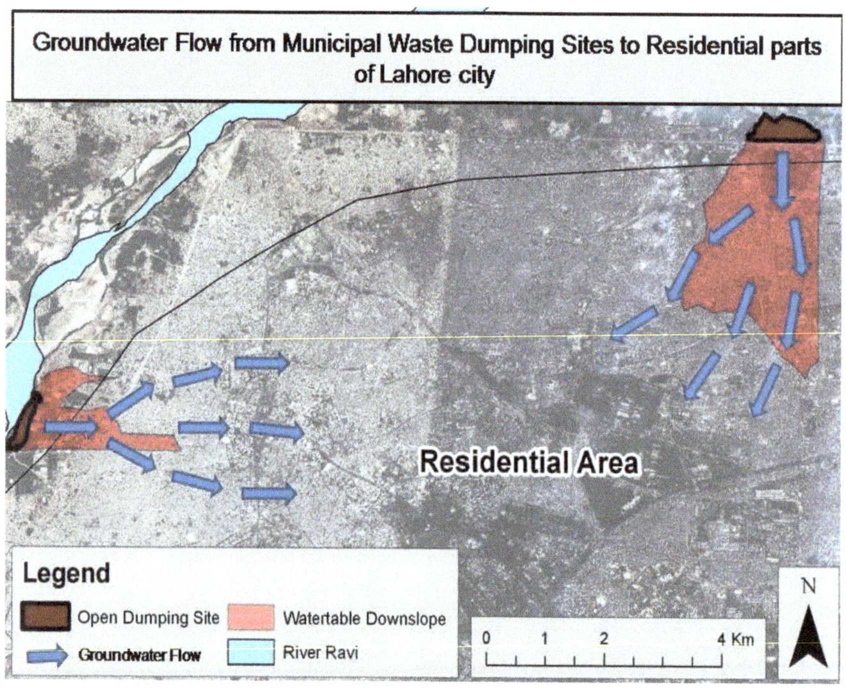

Figure 6. Groundwater flow to the Lahore city.

which has to be ultimately extracted for human usage.

Effects on the drinking water supply units/tube wells

Table 2 lists the details of risk prone tube wells identified in this study. Nine tube wells were located within contamination reaches in Mahmood Booti dumping site (Figure 7) and two tube wells were in the Saggian dumpsite (Figure 8).

The quality parameters for the risk prone tube wells are shown in Table 3. The landfill leachate contains acidic substances that cause decrease in pH value of the leachate (Kareem et al., 2010) which further tends to decrease pH of the ground water if it gets mixed with it. Keeping in view this effect to the pH parameter, Table 3 shows the average measured pH of groundwater is more acidic in the pre-monsoon season than the post-monsoon

Table 2. Tube wells at risk along the dumping sites.

S/N	Serving area	Effecting dumping site	Design capacity (Cusecs)	Distance from landfill (m)	Water depth Nov, 2011(m)	Water sampling depth (m)
1	Madho Lal Hussain	Mahmood Booti	0.112	2565	39.62	137
2	Dy. Yaqoob Colony	Mahmood Booti	0.112	2049	32.92	137
3	Aliya Town	Mahmood Booti	0.112	548	32.69	137
4	Gulshan-e-Shalimar	Mahmood Booti	0.112	1148	32.5	137
5	Angoori Bagh SchemeNo.1	Mahmood Booti	0.056	3172	39.5	137
6	Gosha-e-Angoori	Mahmood Booti	0.112	3190	39.1	137
7	Kotli Pir, Abdul Rehman	Mahmood Booti	0.112	3328	39.8	137
8	Naseer Abad, Pakistan Mint	Mahmood Booti	0.112	2595	39.5	137
9	Mahmood Booti Well Center 3	Mahmood Booti	0.112	50	32.55	137
10	Rafi Abad Darbar	Saggian	0.056	987	23.71	79
11	Islam Pura	Saggian	0.112	2095	25	79

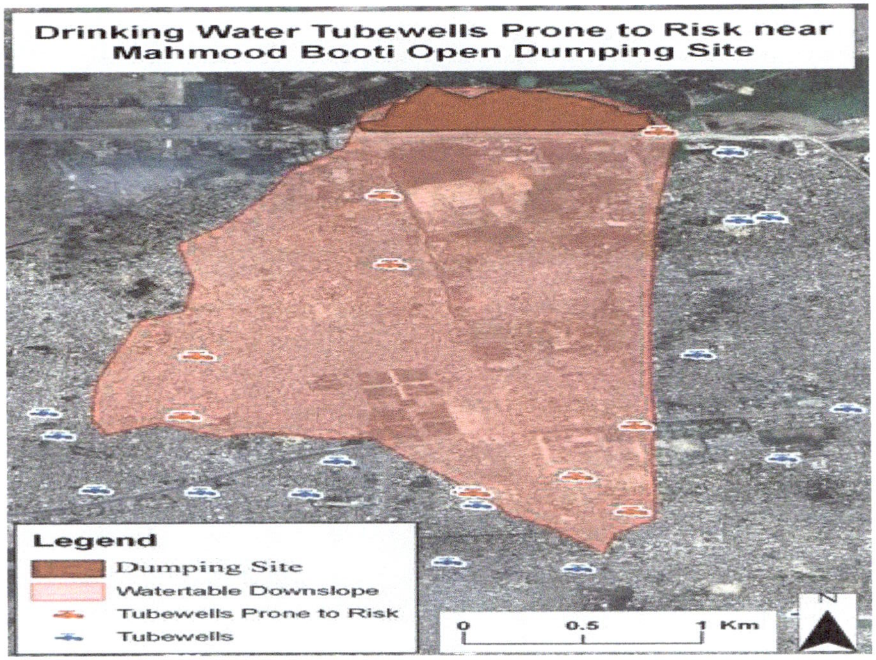

Figure 7. Drinking Water at risk around Mahmood Booti.

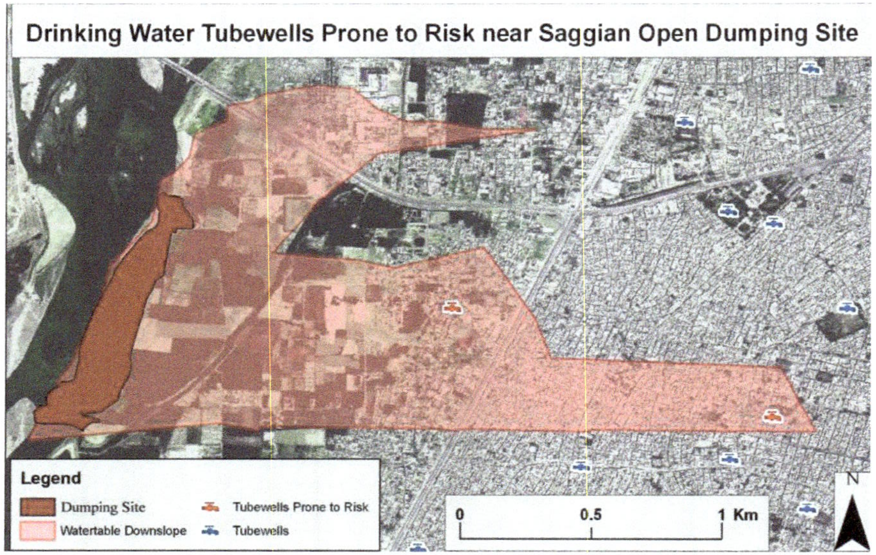

Figure 8. Drinking Water at Risk around Saggian

Table 3. Quality degradation chart of the risk prone tube wells.

S/N	Location	PH		Turbidity (NTU)		Conductivity (ms/cm)		TDS (ppm)		*E. coli*	
		Pre MS*	Post MS*	Pre MS*	Post MS*	Pre MS*	Post MS*	Pre MS*	Post MS*	Pre MS*	Post MS*
1	Madho Lal Hussain	7.6	8.4	0.59	0.08	526	430	331.3	290	Absent	Present
2	Dy. Yaqoob Colony	8.1	8.5	0.16	0.09	518	410	236.3	290	Absent	Present
3	Aliya Town	7.5	8.2	1.07	0.13	514	390	323.8	380	Absent	Present
4	Gulshan -e-Shalimar	8.6	8.6	3.44	0.18	886	380	558.1	390	Absent	Present
5	Angoori Bagh SchemeNo.1	7.5	7.9	1.85	0.15	921	1010	580.2	465	Absent	Present
6	Gosha-e-Angoori	7.8	7.9	3.8	0.15	621	1010	391.2	465	Absent	Present
7	Kotli Pir, Abdul Rehman	7.9	8.2	2.02	0.16	560	560	352.8	390	Absent	Present
8	Naseer Abad, Pakistan Mint	7.9	8.1	1.94	0.21	574	530	361.6	359	Absent	Present
9	Mahmood Booti Well Center 3	7.9	8.1	1.97	0.15	380	450	239.4	310	Absent	Present
10	Rafi Abad Darbar	8	8.8	0.57	0.15	321	260	202.2	285	Absent	Present
11	Islam Pura	7.7	8.5	1.67	0.18	392	290	246.9	210	Absent	Present
Averages		7.8	8.2	1.73	0.15	564.8	520	347.6	348.5	Absent	Present

*MS- Monsoon.

Table 4. Change in co-relation (Pearson's r test) among quality parameters.

Correlation between	Pre-monsoon	Post-monsoon
Electric conductivity/TDS	0.97638	0.82374
Electric conductivity/Turbidity	0.511273	0.05530
TDS/Turbidity	0.589632	0.16959
Electric conductivity/pH	0.316727	0.82572
TDS/pH	0.316727	0.68812

measurement. This trend shows the effect of leachate intrusion to be more prominent in pre-monsoon season. As mentioned earlier, the groundwater recharge of the Lahore district is contributed by two main sources; one is the River Ravi and the other is rainfall. The measurement of pH in the pre-monsoon season is more influenced by base flow which carries the effects of the leachate to the groundwater down slope. In the monsoon season groundwater gets the dominant share of its recharge from rainwater that seeps down. In this way, quality of the groundwater is more influenced by the local circumstances. High value of pH in post-monsoon analysis indicates that the decreased value of pH in the pre-monsoon could possibly be due to the leachate, which is diluted in post-monsoon, affecting the increase in pH because of the local influence.

A similar trend is observed in the measurements of turbidity which shows an average decrease from 1.73 to 0.15 NTU. Landfill leachate tends to increase electrical conductivity (Kareem et al., 2010), therefore its value must be higher in the pre-monsoon as compared to the post-monsoon. The measurements (Table 3) agree with the hypothesis, though the effect has not been found as prominent as it was in the cases of pH and turbidity parameters. According to the cited literature, wherever hydrocarbon degradation takes place, it tends to increase total dissolved solids (TDS) (Atekwana et al., 2004). Kareem et al. (2010) has measured high concentration of TDS in the leachate for the same study area. With reference to his study, pre-monsoon measurement of TDS should be higher than the post-monsoon measurements as in this season effect of leachate is diluted considerably as found for other quality parameters, whereas, behavior of TDS is found somewhat variant from the expected, as it has shown an increasing trend from 347.62 to 348.54 ppm.

This anomaly may also be the result of the local geochemical heterogeneity which is more prominent in the post-monsoon season. However, another interesting behavior has been identified in the relation between electrical conductivity and the amount of TDS. Nowaday's electric conductivity is used as a proxy for the measurement of TDS in the water (Post, 2012), based on the assumption that almost all TDS constituent participate in increasing conductivity of the groundwater (Atekwana et al., 2004). In fact, the groundwater contains both the charged and the uncharged species in proportions that may vary depending upon geochemical heterogeneity. For example, dissolved silica (H_4SiO_4) and organic substances do not contribute to increase electric conductivity (Post, 2012). Any increase in uncharged substances disturbs the correlation between the two quantities. Therefore, electric conductivity cannot be directly used for the measurement of TDS, but it only gives a good approximation of the TDS through the relation given by Lloyd and Heathcote (1985):

$$TDS = K_c \times EC$$

Where EC is electrical conductivity and K_c is correlation factor that may vary depending upon proportion of charged and uncharged constituents (Atekwana et al., 2004).

Decrease in the correlation between electrical conductivity and TDS in the post-monsoon season as shown in Table 4, indicates that base flow contribution to the TDS are dominated by charged species for the contributing leachate (Kareem et al., 2010). Whereas decrease in correlation factor and increased TDS in the post-monsoon season reflect that contribution of local influences to TDS is dominated by uncharged species. The charged species may include inorganic substances which may also contain metals for groundwater pollution (Rivett et al., 2011). Changes in the correlation may also be used as an indicator of raising the metal contents in groundwater due to intrusion of the leachate which is one of the serious issues of groundwater contamination (Jensen et al., 1998; Baumann et al., 2006; Malana and Khosa, 2011).

Increase in the effects of leachate to groundwater has also been assessed in terms of correlation of the measured parameters with distance of sampling depths from dumping sites as shown in Table 5.

With reference to Table 5 almost all the parameters had shown increase in the leachate influence over the time except turbidity. The smallest rise is found in the correlation of TDS with the distance, it may be the result of TDS contribution by factors other than the leachate, which already had been explained. Looking at the influential trends over pH and electric conductivity, the effect is almost doubled in the time gap of sampling. Here, it is very important to keep in mind that the samples

Table 5. Changes in influence of leachate over water quality.

S/N	Quality parameter	2010	2012
1	pH	0.25382	0.41689
2	Turbidity	0.20550	0.02893
3	Conductivity	0.39163	0.63782
4	TDS	0.37129	0.38575

Table 6. Additional water quality parameters.

S/N	Location	T.H	Ca	Mg	Cl	HCO_3
1	Madho Lal Hussain	140	35.2	12.4	18	156
2	Dy. Yaqoob Colony	140	35.2	22.4	19	144
3	Aliya Town	176	40	18.2	16	154
4	Gulshan -e-Shalimar	102	19.2	12.9	70	100
5	Angoori Bagh SchemeNo.1	250	48.8	30.7	42	246
6	Gosha-e-Angoori	120	24	14.4	27	172
7	Kotli Pir, Abdul Rehman	140	26.4	17.7	30	188
8	Naseer Abad, Pakistan Mint	120	24.8	15.3	21	166
9	Mahmood Booti Well Center 3	136	28.8	15.3	14	122
10	Rafi Abad Darbar	140	27.2	17.2	10	92
11	Islam Pura	124	24.8	14.8	18	130
Averages		144.36	30.4	17.39	25.91	151.82
Correlation with distance		0.1811	0.10047	0.2589	0.16906	0.72278

in 2012 had been taken after monsoon that had diluted the effects of base flow but the results are still alarming. Other parameters like total hardness (TH), Calcium (Ca), Magnesium (Mg), Chlorine (Cl), and Bicarbonates (HCO_3) can also be used for groundwater quality measurement (Malana and Khosa, 2011) and their source may also be associated with dumping of MSW (Xing et al., 2013). These parameters had also been analyzed in the first dataset taken from WASA, Lahore and are given in Table 6.

Regarding these additional water quality parameters two observations are important to note. The first one is the tube well installed at Angoori Bagh Scheme No. 1 with comparatively high values of all quality parameters. At the second place, the correlation of bicarbonates (HCO_3) with distance from dumping site is high, which shows contribution of leachate to this quality parameter.

CONCLUSION AND RECOMMENDATION

The Inverse Watershed technique had been found fruitful for the identification of contamination plume under Municipal Waste Dumping sites as the results agree with the reviewed literature. Flow of groundwater under both sites is found towards the residential part of the city, which proves unsuitability of locations of the dumping sites. The groundwater in the down slope region around Mahmood Booti has nine tube wells that are being used for the provision of drinking water to residents. Similarly, the leachate plume detected around Saggian dump contains two tube wells of the same kind.

Hazardous effects of landfill leachate to groundwater has been identified which are more prominent in the pre monsoon season and diluted by the intrusion of rainfall water to it in the post monsoon season. The variation in the correlation factor between electrical conductivity and TDS may possibly be due to the intrusion of metals to the groundwater by the leachate. The quality parameters are well within the range of the drinking water standards for these deep tube wells but the increasing influence of leachate in the down slope regions may turn out to be a big issue in the near future.

Therefore, in future, drinking water supply tube wells should be installed at opposite sides of the groundwater down slope directions for the landfills. As the shallow water is more prone to these risks, so shallow bores in the area for drinking water need to be banned.

ACKNOWLEDGEMENTS

We would like to acknowledge and extend our heartfelt gratitude to Mr. Arif Butt, Director Hydrology, WASA,

Lahore for providing data. We also acknowledge Nano Science and Catalysis Division, National Centre for Physics, Quaid-i-Azam University Islamabad, Pakistan and College of Earth and Environmental Sciences, University of the Punjab, Lahore, Pakistan, for providing us technical and analytical facilities.

REFERENCES

Atekwana EA, Rowe RS, Warkema DD, Legall FD (2004). The Realtionship of total dissolved solid measurements to bulk electrical conductivity in an aquifer contaminated with hydrocarbons. J. Appl. Geophys. 56:281-294.

Basharat M, Rizvi SA (2011). Groundwater extraction and waste water disposal regulations – Is Lahore aquifer at stake with as usual approach. Pakistan Engineering congress, World Water Day. pp. 112-134.

Batool SA, Chohdary MN, (2009). Municipal solid waste management in Lahore City District, Pakistan. Waste manage. 29:1971-1981.

Baumann T, Frushtorfer P, Klein T, Niessner R (2006). Colloid and heavy metal transport at landfill sites in direct contact with groundwater. Water Res. 40:2776-2786.

Biswas AK, Kumar S, Babu SS, Bhattaacharyya JK, Chakrabarti T (2010). Studies of Environmental Quality in and around municipal solid waste dumpsite. Resour. Conserv. Recycling. pp. 129-134.

Butt TE, Lockley E, Oduyemi KOK (2008). Risk assessment of landfill disposal sites- State of the art. Waste Manage. 28:952-964.

Butt TE, Oduyemi KOK (2003). A holistic approach to Concentration Assessment of Hazards in the risk assessment of landfill leachate. Environ. Int. 28:597-608.

Cherry J (1990). Groundwater Monitoring: Some Deficiencies and Opportunities. Hazardous Waste Site Investigationns; Towards Better decision, Proceedings of 10th ORNL Life sciences symposium Gatlinberg, TN, May 1990, Lewis publishers.

Cherry J (1991). Waterloo Center for Ground Water research. University of Waterloo, Ontario, Canada, Personal Communication.

Chian SK, DeWalle FB (1976). Sanitary landfill leachate and their treatment. J. Environ. Eng. 102:411-431.

Demitriou E, Karaozas I, Saratakos K, Zacharias I, Bogdanos K, Diapoulis A (2008). Groundwater risk assessment at a heavily industrialized catchment and the associated impacts on a peri-urban wetland. J. Environ. Manage. 988:526-538.

Din MAM, Jaafar WZW, Obot RMM, Hussain WMAW (2008). How GIS can be a Useful Tool to Deal with Landfill Site Selection. International Symposium on Geoinformatics for Spatial Infrastructure Development in Earth and Allied Scinces.

ESA (2001). Water in a Changing World. Ecological Society of America. J. Issues Ecol. Vol. 9.

Feta D, Pabadopoulos A, Loizidou M (1996). A study on the landfill leachate and its impact on the ground water quality of the greater area. Environ. Geochem. Health. 21:175-190.

Gabriel HF, Khan S (2006). Policy Options for Sustainable Urban Water Cycle Management in Lahore, Pakistan. Proceedings of ERSEC Water Workshop 2006 Beijing, China. pp. 180-202.

Gabriel HF, Khan S (2010). Climate Responsive Urban Groundwater Management Options in a Stressed Aquifer System. IAHS Publications, 338:166-168. ISSN 0144-7815.

Gorsevski PV, Donevska KR, Mitrovski CD, Frizado JP (2012). Integrating multi-criteria evaluation techniques with geographic information systems for landfill site selection: A case study using ordered weighted average. Waste Manage. 32:287-296.

Jensen DL, Ledin A, Christensen TH (1998). Specification of heavy metals in landfill leachate polluted groundwater. Water Resour. 33(11):2642-2650.

Kareem S, Chaudhry MN, Ahmad K, Batool SA (2010). Impact of Solid Waste Leachate on Groundwater and Surface water. General Chem. Soc. Pak. 32(5):606-612.

Lashari B, McKay J, Villholth K (2007). Institutional and Legal Groundwater Management Framework: Lessons Learnt from South Australia for Pakistan. Int. J. Environ. Dev. 4(1):45-59.

Lee GF, Lee AJ (1993). Groundwater Quality Monitoring at lined landfills: Adequacy of Subtitle D Approaches. Report of G. Fred Lee & Associates, El Macero, CA.

Li Y, Li J, Chen S, Diao W (2012). Establishing indices for groundwater contamination risk assessment in the vicinity of hazardous waste landfills in China. Environ. Pollut. 165:77-90.

Limouzin M, Maidment D (2009). Water Scarcity is an Indicator of Poverty in the World. A term project of University of Taxas at Austin.

Lloyd JW, Heathcote JA (1985). Natural in organic Hydro-Chemistry in realtion to groundwater. Clarendon Press, Oxford, England.

López JAR, Hernández JR, Mancilla OL, Diazconti CC, Garrido ML (2008). Assessment of groundwater contamination by landfill leachate: A case in México. Waste Manage. 28:33-39.

Mahini SA, Gholamafard M (2006). Siting MSW Landfills with a Weighted Linear combination Methodology in a GIS Environment. Int. J. Environ. Sci. Technol. 3(4):435-445.

Mahmood K, Rana AD, Tariq S, Kanwal S, Ali R, ALI AH, Tahseen T (2013). Groundwater Levels susceptibility to degradation in Lahore Metropolitan. Sci. Int. 25:123-126.

Malana MA, Khosa MA (2011). Groundwater Pollution with special focus on arsenic, Dera Ghazi Khan-Pakistan. J. Saudi Chem. Soc. 15:39-47.

Nyenje PM, Foppen JW, Muwanga R, Uhlenbrook S (2013). Nutrient Pollution in shallow aquifers underlying pit latrines and domestic solid waste dumps in urban slums. J. Environ. Manage. 122:15-24.

PCRWR (2007). Water quality Monitoring Report (2005-6). ISBN 978-696-8469-18-4, Pakistan Council of Research in Water Resources (PCRWR), www.pcrwr.gov.pk.

Post VEA (2012). Electrical Conductivity as proxy for ground water density in coastal aquifers. Ground Water. 50(5):785-792.

Rivett MO, Ellis PA, Mackay R (2011). Urban groundwater baseflow influence upon inorganic river-water quality: The river tame headwaters catchment in the city of Birmingham, UK. J. Hydrol. 400:206-222.

Santos FAM, Mateus A, Figueiras J, Gonçalves MA (2006). Mapping groundwater contamination around a landfill facility using the VLF-EM method — A case study. J. Appl. Geophys. 60:115-125.

Singh RK, Datta M, Nema AK (2009). A new system for groundwater contamination hazard rating of landfills. J. Environ. Manage. 91:344-357.

World Bank (2006). Pakistan Strategic Country Environment Assessment. Main Report (Volume 1), 36946-PK.

Xing W, Lu W, Zhao Y, Zhang X, Deng W, Christensen TH (2013). Environmental impact assessment of leachate recirculation in landfill of municipal solid waste by comparing with evaporation and discharge (EASEWASTE). Waste Manage. 33:382-389.

Zhang H, Zhang DQ, Jin TF, He PJ, Shao ZH, Shao LM (2011). Environmental and economic Assessment of combined biostabilization and landfill for municipal solid waste. J. Environ. Manage. 92:2533-2538.

Heavy metals in street soil dusts of industrial market in Enugu, South East, Nigeria

Nwachukwu R. Ekere* and Okiri P. Ukoha

Department of Pure and Industrial Chemistry, University of Nigeria, Nsukka, Nigeria.

Analysis of some heavy metals present in soil dusts of Industrial Market in Enugu, South East Nigeria was carried out. Soil dust samples were collected from five different sections in the market and a control sample was collected 500 m away from the market. The samples were analyzed after aqua regia digestion with Flame Atomic Absorption Spectrophotometer for Lead, Iron, Nickel, Copper and Cadmium. The result showed a variation, which indicated that concentration of Fe > Cu > Pb > Cd > Ni. The levels of some of these metals were higher than the recommended allowable limit for such metals in soil by regulating agencies.

Key words: Soil dust, contamination, industrial market, pollution index, enrichment coefficient.

INTRODUCTION

The presence of high concentration of heavy metals in street dusts and soils have become a matter of serious concern of late. Street dust is derived from anthropogenic activities by interaction of natural elements with pollution sources (Turer et al., 2001; Al-Khashman, 2004; Bhattacharya et al., 2011). Reports (Archer and Barratt, 1976; Ayodele and Gaya, 1998; Jaradat and Momani, 1999; Shinggu et al., 2007; Osakwe, 2010) indicate increasing contamination of street dust and soils with heavy toxic metals poses serious threat to healthy living and environmental sustainability. For instance, human exposure to heavy metals has risen dramatically in the last 50 years as a result of an exponential increase in the use of heavy metals in industrial processes and products (Ano et al., 2007). Ultimately, they enter the food chain and affect human and animals. There is consensus among scientists that the increasing rate of heavy metals contamination of soil ecosystem is due to anthropogenic activities such as industrial and energy production, construction and fuel combustion (Murphy, 1981; Nasralla, 1984; Li et al., 2001; Li et al., 2006, 2007; Sharma et al., 2007; Nyangababo and Hamuya, 1986; Sutherland, 2000; White head, 1975). In humans, heavy metals are systemic toxins with specific neurotoxic, nephrotoxic, fetotoxic and teratoxic effects (Murphy, 1981). Heavy metal pollution can affect all environments but its effect is most long lasting in soils due to relatively strong adsorption of many metals on loamy and clay colloid of soil. Many reagents such as Calcium chloride, ammonium nitrate, and ammonium acetate have been employed to extract the mobile or bioavailable forms of heavy metals in single extraction method (Ahumada et al., 1999; He and Singh, 1995; Karezewska, 1996). Sequential extraction method has also been adopted (Narwal et al., 1996). However, acid digestion using aqua regia appears to be most common (Fazeli et al., 2009). Soil dusts may be analyzed for many reasons but this work was aimed at estimating the concentration of some heavy metals in soil dust of Ogbete Industrial market, Enugu.

MATERIALS AND METHODS

Enugu is the capital of Enugu State, South East Nigeria. It has a population of 722, 000 in 2006 (NPC, 2006) and is located between 6° 21′N and 6° 30′ N and 7° 26′ E 7° 30′ E. Enugu urban has a humid tropical climate and the mean monthly temperature lies between 27 and 29°C with rainforest savanna ecotone vegetation. The industrial market covering a space of about 2 km^2 occupies the southern part of the city, popularly called coal camp because early coal miners settled in quarters which presently house some of the small and medium scale industries (SMEs). Found in the industrial market include foundry, motor parts fabrication, aluminum smelting and fabrication, woodworks and brickworks, automobile mechanics,

*Corresponding author. E-mail: nwachukwuekere2006@yahoo.com.

Table 1. Concentration (mg/kg) of heavy metals in the soil dust.

Sample/Identification	Cu	Cd	Fe	Ni	Pb	EC* (µS/cm)	pH
Roadside I	64.0±3.0	2.8±0.2	690.0±0.2	0.4±0.1	31.0±0.1	0.45±0.0	7.9±0.1
Welding works II	23.0±1.1	2.2±0.3	281.1±0.0	0.8±0.2	35.0±0.1	0.32±0.1	8.4±0.0
Battery sales and Brick Cutting III	34.0±0.5	2.3±0.5	752.0±0.2	0.2±0.3	9.5±1.3	1.02±0.2	6.3±0.3
Motor Park IV	48.0±0.3	0.4±0.3	651.0±1.0	0.8±0.1	48.0±0.3	0.24±0.3	7.2±0.2
Automobile Mechanic V	63.0±1.2	0.8±0.1	613.5±2.0	0.5±0.2	87.0±0.1	0.51±0.1	8.1±0.3
Reference Station VI	2.8±0.1	0.2±0.1	125±1.0	0.1±0.0	4.8±0.0	0.21±0.0	6.9±0.0

*EC = Electrical conductivity.

blacksmithing and a host of others. Also, dealers in assorted machines, spare parts, food, and household effects litter the industrial market. It has been established that the industrial market contribute significantly to the environmental degradation of the city especially the underground and surface waters in the area (Chima et al., 2009; Ekere, 2011).

Sampling site description

Samples were taken from coal camp Industrial market, Enugu, Nigeria at five different areas of industrial activities. These areas include (i) Automobile mechanic (ii) road side (iii) Battery and brick works (iv) Motor park and (v) welding works areas. Standard methods were applied for sample collection, preservation and analysis (Roger, 1994). Street soil dusts were sampled using plastic brush and a plastic scoop. The soil dust samples were passed through 62 µm sieve prior to analysis. Ten replicate samples were collected from each site during the study period, which covered four months (December-March). The samples were collected at this period (dry season) to avoid washing away the heavy metals by rain. Another sampling site was selected 500 m from the market to serve as a control.

Dust samples were collected and transferred to clean polyethene bags. Each of the samples was dried in an oven at 100-110°C to drive out moisture until a constant weight was obtained. The soil pH of each dust sample was determined in 1:5 dust: distilled water (w/v) suspension using a pH meter (Kent model) while the electrical conductivities (EC) was measured in a 1:2 dust: distilled water (w/v) suspension using a conductivity meter. Particular like slag stones were separated from the samples and the samples ground into a fine powder. The ground samples were sieved using a 250 µm sieve mesh diameter.

All experiments were performed with analar grade chemicals. One gramme of each sample was digested using aqua regia, 3:1 ratio of HCl to HNO_3 (Ayodele and Gaya, 1994; Xianagding et al., 2001) to incipient dryness. Then 5 ml $HClO_4$ was added and the mixture heated until a homogeneous solution was obtained. During the digestion, the samples were heated in a water bath at 100°C. Calibration standards were prepared from the stock solution of the metals. The digested samples were then analyzed for desired heavy metals using Buck Scientific Model 210 VGP Atomic Absorption Spectrophotometer.

RESULTS AND DISCUSSION

The concentrations of the assayed heavy metals are presented in Table 1. The pH of the soil dust suspension ranged from 6.3 to 8.6. The electrical conductivity (EC) of the sample dusts was relatively high in all the stations indicative of presence of ionic species. The concentration of Iron had a range of 752.0 to 12.5 mg/kg soil dust, with a concentration trend in sampling sites of III > I > IV > V > 11 > VI. Iron had the highest concentration in soil dust from welding work section and lowest in the automobile mechanic section. Iron also had the highest concentration among all the metals in the reference station, some 500 m from the border of the industrial market. Earlier studies (Xianagding et al., 2001; Dara, 1993) found out that natural soils contain appreciable amount of Iron. The levels of iron observed in this study were higher than some reported elsewhere (Ahumada et al., 1999; He and Singh, 1995; Karezewska, 1996). Iron in industrial areas result from construction, filing and other metal works.

The levels of Copper in the samples studied ranged from 64.0 to 2.8 mg/kg. These were significantly higher than values reported elsewhere (Rashad and Shalaly, 2007; Dara, 1993). The presence of Cu in industrial street dust is attributable to smelting, battery, soldering works.

Lead levels in the soil dust studied ranged from 87.0 to 4.8 mg/kg, greatly higher than some earlier reported works (Rashad and Shalaly, 2007; Ayodele and Gaya, 2003). Lead enters dust through the actions of electrical smelting and soldering workers. Lead is noted as toxin and it has adverse effects on humans and animals (Murphy, 1981).

The concentrations of Cadmium ranged from 2.3 to 0. 2 mg/kg. Cadmium presence in the soil dusts results from industrial works such as battery and other electrical works. Nickel levels ranged from 0.8 to 0.1 mg/kg. The element has the lowest concentration of all the metals analyzed. The levels of the metals found in the industrial market soil dust are appreciably higher than those found in the reference station, an indication of pollution of the market soils. The result of this work has far reaching implications to the environment. Heavy metals are widely distributed from the pollution sources into various components of the environment ranging from the humans to air, vegetation and water. For instance, in a study of heavy metal pollution of ground water in the same market (Ekere, 2011), many heavy metals were found to be in excess of the maximum allowable limit. Health risks associated with Cd, Pb, Cu, Ni, and Fe have been

Table 2. C/P index of some metals in the soil dusts.

Sample	Fe	Cu	Pb
I	0.14	1.78	0.37
II	0.06	0.61	0.41
III	0.15	0.94	1.12
IV	0.13	1.33	0.57
V	0.12	1.75	1.02

Table 3. Enrichment Coefficient (EC) of the metals studied.

Sample	Fe	Cu	Pb	Cd	Ni
I	5.52	4.1	1.16	1.62	0.72
II	2.25	3.61	0.87	4.84	3.52
III	6.01	2.02	0.34	1.96	0.34
IV	5.21	3.26	1.90	0.38	1.52
V	4.90	4.50	3.69	0.80	1.00

extensively reviewed and hardly need further stress (JECFA, 1989; Oldiges et al., 1989; Nriagu, 1996; Eaton and Roberton, 1994; Glover-KerKvilet, 1995; Imeggbue et al., 2006; Jkrup, 2003). The contamination/pollution index (C/P) of the three most abundant metals in the soil dusts was calculated as recommended by Lacatusu (2000) thus:

$$C/P = \frac{\text{Conc. of the metal in the soil}}{\text{Target value}}$$

where the target value was obtained by using the standard formulated by the Department of Petroleum Resources (Osakwe, 2010) in Nigeria as cluster abundant values for maximum allowable concentration of heavy metals in soil in mg/kg {5000 for Fe; 85 for Pb; 36 for Cu}. The C/P index values for the dust samples are shown in Table 2.

A C/P value greater than 1 is an indication that the soil is polluted while values less than 1 indicates contamination. Based on the classification, it can be easily seen that the dust samples from the industrial market in locations I, IV and V were polluted with respect to Cu and Pb while sections III and V are polluted with respect to Pb.

In order to evaluate the anthropogenic environmental status of the elements examined, the Enrichment Coefficient was used (Assah et al., 2003; Egbareuba and Odjada, 2002). The Enrichment Coefficient {EC} in given as:

$$EC = \frac{C_1Me/C_1n}{C_2Me/C_2n}$$

Where C_1Me and C_2Me are examined metal content in the examined environment and reference environment respectively while C_1n and C_2n are reference element in the examined environment and reference environment respectively.

Iron was used as the reference element as it in the most naturally abundant element in the soil dust. Five contamination categories are recognized on the basis of EC (Egbareuba and Odjada, 2002) as follows:

i. Deficiency to minimal enrichment for values below 2,
ii. Moderate enrichment for values from 2 to 5,
iii. Significant enrichment for values from 5 to 20,
iv. Very high enrichment for values from 20 to 40,
v. Extremely high enrichment for values above 40.

The result of the analysis is shown in Table 3. Enrichment coefficient is a very veritable tool in assessment of anthropogenic contribution to environmental pollution with heavy metal in literature (WHO, 1989). The EC values obtained for Iron have a range of 2.25 to 6.01 placing the element in the category of moderate enrichment to significant enrichment. Copper with a range of 2.02 to 4.50 has moderate enrichment while lead had minimal enrichment in all samples except V (automobile mechanic) which was moderately enriched. Cadmium and Nickel fall within deficiency to minimal enrichment category. Some of these values were above the range reported in some similar studies but definitely the contributions of anthropogenic activities is evident (Assah et al., 2003; Egbareuba and Odjada, 2002; Akoto et al., 2008; Ogundiran et al., 2012; Amusan et al., 2005).

Conclusion

The soil dusts from the industrial market studied indicated pollution of the soil environment by heavy metals. The values obtained from the contamination/pollution index corroborated this view. Further more, analysis of the enrichment coefficient indicated a range of minimal to significant enrichment for the metal contents studied.

REFERENCES

Ahumada I, Mendoza J, Asear L (1999). Sequential Extraction of heavy metals in soils irrigated with waste water. Commun. Soil Sci. Plant Anal. 30:1507-1519.

Akoto O, Ephraim JH, Dako G (2008). Heavy metals pollution in surface soils in the vicinity of Abundant Railway Servicing workshop in Kumasi, Ghana. lat. J. Environ. Res. 2(4):359-364.

Al-Khashman OA (2004). Heavy metal distribution in dust street dust and soils from work place in Karak Industrial Estate, Jordan. Atmos. Environ. 38:6803-6812.

Amusan AA, Ige DU, Olawales R (2005). Characteristics of Soils and Crop uptake of metals in Municipal waste dumpsites in Nigeria. J. Hum. Ecol. 17(3):167-171.

Ano AO, Odoemelam SA, Ekwueme PO (2007). Lead and Cadmium levels in soils and cassava. (*Manilot Esculenta* Grantz) along Enugu-Port Harcourt Expressway in Nigeria. Electronic J. Environ. Agric. Food Chem. 6(5):2024-2031.
National Population Commission of Nigeria (NPC) (2006). Population and Housing Census Result. NPC, Lagos. P. 26.
Archer A, Barratt RS (1976). Lead in the Environment. Monitoring for Lead. Sci. Total Environ. 96:173-176.
Assah VA, Abimbola AF, Such CE (2003). Heavy metal concentration and distribution on surface soils of the Rassa Industrial Zone I Dounal Cameroon. Arabia J. Sci. Eng. 31(24):147-158.
Ayodele JT, Gaya UM (1998). Zinc and Manganese in Kano municipality street Dust. J. Chem. Soc. Nig. 23:24- 31.
Ayodele JT, Gaya UM (2003). Nickel in Municipal street dust. J. Chem. Soc. Nig. 25(1):15-20.
Ayodele JT, Gaya UM (1994). Determination of lead in street dust to index its pollution in Kano municipal. Spectrum J. 1:92-97.
Bhattacharya T, Chakraborty S, Fadadu B, Bhattacharya P (2011). Heavy Metal concentrations in street and Leaf Deposited Dust in Anand City, India. Res. J. Chem. Sci. 1(5):61-66.
Chima GN, Ogbonna CE, Nwankwo IU (2009). Effects of urban wastes on the quality of Asata River in Enugu, South Eastern Nigeria. Global J. Environ. Sci. 8(1):31-39.
Dara SS (1993). A Textbook of Environmental Chemistry and Pollution Control. S. Chand and Company Ltd. Ran Nagar. New Delhi 110055.
Eaton DL, Roberton WO (1994). Toxicology' in Rosenstick L and Cullen MR eds. Textbook of Clinical Occupational and Environmental Medicine. WB Saunders Company. Philadelphia pp. 116-117.
Egbareuba F, Odjada V (2002). The Pollution Potential and Chemical Interaction Parameter of some agro and industrial wastes on soil. Nig. J. Basic Appl. Sci. 11:177-188.
Ekere NR (2011). Heavy metals contamination of Ground water in Industrial areas of Enugu Nigeria J. Appl. Sci. 14(1):9535-9542.
Fazeli MS, Moosavi MH, Pournia M, Jassemi ZZ (2009). Metals Distribution in Top soils around Industrial Town of Awhaz II. Ahwaz, Iran. J. Appl. Sci. 9(6):1121-1127.
Glover-Kerkvilet J (1995). Environmental Assult on Immunity. Environ. Health Perspect. 103(3):236-237.
He QB, Singh BR (1995). Calcium Availability to Plants as Affected by Repeated Applications of Phosphorous Fertilizers. Acta. Agric. Scand. B. 45:22-31.
Imeggbue CMA, Isirimah NO, Igwe C, William ES (2006). Characteristic, levels of heavy metals in soil profile is automobile Mechanic waste dumps in Nigeria. Environmentalist 26:131-137.
Jaradat OM, Momani AK (1999). Contamination of road side soil, plants and air with Heavy metals in Jordan. A Comparative study. Turk. J. Chem. 23:209-220.
JECFA (1989). Toxicological evaluation of certain food additives and contaminants. World Health Organization. Joint FAO/WHO Expert Committee on food Additives Cambridge, Cambridge Univ. Press (WHO Food Additives series No. 24) pp. 163-219.
Jkrup L (2003). Hazards of heavy metals contamination. Great Britain. British Council Publication. pp. 18-22.
Karezewska A (1996). Metal Species Distribution in Top and Subsoil in Area Affected by Copper Smelter Emissions. Appl. Geo. Chem. 11:35-42.
Li XD, Poon CS, Liu PS (2001). Heavy metal contamination of urban soils and street dusts in Hong Kong. Appl. Geo. Chem. 16:136-1368.
Lacatusu R (2000). Appraising levels of soil contamination and pollution with heavy metals European Soil Bureau Research Report. 4:393-402.
Li MS, Lou YP, Su ZY (2007). Heavy metal concentrations in soils and plant accumulation in a restored manganese mine land in Guangxi. South China. Environ. Pollut. 147:168-175.
Li Y, Wang X, Gou X, Isu YB, Wang B (2006). Risk assessment of heavy metals in soils and Vegetables around non- ferrous metals mining and smelling sites, Baiyin, China. J. Environ. Sci. 18:1124-1134.
Nasralla MM (1984). Lead in Jaddah urabn dust. Environ. Pollut. (series B) 8:133-141.
Murphy CB (1981). Bioaccumulation and toxicity of heavy metals and related trace elements. J. Wat. Pollut. Contr. Fed. 53(6):993-999.
Narwal RP, Singh BR, Salbu B (1996). Association of Cadmium, Zinc, Copper and Nicked with Components in Natural Heavy Metal Rich Soils Studied by Parallel and Sequential Extractions. Com. Soil Sci. Plant. Anal. 30:1209-1230.
Nriagu JO (1996). History of Global metal pollution. Science 272:223-224.
Nyangababo JT, Hamuya JN (1986). The deposition of lead, Cadmium, Zinc and Copper from motor traffic on Bombay road in Kampala city. Inter. J. Environ. Stud. 28:115-119.
Ogundiran MB, Ogundela DT, Afoloyan P, Osibanjo O (2012). Heavy metals levels in forage Grasses, Leachate and Lactating Cows Reared around lead Slag Dumpsites in Nigeria Int. J. Environ. Res. 6(3):695-702.
Oldiges H, Hochrainer T, Glaser U (1989). Long term Inhalation Study with Wistar rats and four Cadmium Compounds. Toxicol. Environ. Chem. 19: 217-222.
Osakwe SA (2010). Distribution of Heavy metals in soils around automobile dumpsites in Agbo and Environs, Delta State, Nigeria. J. Chem. Soc. Nig. 35(1):53-60.
Rashad M, Shalaly EA (2007). Dispersion of heavy metals around Municipal solid waste (MSW) dumpsites, Alexandria, Egypt. Am. Eurasian J. Agric Environ. Sci. 2(3):204-212.
Roger NR (1994). Environmental Analysis. 13th edition. John Wiley & Sons. New York. pp. 25-40.
Sharma RK, Agraval M, Marshal F (2007). Heavy metal contamination of soils and vegetables in sub urban areas of Varansi, India. Ecotoxicol. Environ. Safety 66:258-266.
Shinggu DY, Ogugbuaja VO, Barminas JT, Toma I (2007). Analysis of street dust for heavy metal pollutants in Mubi, Adamawa State Nigeria. Int. J. Phy. Sci. 2(11):290-293.
Sutherland RA (2000). Bed Sediment associated trace metals in the urban stream. Oahu Hawai. Environ. Geol. 39:611-627.
Turer D, Maynard JB, Sansalone JJ (2001). Heavy metal contamination in soils of urban highways. Comparison between runoff and soil concentration at Cincimmati, Ohio. 132:293-314.
White head PJ (1975). Pye Unicam Atomic Absorption Data Book. Ist ed. Pye Unicorn Ltd. England; Cambridge.
WHO (1989). Environmental Health Criteria 3 and 5 CPCB (1997). CPCB News letter 4(iii).
Xianagding Li, Chi-sum P, Pui-sumliu S (2001). Heavy metal contamination of urban soils and street dust in Hong Kong. Appl. Geo. Chem. 16:1361-1368.

16

Assessment of industrial and adaptable building components for a residential layout

Nasibeh Sadafi*, M. F. M. Zain and M. Jamil

Department of Architecture, Faculty of Engineering, National University of Malaysia, 43600, Bangi, Selangor, Malaysia.

From environmental point of view, building waste is a critical issue in present construction. In order to reduce the depletion of valuable materials and natural resources, waste management of constructions needs to be considered in the early stages of design. This requires innovative design strategies that take into account the future demolition and provide flexible and versatile structures. Industrial, flexible and demountable building system (IFD) is trying to create buildings with higher quality, more adaptable and better environmental characteristics. But till date, a recognized method for IFD building assessment has not been defined. This research focuses on IFD component analysis process for a residential terrace house layout. In the first step of the research, a conceptual prototype for the key elements was proposed. Next step was to create a procedure for evaluating the suggested design. The evaluation process contains two parts. First, the layout and design characteristics assessment according to IFD criteria was developed. Second, the stability, seismic and load bearing capacity of the designed structure was characterized by conventional nonlinear static analysis (pushover) through finite element computations with ETABS software. It is proposed that, application of these assessments can testify to the flexibility and strength of the designed layout.

Key words: Industrial, flexible and demountable building system (IFD), building components, modular coordination system, terrace house layout.

INTRODUCTION

Construction could be defined as transformation of materials, resources or components into buildings and infrastructural works (Egmond and Scheublin, 2005). In fact, a building construction process is a complex system that involves required interactions between different parties. On the other hand, in most of developing countries, insufficient cooperation between the parties have caused dysfunctional team works, and lots of opportunities for optimal use of resources as well as innovations in design and construction. Yet, conventional ways of construction ignores the in-built capacity of the building for easy adaptation over time (Paduart et al., 2008; Razaz, 2010). Therefore, any change or renovation of the building will not be easily possible or result in significant cost and waste during its lifecycle (Sunikka and Boon, 2003; Tam et al., 2006). Besides, customers' demands for variable production output, require the construction industry to make the process of production more flexible for achieving higher quality products with variety of kinds (Egmond and Scheublin, 2005). Industrial and flexible system tries to improve the way of building construction in such a way that result in a faster, economical, higher quality and environmental friendly building in comparison with traditional methods (Bon and Hutchinson, 2000; Ball, 2002; Kohler and Hassler, 2002; Zegers and Herwijnen, 2004; Gallant and Blickle, 2005).

However, flexible and reusable construction system can only be attained if the design process considers the general construction system and construction detailing at the same time (Paduart et al., 2008). Therefore, attention has to be paid to the integration and technical requirements of the elements and the connections as well. This research project focuses on analysis process of an IFD terrace house layout. The first step is to define the

*Corresponding author. E-mail: nsadafi@ eng.ukm.my.

criteria for flexible and dismountable building elements. In order to objectively assess whether the design meets these set of criteria, an evaluation method was developed. The evaluation method helps to recognize where improvements can be made. Furthermore, ETABS software has been applied to assess the structural design of the building. The analysis methods have helped to draw the results and validation of the prototype design.

ASPECTS OF IFD SYSTEM

Industrial, flexible and demountable systems (IFD) imply to easy change and adaptation of buildings while reducing resource depletion and construction cost (Bon and Hutchinson, 2000; Ball, 2002; Kohler and Hassler, 2002; Gallant and Blickle, 2005). Such an approach focuses on the long-term performance of building structures and materials, which attempts to introduce flexible building concepts using industrially produced and demountable systems (Durmisevic, 2006).

Industrial way of construction

Industrial way of construction is to produce building elements in a non-project related manufacturing, under controlled circumstances and in a repeatable process (Zegers and Herwijnen, 2004). International Council for Research and Innovation in Building and Construction, Work group 24 (CIB W24) has defined the Industrialized Building as a building technology in which modern systematized method of design, production planning and control as well as mechanized and automated manufacture are applied. In this case, the elements can be used in several buildings with different characteristics to reduce the cost of manpower and time consuming activities. The quality of the building parts will be controlled in the manufacturing and assembly process either in the factory or the building site. Furthermore, the repeatability of the process makes the manufacturing easier and faster (Hamid et al., 2008). However, industrialized building is not necessarily equal to mass production. Industrial production of the components will increase the opportunities for flexible application.

Flexible design

Flexible building parts require adapting to change according to users' needs easily. During the life of the building, changes will be required according to users demand or to facilitate other functionalities (Slaughter, 2001). Rather than this, new technologies or regulations might be introduced. In fact change is an obvious and inevitable characteristic of built environment and requires sufficient attention at the initial design phase (Brand, 1994; Greden, 2005).

So, it is necessary to provide the building that would be able to change the layout and using materials to meet the new requirements. The importance of this trend is that extending the useful life of existing buildings and constructed systems through accommodating change easily, fast, and inexpensively. It will also improve the value of any property by making it as an investment opportunity (Slaughter, 2001; Morgan and Stevenson, 2005). On the other hand, it supports the key concept of sustainability by lowering material, transport and energy consumption as well as waste and pollution (Gregory, 2004; Bullen, 2007).

Demountable system

Non-recyclable building materials are major part of present waste while large parts of the recycled materials are limited to low quality use or even land fill, because of the high amount of contaminants (Durmisevic, 2006). This is because in current construction, components and structures are not designed to be separated or applied in new buildings. In order to overcome the waste issues, more productive design approaches need to be taken that would simplify dismantling of high value materials for reuse and recycle (Chini, 2002). Demount-ability enables the building parts with various life spans to be separated with little damage to other building parts and suited to be reused or recycled. Indeed, design for dismantling encourages the application of recyclable materials as well as simplifying the separation of materials before or after demolition (Dorsthorst and Kowalczyk, 2002). Therefore, it will extend the life of the building as a whole.

Innovation in the construction industry defines the process of development and application of technologies in a new or improved service to increase productivity and user's satisfaction (Egmond and Scheublin, 2005). However, a higher degree of prefabrication needs more work on design and planning process of the whole construction as well.

Description of the designed IFD terrace house

A two story terrace house has been proposed for the prototype unit of study. The lot size was considered as 18600-7800 mm while the interior architectural design attempts to create configuration that encourages the possibility of future adaptations for the spaces use (Figure 1). Therefore, the unusable or single function spaces like circulation and fixed element have been minimized. The plans, section, and elevation have been framed from a modular grid which is based on the available standard sizes (Ministry of Housing and Local Government Malaysia, 2009). The grids will allow simply reducing or expanding the design based on the site restrictions. The size and location of the fenestrations and doors have also been defined according to modular

Figure 1. Prototype terrace house, design concept plans, section and elevation.

design rules and can be changed flexibly.

Partition walls were considered as, 600 mm × 400 mm ×100 mm, concrete panels. The constructability of the walls using the panels is an important factor considered in the development of the panels. Since, the main walls in front and rear façade have considered as non load bearing, the main requirements for the interlocking panels are the stability and transferring the weight load to the beams and columns. The panel dimensions followed modular design ruels, accordingly, other spaces in the house also are in conformity with modular dimensions, therfore encouraging the application of other modularly coordinated components such as doors and windows. Since the typical room height of 2800 mm has been adopted, the application of half or broken block for fitting into the space will not be required. Consequently, no wastage of materials will be provided and exact number of panels that will be required for a specified house could be apprised from the architectural drawings.

The following features are sought to be included for the component adoption:

(i) Efficient interlocking mechanism in various directions.
(ii) Meeting modular coordination requirements, that is, the dimensions are fit into the modular dimensions and horizontal planning dimensions of 3 M and vertical planning dimension of 1 M, where 1 M = 100 mm.
(iii) Dry and fast construction with minimum in-situ casting.

Table 1. Description of the structural elements for the pattern IFD terrace house.

S/N	Element	Description
1	Main load carrying system	Moment resisting frame
2	Foundation	Strip footing with pedestal
3	Pedestal (1)	Square column (450-450-1200 mm) Component code: STR-KP-T-SC-450
4	Pedestal (2)	Square column (600-300-1200 mm) Component code: STR-KP-T-SC-600
5	Roof system	Light weight prefab panels
6	Floor system	Hollow core slab Component code: STR-KP-L-HCS-150
7	Beam (1)	Rectangular Beam (150-500 mm) Component code: STR-KP-R-RB-150-R
8	Beam (2)	Rectangular Beam (300-500 mm) Component code: STR-KP-R-RB-300-R
9	Column(1)	Square column (150-600 mm) Component code:STR-KP-T-SC-150
10	Column(2)	Square column (300-300 mm) Component code:STR-KP-T-SC-300
11	Panel(1)	Non-load bearing Concrete panel (600-100-400 mm) Removable *
12	Panel(2)	Load bearing concrete panel Fixed

*These panels are considered non-load bearing in order to maintain the ability of dismantling and rearranging the wall panels.

(iv) Reduce formwork for construction, so make it more environmentally friendly.

Table 1 is presenting the main structural elements of the designed housing unit. Except the proposed panels, other elements' characteristics are according to standards available in 'Modular Design Guide' and components available in IBS catalog booklet (Ministry of Housing and Local Government Malaysia, 2009).

EVALUATION METHOD

Few of the existing modern buildings have been deliberately designed for flexibility which results in difficult process for assessment of that flexibility over time (Davison et al., 2006). Basic principles of the prototype design were to apply high degree of industrialization during construction, flexibility during the installation, and demountability at demolition time.

First step of evaluation

Technical requirements of IFD building design parameters were identified (Gassel, 2003; Zegers and Herwijnen, 2004; Holtz, 2006).Accordingly, the developed prototype has been assessed for its IFD characteristics. The key design parameters are presented in Table 2.

Simulation

In the next step of evaluation, a simulation analysis was run to assess the house layout design. ETABS2000 Extended 3D Analysis of Building Systems (Version 9.0.4) has been used to analyze the structural design of the building. ETABS is well known structural software that utilizes Finite Element Method for analysis of common structural systems (CSI, 2005). It is equipped with steel and concrete design modules which had been used for design of main load carrying elements. Graphical representation of model and loading is also provided (Ghoulbzouri et al., 2009). It is used in our analysis for its relative ease of use, detailed documentation, flexibility and vastness of capabilities.

Applied code

For concrete structure and foundation design, UBC97 has been employed.

Material properties of the main structure (columns, beams and pedestals)

(I) Main compressive strength of concrete = 10 MPa.
(II) Weight per unit volume = 1200 kg/m^3.

Table 2. Definition of IFD characteristics for the panel system.

IFD criteria	Design characteristics
Industrial	
Standardize parts	The whole layout consists of subparts that are manufactured in series
Modular system	All the dimensions are according to Modular system coordination
Reduce number of parts	Consist of small number of parts
Simple assembly protocol	The parts can be assembled on site by means of simples actions and lightweight equipments
Reduce waste	Produces little waste during manufacturing and assembly on site
Changeable	Standard components can be changed during the service life
Flexible	
Freedom of design	Small and changeable parts provide free and adaptable design
Adaptable during assembly	Is not depending on a strict assembly planning
Independence of disciplines	Installation process, bearing structure, outer shell, and interior finishing can be performed independently but combined at the end
Changing of layout	Possibility of changing the layout with little disturbance to other parts of the building
Layout freedom	Open and free interior spaces for future adaptation
Adjustability of building parts	Bearing structure: prefab elements have limited adjustability Installation : dry connections make the installation practically adjustable Outer shell : light panels with dry connections Interior finishing : modular design and adjustable
Demountable	
Reuse from other buildings	The prefab components can be used from other buildings without alteration
Dry connections	Application of dry connections for joining the panels, column, roof, and floor
Demounting of parts	Can be demounted with little disturbance to the other parts
Demounting without waste	Demounting will not cause waste production
Reuse of materials	Elements' materials can be used as new raw materials
Reuse of building parts	One building component can be reuse in the other buildings

(III) Mass per unit volume = 1200 kg/m^3.
(IV) Modulus of Elasticity = 5800 MPa.

Loading

The applied loads have been calculated for two story residential building construction. The loads have been determined according to project specification for dead, live and earthquake loads.

a) Dead load

Weight of the floor is taken as 425 kg/m^2, which is according to the standard specifications of prefabricated concrete slabs.
(I) Weight of roof is taken as 100 kg/m^2 (Light ceiling is considered).
(II) Distributed partition walls without openings' weight have been considered as 420 kg/m (for 2.8 m Height walls) and opening effects has been considered in walls that have window. These loads are assigned in the parts that were needed.
(III) Framing and external walls weight was computed in ETABS2000 program.

b) Live load

(I) Live load of floor is taken as 200 kg/m^2.

c) Earthquake load

Earthquake coefficient has been considered as 0.08. ETABS has the ability to calculate the shear stress according to the considered earthquake coefficient, DL, LL, and elements' weight. This stress will be assigned between the nodes by ETABS.

Modeling and analysis

a) Framing and element section assignments

The grid lines, beams and column elements have been defined and drawn according to the architectural drawings in the first step of modeling. Afterwards, the

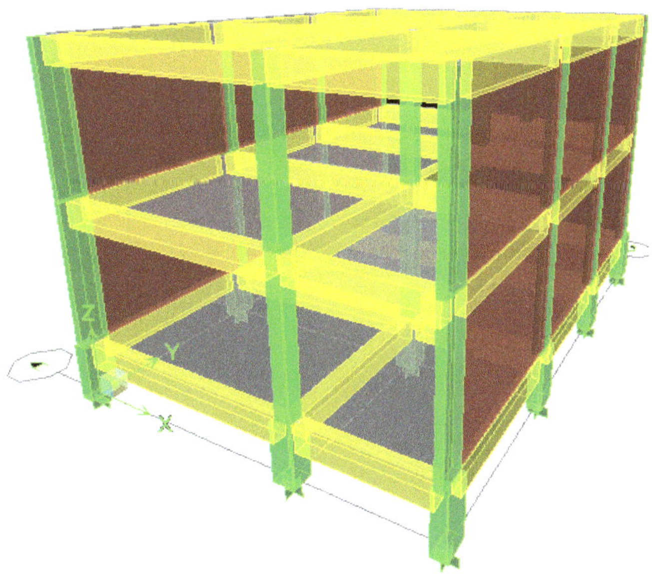

Figure 2. General 3D view of the model.

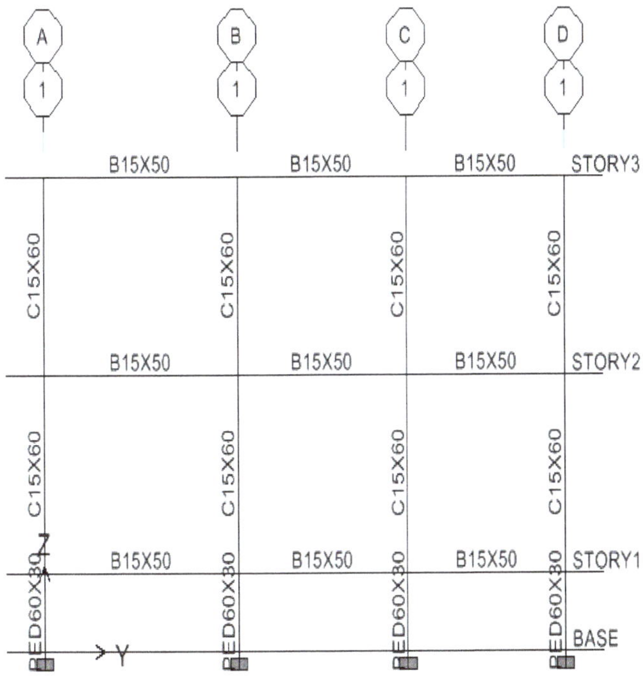

Figure 3. Framing view axis 1-1.

considered sections have been defined and assigned to each related element. The graphical representations of the model are shown in Figures 2-5.

b) Loading and load combinations according to UBC97 Code

In this step, the assumed loads have been assigned to each element (except the elements' self weight that will be considered by ETABS). The load combinations for the three main loads; Dead Load (DL), Live Load (LL), and Earthquake Load (EQ) at X and Y direction, have been determined according to UBC97 Code as follows:

(I) 1.4DL
(II) 1.4DL +1.7 LL
(III) 1.32 DL + 1.1 EQX + 0.55 LL
(IV) 1.32 DL - 1.1 EQX + 0.55 LL
(V) 1.32 DL + 1.1 EQY + 0.55 LL
(VI) 1.32 DL - 1.1 EQY + 0.55 LL
(VII) 0.99 DL + 1.1 EQX
(VIII) 0.99 DL + 1.1 EQX
(IX) 0.99 DL + 1.1 EQY
(X) 0.99 DL - 1.1 EQY

c) Analysis and Design

The concrete structure has been analyzed by ETABS2000 Ver. 9.0.4. The analysis creates reliable estimation of the internal forces, stress, deformations, and behavior of the structure under the appointed loads or movements and consequently guides for safe design of the structures. The three main outputs include the "Shear Forces/Stress", "Moment Forces/Stress" and "Axial Forces/Stress". Each force is available for a special load case, combination or envelope of combinations. Some graphical outputs of the sample's frame under envelope of load combinations are shown in Figure 6.

In the next step, the internal forces were extracted and applied for the design and checking of the sections and structural elements according to the UBC97 Code. Subsequently, the capacity for each column and applied stress were compared. The comparison is produced in graphical and numerical method using capacity ratio of the section. If this ratio is greater than 1 it means that the section is over stressed and needs strengthening. The results from analysis illustrated that the capacity ratio for all columns in the model were smaller than 1. Figure 7 displays one sample section for the capacity ratio of the model.

Furthermore, results from Column P-M-M Interaction Ratios shows that the ratios of existing loads to the capacity of the columns are less than 1.0 in all the stories ETABS controls the capacity of beams according to the internal stresses. The required reinforcements according to the defined beam size and materials (concrete and bars) specifications would be recognized. For this study the results were be compared with prefabricated elements and the optimum sections were chosen accordingly (Figure 8). It means that the assigned prefabricated columns and beams have the capacity to tolerate the existing loads.

Member deflections and story drift are other important parts of the structural design which were checked

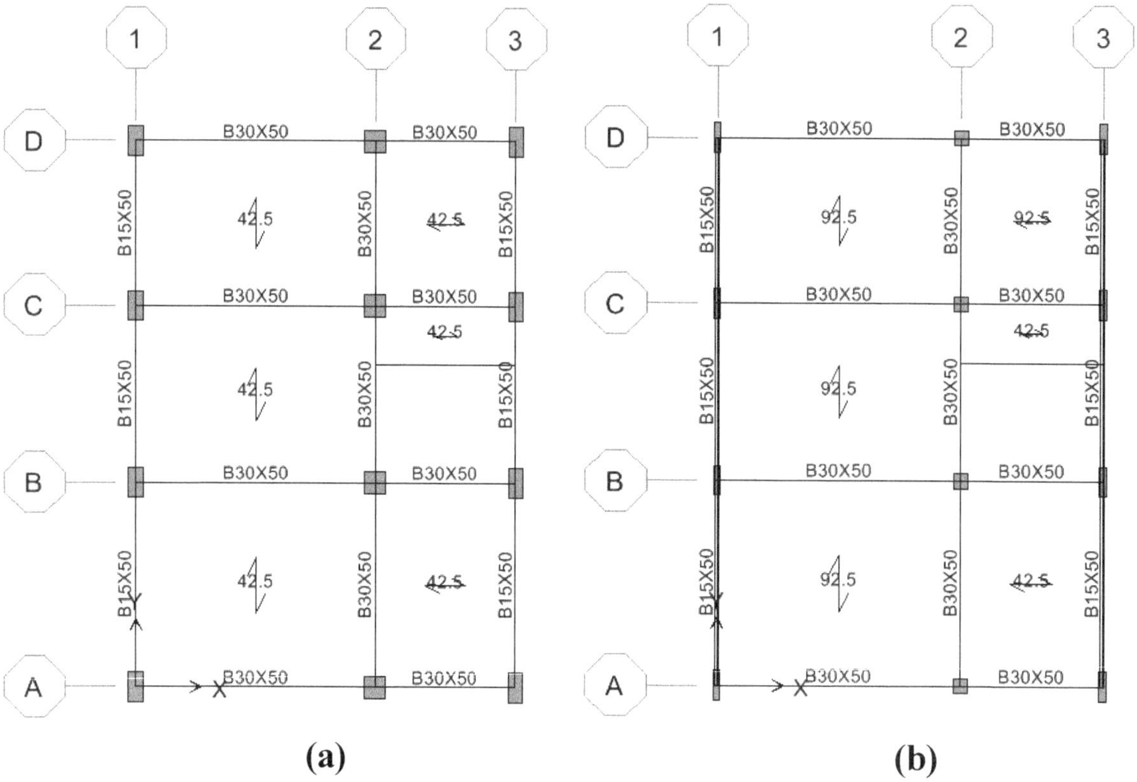

Figure 4. First (a) and second (b) floor dead load.

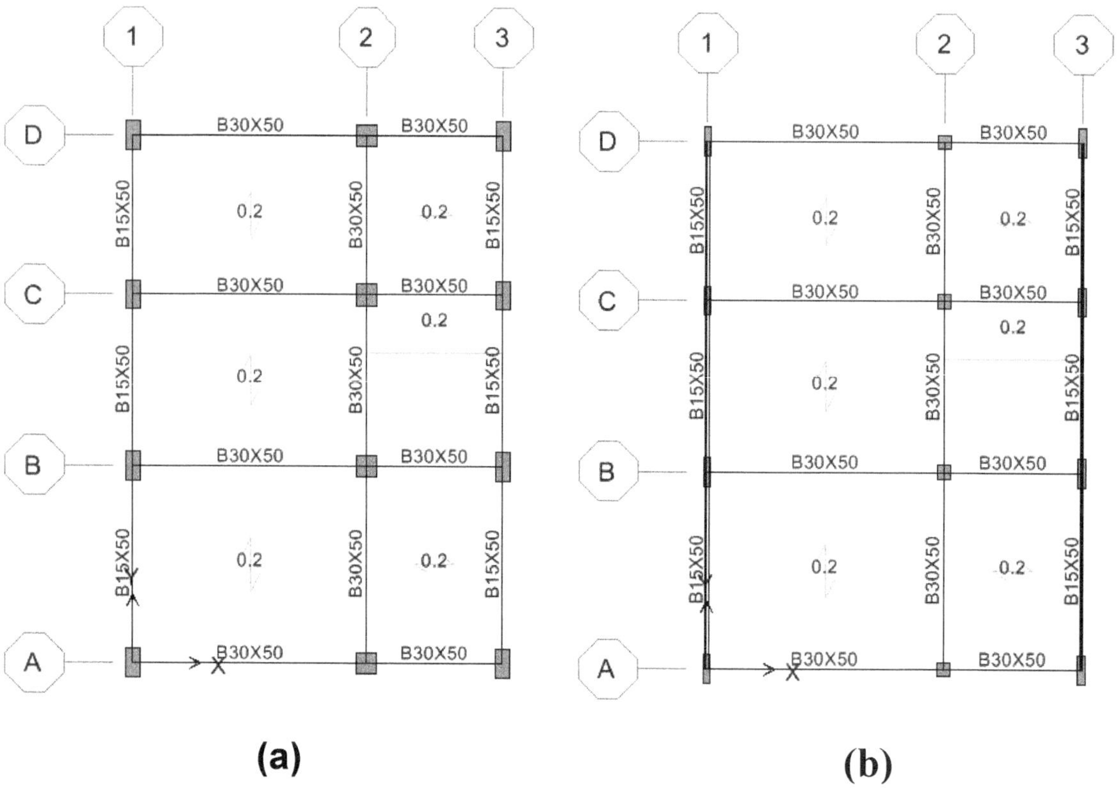

Figure 5. First (a) and second (b) floor live Load.

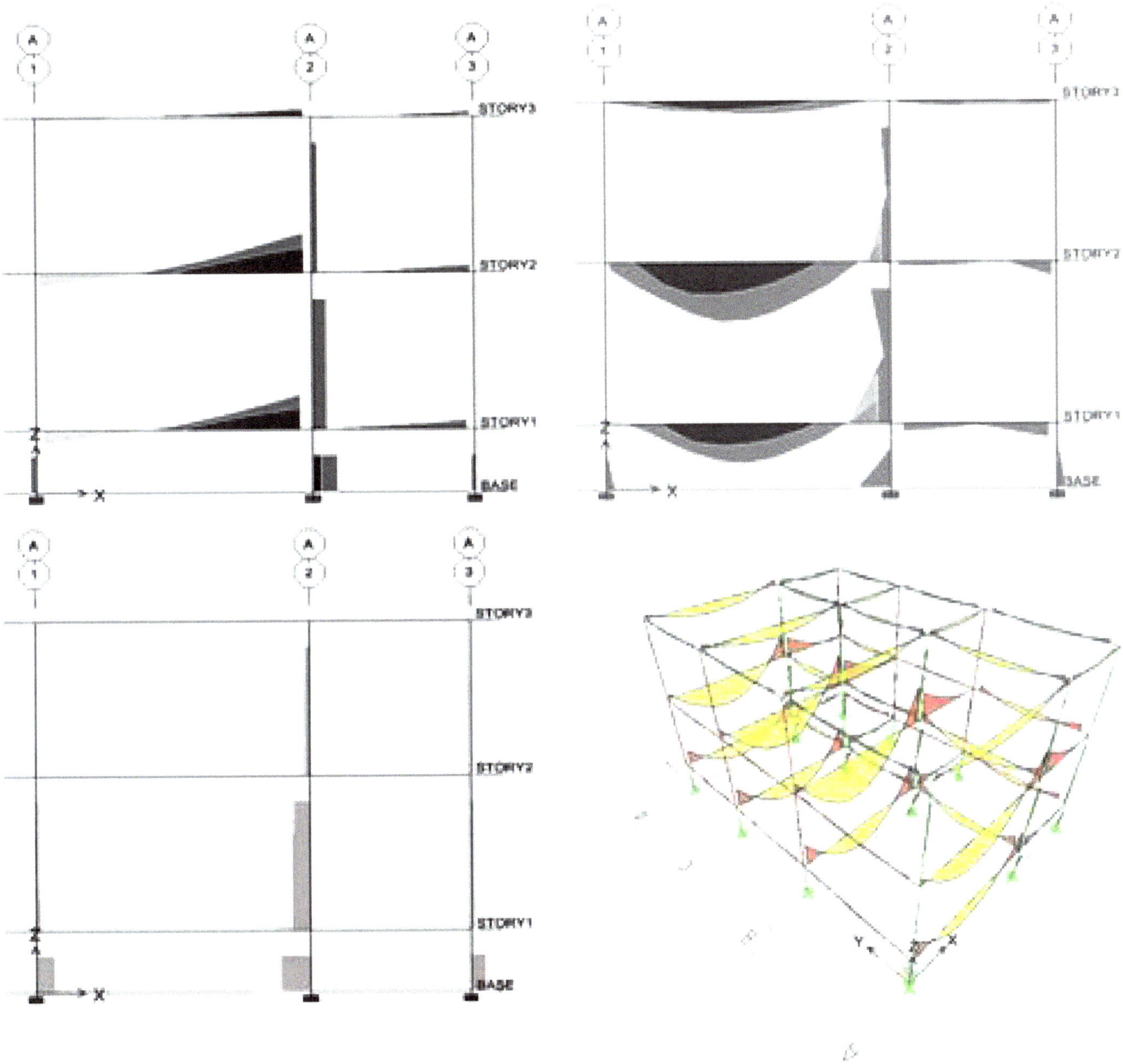

Figure 6. Graphical analysis of the structural frame under load combination.

according to the designing code. Deformed shape of the structure after applying the loads is illustrated in Figure 9. Subsequently, the vertical and horizontal replacements of the beams and columns have been controlled according to ETABS output results and UBC97 criterion.

Figure 10 shows the support reaction driven from the ETABS output files. This evaluation will be applied for future footing design.

Finally, Beam/Column Capacity Ratio was checked in order to obtain the optimum sections.(6/5) Beam/Column Capacity Ratios, shows that the ratio of beams capacity to columns capacity is less than 1.0 in all stories. Figure 11 displays two sample elevations for clarification.

d) Previous studies

Various researchers have applied ETABS software for analyzing the structural elements of the medium and high rise building. Results from similar studies reveal the

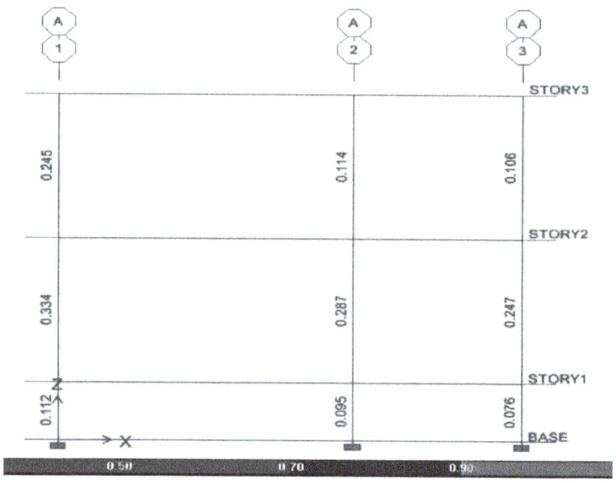

Figure 7. Section from Column's internal forces in the double story sample building.

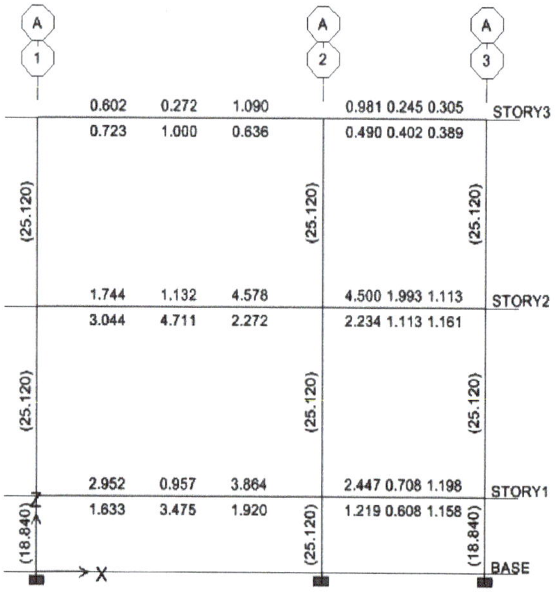

Figure 8. Section from longitudinal reinforcing.

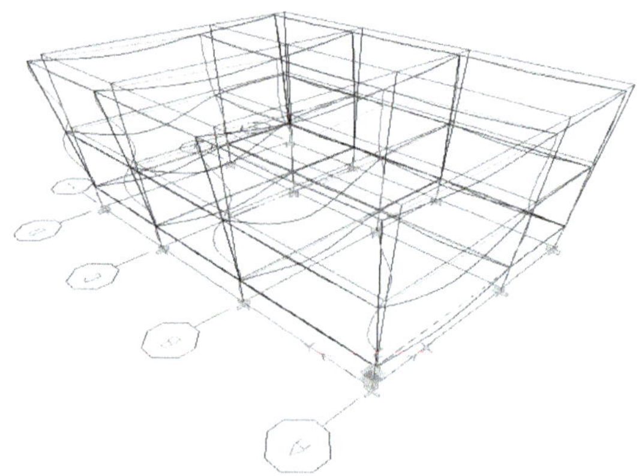

Figure 9. Deformed shape of the structure.

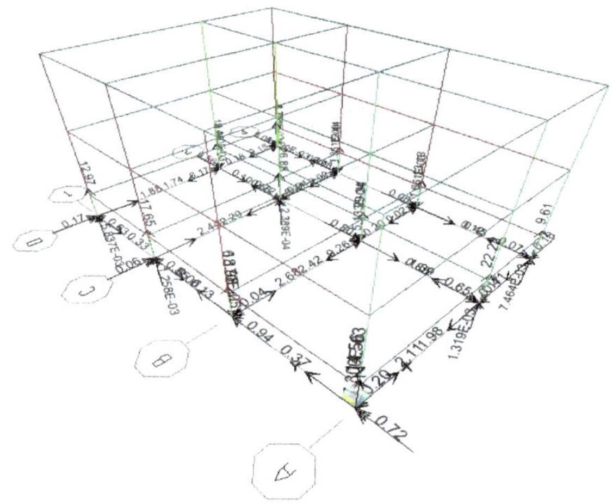

Figure 10. Support reactions for the columns.

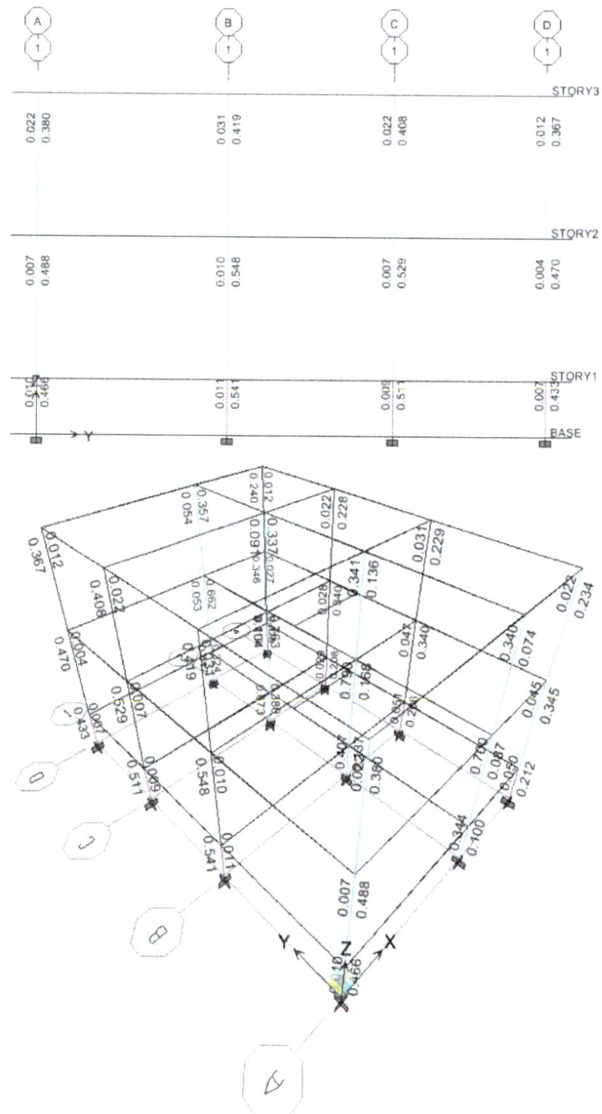

(a)

(b)

Figure 6. (6/5) Beam/Column Capacity Ratios.

applicability of the software and supports the findings from current study.

Kilar and Fajfar, (1996) have analyzed the structure of the seven story reinforced concrete frame-wall building using ETABS. They discussed that asymmetric structures require larger displacements and ductility to develop the same strength as in the symmetric structure remarkably at the flexible side of the building. In fact, at the same displacement, the strength of the symmetric structure is bigger than the strength of the asymmetric one.

Henry (2006) analyzed the flexural and shear capacity of a 5 story office building with ETABS. The study revealed that the critical shear stress from moment transfer in the interior columns were exceed. Consequently, the interior columns under greater lateral loads were upsized to intensify the shear perimeter and reduce shear stresses.

Conclusions

Passive design of the house and application of new construction techniques will make it suitable for current and future requirements of the occupants and the community. This objective can be achieved by using durable materials and fabricated systems, as well as designing flexible and cost-effective layouts to fulfill changing requirements of the occupants and their life style.

This paper presents the process for development and assessment of IFD building components. In an attempt to evolve a new and innovative system, a number of criteria have been studied for the theoretical performance as well as simulation analysis for the structural performance. The main features of innovative layout design are:

- The dimensions satisfy the modular coordination requirements.
- The simple shape of the components will result in easy production and easy assembly of the building.
- The interlocking and connection mechanism will help for efficient assembly.
- Self-aligned construction system ensures accurate and fast construction.

Feasible IFD design requires encouragements such as specific assessment tools for measuring deconstructability of the building elements as well as governmental policy supports for reducing cost and environmental effects. The offered design needs still further development and the detailing principles that will be set up in the next steps of the study.

ACKNOWLEDGEMENT

The authors would like to thank National University of Malaysia, for financial support under GUP project TK-08-16-062.

REFERENCES

Ball R (2002). Re-use potential and vacant industrial; premises: revisiting the regeneration issue in stoke-on-trent. J. Property Res., 19(2): 93-110.

Bon R, Hutchinson K (2000). Sustainable construction: some economic challenges. Building Res. Inform., 28(5/6): 310-4.

Brand S (1994). How buildings learn: what happens after they're built. New York: Viking Press.

Bullen P (2007). Adaptive reuse and sustainability of commercial buildings. Facilities, 25(1/2): 20-31.

Chini AR (2002). Anticipating and responding to deconstruction through building design. Design for Deconstruction and Material Reuse, CIB Publication 272, Proceeding of the CIB task group 39- Deconstruction meeting.

CSI (2005). Analysis Reference Manual for SAP2000®, ETABS® and SAFE™. Berkeley, California, USA, Computers and Structures, Inc.: 415.

Davison N, Gibb AG, Austin SA, Goodier CI (2006). The Multispace adaptable building concept and its extension into mass customisation. Int. Conf. on Adaptable Building Structures.

Dorsthorst BJHt, Kowalczyk T (2002). Design for recycling. Design for Deconstruction and Material Reuse, CIB Publication 272, Proceeding of the CIB task group 39- Deconstruction meeting.

Durmisevic E (2006). Transformable building structures. Delft University of Technology.

Egmond ELCV, Scheublin FJM (2005). Successful Industrialization, innovation and prefabrication in construction. COMBINING FORCES - Advancing Facilities Management and Construction through Innovation.

Gallant BT, Blickle FW (2005). Managing redevelopment of brownfields with major structures. Environ. Prac., 7(2): 97-107.

Gassel FV (2003). Experiences with the Design and Production of an Industrial, Flexible, and Demountable (IFD) Building System, NIST special publication SP.

Ghoulbzouri AE, Khamlichi A, Bezzazi M, Mourabit T, López-Almansa F (2009). Reliability analysis for seismic performance assessment of reinforced concrete buildings 3rd International Conference on Integrity, Reliability and Failure.

Greden VL (2005). Flexibility in Building Design: A real options approach and valuation methodology to address risk. Massachusetts Institute of Technology, pp. 15-16.

Gregory J (2004). Rehabilitation-new ways for older housing. New South Wales, Department of Housing, pp. 30-35.

Hamid Z, Kamar KAM, Zain M, Ghani K, Rahim AHA (2008). Industrialized Building System (IBS) in Malaysia: the current state and R&D initiatives. Malaysia Construction Res. J., 2(1): 1-13.

Henry J (2006). Structural Emphasis, Depth analysis: Concrete design Signal Hill Professional Center, pp. 13-29.

Holtz AJ (2006). London Prototype House: A Flexible Design Alternative for Accomodating Change. PLEA- The 23rd Conference on Passive and Low Energy Architecture.

Kilar V, Fajfar P (1996). Simplified push over analysis of building structure. Eleventh world conference of earthquake engineering.

Kohler N, Hassler U (2002). The building stock as a research object. Building Res. Information, 30(4): 226-36.

Morgan C, Stevenson F (2005). Design for Deconstruction SEDA Design Guides for Scotland: No. 1, Scottish Executive and the Scottish Ecological Design Association, Edinburgh (available at: www.seda2.org/dfd/index.htm).

Paduart A, Debacker W, Henrotay C, Asnong K, Wilde WPD, Hendrickx H (2008). Technical detailing principles for the design of adaptable and reusable construction elements in temporary dwellings. WIT Transactions on Ecology and Environment, Vol 109.

Razaz ZE (2010). Design for dismantling strategies. Building Appraisal, 6(1): 49-61.

Slaughter ES (2001). Design strategies to increase building fexibility. Building Res. information, 29(3): 208-217.

Sunikka M, Boon C (2003). Environmental policies and efforts in social housing: the Netherlands. Building Res. Information, 31(1): 1-12.

Tam VWY, Tam CM, Zeng SXCY, Ng W (2006). Towards adoption of prefabrication in construction. Building Environ., 42(2007): 3642-3654.

Zegers SFAJG, Herwijnen FV (2004). Development of an Industrial, Flexible and Demountable (IFD) floor system. Plea2004- The 21th Conference on Passive and Low Energy Architecture.

17

Organic matter and trace metals levels in sediment of bonny river and creeks around Okrika in Rivers State, Nigeria

Marcus A. C.[1], Okoye C. O. B.[2] and Ibeto C. N.[3]

[1]Department of Chemistry, Ignatius Ajuru University of Education, P. M. B. 5047, Rumuolumeni, Port Harcourt, Rivers State, Nigeria.
[2]Department of Pure and Industrial Chemistry, University of Nigeria, Nsukka, Enugu State, Nigeria.
[3]National Centre for Energy Research and Development, University of Nigeria Nsukka, Enugu State, Nigeria.

Analysis was carried out on the sediment of Bonny River and creeks around Okrika in order to determine the concentrations of organic matter, lead, nickel, vanadium and cadmium as well as their probable sources and diagenesis. Samples for trace metal analysis were prepared by acid digestion and analysed by atomic absorption spectrophotometry. Organic matter content was determined by the method of Walkey and Black. Mean concentrations (µg/g, dry wt) of trace metals were: Pb(23.22 ± 13.43), Ni (57.19 ± 16.93), V (0.06 ± 0.05), Cd (0.24 ± 0.32) while organic matter (% dry wt) were: TOC(1.14 ± 0.40) and TOM(1.990 ± 0.70) and revealed anthropogenic trace metal enrichment. The large standard deviations indicate temporal and spatial variations due largely to anthropogenic waste input. There was a weak association for Pb/organic matter (/r/=0.58, p<0.01) in the wet season suggesting a common source, most likely from land-derived wastes brought in by runoff water.

Key words: Organic matter, trace metals, anthropogenic enrichment, Bonny River.

INTRODUCTION

Trace metals in sediments can play a major role in the pollution scheme of a river system. Sediments are repositories for physical debris and sink for chemical contaminants. They can therefore be used to detect pollutants that escape water analysis and also provide information about the critical sites of the river system. River sediments are a major potential sink for hydrophobic pollutants in the aquatic environment (Kreina et al., 2003; Schorer, 1997). The organic matter content of river sediment has been shown to be an important factor in determining the extent of sorption (Calmano and Fôrstner, 1996). It has been assumed that the efficacy of inorganic exchange sites of clay and its associated organic matter are responsible for the amount and behaviour of the sorbed substances (Karickhoff and Brown, 1978). Bonny River is a recipient of effluent/wastewater from the Port Harcourt Refining Company (PHRC). The company discharges directly into Ekerekana creek, one of the creeks associated with Bonny River. From this discharge outfall, the waste spreads into other creeks and down Bonny River. The studied area is a riverine and intertidal wetland on the north bank of Bonny River, about 35 miles (56 km) upstream from the Bight of Benin in the Eastern Niger

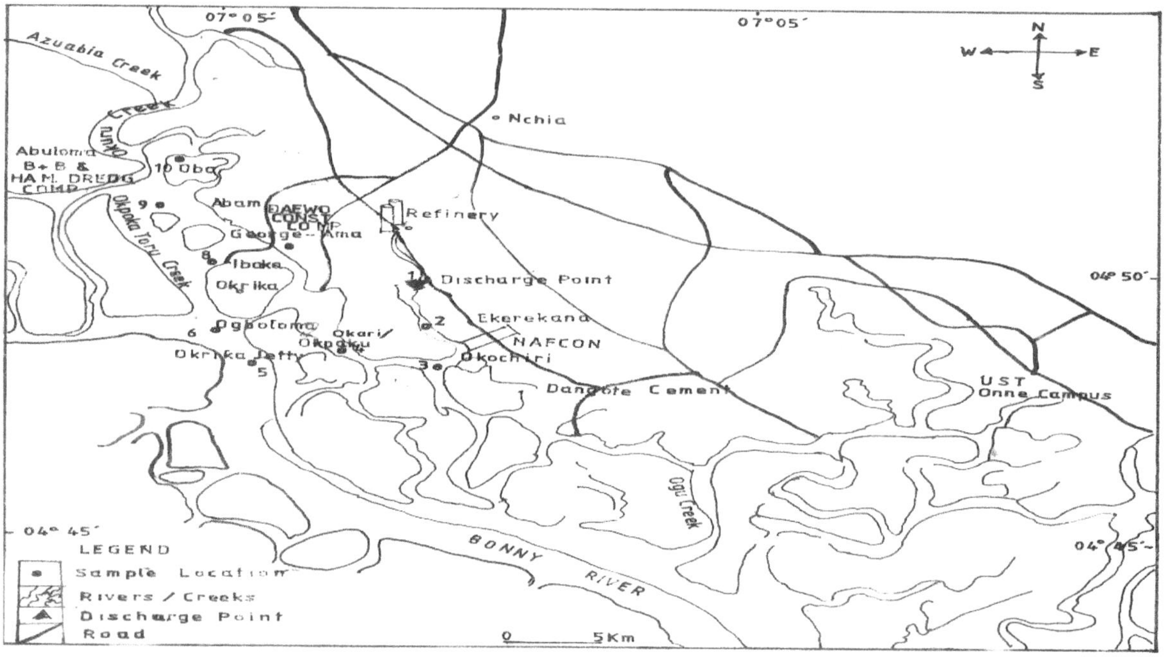

Figure 1. Map of Bonny River and Creeks around Okrika showing sampling locations.

Delta of Nigeria. This area like most parts of the Niger Delta has suffered extensive environmental degradation due to oil exploration and exploitation. The oil industry has attracted many other industries and economic activities and urban population leading to a considerable volume of waste production. Unfortunately the wastes are discharged indiscriminately. Direct and indirect discharge of wastes including domestic, refinery and other kinds of industrial wastes and from land-derived sources, could have great impact on the quality of water in Bonny River and its associated creeks. Many municipal and industrial wastes are rich in trace toxic metals (Yusuf et al., 2003). Rainfall in the coastal belt of the Niger Delta usually starts from March, reaching the peak between June/August and ending in November. During the rains, runoff water collects a lot of wastes from diffuse non-point sources from far and wide into the aquatic system.

Oil prospecting activities and sewage are considered the major sources of organic pollution that can adversely affect fish and other aquatic life principally through dissolved oxygen (DO) depletion, when it occurs in the aquatic system at concentrations higher than normal. Apart from depleting DO, decomposition brings about hazardous consequences on the aquatic life (Saad and Badr, 2012). Organic pollution causes various changes in the chemical and physical properties of water such as pH, turbidity, sedimentation rates, organic matter content, decomposition products, nutrient salt supply and other biologically active materials. This paper therefore examines organic matter, lead, nickel, vanadium and cadmium contents of sediment of Bonny River and creeks around Okrika in Rivers State, and evaluates their sources and seasonal diagenesis.

MATERIALS AND METHODS

Sample collection

Sampling was carried out from 10 designated locations in the creeks and rivers as shown in Figure 1. Surface sediment samples were collected at low tide by the grab method using Eckman grab sampler from 3 to 4 points at each location (APHA, 1975). This was done at two-monthly intervals between December 2008 and October, 2009, covering dry and rainy seasons of the year. The samples for trace metals were put in polythene bags previously washed in dilute HCl while those for organic matter were put in aluminium foils and stored in the laboratory by freezing (APHA, 1975).

Sample treatment and analysis

Sediment samples were allowed to thaw and air-dried at ambient temperature. They were sieved through 0.5 mm sieve and 2 g samples were digested using 25 ml 1:3:1 mixture of $HClO_4$, HNO_3 and H_2SO_4 acids in water bath. To the digest, 10 ml deionized water was added and decanted into 50 ml standard flask and after rinsing, the solution was made up to the mark with deionized water. Buck Scientific atomic absorption spectrophotometer model 200A and air-acetylene flame were used for trace metal analyses. Accuracy of the method was checked using a standard sediment sample PACS-2. For each batch of elemental analysis, an intra-run Quality Assurance Standard (1 mgl^{-1}, Multi-Element Standard Solution, Fisher Scientific) was analysed to check for reading deviation and accuracy of every 10 samples (Cantillo and Calder, 1990). For organic carbon, 1 g dried, sieved sediment was weighed into

Table 1. Results of recovery analysis using standard sediment material (PACS-2).

Value type	Lead	Nickel	Vanadium	Cadmium
Certified value	90.70	183.00	39.50	364.00
Measured value	86.30 ± 4.40	174.00 ± 9.00	37.40 ± 2.10	340.00 ± 24.00
Recovery (%)	94.38	95.08	94.68	93.41

Table 2. Average bimonthly (n=6) concentrations of trace metals (µg/g, dw) and organic matter (%, dw) in sediment of Bonny River and creeks around Okrika.

S/N	Parameter	Sampling locations										Ov.mean±SD
		PRE	EKC	OKC	OOC	OBR	OGR	GAC	IBC	OTR	OAC	
1	Lead	4.21	2.17	33.82	19.06	37.44	26.85	11.69	27.81	29.03	40.10	23.22 ± 13.43
2	Nickel	51.87	46.47	62.41	86.58	44.03	46.75	31.51	81.00	58.19	63.13	57.19 ± 16.93
3	Vanadium	0.02	0.07	0.02	0.01	0.15	0.07	0.13	0.03	0.02	0.08	0.06 ± 0.05
4	Cadmium	0.04	0.44	0.20	0.06	1.01	0.09	0.02	0.02	0.25	0.17	0.24 ± 0.32
5	TOC (%)	0.48	0.75	1.43	1.01	1.40	1.20	0.97	0.80	1.54	1.77	1.14 ± 0.40
6	TOM (%)	0.83	1.23	2.47	1.74	2.42	2.07	1.68	1.39	2.65	3.073	1.99 ± 0.70

PRDC - Port Harcourt Refinery Discharge Channel, PRE - Port Harcourt Refinery Effluent/Wastewater Outfall, EKC - Ekerekana Creek, OKC - Okochiri Creek, OOC - Okari/Okpaku creek, OBR - Okrika/Bonny River, OGR - Ogoloma River, GAC - George Ama Creek, IBC-Ibaka Creek, OTR - Okpoka-Toru or Okpoka River, OAC - Oba Ama Creek, OV. mean - Overall mean, dw -dry wt.

250 ml conical flask and digested with 10 ml 0.5 M $K_2Cr_2O_7$ and 20 ml conc H_2SO_4, for 30 min (Walkey and Black, 1934). 100 ml deionized water and 3 drops of Ferroin indicator were added and titrated with 0.25 M $FeSO_4$ solution. Organic matter was obtained by multiplying organic carbon values by 1.724.

RESULTS AND DISCUSSION

Results of the recovery analysis as shown in Table 1 were in the range of 93.41 to 95.08%, for all the trace metals which shows good recovery and accuracy. The average concentrations of trace metals and organic matter in the sediment are presented in Table 2. The mean values of trace metals (µg/g, dry weight) and organic matter (%) in the year were: Pb (23.22 ± 13.43), Ni (57.19 ± 16.93), V (0.06 ± 0.05), Cd (0.24 ± 0.32), TOC (1.14 ± 0.40) and TOM (1.99 ± 0.70). The detection limits (µg/g) for all the trace metals are as follows: lead (0.005), nickel (0.002), vanadium (0.002) and cadmium (0.002). As shown in Table 2, the levels of nickel and lead were appreciably high when compared with the other two metals. Lead showed average bimonthly concentrations (n=6) of 2.17 to 40.10 and overall mean of 23.22 ± 13.42, while average bimonthly concentrations of nickel were; 31.51 to 86.57 and overall mean was 57.19 ± 16.92. Their occurrences at appreciable levels in most samples suggest widespread contamination arising from many diffuse non-point sources other than refinery effluent waste water. Vanadium and cadmium were very low with mean values of 0.06 and 0.24 µg/g, dry wt, respectively. Organic matter was also very low 0.48 to 1.77%. Generally, all the analytes occurred at low levels and this is attributable to high self purification capacity. Rivers are "open" systems with continual renewal of water, and therefore have high capacity for self-purification due to oxygen diffusion at the surface, together with dilution, which enables the water to assimilate polluting discharges. Consequently, in the sediment, all trace metals investigated recorded low concentrations.

The mean seasonal levels of the metals and organic matter are given in Table 3. Both the metals and organic matter had higher values in the wet season except nickel, but the seasonal levels were only significant ($p<0.05$) for lead and TOM. The absence of significant seasonal differences ($p>0.05$) in the concentrations of trace metals except lead and TOM, may be explained as resulting from high flushing and dilution rates during the rains as well as "solution effect" which according to Welcome (1986) and King and Nkata (1991), a process whereby ions bound in previous semi-dry land by decaying macrophytes get dissolved as water levels increased with inundation of fringing swamps and riparian zones.

Table 4 shows the levels of trace metals (ppm, dry weight) in sediment compared with Netherlands's class limits. The average percentage organic matter is about 3.32%, much below the 10% maximum stipulated for the use of this table. It is clear that all metal levels fall within Classes 1 and 2 which are within acceptable levels inspite of heavy waste loadings from human and sundry sources. However, there is concern with regards to nickel which falls within Class 2, the slightly polluted class.

Table 3. Mean seasonal concentrations of trace metals (µg/g, dw) and organic matter (%, dw) in sediment of Bonny River and creeks around Okrika.

S/N	Metals	Dry season	Wet season
1	Lead	(0.075 - 69.525) 19.472 ± 19.146	(0.069 - 67.356) 29.965 ± 22.074
2	Nickel	(0.075 - 105.775) 60.027 ± 27.827	(0.500 - 102.525) 54.360 ± 29.992
3	Vanadium	(BDL - 0.275) 0.117 ± 0.052	(BDL - 0.875) 0.331 ± 0.226
4	Cadmium	(BDL - 0.863) 0.247±0.305	(BDL - 0.650) 0.512 ± 0.699
5	TOC (%)	(0.341 - 2.141) 1.048 ± 0.418	(0.040 - 3.202) 1.189 ± 0.691
6	TOM (%)	(0.588 - 3.691) 1.747 ± 0.739	(0.641 - 5.520) 2.095 ± 1.142

dw - dry wt.

Table 4. Levels of trace metals (µg/g, dw) in sediment of Bonny River and creeks around Okrika compared with Netherlands's class limits for sediments with organic matter content less than 10%.

Metal	Bonny River and Okrika Creek sediment	Class limits in Netherlands			
		Class 1	Class 2	Class 3	Class 4
Lead	23.22	<530	-	-	>530
Nickel	57.19	<35	35 - <90	45 - < 210	>210
Vanadium	0.06	-	-	-	-
Cadmium	0.13	<2	2 - <7.50	7.50 - > 12	>12

Notes: Class 1 = Excellent; Class 2 = Acceptable; Class 3 = slightly polluted; Class 4 = polluted. Source: EWA (2005).

Matrices of correlation coefficient /r/ in sediment given in Tables 5a and b showed low levels of association between the metals and signify both seasonal and spatial variations. Only the pairs of Pb/TOC and Pb/TOM showed any reasonable association ($p<0.01$) in the wet season. The correlation between TOC, TOM and Pb in the wet season suggests that an important source of lead in the water is organic loading by runoff water. The spatial and seasonal variations leading to lack of associations in the metal-metal correlation matrices could only be ascribed to substantial anthropogenic input of metals into the water from direct waste discharge from domestic and industrial sources and many diffuse non-point sources.

Nickel values determined in the present study compares well with values reported for Iko river sediment in Akwa Ibom State (Benson and Etesin, 2008), but higher than those of Lagos lagoon (Okoye et al., 1991). The levels of lead, nickel and cadmium were all higher than those previously obtained in Niger Delta and Lagos Lagoon. Apart from direct discharge of industrial and domestic wastes, the immense volume of storm water runoff and river waters entering the Bonny River and creeks around Okrika could transport metals originating from wastes discharge on land and in small streams, especially during the rains, thus resulting in the higher levels obtained for most of the metals in the sediment in the wet season.

Conclusion

The sediments of Bonny River and creeks around Okrika

Table 5. Correlation matrices of trace metals and organic matter in sediment of Bonny River and creeks around Okrika for (a) dry season and (b) wet season.

Metal	Pb	NI	V	Cd	TOC	TOM
			(A)			
Lead	1					
Nickel	-0.20297	1				
Vanadium	0.294878*	-0.35906**	1			
Cadmium	0.258518	-0.21362*	-0.07992	1		
TOC	0.504208	-0.07938**	0.027149	0.12231	1	
TOM	0.472626	0.001751**	0.030523	-0.11363	0.936481**	1
			(B)			
Lead	1					
Nickel	-0.19358	1				
Vanadium	0.285027	-0.48197**	1	.		
Cadmium	a.	a.	a.	a.	.	.
TOC	0.569866**	-0.10655	0.054137	.a	1	
TOM	0.581137**	-0.04885	0.036772	.a	0.961787**	1

are enriched with trace metals mainly due to direct input of industrial and domestic wastes and indirect input via tributary rivers and runoff waters. Anthropogenic metal inputs are the major sources of trace metals in the sediment. However, rapid sedimentation appeared to have played major role in accumulation of nickel in the bottom sediment.

REFERENCES

APHA (1975). American Public Health Authority Standard methods 14th edition. Assoc. Wat. Poll. Fed. (APHA-AWWA-WPCF). Washington D.C. USA.

Benson BU, Etesin UM (2008). Metal contamination of surface water, sediment and *Tympanotus fiscatus var. radular* of Iko River and environmental impact due to Utapete gas flare station, Nigeria. Environmentalist 28:195-202.

Calmano W, Fôrstner U (1996). Sediments and Toxic Substances. Springer, Berlin. ISBN 3-540-60051-5. XII, 335 pp.

Cantillo A, Calder J (1990). Reference material for marine science. Fresenius J. Anal. Chem. 338:380-382.

EWA (2005). Official Publication of the European Water Association (EWA). Netherlands. P. 2.

Karickhoff SW, Brown DS (1978). Paraquat sorption as a function of particle size in natural sediments. J. Environ. Qual. 7:246-252.

King RP, Nkanta NA (1991). The status and seasonality in the physicochemical hydrology of a Nigerian rain forest pond. Jpn. J. Limnol. 52:1-12.

Kreina A, Petticrew E, Udelhoven T (2003). The use of fine sediment fractal dimensions and colour to determine sediment sources in a small watershed. Catena 53:165-179.

Okoye COB, Afolabi OA, Ajao EA (1991). Heavy metals in the Lagos lagoon sediments. Inter. J. Environ. Stud. 37:35-41.

Saad MAH, Badr NBE (2012). Dissolved Oxygen and Organic Matter in a Harbor and a Bay in Egypt receiving Organic Pollutants from different sources. CM2005/Z: 13. Accessed from http://www.ices.dk/products/CMdocs/2005/Z/Z1305.pdf on 12th of December, 2012.

Schorer M (1997). Pollutant and organic matter content in sediment particle size fractions Freshwater Contamination (Proceedings of Rabat Symposium S4, April-May 1997). IAHS Publ. 243:59.

Walkey A, Black LA (1934). An examination of the D*agtjareff* method for determination of soil organic matter and a proposed modification of the chromic acid titration method. Soil Sci. 37:29-38.

Welcome RL (1986). The Niger River system. Longman Group Ltd, London. pp. 9-23.

Yusuf AA, Arowolo TA, Bamgbose O (2003). Cadmium, Copper and Nickel levels in vegetables from industrial and residential areas of Lagos City, Nigeria. Food Chem. Toxicol. 41:375378.

Econometric modeling and projection of production, import and export of particle board industry in Turkey

Yildiz Cabuk, Selman Karayilmazlar, S. Murat Onat and Rifat Kurt*

Department of Forest Industrial Engineering, Faculty of Forestry, Bartin University, 74100 Bartin-Turkey.

Although wood is one of the oldest materials used by humans, the production of wood-based panel products has a very short history. The production of panel products constitutes one of the most important developments in forest products industry. In this study, models are established and projections are developed for production, import and export of Turkish Particle board industry by econometric method. As a result of regression analysis performed, it is seen that the Particle board production above 2.5,000,000 m^3 in 2006 shall exceed 4.5,000,000 m^3 in 2021 and the export of around 175000 m^3 shall exceed 250,000 m^3 and import of 175,000 m^3 shall reach 350000 m^3.

Key words: Particle board trade, production, import, export, regression analysis.

INTRODUCTION

The first particle board plant was established in Istanbul in 1955 and the second one was integrated in a plant in Halkali in 1960. In 1972 another plant was built in Isparta for manufacturing three layer particle board from pine chips. From 1970, the building sector grew very rapidly and furniture demand increased significantly resulting in rapid developments in particle board industry and in 1982 the number of plants reached 27 and the production level was 1,600,000 m^3. In 1993, the number of plants was 28 having annual capacity of 2,000,000 m^3. In the later years as parallel to developments in the world, particle board industry in Turkey grew rapidly having 30 plants and 2,500,000 m^3 annual capacities.

Four of the plants were previously owned by the state and they were privatized. The majority of the plants were using standard horizontal press method. These facilities were established according to Siempelkamp, Bison Böhre, Fahrni and Pagnomi machinery and systems. In addition, there is one plant established based on vertical extrusion machinery and system of OKAL type (DPT, 1995). In Turkey, there is also one OKAL type particle board plant, two particle-cement board plants and others produce standard platen type particleboard. Almost all of the production is sold domestically. The major consumers are furniture manufacturers (70%), decorators (15%) and others (Akbulut, 2000).

In 1997, construction business was boomed and particle board production reached 1.7 million m^3. But for 1998 and 1999 the demand for particle board decreased. Domestic sales were raised by 20% in 2000. Twenty-four percent of the export for the same year was carried out to Bulgaria, 17% to Macedonia, 15% to Northern Cyprus and 11% to Georgia. 55% of the total exported particleboards were laminated type (Yaman, 2002).

Particle board industry in Turkey has shown significant improvements in recent years, in parallel with the development and growth of especially the furniture industry. This development has been seen as improvement of much of the existing investments product structure in the form of expansion to bring new features in the sector rather than the formation of new investments. Presenting laminated particle board instead of similar products like medium density fiberboard (MDF) on the market brought significant ease of use for small and medium-sized businesses (Koc and Aksu, 1999).

As it can be seen in Table 1 and Figure 1, the average particle board production between the years of 1982 to 1983 was about 330,000 m^3, between 1984 and 1991 it

*Corresponding author. E-mail: rkurt@bartin.edu.tr.

Table 1. The production, export and import amounts of Particle board panels in Turkey (m^3) (FAO, 2008).

Year	Production	Export	Import	Year	Production	Export	Import
1982	330000	0	0	1995	1243000	51000	44000
1983	330000	0	0	1996	1193000	30000	17000
1984	546000	0	0	1997	1728000	31000	25000
1985	636000	0	0	1998	1525000	27000	19000
1986	636000	0	0	1999	1643000	30000	14000
1987	636000	0	0	2000	1884000	32000	219000
1988	636000	0	0	2001	1664000	64000	33000
1989	636000	0	0	2002	1999000	99000	77000
1990	636000	0	0	2003	2264000	111800	156000
1991	636000	0	0	2004	2700000	172419	272791
1992	947000	10954	10839	2005	2890000	284227	270736
1993	883000	3000	44300	2006	2750000	178000	179000
1994	898000	32600	64000				

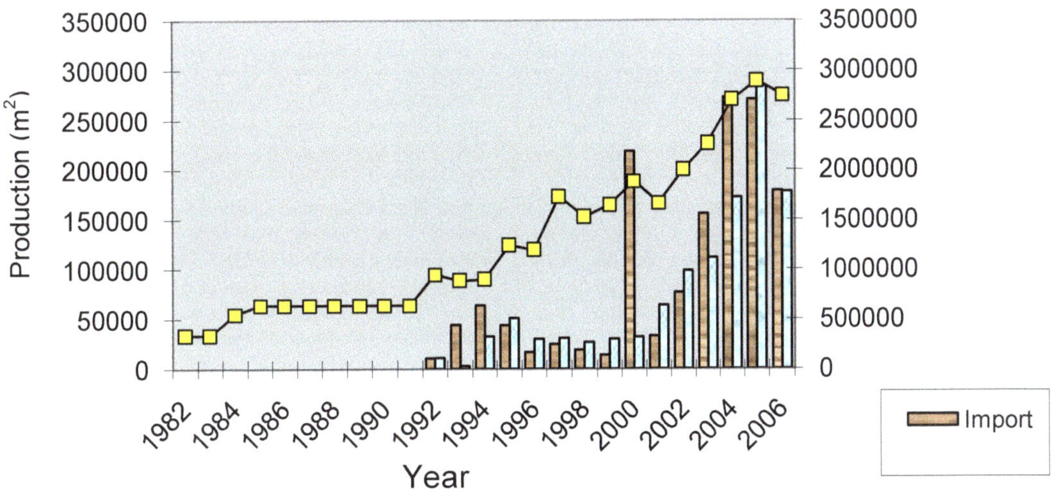

Figure 1. The production, export and import amounts of particle board panels in Turkey.

was 625,000 m^3, between 1992 and 1994 it was 910,000 m^3, between 1995 and 2001 it was 1,555,000 m^3, in 2002 it was 2,000,000 m^3 and 2003 to 2006 it was 2,650,000 m^3. Exports began in 1992, and remained until year 2000 as 27,500 m^3/year. In 2001, it was around 64,000 m^3, from 2002 to 2003 it was 105,000 m^3 and from 2004 to 2006 it was approximately 210,000 m^3. In the same way, the average of 1992 to 1999 years imports was around 30,000 m^3. In 2000 it went up to 219,000 m^3 and in 2001 it fell to 33,000 m^3. In 2003, it increased again and reached to 156,000 m^3. Since 2004, import was, on average, 240,000 m^3.

On the other hand, in 1982 particle board production amount was 330,000 m^3 as the minimum, in 1992, there was an increase of 187% (2.87-fold) to 947,000 m^3, 1,243,000 m^3 in 1995, 82% (1.82-fold) increase in 2003 to 2,264,000 m^3, and 21.5% increase in 2006 when reached 2,750,000 m^3.

Exports started in 1992 as 10,954 m^3, and in 2003, it was 111,800 m3 with 920% increase. In 2005, it reached 284,227 m^3 with 154% (2.54-fold) increase. However, in 2006 the amount of export was realized as 178,000 m^3 compared to 284,227 m^3 in 2005 meaning 37.37% of decline. Similarly, in 1992, the amount of imports was 10,839 m^3. In 2000, there was 1920% or about 20-fold increase as 219,000 m^3. In 2003, it then decreased by 29%, or to 156,000 m^3. The amount of imports increased to 272,791 m^3 in 2004, and fell again in 2006; it was declined to 179,000 m^3 (about 34.4%).

MATERIALS AND METHODS

Model building and regression analysis

Basic econometric method used in this study is multiple regression modeling. By applying this method in the present study, the aim is

Table 2. Population, GNP and GDP of Turkey (TUIK, 2008).

Year	Population (000)	Per capita (TL)	GNP ($)	Per capita (TL)	GDP ($)	Year	Population (000)	Per capita (TL)	GNP ($)	Per capita (TL)	GDP ($)
1982	46,688	227,293	1,375	224,730	1,360	1995	61,644	127,423,385	2,759	125,923,952	2,727
1983	47,864	291,096	1,264	290,528	1,261	1996	62,697	238,896,076	2,928	235,611,117	2,888
1984	49,070	451,758	1,204	448,281	1,195	1997	62,480	470,442,977	3,079	461,522,054	3,021
1985	50,306	702,706	1,330	697,640	1,320	1998	63,459	843,358,573	3,255	822,976,986	3,176
1986	51,433	995,174	1,462	993,124	1,459	1999	64,345	1,216,609,421	2,879	1,203,124,428	2,847
1987	52,561	1,427,282	1,636	1,421,623	1,629	2000	67,461	1,861,759,072	2,965	1,846,747,873	2,941
1988	53,715	2,404,824	1,684	2,405,743	1,685	2001	68,618	2,571,977,513	2,123	2,600,082,172	2,146
1989	54,893	4,196,709	1,959	4,141,220	1,933	2002	69,626	3,950,138,827	2,598	3,986,643,746	2,622
1990	56,203	7,066,839	2,682	6,993,580	2,655	2003	70,712	5,044,135,199	3,383	5,087,720,980	3,412
1991	57,305	11,070,462	2,621	10,995,846	2,603	2004	71,789	5,974,903,440	4,172	5,996,900,319	4,187
1992	58,401	18,897,021	2,708	18,721,735	2,682	2005	72,065	6,749,476,615	5,008	6,760,596,160	5,016
1993	59,491	33,573,525	3,004	33,313,730	2,981	2006	72,974	7,890,261,766	5,477	7,897,637,938	5,482
1994	60,576	64,182,233	2,184	63,860,757	2,173						

to show relationship of one dependant variable and multiple independent (explanatory) variables over some certain past period and, accordingly, to make projections on present and future quantity of a dependant variable at an acceptable confidence level.

For establishment of the most appropriate regression models for the projection operations, wherein the particle board production, import and export were dependant variables, and the industrial wood sales (m^3) from general directorate of forestry (OGM), gross national product per capita (TL and USD, separately), population, number of buildings by area (m^2) as per the occupancy permit; number of buildings constructed as per the occupancy permit, inflation rate (on annual consumer price index (CPI) and producer price index (PPI) basis), exchange rates (USD), economic growth rate, construction materials price index, gross domestic product (GDP) per capita, timber sales (m^3) by general directorate of forestry (TL and USD, separately) were used as independent variables, all of which are considered to be effective in the production, import and export quantities of the forest industry products. Parameters of the econometric modeling rest on time series of past 25-years and projection was made for the next 15 years around on basis of a variety of reasonable assumption and scenarios.

The data in question were obtained either by direct access to or via websites of Turkish Statistics Institute (TÜİK), Undersecretariat of Foreign Trade (DTM, 2008), state planning organization (DPT 2006), Export Development Center, Ministry of Industry and Trade (IGEME, 2008), World Food and Agricultural Organization (FAO), Forest Stewardship Council (FSC, 2008), and Turkish General Directorate of Forestry (OGM, 2008). Furthermore, some informations and documents of the organizations operating in the sector, the records of Turkish Association of Chambers and Exchanges (TOBB, 2001, 2007) and particle board Industrialists Society and websites of the organizations and enterprises having direct or indirect relation to the subject of the study were all used.

Twenty-five years (1982 to 2006) data on the aforementioned independent and dependant variables are organized in independent variables (Tables 2 to 4) and transferred to the computer environment for multiple regression analysis to be conducted at SPSS statistical package program. Information about calculations made for missing or unavailable data are given under the tables. It is seen that export and import figures for the years 1982 to 1991 given in Table 1 are zero. This situation does not mean that no data was found for the said years, but shows the real status. In other words, zero values for some years show that import and export of Turkey was taken as zero as they are actually or very small or negligible level.

Furthermore, the economic growth rate (%), one of the independent variables given in Tables 2 and 4 has caused some difficulties with the fixed prices and gross national product (GNP) per capita (TL) in all models. It was considered that the reason is that both it causes multiple linear connection when the economic growth rate (%) that may replace these variables is used together with current prices and GNP per capita ($) and it is expressed by very high figures, thus the coefficients in the equations appear as zero. Consequently, when searching appropriate model, the said variables were not used together, but individually and the significant and valid variable out of them has taken its place in the model.

RESULTS AND DISCUSSION

After building the most appropriate regression models for projection (3 distinct models for particle board production, import and export), forecast values for independent variables applicable for each model for the next 15 years were obtained (year) based on the time series and projection values were calculated on basis of these figures.

Regression analysis results of particle board industry (production-import-export)

Particle board production

As it may be seen in the summary Table 5 given that both regression models, one built with one independent variable (Population), and the other with two independent variables (Population and CPI) are valid and significant, that is, usable for projections. The reason is that it indicates that the coefficient of determination r^2 is quite high in both regression models and F statistical values are significant when the models are valid or when the relationship between the dependant variable and

Table 3. The industrial wood and log sales by General Directorate of Forestry, number of buildings by area and number of buildings constructed as per the occupancy permit and exchange rates ($) of Turkey (OGM, 2008; TUIK, 2008).

Year	Log (000 m^3)	Industrial Wood (000m^3)	Buildings Number of building	Permits Area	Annual Exchange rates ($)	Year	Log (000m^3)	Industrial Wood (000m^3)	Buildings Number of building	Permits Area	Annual Exchange rates ($)
1982	4,066	5,821	*45,995	22,945,123	164.07	1995	3,578	8,046	137,905	83,956,863	46,558.58
1983	3,945	6,665	58,968	25,554,984	228.14	1996	3,172	7,528	126,722	78,477,686	83,043.91
1984	4,078	7,596	63,153	28,887,793	369.75	1997	2,845	6,974	126,956	83,388,824	165,170.83
1985	3,892	7,407	71,844	37,251,360	522.91	1998	2,817	7,051	116,235	78,568,789	264,183.08
1986	3,746	7,570	102,888	55,624,440	676.56	1999	2,833	7,066	92,469	62,761,914	427,202.08
1987	3,687	7,251	138,155	70,912,137	866.08	2000	3,007	7,329	79,140	61,694,941	628,804.5
1988	3,572	7,447	139,995	67,861,304	1,448.46	2001	2,738	6,778	77,430	57,449,494	1,245,609.58
1989	3,393	7,460	136,015	62,923,939	2,137.81	2002	3,297	8,005	47,242	36,187,021	1,517,018.41
1990	3,310	6,581	123,304	60,083,035	2,634.47	2003	2,827	7,320	53,843	45,516,030	1,493,827.91
1991	3,159	6,513	121,486	61,447,817	4,264.53	2004	3,065	8,253	75,495	69,719,611	1,421,467.33
1992	3,353	6,897	137,990	73,062,016	6,994.97	2005	2,936	8,100	114,254	106,424,587	**1,344,966.66
1993	3,199	7,010	147,033	85,080,806	11,193.6	2006	3,480	9,299	114,204	122,909,886	**1,433,958.33
1994	2,939	6,712	143,281	81,715,801	30,266.88						

*The calculation is based on 22% being the average of three year increase on the number of buildings. **, The US$ and Turkish Lira exchange rates were ignored for 2005-2006 US$ rates.

Table 4. Annual CPI, PPI, economic growth rate and construction materials price index of Turkey (TUIK, 2008).

Year	The base year 1978 CPI (%)	The base year 1981 PPI (%)	Economic growth rate (%) constant prices	Economic growth rate (%) current prices	Construction materials price index (1968=100)
1982	410.29	127.05	0.6	29.0	3882
1983	539.00	165.68	1.7	28.1	5441
1984	799.95	249.13	4.5	55.2	7878
1985	1159.63	356.79	1.7	55.5	12525
1986	1560.98	462.25	4.4	41.6	16916
1987	2167.51	610.40	7.5	43.4	23075
1988	3800.95	1027.30	-0.7	68.5	38744
1989	6447.44	1741.99	-0.6	74.5	62699
1990	10547.15	2741.10	6.8	68.4	91729
1991	17503.32	4260.36	-1.6	56.7	152580
1992	30052.64	7051.58	4.4	70.7	246594
1993	50392.45	11545.97	6.2	77.7	406756
1994	106102.03	25212.55	-7.8	91.2	887488
1995	206323.49	47528.46	6.1	98.5	1511717
1996	366475.34	84934.70	5.3	87.5	2765327
1997	672724.15	153300.04	8.7	96.9	5104892
1998	1225733.19	260825.50	2.3	79.3	8538854
1999	1943577.71	398121.90	-7.4	44.3	12277603
2000	2960721.26	600952.65	1.4	53.0	18851834
2001	4545059.66	998582.63	-11.1	38.1	31567385
2002	6733431.01	1510984.00	6.4	53.6	45494981
2003	8506320.48	1871847.92	4.2	27.7	56359182**
2004	9208409.60	2099693.40	8.2	18.5	63218094**
2005	10136772.60	2260856.62	7.2	13.0	68066921**
2006	11657288.49*	*2599985.11	4.6	16.9	78276959**

*, The increase rate of the last three year was found as 15% and 2006 values were calculated according to this rate. **, PPI was calculated according to last four years increase rates (23.88, 12.17, 7.67 and 15%) respectively.

Table 5. Model summary[c].

Model	R	R^2	Adjusted R^2	Std. error of the estimate
1	0.941[a]	0.885	0.880	272344.38004
2	0.973[b]	0.948	0.943	187889.48198

[a], Predictors: (Constant), population. [b], Predictors: (Constant), population and CPI. [c], Dependent variable: particle product.

Table 6. ANOVA[c].

Model		Sum of squares	df	Mean square	F	Sig.
	Regression	13123024949215.630	1	13123024949215.630		
1	Residual	1705943610784.365	23	74171461338.451	176.928	0.000[a]
	Total	14828968559999.990	24			
	Regression	14052314496387.000	2	7026157248193.500		
2	Residual	776654063612.994	22	35302457436.954	199.027	0.000[b]
	Total	14828968559999.990	24			

[a]Predictors: (Constant), population. [b]Predictors: (Constant), population and CPI. [c]Dependent variable: particle product.

Table 7. Coefficients[a].

Model		Unstandardized coefficients		Standardized coefficients	t	Sig.
		B	Std. error	Beta	B	Std. error
1	(Constant)	-4194805.932	414793		-10.113	0.000
	Population	90.774	6.824	0.941	13.301	0.000
	(Constant)	-2357852.321	458343.61		-5.144	0.000
2	Population	56.790	8.126	0.589	6.988	0.000
	CPI	0.090	0.018	0.432	5.131	0.000

[a]Dependent variable: particle product.

independent variable is significant at α=0.05. However, in this case of projection, the regression model with two independent variables (Population and CPI) shall be used. Here, r^2=0.948 is a very high coefficient of determination. This figure indicates that the selected independent variables express the particle board production around 95%, demonstrating that the structure of the linear model is appropriate. Below other results of the regression model such as analysis of variance (analysis of variance) (Table 6), coefficients (Table 7) and dispersion graphic (Figure 2) are given.

As it may be seen from Table 7, regression equation for the particle board production would be as follows (Model 2):

Y = - 2357852.321 + 56.79 POPULATION + 0.090 CPI.

Particle board import

As it may be seen from the summary Table 8, the regression model with one independent variable (CPI) is valid and significant (Table 8). Other ten independent variables included in the model for the regression analysis were not included in the regression equation as a result of the operation. The reason is that here stepwise method was applied in the multiple regression analysis. Consequently, it is seen that with the method eliminating the multiple linear connection (a condition deteriorating the model caused by high correlation among the independent variables), the other independent variables remain outside the model without the variable(s) affecting the particle board import most and overloading the model. As a result, the resulting coefficient of determination r^2 is relatively high, and F statistical values show the model is valid and relationship between the dependant variable and independent variable is significant at significance level of α= 0.05. Here r^2 = 0.615 is a coefficient of determination which can be considered relatively high.

Coefficient of correlation (R) of 0.78, as a measure of

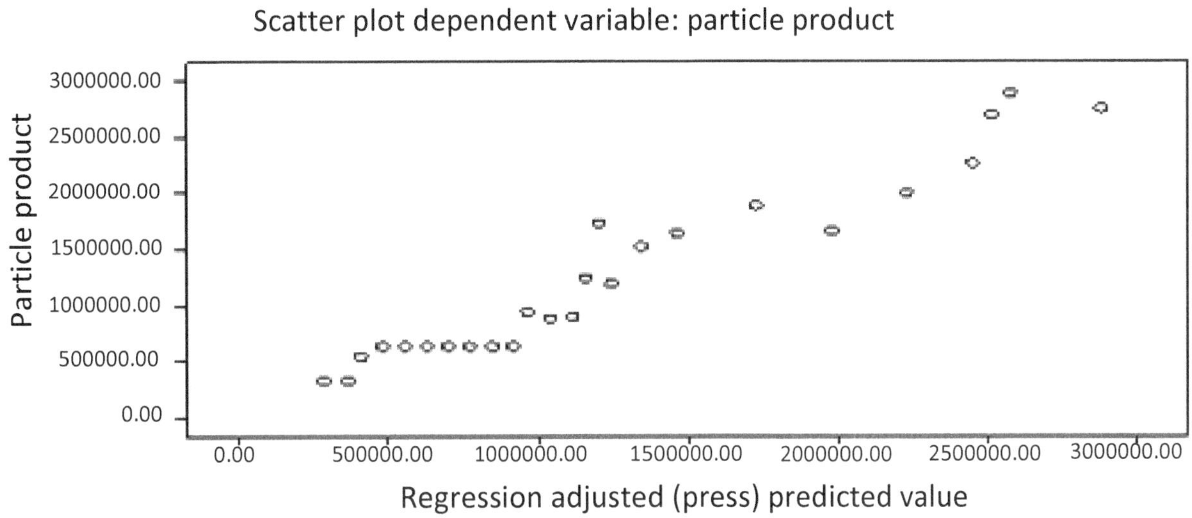

Figure 2. The scatter diagram of particle board production.

Table 8. Model summary[b].

Model	R	R^2	Adjusted R^2	Std. error of the estimate
1	0.784[a]	0.615	0.586	61873.57933

[a], Predictors: (Constant), CPI. [b], Dependent variable: Particle import.

Table 9. ANOVA[b].

Model		Sum of squares	df	Mean square	F	Sig.
1	Regression	79553455412.729	1	79553455412.729	20.78	0.001[a]
	Residual	49768417648.204	13	3828339819.093		
	Total	129321873060.933	14			

[a], Predictors: (Constant), CPI. [b], Dependent variable: Particle import.

the linear relationship between the variables, is about the level 0.60 to 0.70 which is recognized in the literature. This figure indicates that the selected independent variable express particle board import around 62% and the structure of the linear model built is suitable. Subsequently, other results of the regression model such as ANOVA (Table 9), coefficients (Table 10) and dispersion graphic (Figure 3) are given. As it may be seen from Table 10, regression equation for the particle board import shall be as follows (Model 1): Y= 26918.227 + 0.018 CPI.

Particle board export

As it may be seen in the Table 11, both regression models, one built with one independent variable (PPI), and the other with two independent variables (PPI and number of building) are valid and significant, that is, usable for projection. The reason is that it indicates that the coefficient of determination r^2 is quite high in both regression models and F statistical values are significant when the models are valid or when the relationship between the dependant variable and independent variable is significant at $\alpha=0.05$. However, in this case of projection, the regression model with one independent variable (PPI) shall be used.

Due to the fact that determination coefficient was found very low for projections for number of buildings. Therefore, it was excluded from the model. Here, r^2 =0.807 is a relatively high coefficient of determination. This figure indicates that the selected independent variables express the particle board export around 81%, demonstrating that structure of the linear model is appropriate. Subsequently, other results of the regression model such as ANOVA (Table 12), coefficients (Table 13) and dispersion graphic (Figure 4) are given. As it may be seen from Table 13, regression equation for the particle

Table 10. Coefficients[a].

Model		Unstandardized coefficients		Standardized coefficients	t	Sig.
		B	Std. Error	Beta	B	Std. error
1	(Constant)	26918.227	22077.077		1.219	0.244
	CPI	0.018	0.004	0.784	4.559	0.001

[a]Dependent variable: Particle import.

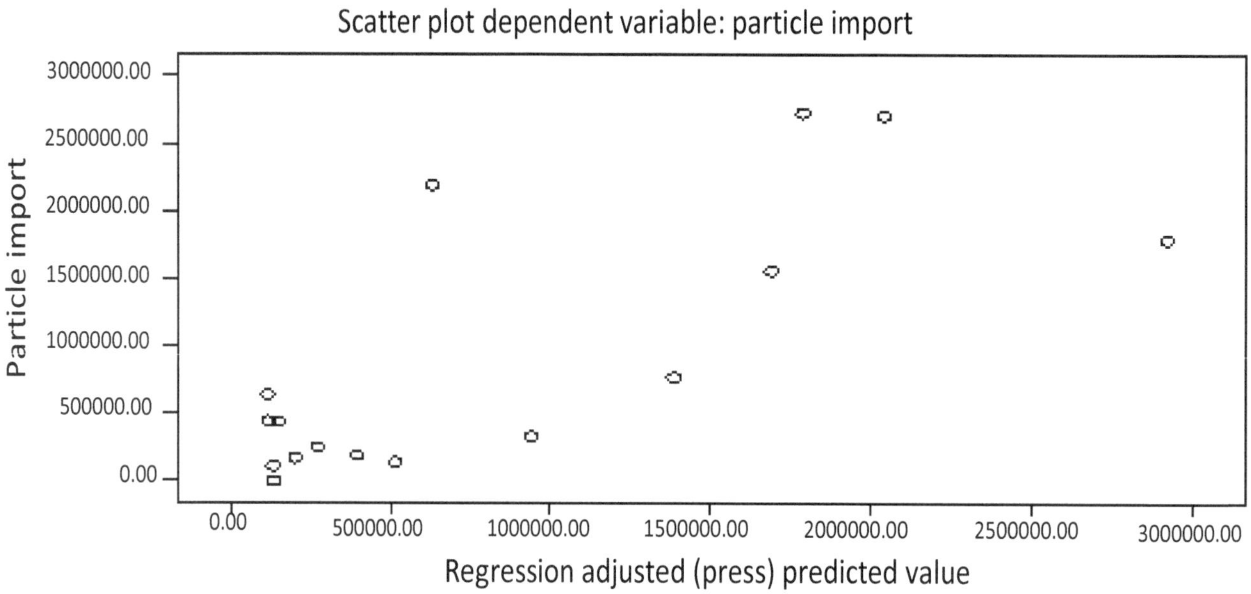

Figure 3. The scatter diagram of particle board import.

Table 11. Model summary[c].

Model	R	R^2	Adjusted R^2	Std. error of the estimate
1	0.898(a)	0.807	0.792	36095.25029
2	0.932(b)	0.869	0.847	30910.49550

[a], Predictors: (Constant), PPI. [b], Predictors: (Constant), PPI and number of building. [c], Dependent variable: Particle export.

Table 12. ANOVA[c].

Model		Sum of squares	df	Mean square	F	Sig.
1	Regression	70719750322.554	1	70719750322.554	54.280	0.000[a]
	Residual	16937272216.779	13	1302867093.598		
	Total	87657022539.333	14			
2	Regression	76191517753.570	2	38095758876.785	39.872	0.000[b]
	Residual	11465504785.763	12	955458732.147		
	Total	87657022539.333	14			

[a], Predictors: (constant), PPI. [b], Predictors: (Constant), PPI and number of building. [c], Dependent variable: particle export.

Table 13. Coefficients[a].

Model		Unstandardized coefficients		Standardized coefficients	t	Sig.
		B	Std. error	Beta	B	Std. error
1	(Constant)	12386.331	12809.764		.967	0.351
	PPI	0.075	0.010	0.898	7.368	0.000
2	(Constant)	-78442.205	39508.070		-1.985	0.070
	PPI	0.090	0.011	1.074	8.414	0.000
	Numberbuild	0.737	0.308	0.305	2.393	0.034

[a], Dependent variable: particle export.

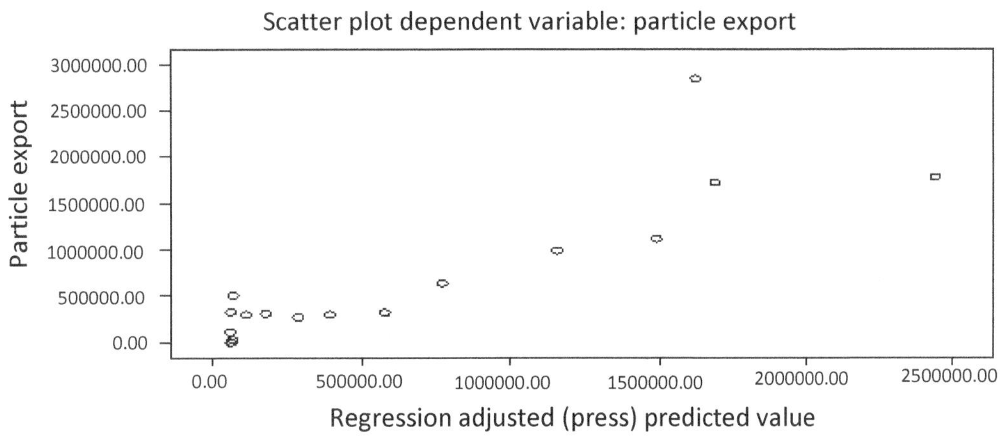

Figure 4. The scatter diagram of particle board export.

Table 14. The estimated values of the independent variables between the years 2007 and 2021 (population, OGM wood sales, foreign exchange).

Year	Population (00) person	OGM wood sales (m^3)	Foreign exchange (USD$)	Year	Population (00) person	OGM wood sales (m^3)	Foreign exchange (USD$)
2007	74,609.64	7,970.756	1,286,324.672	2015	83,443.24	8,379.124	1,828,467.128
2008	75,713.84	8,021.802	1,354,092.479	2016	84,547.44	8,430.170	1,896,234.935
2009	76,818.04	8,072.848	1,421,860.286	2017	85,651.64	8,481.216	1,964,002.742
2010	77,922.24	8,123.894	1,489,628.093	2018	86,755.84	8,532.262	2,031,770.549
2011	79,026.44	8,174.940	1,557,395.900	2019	87,860.04	8,583.308	2,099,538.356
2012	80,130.64	8,225.986	1,625,163.707	2020	88,964.24	8,634.354	2,167,306.163
2013	81,234.84	8,277.032	1,692,931.514	2021	90,068.44	8,685.400	2,235,073.970
2014	82,339.04	8,328.078	1,760,699.321				

board export shall be as follows (Model 1):

Y= 12386,331 + 0.075 PPI.

Calculation of the estimated value of the independent variables in the projection models

In the estimated values of the independent variables (Tables 14 to 17), the independent variables of Population, OGM wood sales, foreign exchange, CPI, PPI, price index, building area, number of buildings, GNP and economic growth are projected by years (x), using the data for the period of 1982 to 2006 by help of regression analysis. For the said projection, the following regression equations were found and these equations were used for the calculations (Table 18).

Table 15. The estimated values of the independent variables between the years of 2007-2021 (CPI, PPI, construction materials price index).

Year	CPI	PPI	Construction materials price index	Year	CPI	PPI	Construction materials price index
2007	13,886,464	1,719,991	52,165,111.15	2015	17,214,102	2,459,845	74,568,501.25
2008	14,302,418	1,812,472	54,965,534.91	2016	17,630,057	2,552,327	77,368,925.01
2009	14,718,373	1,904,954	57,765,958.68	2017	18,046,011	2,644,809	80,169,348.77
2010	15,134,328	1,997,436	60,566,382.44	2018	18,461,966	2,737,291	82,969,772.53
2011	15,550,283	2,089,918	63,366,806.20	2019	18,877,921	2,829,773	85,770,196.30
2012	15,966,238	2,182,400	66,167,229.96	2020	19,293,876	2,922,255	88,570,620.06
2013	16,382,192	2,274,882	68,967,653.72	2021	19,709,831	3,014,736	91,371,043.82
2014	16,798,147	2,367,363	71,768,077.49				

Table 16. The estimated values of the independent variables between the years of 2007 to 2021 (building area, number of building, GNP).

Year	Building area	Number of building	GNP	Year	Building area	Number of building	GNP
2007	89,153,950.80	102,594.396	4,301.642	2015	104,130,998.7	101,926.284	5,330.618
2008	91,026,081.78	102,510.882	4,430.264	2016	106,003,129.6	101,842.770	5,459.240
2009	92,898,212.77	102,427.368	4,558.886	2017	107,875,260.6	101,759.256	5,587.862
2010	94,770,343.75	102,343.854	4,687.508	2018	109,747,391.6	101,675.742	5,716.484
2011	96,642,474.73	102,260.340	4,816.130	2019	111,619,522.6	101,592.228	5,845.106
2012	98,514,605.71	102,176.826	4,944.752	2020	113,491,653.6	101,508.714	5,973.728
2013	100,386,736.7	102,093.312	5,073.374	2021	115,363,784.6	101,425.200	6,102.350
2014	102,258,867.7	102,009.798	5,201.996				

Table 17. The estimated values of the independent variables between the years of 2007 to 2021 (economic growth %).

Year	Economic growth (%) (current prices)	Year	Economic growth (%) (current prices)
2007	46.574	2015	41.070
2008	45.886	2016	40.382
2009	45.198	2017	39.694
2010	44.510	2018	39.006
2011	43.822	2019	38.318
2012	43.134	2020	37.630
2013	42.446	2021	36.942
2014	41.758		

Particle board production, export and import projection values in Turkey

In Table 19, Turkish particle board production, export and import projection values are given for the period of 2007 to 2021. These values were obtained by putting in place the estimated values of the valid and significant independent variables build for these equations for the period between 2007 to 2021 in the equation found as a result of regression analysis conducted for the particle board production, export and import values previously for the period of 1982 to 2006. In the projection, the regression models were used and the results are shown in Table 19.

CONCLUSION AND RECOMMENDATIONS

There are more than 30 particle board plants having total annual production capacity close to 3 million m^3, each with average capacity around 75 thousands m^3 and capacity usage rate around 80%. Production in 1980,

Table 18. Regression equations used for the estimation of the independent variables.

$Y_{Population} = -2141520 + 1104.200.x$ ($r^2 = 0.99$)	$Y_{CPI} = -820934779 + 415954.780.x$ ($r^2 = 0.66$)
$Y_{OGM} = -94479 + 51.046.x$ ($r^2 = 0.29$)	$Y_{PPI} = -183891069 + 92481.844.x$ ($r^2 = 0.66$)
$Y_{Pricet\ Indx} = -5568285378 + 2800423.762.x$ ($r^2 = 0.66$)	$Y_{E.Growth} = 1428 - 0.688.x$ ($r^2 = 0.04$)
$Y_{B.Area} = -3668212931 + 1872130.982.x$ ($r^2 = 0.33$)	$Y_{GNP} = -253843 + 128.622.\ x$ ($r^2 = 0.74$)
$Y_{Number\ Build.} = 270207 - 84.x$ ($r^2 = 0$)	$Y_{Foreign\ exch.} = -134723663 + 67767.807.x$ ($r^2 = 0.7$)

Table 19. Particle board production, export and import projection values in Turkey (m^3).

Year	Production	Export	Import	Year	Production	Export	Import
2007	3,129,011	141,386	276,875	2015	3,930,158	196,875	336,772
2008	3,229,154	148,322	284,361	2016	4,030,302	203,811	344,259
2009	3,329,298	155,258	291,849	2017	4,130,445	210,747	351,746
2010	3,429,441	162,194	299,336	2018	4,230,589	217,683	359,234
2011	3,529,585	169,130	306,823	2019	4,330,732	224,619	366,721
2012	3,629,728	176,066	314,311	2020	4,430,876	231,555	374,208
2013	3,729,872	183,002	321,798	2021	4,531,019	238,492	381,695
2014	3,830,015	189,939	329,285				

around 300 thousands m^3, increased to 600 thousands m^3 between the years 1985 to 1990 and to around 900 thousands m^3 between the years 1990 to 1995, with an average increase of 300 thousands m^3 for each five years, resulting to 2 million 740 thousands m^3 in recent years. Although no significant difference is seen between import and export in the Turkish Particle board foreign trade, the import seems more weighted. Import and export quantities have realized above 150 thousands m^3 by 2003.

In this study, made for the said purposes, the following results were obtained and evaluations made: In the regression analyses performed for projection of particle board production, import and export, the nine independent variables used include round timber and industrial wood sales by the general directorate of forestry (m^3), gross national product per capita (thousand person), building area as per the occupancy permit (m^2), inflation rate, exchange rates, population, economic growth and construction materials price index.

All possible models for particle board production, import and export projections and their combinations were tried and the most appropriate regression models were searched and thus regression models were formed. As period up to the year 2021 was target in the projection of particle board production, import and export quantities made, the estimated values of the independent variables significant and valid for the models built ware calculated by a separate regression analysis, proceeding to the projection operation.

As a result of different regression model trials, the independent variables of population and CPI have provided sufficient explanation ($r^2 = 0.948$) for the particle board production and were used as an estimation tool for the projection of particle board production.

Similar operations in the particle board import and export projection were performed by using same data and changes in the period. It is seen from the results of the regression analysis that CPI by itself as a significant and independent variable provides explanation ($r^2 = 0.615$) for import, and that the variable PPI in the particle board export projection provide explanation as significant variable (r^2=0.807) and can be used as projection tool for export.

When examining the particle board production, export and import estimated figures, the following results appear: From the regression analysis, it is seen that the particle board production above 2.5 million m^3 in 2006 shall exceed 4.5 million m^3 in 2021 and the export of around 175 thousands m^3 shall exceed 230 thousands m^3 and import of 175 thousand m^3 shall reach to 380 thousands m^3.

That the foreign trade enjoys an important place in the industrialization policies shows the need to develop particle board industry and give important to this sector. For this reason, it has become very important to examine changes to occur in the production and foreign trade structure of particle board industry in Turkey over time, determine short and long term development, strategy and policies for the particle board industry and perform realistic projections about production-import-export in future and hence this study has been a very new, important and comprehensive one in filing the gap of search mentioned above with the production, import and export projections for the particle board industry with a confidence level and acceptable error extent. By this study, the relations explaining production and foreign trade of the particle board industry in Turkey have been set forth and projection data were obtained by scientific data.

ACKNOWLEDGEMENT

This research is part of a project supported by the Scientific and Technological Research Council of Turkey (TUBİTAK) (Project Number: 108O416).

REFERENCES

Akbulut T (2000). Particle board industry. Laminate Furniture Decoration Art Des. J. 7:112-119.
DPT (1995). 7th Five-year Development Plan, Special forest products industry commission report, Ankara-Turkey. p. 445.
DPT (2006). 9th Five-year Development Plan, Specialized commission report of wood products and furniture (2007- 2013), Ankara-Turkey.
DTM (2008). Undersecretariat Of The Prime Ministry For Foreign Trade of Turkey, Website, http://www.dtm.gov.tr/dtmweb/index.cfm.
FAO (2008). Food and Agriculture Organization of the United Nations, Website, http://faostat.fao.org/site/626/default.aspx#ancor.
FSC (2008). Forest Stewardship Council, Website, http://www.fsc.org/.
IGEME (2008). Export Development Center, Ministry of industry and trade, Website, http://www.igeme.org.tr/.
Koc H, Aksu B (1999). Fiberboard foreign trade of Turkey. Laminate Furniture Decoration Art Des. J. 3:82-85.
OGM (2008). General Directorate of Forestry, Forestry statistics, Website, http://www.ogm.gov.tr/.
TOBB (2001). Records of Union of Chambers and Commodity Exchanges of Turkey, Ankara-Turkey.
TOBB (2007). Union of Chambers and Commodity Exchanges of Turkey, Industry database, Website, www.tobb.org.tr.
TUIK (2008). Turkish Statistics Institute, Website, http://www.tuik.gov.tr/VeriBilgi.
Yaman A (2002). Developments and prospects in wood and sheet industry. Laminate Furniture Decoration Art Des. J. 18:92-107.

Regional variation of columnar refractivity with meteorological variables from climate monitoring satellite application facility (CM SAF) data over Nigeria

I. Emmanuel, B. Adeyemi and K. D. Adedayo

Department of Physics, Federal University of Technology, P. M. B. 704, Akure, Ondo State, Nigeria.

Data from satellite application facility on climate monitoring (CM SAF) have been used to estimate and study seasonal variation of columnar refractivity in relation with relative humidity and temperature for twenty six stations categorized into four climatic regions (coastal, Guinea savannah, midland and sub Sahel regions) over Nigeria. Mean, standard error, maximum, minimum values and ranges of these variables were also determined for different regions at each atmospheric level. The results have revealed that the monthly variation of refractivity along side with meteorological factors (relative humidity and temperature) from one region to another over Nigeria are influenced by seasonal variation, geographical location and environmental situation of each region. It is also noticed that the range of relative humidity and temperature are wider in the northern part (56.19 and 72.51% and 5.64 and 5.90 K respectively) compared to the southern part (17.85 and 25.27% and 3.19 and 5.44 K respectively) of the country. This effect was observed in the variation of refractivity also. Effect of inter tropical discontinuity (ITD) was recorded in the southern variation of relative humidity and refractivity.

Key words: Columnar, refractivity, meteorological, variation.

INTRODUCTION

The structure of the radio refractive index, *n*, at the lower part of the atmosphere is a very important parameter in the planning of the communication links. It is defined as a ratio of the radio wave propagation velocity in free space to its velocity in a specified medium (Freeman, 2007). Radio–wave propagation is determined by changes in the refractive index of air in the troposphere (Adediji and Ajewole, 2008). Changes in the value of the radio refractive index in the troposphere can curve the path of the propagating radio wave. The atmosphere radio refractive index depends on air temperature, humidity, atmospheric pressure and water vapour pressure. Even small changes in any of these variables can make a significant influence on radio-wave propagation, because radio signals can be refracted over the whole signal path (Priestley and Hill, 1985).

The refractive index is responsible for bending of propagation direction of the electromagnetic wave (Guanjun and Shukai, 2000). Previously worked with Adeyemi and Emmanuel (2011) and Adeyemi (2006) showed that seasonal variation affects radio refractivity in Nigeria. Earlier research work carried out on refractivity in Nigeria was based on radiosonde data (Willoughby et al., 2002) for view available locations and data logger data (Adediji and Ajewole, 2008). This research work covers all the geographical areas/zones in Nigeria.

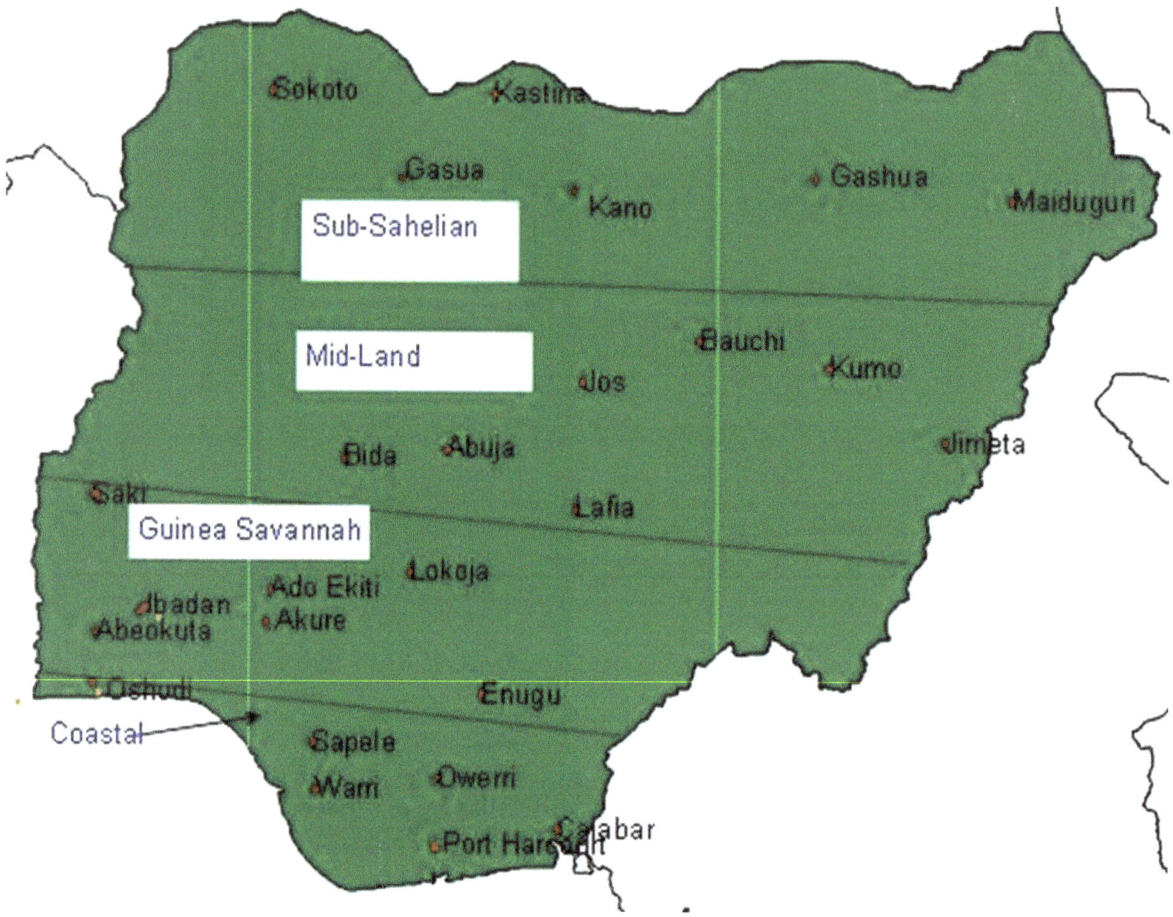

Figure 1. Map of Nigeria showing study locations.

Nigeria lies wholly within the tropical zone, there are wide climatic variations in different areas of the country. Southern Nigeria is characterized by long rainy season starting in March and it lasts until the end of July, with a peak period in June over most parts of southern Nigeria. A short dry season is experienced in August for 3 to 4 weeks (Adeyemi and Aro, 2004). However, the real dry period known as the "August break" is generally observed in the last two weeks of August in most parts of southern Nigeria. This is followed by brief wet season from early September to Mid-October, with a peak period at the end of September. The long dry season period starts from late October and last to early March with peak dry conditions between early December and late February. This period witnesses the prevailing influences of the dry and dusty northeast winds, as well as the 'Harmattan' conditions.

The climatic conditions in the northern part of Nigeria exhibit only two different seasons, namely, a short wet season and a prolonged dry season. The main goal of this paper is to estimate refractivity and study its seasonal variation along side with the temperature and relative humidity at three different atmospheric levels (column) over Nigeria using satellite retrieved data.

DATA ANALYSIS TECHNIQUES

Atmospheric data used in this analysis were obtained from the Department of Satellite Application Facility on Climate Monitoring (CM-SAF), DWD Germany. The CMSAF focuses on the atmospheric part of the Essential Climatic Variables defined within the framework of the Global Climate Observing System (GCOS). The CMSAF operationally applies the international ATOVS processing package (IAPP) to retrieve humidity and temperature from ATOVS observations on-board NOAA - 15, 16 and 18 satellites. The profiles are vertically integrated and averaged to provide temperature and humidity for 5 layers (that is, 925, 775, 600, 400 and 250 mbar). The data used were obtained between the period of 2004 to 2006 for twenty six stations, classified into four climatic regions based in Nigeria (Figure 1), (Olaniran and Sumner, 1989). These regions are:

(i) Coastal region, ((Lat 4.0° to 6.5°, Lon 3.7° to 7.1°) dominated by tropical maritime (mT) air for most of the year;

(ii) Guinea savannah region, (Lat 6.6° to 7.8°, Lon 3.9° to 7.5°), where mT air dominates for about 7 months and tropical continental (cT) air for the remaining 5 months;

(iii) Midland, (Lat 8.0° to 10.8°, Lon 3.9° to 9.8°), which is predominantly highland, where the cT air mass dominates, but

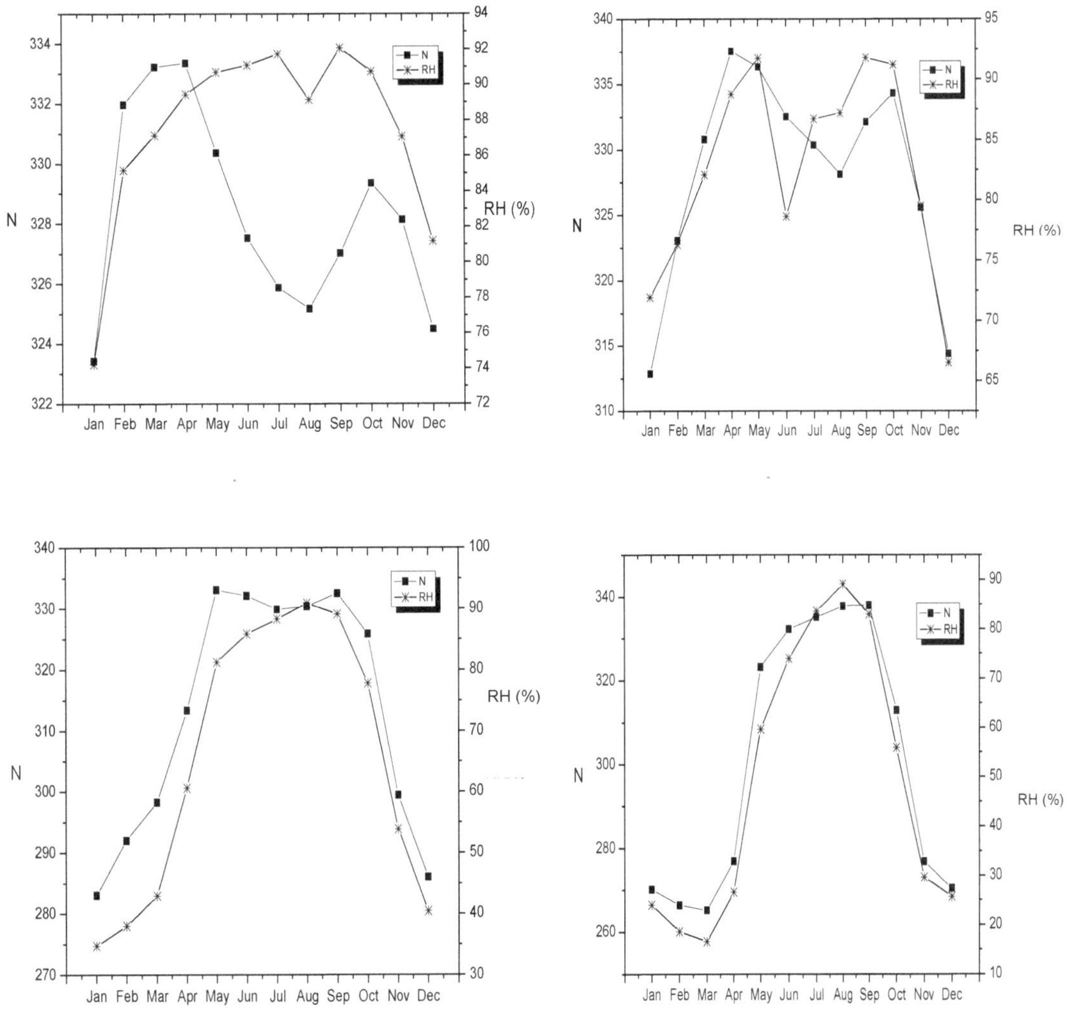

Figure 2. Monthly variation of Refractivity and Relative Humidity at Low Level in (a) Coastal (b) Guinea savannah (c) Midland and (d) Sub sahel regions.

where the topography effectively extended the length of the humid period, due to localized convection and orographic effects (Olaniran, 1983b; Adeyemi, 2004);
(iv) Sahelian region, (Lat 11.0° to 13.0°, Lon 3.7° to 13.333°), where the cT air mass predominates, and mT air mass invades for only 3 to 5 months at most (Olaniran, 1983b; Adeyemi, 2004).

Refractivity value, N, is deducted from the atmospheric data, using equation below.

$$N = (n-1)\times 10^{6} = \frac{77.6}{T}\left[P + \frac{4810\, e}{T}\right] \quad (1)$$

Where: P = Atmospheric pressure (hpa), e = Water vapour pressure (hpa) and T = absolute temperature (K), (Oyedum and Gambo, 1994).

$$e = \frac{He_s}{100} \quad (2)$$

Relative humidity (RH), temperature and refractivity were further grouped into three different atmospheric levels (columns), such as, between surface and 925 mbar have been labeled as low level; between 775 and 600 mbar as mid level and between 400 and 250 mbar as upper level.

RESULTS AND DISCUSSION

Monthly variations of relative humidity and radio refractivity

Figures 2 to 5 show the monthly variation of refractivity, N and relative humidity, RH at different atmospheric levels over Nigeria. From Figure 2a, RH increases with increase N, from January to April in coastal region at low level. A little dip is noticed in August both in RH and N, though that of N is conspicuous. The values of RH were high

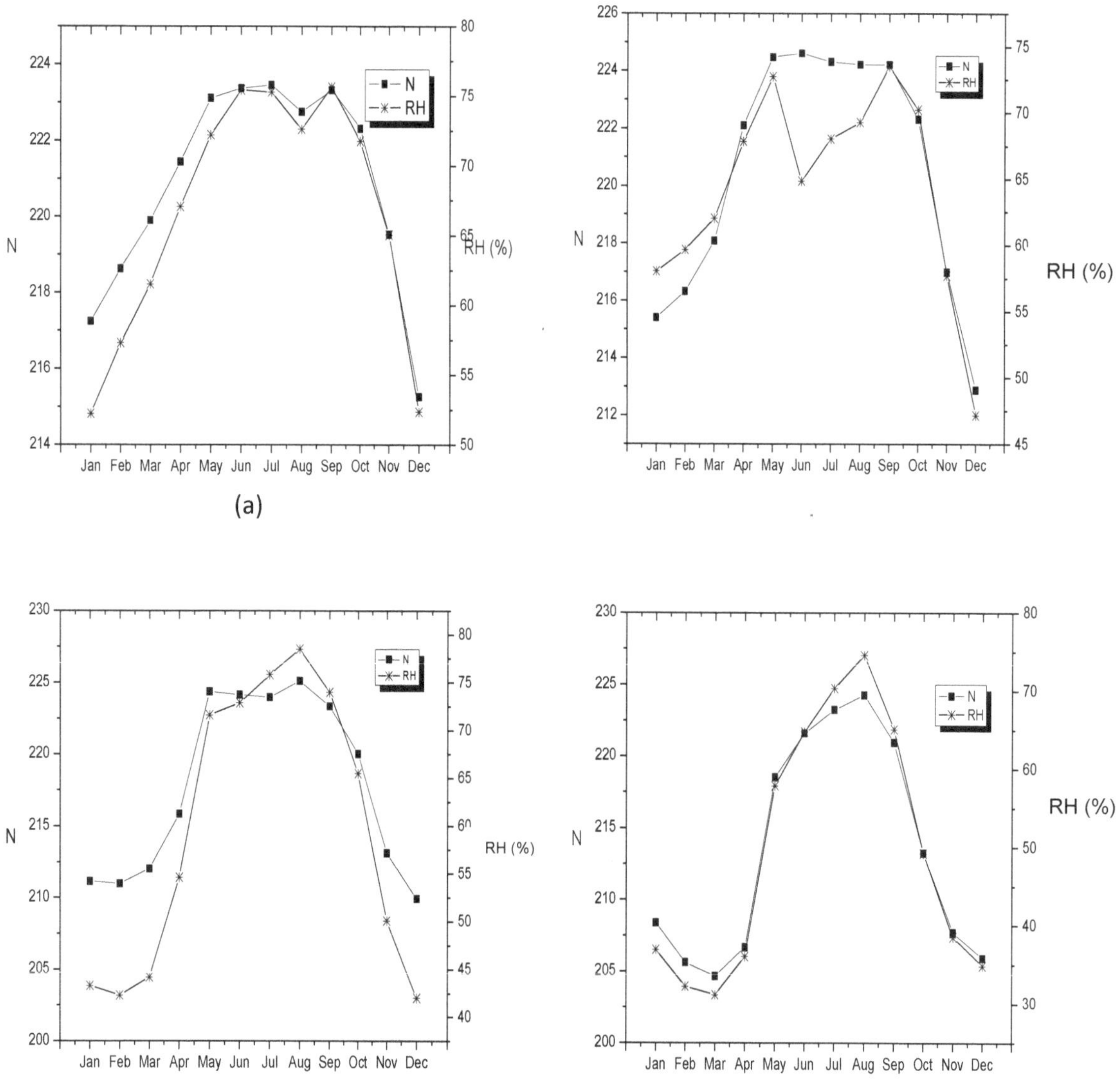

Figure 3. Monthly variation of Refractivity and Relative Humidity at Mid Level in (a) Coastal (b) Guinea savannah (c) Midland and (d) Sub sahel regions.

during the rainy season with an average value of 150% and low during the dry season with an average value of 144.97%, likewise N with average values of 328.4 and 327.2 during the rainy and dry season respectively. The variation of RH and N in the Guinea savannah region, (Figure 2b) is similar to that of the coastal region. It rises from January and reaches its peak in May. Two peaks with a dip in between were noticed here. One in May and the second in September/October. The value of RH was also higher during the rainy season and low during the dry season with average values 87.99 and 75.26% respectively. The dip in between the peaks was noticed between June to August. The average values of N during the dry and rainy seasons are 324.9 and 324.7 respectively. The variation of RH and N at midland and sub Sahel regions are similar. RH and N have the same pattern; they were high during the rainy months of May – September and low in the dry months of October to April in both regions (Figure 2c and d). At midland, their average values of RH during these periods are 69.50 and 56.92%, respectively and for N is 329.3 and 291.5 respectively; at sub Sahel region, it is 67.36 and 22.87%, respectively, while that of N are 321.9 and 269.4, respectively.

At mid level, (Figure 3), in coastal and guinea savannah regions, the monthly variation of RH and N follow the same pattern. Both RH and N have two peaks in these regions, these peaks were noticed in June/July

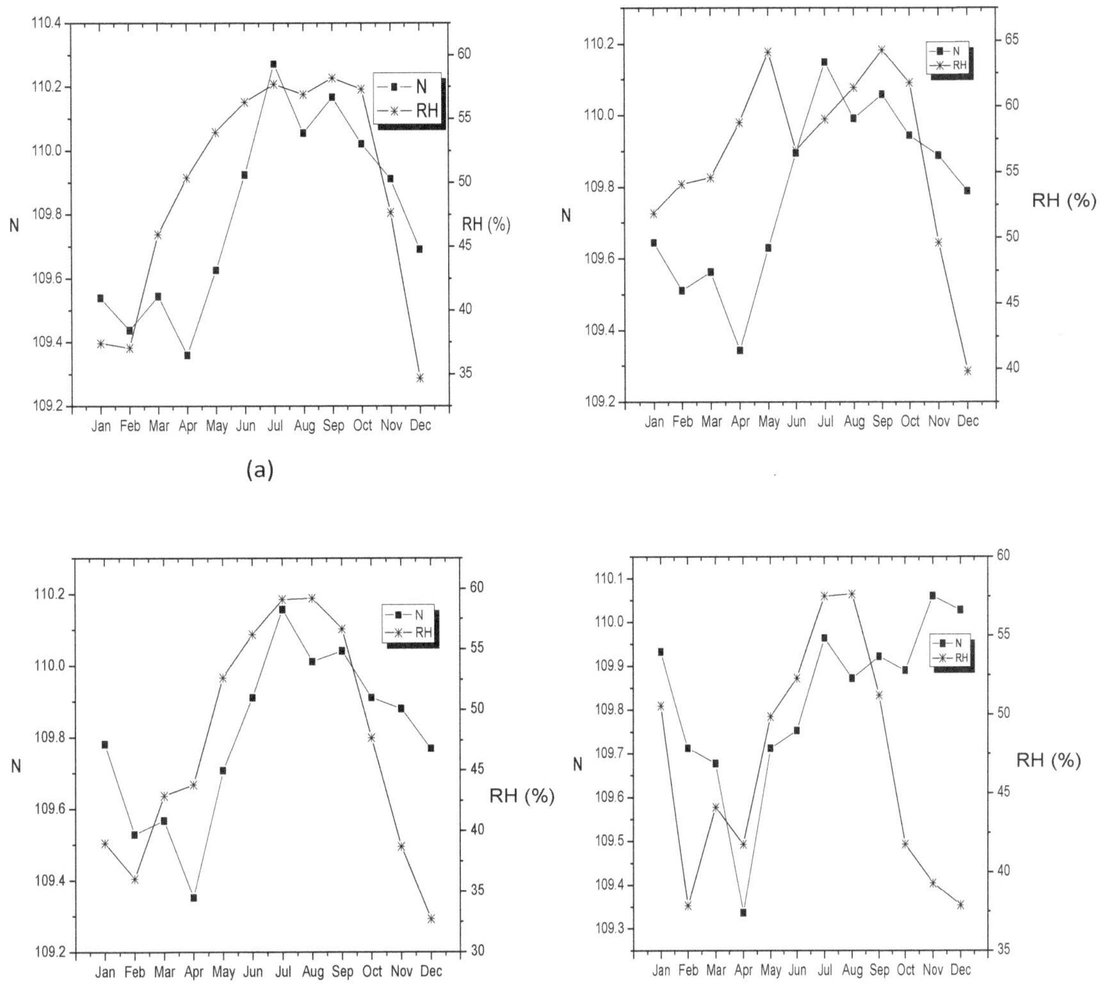

Figure 4. Monthly variation of Refractivity and Relative Humidity at Upper Level in (a) Coastal (b) Guinea savannah (c) Midland and (d) Sub sahel regions.

and September with the dip in the coastal region; and May and September in guinea savannah with dip in June are extending into August here (Figure 3a-b). Here the values of RH and N were high during the rainy season and low during the dry season (Table 1). Also variation at midland and sub Sahel regions follows the same pattern both in RH and N. Here RH and N rise at the same time and attain their peak at the same time. The values were high during the rainy and low during the dry seasons (Table 1). At the upper level (Figure 4) the monthly variation of RH was different from that of N in all the regions. The variations are summarized in Table 1.

Monthly variations of temperature and radio refractivity

Figures 5 to 7 show the monthly variation of temperature, T (in Kelvin) along side with radio refractivity for the three atmospheric levels (low level, mid level and upper level). At low level, the average values of T, were higher during the dry months of November – March with an average value of 292.18 K, and slightly dropped till it gets to minimum values in the months of July/August with an average value of 289.50 K in the coastal region. The average value during the rainy months here is 290.47 K. In Guinea savannah, variation of T follows the same pattern as that of coastal regions. The values were higher during the dry season and low during the rainy season. Their values during these periods are 293.60 and 291.49 K. The dip is noticed in August with an average value of 289.84 K. In midland region, T gradually increases from January and attain its maximum in March/April. The value suddenly drops to minimum values in August with a value of 291.14 K and partially rises again. Its average values during the rainy and dry season are 292.26 and 294.39 K

Table 1. Analysis result of relative humidity over Nigeria.

Region	Low level					Mid level					Upper level				
	Mean RH	Max RH (%)	Min RH (%)	ΔRH (%)	SE	Mean RH	Max RH (%)	Min RH (%)	ΔRH (%)	SE	Mean RH	Max RH (%)	Min RH (%)	ΔRH (%)	SE
Coastal	87.47	92.07	74.22	17.85	1.51	66.54	75.65	52.22	23.43	2.54	66.54	58.17	34.69	23.48	2.55
Guinea savannah	82.69	91.77	66.5	25.27	2.39	64.26	79.5	47.16	32.34	2.21	64.26	64.28	39.81	24.47	2.02
Midland	65.31	90.91	34.72	56.19	6.47	59.57	78.49	42.02	36.47	4.27	59.57	59.21	42.02	17.19	2.74
Sub sahel	48.82	89.08	16.57	72.51	8.15	49.32	74.56	31.15	43.41	4.70	49.32	57.63	37.84	19.79	2.09

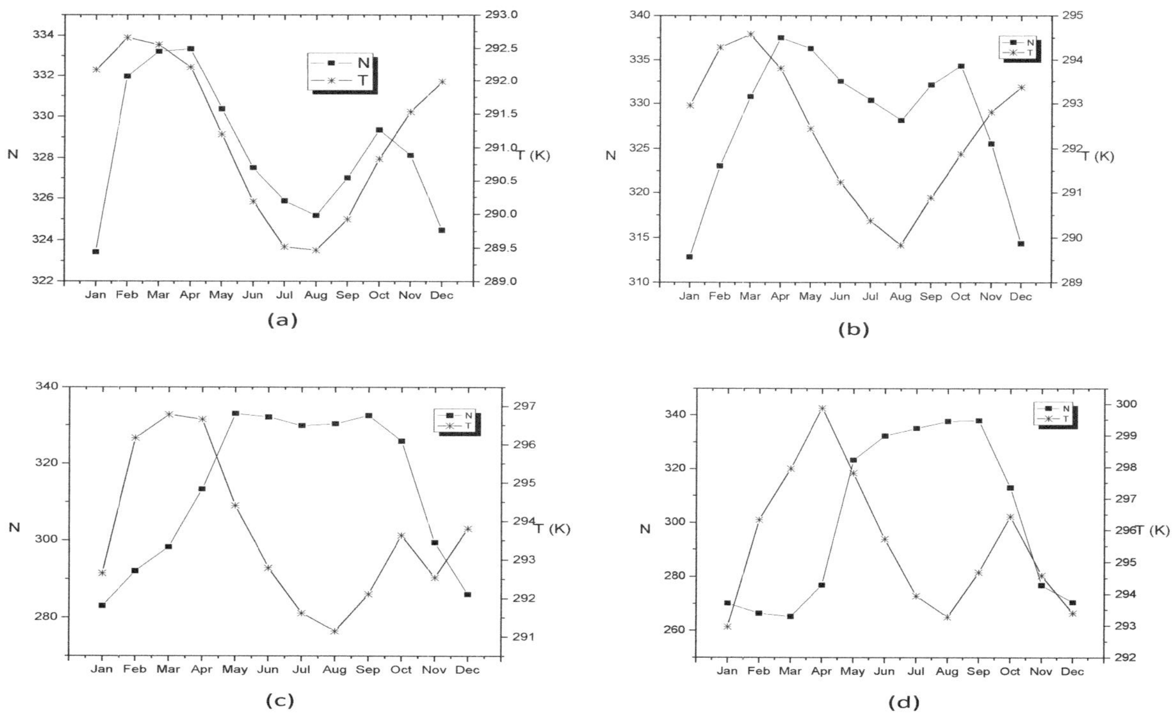

Figure 5. Monthly variation of Refractivity and Temperature at Low Level in (a) Coastal (b) Guinea savannah (c) Midland and (d) Sub sahel regions.

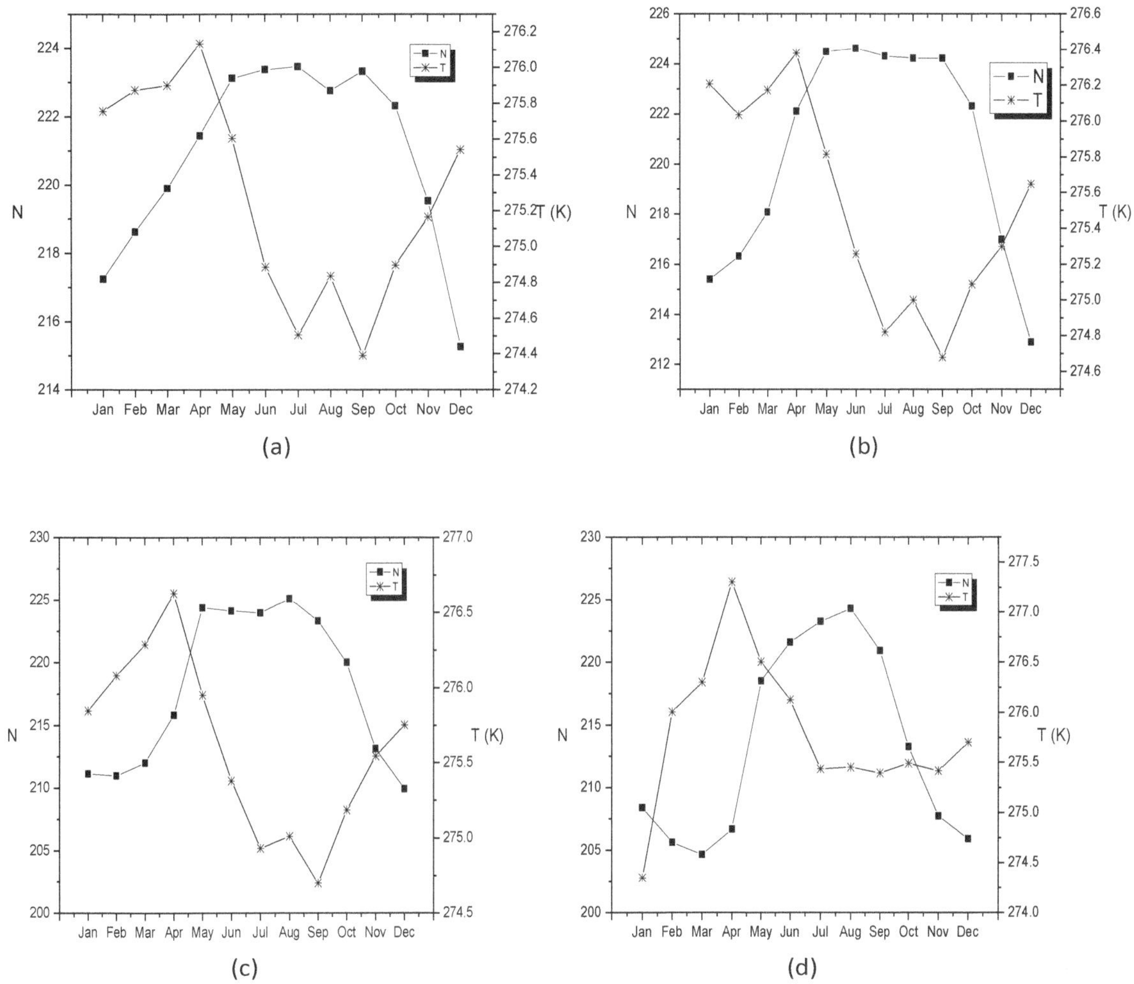

Figure 6. Monthly variation of Refractivity and Temperature at Mid Level in (a) Coastal (b) Guinea savannah (c) Midland and (d) Sub sahel regions.

respectively. The variation in the sub Sahel region is a little bit similar to that of midland region. Here two peaks are discernible; the major one in April with an average value of 299.87 K and minor in October with an average value of 296.43 K. The dip in between is noticed in August with an average value of 293.28 K. At mid level (Figure 6) monthly variations of T in coastal and Guinea savannah regions are similar. The variation does not have a particular pattern (Figure 6a and b). The behavior of T, at midland region is shown in Figure 6c. The values of T rise gradually from January to a peak in April and started descending until it reach its minimum in July. It partially rises in August and fall the remaining months of the year. In the sub Sahel region, the value of T is low in January, and it gradually rises to a maximum in April before it falls and reaches a minimum value in July. The value of T is almost constant for the remaining of the year. Figure 7a to d shows the monthly variation of T and N at the upper level, the behavior of T is almost the opposite of N in all the regions. This shows greater dependence of N on T at this level.

Statistic variation of relative humidity, temperature and refractivity

Tables 1, 2 and 3 show the statistical analysis results of relative humidity, temperature and refractivity, respectively. The result includes mean value, maximum value (max), the minimum value (min), differences (ΔRH) and standard error (SE) of the above parameters. Mean value of relative humidity is higher in the southern Nigeria (coastal and Guinea savannah regions) than the northern part (midland and sub Sahel regions) at low level and mid level (Table 1). Higher range of relative humidity (ΔRH) is also recorded in the northern part. This may be attributed to high and low rainfall witness in the southern regions and Northern regions, respectively. This is in line with

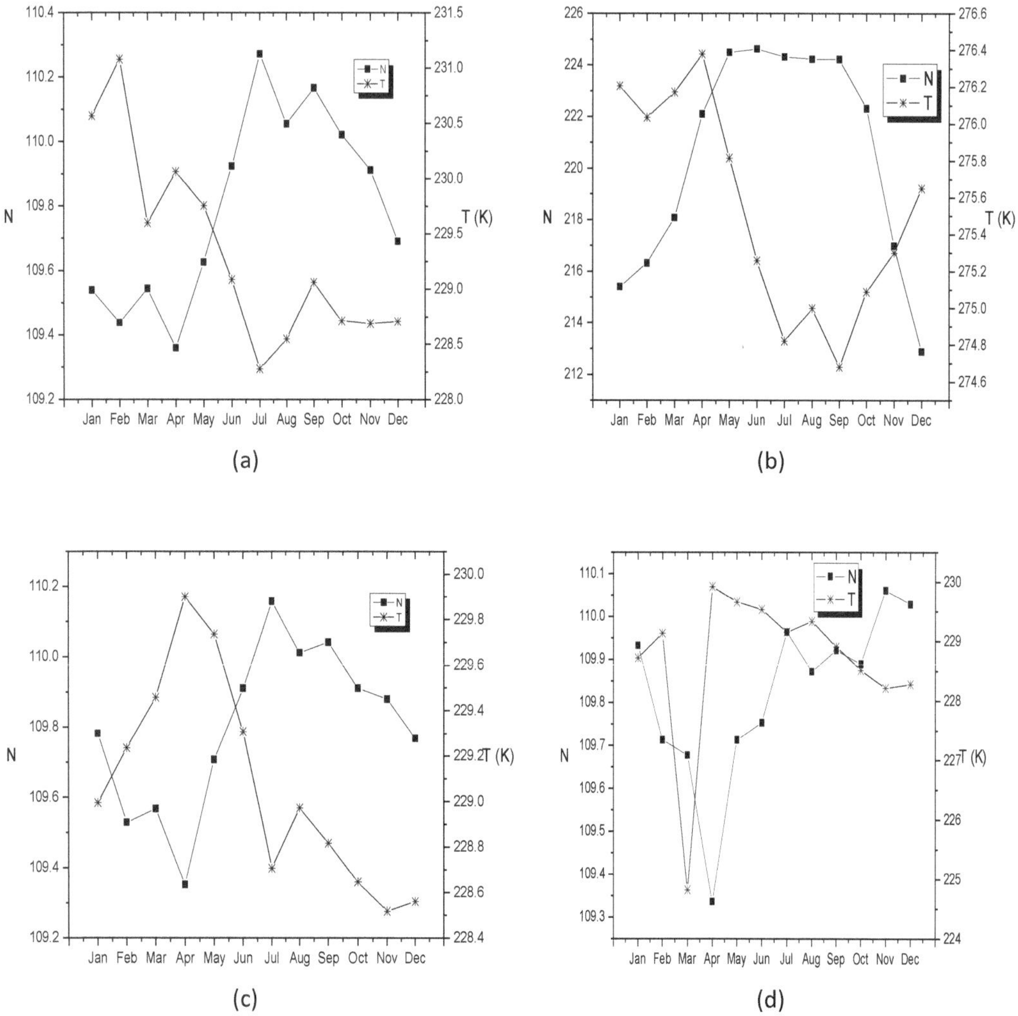

Figure 7. Monthly variation of Refractivity and Relative Temperature at Upper Level in (a) Coastal (b) Guinea savannah (c) Midland and (d) Sub Sahel regions.

finding of Olusina and Odumade (2012). Temperature values are higher in the northern regions than the southern regions (Table 2). From Table 3, high values in variation of refractivity were noticed in the northern regions.

Conclusion

Using satellite data from CM – SAF, it has been established that monthly variation of refractivity is majorly influenced by relative humidity and temperature which also vary geographically and seasonally. At low level and mid level, relative humidity and refractivity decreases northward while temperature increases northward. The dip noticed (in August) in monthly variation of relative humidity and refractivity in coastal and Guinea savannah regions at low level may be associated to the August break (a short period of dryness). This break (known as the intra-monsoonal period) is believed to be a consequence of several factors such as coastal upwelling and the northern advance of the subtropical high pressure systems of the southern Atlantic ocean or because of the circulation aloft which becomes divergent and subsident due to frequent occurrence of inversions and isothermals in the upper atmosphere along the coast when the weather zone – E makes its appearance a short way inland from the coast (Adedokun, 1986; Balogun and Adedokun, 1985; Balogun, 1981; Olaniran and Sumner, 1989; Trewartha, 1970). The values of relative humidity and refractivity are partially low during the dry season and high during the rainy season in southern regions and much lower in the dry month at northern regions.

There is a unique variation in the ranges of these parameters (relative humidity, temperature and refractivity) over the four regions in the country.

Table 2. Analysis result of temperature over Nigeria.

Region	Low level					Mid level					Upper level				
	Mean	Max T (K)	Min T (K)	ΔT (K)	SE	Mean	Max T (K)	Min T (K)	ΔT (K)	SE	Mean	Max T (K)	Min T (K)	ΔT (K)	SE
Coastal	291.18	292.65	289.46	3.19	0.34	275.29	276.13	274.39	1.74	0.17	229.34	231.08	228.71	2.37	0.25
Guinea savannah	292.21	292.95	287.51	5.44	0.44	275.53	276.21	274.68	1.53	0.17	229.04	229.98	228.53	1.45	0.14
Midland	293.69	296.78	291.14	5.64	0.56	275.61	276.63	274.7	1.93	0.17	229.07	229.73	228.51	1.22	0.13
Sub Sahel	295.58	299.87	293.97	5.90	0.63	275.79	277.3	274.35	2.95	0.21	228.68	229.91	228.21	1.7	0.38

Table 3. Analysis result of refractivity over Nigeria.

Region	Low level					Mid level					Upper level				
	Mean N	Max N	Min N	ΔN	SE	Mean N	Max N	Min N	ΔN	SE	Mean N	Max N	Min N	ΔN	SE
Coastal	328.32	333.35	323.42	9.93	0.97	220.86	223.45	215.26	8.19	0.79	109.80	110.27	109.36	0.91	0.08
Guinea savannah	328.16	337.51	312.85	24.66	2.30	220.49	224.47	212.88	11.59	1.23	109.78	110.15	109.34	0.81	0.07
Midland	312.99	333.05	283.00	50.05	5.75	217.83	223.97	210.97	13.00	1.79	109.80	110.16	109.35	0.81	0.07
Sub Sahel	300.42	337.97	265.18	72.79	9.13	213.40	224.26	204.66	19.60	2.24	109.82	110.06	109.33	0.73	0.06

It was observed that the ranges of the parameters increases northward at low level and mid level. This may be associated with the seasonal variation in these regions. As the southern part (coastal and Guinea savannah regions) of the country is characterized with long rainy season and short dry season, whereas the northern part (midland and sub Sahel regions) is characterized with short rainy season and long dry season (Imo and Ekpenyong, 2011). The ranges of relative humidity at low level are 17.85, 25.27, 56.19 and 72.51% in coastal, Guinea savannah, midland and sub Sahel regions, respectively at low level. For the temperature at this level, the differences are 3.19, 5.44, 5.64 and 5.9 K, respectively. This may be accounted for the variation observed in refractivity at this level. From the foregoing, the behaviour of radio wave propagation will vary from one region to the other.

ACKNOWLEDGEMENTS

The authors wish to express their gratitude to the officials and management of the Department of Climate Monitoring SAF, Deutscher Wetterdienst, Offenbach, Germany for providing the data. The efforts of Dr. Joerg Schulz, who incidentally was the host of the corresponding author during his visit to CM-SAF, and Mr. Markus Jonas in making sure that the data were available in ASCI format is also greatly appreciated.

REFERENCES

Adediji AT, Ajewole MO (2008). Vertical profile of radio refractivity gradient in Akure Southwest Nigeria. Progress Electromagn. Res. 14:157-168.

Adedokun JA (1986). On a relationship for estimating precipitable water vapour aloft from surface humidity over West-Africa. J. Climatol. 6:161-172.

Adeyemi B, Aro TO (2004). Variation in surface water vapour density over four Nigeria stations. Nig. J. Pure

Appl. Phys. 3:46-55.
Adeyemi B (2004). Tropospheric radio refractivity over three radiosonde stations in Nigeria. Ife J. Sci. 6(2):167-176.
Adeyemi B (2006). Surface water vapour density and tropospheric radio refractivity linkage over three stations in Nigeria. J. Atmos. Solar-Terr. Phys. 68:1105-1115.
Adeyemi B, Emmanuel I (2011). Monitoring tropospheric radio refractivity over Nigeria using CM – SAF data derived from NOAA 15, 16 and 18 satellites. Indian J. Radio Space Phys. 40:301-310.
Balogun EE, Adedokun JA (1986). Note and correspondence on the variations in precipitable water over some West African stations during the special observation period of WAMEX. Am. Meteorol. Soc. 1(14):772-775.
Freeman RL (2007). Radio system design for telecommunications, 3rd ed., Wiley–Interscience, New York.
Guanjun G, Shukai L (2000). Study on the vertical profile of refractive index in the troposphere. Int. J. Infrared Millimeter Waves. 21(7):1103-1112.
Imo JE, Ekpenyong N (2011). Extreme climatic variability in northwestern Nigeria: An analysis of rainfall trends and patterns. J. Geogr. Geol. 3(1):51-62.
Olaniran OJ, Sumner GN (1989). A study of climate variability in Nigeria based on the onset, retreat, and length of the rainy season. Int. J. Climatol. (UK). 9:253-269.
Olaniran OJ (1983b). The onset of rainfall and the start of the growing season in Nigeria. Niger. Geogr. J. 26.
Olusina JO, Odumade OM (2012). Modeling climatic variation parameters of Nigeria using the statistical downscaling approach. FIG working week 2012, knowing to manage the territory, protect the environment, evaluate the cultural heritage, Rome, Italy, pp. 6-10.
Oyedum OD, Gambo GK (1994). Surface radio refractivity in northern Nigeria." Nig. J. Phys. 6:36-41.
Priestley JT, Hill RJ (1985). Measuring high–frequency refractive index in the surface layer. J. Atmos. Oceanic Technol. 2(2):233-251.
Trewartha GT (1970). The Earth's Problem Climates. The University of Wisconsin Press. pp. 91-110.
Willoughby AA, Aro TO, Owolabi IE (2002). Sesasonal variations of radio refractivity gradient in Nigeria, J. Atmos. Solar-Terr. Phys. 64:417-425.

20

Performance of community based organizations in managing sustainable urban water supply and sanitation projects

G. E. Kifanyi[1] , B. M. B. Shayo[2] and J. M. Ndambuki[1]

[1]Department of Civil Engineering, Tshwane University of Technology, Private Bag X680, 0001 Pretoria, South Africa.
[2]Department of Civil Engineering, Mbeya University of Science and Technology, P. O. Box 131 Mbeya, Tanzania.

The advocacy of climatic change and ever increasing poverty has its impact on environmental degradation. Most of underdeveloped countries have experienced this impact through technical, economic, institutional and social constraints on water supply and sanitation infrastructures. The impact has much effect in urban areas (informal settlements) due to the nature of such settlements. Lack of a sustainable approach and an appropriate technology for the provision of water and sanitation has resulted in decreased coverage in terms of percentage of the population with access to safe water and adequate sanitation, thus necessitating the need for appropriate and sustainable water and sanitation systems. New strategies for water supply delivery, particularly the community participation has emerged and some Community Based Organizations (CBOs) have made a significant contribution to the development of community participation approaches. This paper analyses the effectiveness of community participation, appropriate technologies and the institutional arrangements for the overall performance. Tanzanian CBOs managing water supply and sanitation systems were studied. Results revealed that CBOs are effective in cost recovery as well as effective water and sanitation provision in low-income earners areas.

Key words: CBOs, community participation, environmental degradation, water supply and sanitation.

INTRODUCTION

There is a direct linkage between livelihood and well being of human beings and water supply and sanitation services. Improved water supply and sanitation ultimately contribute towards nation building and prosperity by enhancing the health status of the common mass and thus, their economic productivity.

Despite significant progress recorded in the extension of basic water supply and sanitation services across the developing countries, climatic change and the ever increasing population and poverty has increased its impact on environmental degradation. Most of undeveloped countries have experienced this impact through technical, economical, institutional and social constraints on water supply and sanitation infrastructures. The impact has much effect in urban areas (informal settlements) due to the nature of such settlements. Lack of a sustainable approach and an appropriate technology for the provision of water and sanitation has resulted in decreased coverage in terms of percentage of the population with access to safe water and adequate sanitation, thus necessitating the need for appropriate and sustainable water and sanitation systems. Moreover, the high rate of growth of urban populations in Africa and Tanzania in particular, has resulted partially from an increase in the number of urban households, brought about by changes in standards of living and attitudes towards dependence. The rate of growth in number of households across Africa averaged 3.1% between 1985 and 2000, and is set to continue at this rate until 2030 (Thompson et al., 2001). The consequent increase in demand for basic housing and services for urban populations, as well as skewed distribution of investment towards affluent suburban developments, has resulted in rapid expansion of unplanned and unserviced settlements. Overcrowding exacerbates rate of transmission of infectious diseases, such as gastro-intestinal infections, and respiratory diseases such as tuberculosis, commonly associated with poor water quality, poor sanitation facilities, poor ventilation and air pollution (Esrey et al., 1990).

Population growth in Tanzanian urban areas has been increasing rapidly since 1960s. The 1978 and 1988 national census shows that a bigger share of urbanization results from rural-urban migration (Kaliba, 2002; Kessy, 2002). The resulting fast growth in population in urban areas increases pressures on resources to satisfy the increased demand for clean water and sanitation facilities. The increased demand for urban housing has been accompanied by inadequate planning enforcement that has resulted in increased informal settlements with inadequate water and sanitation services. Unsafe water and poor sanitation are the primary causes for the vast majority of water borne and primarily diarrhoeal diseases. Every year, unsafe water, coupled with lack of basic sanitation, kills at least 1.6 million children under the age of five years; more than eight times the number of peoples who died in the Asian tsunami of 2004 (WHO/UNESCO, 2006). Water borne diseases also inflict significant economic burden through the loss of productivity in the workforce and through increasing national health care costs. As a consequence of these pit falls, a billion people were locked in a cycle of poverty and disease (UNICEF/WHO, 2004).

This paper aims to address the effectiveness of community participation, appropriate technologies and institutional arrangements for the overall CBOs' performance in Tanzania. Through literature reviewed and case study, Tanzanian CBOs performance in managing sustainable urban water supply and sanitation projects was analyzed. The study was based on Hanna Nassif Community Development Association (HNCDA) in Dar Es Salaam, Tanzania (a Community Based Organization (CBO) Project).

METHODOLOGY

The study was carried out through physical observations, documentary analysis, quantitative and qualitative methods. Data for the study was collected through a focused survey group of randomly selected CBOs practitioners and key stakeholders. Interview/questionnaires supported with semi structured interview techniques were used to collect data. The interview with water kiosks operators and public water standpipe operators, water users who were found fetching water at water collection points and head of households were carried out to evaluate the satisfaction of the users and the performance of the CBOs. Findings from the 1994 and 1998 baseline survey studies were used as a basis for comparing the trend of water supply and sanitation services. The 1998 baseline study indicated that the number of registered households at Hanna Nassif settlement area were 1897. These registered households were used as a basis for selection of the households for the study. A total of 96 households (about 5.1% of the total registered households) were randomly selected for interview and physical observation of sanitation systems. In Hanna Nassif settlement area, there were 10 water kiosks and 20 public water standpipes. Interview was carried out with 8 water kiosks operators and 17 standpipes operators, HNCDA Secretary, 3 HNCDA Committee members, HNCDA treasurer, and 270 water users who were found fetching water at the water collection points. A total of 396 respondents were interviewed. Tables 1 and 2 summarize the interview/ questionnaires distribution patterns.

Existing conditions of water supply and sanitation services

The problem of water and sanitation in urban and rural areas of developing countries including Tanzania is a major concern. The available water sources throughout the world are becoming depleted and this problem is aggravated by the rate at which populations are increasing, especially in developing countries. Currently, some 30 countries are considered to be water stressed, of which 20 are absolutely water scarce. It is predicted that by 2020, the number of water scarce countries will likely approach 35 (Rosegrant et al., 2002). It has been estimated that, one-third of the population of the developing world will face severe water shortage by 2025. Furthermore, the limited water resources as well as the lack of safe drinking water and sanitation are the most serious challenges of the twenty first century. Over 1 billion people lack access to clean water with nearly all of them living in developing countries. Yet, 2.6 billion people, 40% of the world population, half the developing world lack even a simple improved latrine (UNICEF/WHO, 2004; Elimelech, 2006).

Water supply

According to Antonio (2005), more than 1.2 billion people in the world still lack access to safe drinking water. Barney (2005) noted that, over the next 30 years, virtually all of the world's population growth is expected to be concentrated in urban areas in the developing countries, with its attendant socio-economic and environmental impact. This portrays that developing countries are facing great challenges in meeting community water supply needs.

In Tanzania, increasing levels of poverty, population growth and lack of a sustainable housing policy mean that urban growth is often absorbed into informal settlements. These areas are characterized by lack of basic infrastructure like water (AFTU 1&2, 2002). Many urban residents cannot afford housing, and authorities themselves have few resources with which to improve or maintain infrastructure and services. Consequently housing, health, water quality and environmental conditions in the growing informal settlements of Tanzania's cities are extremely poor (Kessy, 2002).

The National Water Policy of Tanzania stipulates the need to have water resources with acceptable quality. Increased human activities including poor land use practices, as well as uncontrolled abstractions and pollution of water body impact on the quality and quantity of the available water (URT, 2002). Tanzania has a population of more than 38 million, with 25% of this population living in urban areas, while water supply extends to some 85% of the

Table 1. Interview/questionnaires distribution.

S/No	Category	Category population	Total respondents Interviewed
1	Kiosk Operators	10	8
2	Water standpipe Operators	20	17
3	HNCDA Committee members	4	3
4	HNCDA Treasurer	1	1
5	HNCDA Secretary	1	1
6	Head of Households (Hanna Nassif Registered houses)	1897	96
	Total	1933	126

Table 2. Interview distribution at water collection points.

Kiosk	Water collection point	Number of water collection points	Targeted number of interviewees	Total respondents interviewed
1	Kiosks	10	150	97
2	Public water standpipes	20	250	173
Total		30	300	270

urban population service is often poor and at best erratic. The urban poor do not receive adequate services although in some areas this situation is beginning to improve (UWSA, 2001).

The Tanzania urban water and sanitation sector like many similar institutions is undergoing decentralization. This process means a shift away from organizational structures where the district water engineer was responsible technically to the ministry but administratively to the local municipality, which in turn was responsible for managing cost recovery. Since 1994, a new scenario has emerged in Tanzania and now all the responsibility is devolved to the Urban Water and Sewerage Authorities (UWSA). With the support of legislation, the UWSA have significant autonomy linked to the achievement of key performance indicators.

The impact of decentralization is most apparent in the obligations of the Ministry of Water and the UWSAs. The traditional ministry role of 'provider of services' is now one of regulator of policy and facilitator of service expansion through fund allocation and UWSA performance monitoring. UWSAs is responsible for provision of services in quantity and quality standards required, set water and sewerage tariffs, collect revenue and plan and execute new projects for water supply. Table 3 shows the main source of drinking water for Tanzanian urban households in percentage. Table 3 reveals that few people have access to water supply (only 34.1% of the urban households is served by utility directly). Despite the many opportunities UWSAs face, many hurdles including the fact that not all customers pay their water bills, current supplies are intermittent, unaccounted for water (UFW) is in excess of 40% and leakage is also high (UWSA, 2001).

Typically, the urban poor are not seen as a viable customer base as the authority is pre-occupied by an attempt to recover costs for other market segments. Most of the poor segment in urban slum areas makes use of unsafe water for domestic purposes (UN-HABITAT, 2009). The water shortage drives most of the urban poor to fetch unsafe water for domestic use. Figure 1 shows the situation where most of urban poor fetch water for domestic purposes.

Sanitation

In developing countries such as Tanzania, rapid population growth and urbanization is creating an added demand for housing, infrastructure services including sanitation services. Providing sanitation services especially for the poor who are living outside the designated residential areas like illegal settlements or slums (informal settlements) is a challenge. The World Bank (2000) report estimates that almost 26% of the global urban population, over 400 million people, lack access to the simplest latrine.

Sanitation coverage in developing countries (49%) is only half that of the developed world (98%). In Sub-Saharan Africa, the coverage is a mere 36%, and over half of those are without improved sanitation. Similarly, nearly 1.5 billion people live in China and India without access to improved sanitation services (WHO/UNESCO, 2006). The number of deaths attributable to poor sanitation and hygiene alone may be as high as 1.6 million a year. Statistics on wastewater treatment reveal that almost 85% of global wastewater is discharged without treatment leading to serious impacts on public health and the receiving water environment.

The situation is even worse in areas of low-income settlements. Septic tanks and feeder networks regularly discharge effluent into street gutters, open streams or drainage canals. This creates unpleasant living conditions, public health risks and environmental damage (GHK, 2002). Sanitation coverage refers to the percentage of the population that has access to safe collection and disposal of wastewater. In the Southern Sahara countries, the coverage increased from 28% in 1980 to 36% in 1990. In 2003, the access to adequate excreta disposal was averaged to only 35% of the population. About 21.7% of the population had access to flush toilets, 66.8% to pit latrines and 11.6% use other options including open defecation (WHO, 2003). According to Ngware and Kironde (2000), only 6% of Dar-es Salaam's population has access to the choked sewers while 8% use septic tanks and 80% use pit latrines. Moreover, in informal settlements, the latrines are constructed in wrong locations due to lack of space. This situation has also been observed in Kinondoni Hanna Nassif Dar-es Salaam, Tanzania (Figure 2).

Summary results of review of existing conditions of water supply and sanitation services in Tanzania

Despite many opportunities, Tanzania has a severe

Table 3. Main source of drinking water for Tanzanian urban households (% of households).

Source	Households (%)	Improved/ unimproved	Utility/ non-utility
Piped into dwelling/yard/plot	18.6	All improved sources: 77%	Households served by utility directly: 34.1%
Public tap	15.5		
Neighbour's tap	32.8		
Protected well / borehole	10.1		
Unprotected well / borehole	8.2	All unimproved sources: 23%	Households not served by utility: 65.9%
Surface sources (river, lake, etc)	4.7		
Vendor / tanker truck	7.2		
Springs	1.6		
Other	1.5		

Source: (TDHS, 2004).

Figure 1. A woman fetching water from a leaking pipe in a grubby area found at Bagamoyo town (a town near Dar es Salaam City). Source: UN-HABITAT (2009).

problem of safe water supply and access to improved sanitation services. Few people have access to water supply, with only 34.1% of the urban households being served by utility directly. Moreover, Tanzanian Urban Water and Sewerage Authorities (UWSAs) face many hurdles such as customers who do not pay their water bills, current supplies are intermittent, unaccounted for Water (UFW) in excess of 40% and high leakage of water. Furthermore, in sanitation sector the situation in Tanzania is even worse than water supply. For example, a survey conducted in 1996 revealed that only 6% of Dar es Salaam's population has access to the choked sewers

Figure 2. In house latrines found in Kinondoni Hanna Nassif Dar es Salaam, Tanzania.

while 8% use septic tanks and 80% use pit latrines. However, the introduction of Community Based Organizations (CBOs) in Tanzania in upgrading informal settlements may improve the situation.

Performance of community based organizations (CBOS) in managing sustainable urban water supply and sanitation projects

Overview

Historically, the performance of urban water and sanitation systems in developing countries including Tanzania remains below expectation. It should be recognized that urban water management poses extraordinary complex problems that cannot be solved by individual stakeholders. The failing of systems particularly in developing countries has been partly the result of a top-down approach with limited involvement of stakeholders (Khatri and Vairavamoorthy, 2007).

At the root of many failures is the lack of public participation, including participation of organized forms of the public such as CBOs and NGOs. Some would argue it is the 4th pillar of sustainable development. Participation is not happening due to governance failures as well as lack of awareness. People still do not understand the concept of sustainability and do not know how they can contribute to its achievement. This leads to gaps and divergences in governments, CBO and NGO efforts to raise awareness among citizen groups.

Conventional wisdom dictates that without community participation, there is little likelihood of sustainability being realized (Narayan, 1995; Oyesiku, 1998). This is in part a pragmatic recognition of Governments' inability to deliver services, but in part an ideological proposition which values concepts such as 'empowerment', and 'capacity building' for their own sake. Even from a strictly practical approach, a number of the issues mentioned earlier illustrate the need for capacity building at the community level as well as at the level of Government or NGO, CBO.

Sustainability of facilities was reported by Overseas Development Institute (ODI) (2004), to be a major concern in the developing countries. Performance on sustainability is often gauged by looking at the number and proportion of functioning and non-functioning facilities. It also further noted that a functioning facility requires attention to a range of managerial, social, financial, institutional and technical issues.

In developing countries (including Tanzania), a significant number of projects, including those in the water and sanitation sector, fail to deliver benefits to society over the long term (Antonio, 2005). Part of the cause of this failure lies in poor understanding of the issues of impact and sustainability. Abrams (1998) pointed out that, "if the water flows, then all of the many elements which are required for sustainability must have been in place. There must have been money for recurring expenses and for the occasional repair, there must have been acceptance from the consumers of the service, the source supplying the service must have been adequate, the design must have been properly done, and there must have been sound construction".

To quote from the work of Abrams (1996) on the review of the African domestic water and sanitation: "It is generally agreed that community engagement and empowerment is the solution to the sustainability of water supply and sanitation services". The hallmarks of empowerment and capacity building are factors such as transparency, partnership, flexibility, respect, and empathy.

Education in health and hygiene, training in maintenance and the handling of cash, and involvement of women in community institutions and decision making, are key activities needed to create local capacity to manage. On the part of Governments and NGOs/CBOs, listening and learning from the community, developing respect for existing methods of organization, problem solving, conflict resolution, and decision making, are essential components of such capacity building work. This calls for a fundamentally new approach to water, sanitation and human settlements whose aim is to satisfy the water needs while protecting the eco-systems.

Governments need to support and reinforce the efforts of the NGO/CBO community in this respect. This would lead to greater participation of the public to promote sustainable development at local levels. For example, in the United Kingdom, issues of sustainability are now being included in the school curriculums via for example, the sustainable Design Awards for 16-18 year olds and the sustainable Technology Education Project for 11-16 year olds. Such programmes should be encouraged, replicated and amply funded to raise awareness (UN, 2004).

Two different kinds of partnerships have been tried to increase effectiveness in poverty reduction, and sustainable development. The first is through working with the private sector, especially for urban infrastructure (including provision for water and sanitation) and in some cases housing finance. The second was working with the urban poor, including their community organizations and local NGOs. For extending and improving provision for water and sanitation, privatization proved disappointing; it did not succeed in bringing in large new sources of capital for investment from private markets (which had been one of the justifications for promoting it). Private companies were interested primarily in large cities with sizable middle classes rather than the smaller, poorer urban settlements where most of those without adequate provision actually live.

In Tanzania, CBOs and NGOs are observed currently participating fully in development and capacity building of urban and rural areas' projects, which central Government and local Governments are embarking upon, particularly with assistance of UNICEF, IMF, World Bank and other international organizations financial supporters. CBOs and NGOs participation coupled with Governments and International supports appeared to be the right step at right time in the right direction in quest to ensuring better access to basic sanitation and safe drinking water.

Performance trend of the Dar-es- Salaam (DSM) water supply

The government of Tanzania like any other developing country is unable to provide and maintain the most needed services to the urban community. This situation calls for more involvement and participation of the community aiming at stimulating the responsibility and willingness of community to operate and maintain their water systems. In DSM, the demand for clean water has in most cases outstripped the supply because the supply has been faced with inadequate and unreliable water sources. Other problems include unreliable power supply, dilapidated main water supply pipes from Lower Ruvu, and Upper Ruvu, characterized by leakages, problems of distribution systems, spaghetti of the subscribers' distribution pipes as well as illegal connections.

In 1984, the Tanzania government formed National Urban Water Authority (NUWA), in order to redress the problems of inadequate water supply throughout all urban areas in the country. NUWA was expected to start with DSM city and gradually extend its jurisdiction to all urban areas in the country. In 1997, the water demand in DSM was estimated to be 409,500 m^3 while the supply was 191,000 m^3 (Ngware and Kironde, 2000). Therefore, the supply was only 47% of the demand. It was then the same year the NUWA was transformed and renamed as DAWASA to operate in DSM, and was further given mandate to take over sewerage services in the city. In 2003, DAWASA was privatized to a foreign company- the Dar-es-Salaam City Water Company.

DAWASA was privatized by the Government of Tanzania so that to fulfill loan conditions of US$98.76 million (that is, 164.6 billion Tanzanian Shillings (TZS), 1 Tanzanian shilling = 0.0006 US dollars) from the World Bank, International Development Agency (IDA), the African Development Bank (AfDB), and the European International Bank (EIB) for rehabilitation of water supply systems for the Dar-es-Salaam City, Kibaha Town and Bagamoyo (URT, 2004). It was expected that major rehabilitation of the water supply system of the city would be covered both in Lower and Upper Ruvu sources. Table 4 shows the water connection and distribution by City water in Kinondoni Municipal.

The Government of Tanzania expected the water supply service provision within the city to improve after the rehabilitation work, whereby leakages were expected to decrease from the 55 to 28% and revenue collection to

Table 4. Water connection and distribution by City Water in numbers at Kinondoni Municipal.

Domestic	Institutional	Commercial	Industrial	Total
5311	63	1056	6	6436

Source: (URT, 2004).

Table 5. Perceptions of trends in water supply over the last five years.

Indicator	Dar es Salaam (%)		Other urban (%)		Rural areas (%)		All areas (%)	
	I ↑	D ↓	I ↑	D ↓	I ↑	D ↓	I ↑	D ↓
Dry season shortages	20	33	27	26	20	29	21	29
Cleanliness	15	17	17	14	7	22	11	19
Queuing time	12	26	17	14	12	13	12	17
Cost	8	32	14	16	12	8	11	16
Distance	17	13	17	11	9	14	13	13

I ↑, Improved; **D** ↓, deteriorated. Source: Views of the People Survey (2007).

increase from 45 to 90% (URT, 2004). However, people perceived differently. Table 5 shows the perceptions of trends in water supply. The survey data indicators in Table 5 revealed that more efforts are needed to improve the water supply in urban as well as rural areas. However, results from the Jan et al. (1993) on developing, managing and maintaining community water supplies and sanitation projects suggests that water and sanitation targets can be achieved through empowering individuals, households, and communities to take charge of their development needs. Hence, full involvement of communities in all stages of project implementation and management is the correct pragmatic approach for the improvement of water supply delivery and sanitation services in urban as well as rural areas.

CASE STUDY: HANNA NASSIF COMMUNITY DEVELOPMENT ASSOCIATION (HNCDA)

Evolution of the Hanna Nassif Community Development Association (HNCDA)

Hanna Nassif ward is in Kinondoni Municipality Dar Es Salaam, Tanzaina, with three sub-wards; Hanna Nassif, Mkunguni, and Kisutu. The upgraded area covers a part of the Hanna Nassif sub ward and another part of the Mkunguni sub ward. According to the 2002 census report, the upgraded settlements cover an area of 45 ha, with a population of twenty thousand (20000) people. The need for establishing a community based organization in Hanna Nassif settlement stemmed from the commonly felt problems, that is, flooding, poor road and drainage networks, inadequate water supply and sanitary services.

Hanna Nassif was earmarked for upgrading since the second phase of the National Sites and Services and Squatter Upgrading Programme phase II and III that commenced in 1976 and 1981 respectively (Kessy, 2002). The programme was financially supported by the World Bank, but due to some reasons the efforts to upgrade Hanna Nassif could not succeed until 1991, when the move to improve basic infrastructure through community based approach was initiated by the community itself. This was a proper approach since community approach sought to empower local organizations and individuals through an atmosphere of dignity and participation, and an orientation to achieving durable results (O'Regan and Conway, 1993).

The primary objective of infrastructure improvement at Hanna Nassif was to improve the living environment by empowering the community to pioneer the development of basic infrastructure in collaboration with other stakeholders (Kessy, 2002). In recent years, many development agencies have focused on the promotion of participatory approaches to encourage bottom-up planning and empowerment of communities so that they take more control of development activities. It was under this effort that the Hanna Nassif community based infrastructure upgrading was promoted by the National Income Generation Programme (NIGP) (Salewi, 2004). This is in line with the National Water Policy on the issue of water for low-income groups and community user groups (the goal is to improve water and sanitation services in low income urban and peri-urban areas) (URT, 2002).

Performance trend of Hanna Nassif water supply and sanitation services

Performance of water supply services

Water Supply Condition: The Hanna Nassif water supply depends on the water main pipe through the

Table 6. Water supply connection in Hanna Nassif settlement.

Type of connection	1994 Total houses connection (%)	1998 Total houses connection (%)	Remarks
House connection	18.8	20	In house water network
Plot connection	39.4	8	Single tap
No water connection	41.8	72	Were buying water within settlement area

Source: UCLAS/ILO (1994; 1998).

settlement to Kinondoni area. The situation of water supply at Hanna Nassif settlement area is relatively better compared to the situation it was before the commencement of the project, out of 96 head of households interviewed, 85 (88.5%) head of households agreed that the project has reduced water shortage and improved drainage system. The same pattern of result was portrayed from the interview conducted with water users who were found fetching water at water collection points, out of 270 respondents, 246 (91.1%) respondents said that water shortage was reduced and drainage system was relatively improved. They pointed out that before commencement of the project pit latrines flooding was a major problem during rainy seasons. Construction of drainage system has reduced flooding problem. Moreover, the 1994 and 1998 trend of water supply was almost the same. Findings from the 1994 and 1998 baseline studies indicated that the Hanna Nassif settlement area did not experience frequent water shortages either from rationing or low pressure. The 1994 survey revealed that 63.3% of the total respondent households experienced a water shortage only once a month. The results from the 1998 survey revealed almost the same about 60% of the interviewed households experienced a water shortage within the same period. Moreover, findings of the 1994 and 1998 baseline studies showed that majority of the households get their water for domestic use from within the settlement. Furthermore, housing registration survey which was carried out on May 1998 revealed that out of 1897 houses, 20% (373) had private water connection, 8% (149) houses had plot connection and the remaining 72% (1375) had no water connection. People living in houses with no water connection were buying water within settlement area. Table 6 summarizes the water supply connection in Hanna Nassif settlement between 1994 and 1998.

Water Consumption: The HNCDA study indicated that water consumption within the Hanna Nassif settlement was gradually increasing. There was relatively small change of water consumption compared to the water consumption patterns revealed in the 1994 and 1998 baseline studies. The study showed that about 27% of the 396 total respondents interviewed were using less than 5 buckets per day or below 100 Lt per day, 46.2% between 101-200 Lt per day, 18.3% between 201-250 Lt per day and the remaining 8.5%, were using above 250 Lt per day. Findings from the 1994 and 1998 baseline studies portrayed almost the same trend of water consumption. The 1994 survey study revealed that 47.9% of the sample population used below 100 Lt per day, 40.5% between 101-200 Lt per day, 7.2% between 201-250 Lt per day and 4.5% used above 250 Lt per day. The 1998 water consumption pattern indicated that 34% of the sample population was using below 100 Lt per day, 43% between 101-200 Lt per day, 15% between 201-250 Lt per day and the remaining 8% were using above 250 Lt per day. Table 7 shows water consumption pattern within the settlement between 1994 and 1998.

Performance of sanitation services

The HNCDA study revealed that most of the houses in Hanna Nassif settlement area use pit latrines, out of 96 households which were visited, 79 (82.3%) households use pit latrines, 15 (15.6%) households use flush toilets and the remaining 2 (2.1%) households had no toilet facilities. Table 8 shows the summary of the observation results.

Moreover, results from the interview which was conducted with water users who were fetching water at water collection points reflected almost the same pattern; out of 270 respondents, 238 (88.2%) said that they use pit latrines, 29 (10.7%) said they use flush toilets and the remaining 3 (1.1%) respondents said that they had no toilets; however, they further pointed out that their toilets were demolished by flood. Table 9 summarizes the results of the interview conducted at the water collection points.

The 1998 baseline study showed that out of 1897 households that were registered, only 128 (6.7%) were using water closet and septic tanks sanitation system, 1758 (92.7%) had pit latrines and the remaining 11 (0.6%) had no toilets. Moreover, the 1998 baseline study revealed that one among the problems facing residents of Hanna Nassif was that of flooded latrines, about 11% of the sample households had been experiencing the problem. However, recent observations in the Hanna Nassif settlement revealed that this problem has been reduced as a result of improvement on the drainage system.

Table 7. Water consumption pattern.

Amount of water consumption (L)	Year 1994 (%)	Year 1998 (%)
Below 100	47.9	34
101-200	40.5	43
201-250	7.1	15
Above 250	4.5	8

Source: UCLAS/ILO (1994; 1998).

Table 8. Use of sanitation systems at Hanna Nassif settlement.

Sanitation system	Number of households	Percentage
Pit latrine	79	82.3
Closet	15	15.6
No Toilet	2	2.1
Total	96	100

Table 9. Summary of results of the interview conducted at water collection points.

Type of sanitation system	Number of households	Percentage
Pit latrine	238	88.2
Flash Toilet	29	10.7
No Toilet	3	1.1
Total	270	100

Institutional arrangements

The source of the Hanna Nassif community water supply is from the City Water main distribution pipe along Kawawa Road. There are 10 public water kiosks, and 20 standpipes. The land on which the kiosks and the standpipes are installed was obtained from individual landlords. The landlords share the net revenue collected each month as partners in the service (that is, 50% of total collections less operational costs and the City Water bill).

The water supply serves the entire population of the Hanna Nassif sub ward and part of the Mkunguni sub ward. Many respondents of this study were satisfied by the service, but lamented that the HNCDA has raised the price of 20lt bucket dramatically from 0.006 to US$0.012 (that is, 10/= to 20/= Tanzanian shillings (TZS.)). The respondents were comparing the price with those charged by some religious institutions at the area who own deep wells and sell water at a price of 0.006 US$ (that is, 10/=TZS) per 20lt. bucket.

The distribution of water from the main pipes is not reliable due to leakages and the dilapidated condition of the pipes, which has forced the City Water Company to introduce an allocation system of supply to most of the city areas. The introduced water allocation system has caused the community standpipes and some kiosks to remain dry in some days when the area is not supplied from the mains. The kiosks are installed with reservoir five tanks with capacity of 5,000lt, and the remaining five kiosks with tanks of 10,000lt capacity. However, this reserve cannot supply water to the community satisfactorily for a day, where as the allocation can sometimes extend for two or more days.

Since the Hanna Nassif Community water supply depends totally on the supply of the City Water Company, the performance of the Community Water Supply to a greater extent depends on the performance of the City Water. Therefore, the Institutional arrangement with the Community Water Supply Management to a greater extent influences the performance and the assurance of the sustainability of the service to the settlements. The government of Tanzania as in many developing countries is unable to provide and maintain the urban water supply systems. But since water and sanitation are critical components of development, the Central Government, Local governments, and Government Development Partners have a role to play in order to encourage and support the Communities, NGOs and CBOs in developing and maintaining water supply and Sanitation systems. This is critical especially in low-income urban communities.

Figure 3. A woman fetching water at a water kiosk.

Performance of the actors

The performance of the actors in community-based water supply depends on the efficiency of the management team. The community participation in community based water services aims at stimulating the responsibility and willingness of community to operate and maintain their water and sanitation systems. They achieve this not only by training those involved in the day to day operations, but also by participatory planning to tailor services by organizing water committees to represent their community. The organization of water committee should focus on roles and responsibilities of the most affected, for example, the women and the poor, so that they would not be excluded from the use of the system and from decision-making. The HNCDA water committee has only 4 members that include two women. This committee is supposed to administer 20,000 people on water issues. More involvement of women especially in decision-making could help in the improvement of the service provision. Figure 3 shows a woman fetching water at a water kiosk.

Accountability

The HNCDA's water supply service has three stages of accountability:
(a) Operators are answerable to the HNCDA water committee.
(b) The water committee is accountable to the executive committee.
(c) The executive committee is held accountable to the members of the HNCDA.

This set up needs frequent reporting of the progress from the operators up to the executive committee and from the committee to the general meeting. Apart from the reporting system, there is a need to have an organ within the system which is directly accountable to the users of the water. Lack of accountability to the users can lead to complaints that are detrimental to the overall performance of the water supply service. Financial accountability is also another issue of concern in Hanna Nassif community water supply. When interviewed, the secretary of the executive committee admitted that the HNCDA has no culture of frequent financial reports by the treasurer and/or the water committee to its members. The HNCDA constitution provides for annual financial reports in each general meeting. Frequent community meetings with an oral report given by the treasurer followed by questions and answers could help to minimize questions on financial accountability.

Financial records and transparency

In Hanna Nassif, the kiosks and standpipes operators are not trained to manage financial recording and reporting. May be that is why the water points' financial records, are not kept by the operators. All operators who were interviewed in this study said that they are only obliged to collect the daily sales at the water point, and send the sum to a selected member of the water committee, who in turn sends the money to the treasurer.

According to the operators interviewed in this study, daily collections per water point ranges from 1.2 to US$3.00 (that is, 2,000/= TZS to 5,000/= TZS) for the kiosks while the sales at standpipe ranges from 1.2 to US$1.8 (that is, 2,000/= TZS to 3,000/= TZS). Therefore this could be averaged to US$63.00 (that is, 105,000/= TZS) per month per kiosk and US$45.00 (that is, 75,000/= TZS) per month per standpipe. The interview with the HNCDA Treasurer showed that total sales per month is averaged at US$36.00 (that is, 60,000/= TZS) per kiosk and US$24.00 (that is, 40,000/= TZS) per standpipe per month. She said that the HNCDA pays US$236.06 (that is, 393,440/= TZS) monthly bill to the City Water, and she averaged a balance of US$90.00 (that is, 150,000/= TZS) as revenue after deducting all costs.

Satisfaction of the users

Most of the customers interviewed were women found at the water points, who indicated satisfaction with the community water supply. They only lamented on the dramatic raise in price per 20lt. bucket. Responding on the question of the increase in price, the HNCDA secretary said the increase was due to the City Water Tariffs that are US$0.00048 (that is, 80 cents TZS) per liter which amounts to US$0.0096 (that is, 16 /= TZS) per 20lt. Before privatization of water distribution services, DAWASA was charging all community water supplies a subsidized price per liter, US$0.00024 (that is, 40 cents TZS) per liter which amounted to US$0.0048 (that is, 8/= TZS) per 20lt. And the HNCDA was selling at US$0.006 (that is, 10/= TZS) per 20lt.

CONCLUSION AND RECOMMENDATIONS

In recent years, many development agencies in developing countries including Tanzania have focused on the promotion of participatory approaches to encourage bottom-up planning and empowerment of communities so that they take more control of development activities, affecting their lives. Governments' inability largely due to lack of resources (financial) to maintain water and sanitation infrastructure has been the major factor contributing to the need for promotion of community participation in development programmes. Involvement of CBOs in managing water supply and sanitation projects provide an indirect economic impact. The socio-economic impact is attributed in the improvement of living standards of the society. The Hanna Nassif Community Development Association (HNCDA) was meant for infrastructure upgrading. The HNCDA study revealed that CBOs can contribute in improving water supply and sanitation systems, especially in informal settlements where most of households are low-income earners. However, there are numerous risks that threaten the sustainability of CBOs. These problems include; technical maintenance of the systems, management, cost recovery, planning, transparency in financial matters and in decision-making, as well as communications between committees and the community. Despite of these challenges, it has been revealed that full involvement of communities in all stages of project development, implementation and management lead to successful water supply and sanitation project success. Moreover, it is obvious that this approach does not divest Governments and NGOs of their responsibility for continuing and evolving support of the programmes which the communities promote. As communities change, and the needs of their water and sanitation systems change, the appropriate type of support (education, training, financial subsidy, technical assistance, maintenance, and even rehabilitation) should evolve. Continuous support to community participation and specifically institutional, legal, and contractual links between communities, Governments and NGOs should be developed.

REFERENCES

Abrams LJ (1996). Water Supply and Sanitation Development in Africa, Africa Sector Review-Report, UN Water Supply and Sanitation Collaborative Council, Working Group on Water Supply and Sanitation Development in Africa.

Abrams LJ (1998). Understanding sustainability of local water services. Accessed from http://www.thewaterpage.com/sustainability.htm on January 2011.

Antonio GF (2005). The European response to the challenges of Water and Sanitation in Developing Countries. In TRIBUNE, the document of European Commission.

Barney C (2005). Urbanization in Developing Countries: Current trends, future Projection, and Key Challenges for Sustainability, J. Technol. Soc. 28:63-80.

Elimelech M (2006). The Global Challenge for adequate and Safe Water, J. Water Supply, Res. Technol.-AQUA, 55(1):3-8.

Esrey S, Potash JB, Robert SL, Shiff C (1990). Health benefits from improvements in water supply and sanitation, Survey and analysis of the literature on selected diseases. US Agency for International Development (USAID), Washington DC, USA, Water and Sanitation for Health (WASH) Technical /66/Rep.

GHK (2002). Effective Strategic Planning for Urban Sanitation Services: Fundamentals of Good Practice, GHK Research and Training, UK. Accessed from http://www.ghkeurope.com/products/ prd02.asp?id=4 on March 2011.

Jan D, Gerry G, Michad W (1993). Developing and Managing Community Water Supplies. Oxfam Publication Oxford UK, Oxfam Development Guidelines P. 8.

Kaliba RM (2002). Participatory Evaluation of Community Based Water and Sanitation Programs: The Case of Central Tanzania. PhD Thesis, Department of Agricultural Economics, Kansas State University, Manhattan.

Kessy AG (2002). Community Participation in Urban Infrastructure Provision, Research Series P. 33, Dortmund.

Khatri KB, Vairavamoorthy K (2007). Challenges for Urban Water Supply and Sanitation in the Developing Countries: Discussion Draft

Paper for the Session on Urbanasation, UNESCO, Delft, Netherlands.
Narayan D (1995). The Contribution of People's Participation: Evidence from 121 Rural Water Supply Projects: Environmentally Sustainable Development Occasional Paper Series, No. 1, World Bank, Washington DC, USA.
Ngware SSA and Kironde JML (2000). Urbanizing Tanzania: Issues, Initiatives, and Priorities, DUP (1996) Ltd, Dar es Salaam University Press, Tanzania.
O'Regan F, Conway M (1993). From the bottom – up: Towards a strategy for Income and Employment Generation among the Disadvantaged, Aspen Institute, Washington DC, USA.
Overseas Development Institute (ODI) (2004). From Plan to Action: Water Supply and Sanitation for the Poor in Africa.
Oyesiku OO (1998). Modern Urban and Regional Planning Law and Administration in Nigeria, Kraft Books Limited, Ibadan.
Rosegrant MW, Cai X, Cline SA (2002). Averting an Impending Crisis, Global water Outlook to 2025: Food Policy Report, International Water Management Institute (IWMI), Colombo, Sri Lanka.
Salewi KW (2004). Poverty Alleviation through Infrastructure Provision and Development, NIGP, Dar es Salaam, Tanzania.
TDHS (2004) Tanzania Demographic and Health Survey 2004/05.
The World Bank AFTU 1&2 (2002). Country Assessment Report of Tanzania on: Upgrading Low Income Urban settlements.
Thompson J, Porras IT, Tumwine JK, Mujwahuzi MR, Katui-Katua M, Johnstone N, Wood L (2001). Drawers of Water II: International Institute for Environment and Development, London, UK.
UCLAS (1994). Baseline study, University College of Lands and Architectural Studies (UCLAS)/ILO, Tanzania.
UCLAS (1998). Baseline study, University College of Lands and Architectural Studies (UCLAS)/ILO, Tanzania.
UN (2004). Commission on Sustainable Development, Discussion papers submitted by major groups: Discussion paper on water, sanitation and human settlements submitted by non-governmental organizations, United Nations Economic and Social Council, Item 3 of the provisional agenda, Twelfth session 14-30 Apr. 2004. Accessed from http://sdin-ngo.net/csd/pdf/csd12-dicussionpaper-ngos.pdf on November 2012.
UN-HABITAT (2009). Tanzania: Bagamoyo Urban Sector Profile.
UNESCO/WHO (2006). Meeting the MDG drinking Water and Sanitation target: The urban and rural challenge of the decade, United Nations Children's Fund and World Health Organization.
UNICEF/WHO (2004). Meeting the MDG drinking Water and Sanitation target: A midterm assessment of progress, United Nations Children's Fund and World Health Organization.
URT (2002). National Water Policy, United Republic of Tanzania.
URT (2004). Water Supply Development report, United Republic of Tanzania, Kinondoni Municipal Council, Dar es Salaam, Tanzania.
UWSA (2001). Draft Customer Services Guidelines, Urban Water and Sewerage Authorities, Tanzania.
Views of the People (2007). Research and Analysis Working Group, United Republic of Tanzania. "Views of the People 2007: Tanzanians Give their Opinions on: Growth and Reduction of Income Poverty, Their Quality of Life and Social Well-being and Governance and Accountability." Dar es Salaam, Tanzania.
WHO (2003). Water Supply and Sanitation Sector Monitoring Report, World Health Organization.
World Bank (2000). Entering the 21st Century, World Development Report 1999/2000. Oxford University Press, New York.

Environmental production of fashion colors from natural dyes

Khadijah Q.[1] and Heba M.[1,2]

[1]Department of Textile Printing, Dyeing and Finishing, Faculty of Applied Arts, Helwan University, Egypt.
[2]Department of Fashion Design, Faculty of Art and Design, King Abdul Aziz University, Saudi Arabia.

Dyeing is one of the most delightful textile arts which are an important branch in fashion design. Natural dye is a fascinating phenomenon that enticed researchers to their chemistry and production of fantastic novel hues to explore the point that no other dyes provide a better opportunity teaching how to protect and respect the environment. In this research, silk fabric was dyed by a mixture of different percentages of cochineal, curcumin and indigo powders in water-acetone co-solvent to possess an ecological acceptable for dye extraction. Extraction and dyeing processes were carried out together in one step at different pH values using sonic energy regarding the promising technique in saving energy. Silk dyeing exhibited a variety of shades according to the evaluated colorimetric data which could be attributed to their resonating structures, where the pH values play important parameter in this phenomenon and control the adsorption capacity of dyes onto silk. So, this research give the chance to produce different hues from the traditional natural dyes to improve the natural dyeing cultural heritage to meet the environmental future demands technology of high quality fantastic dyed patterns through an economical point of view.

Key words: Natural dyes, sonic energy, silk, eco-friendly dyeing.

INTRODUCTION

Coloring is one of the most delightful arts and the most important branches of manufacture; before the mid-19th century professional dyers of fine silks and woolens had to rely on such homely substances as dried insects, roots and leaves of plants, concluding that natural dyes are an education in ecology and ethics, where no other dyes provide a better opportunity teaching how to protect and respect the environment (Siva, 2007).

Nowadays, textile dyeing processing industry is one of the major environmental polluters; synthetic dyes have been in use globally in textile industries due to their availability, lower prices and wider ranges of bright shades with considerably improved color fastness properties in comparison with natural dyes (El-Nagar et al., 2005). According to the application of synthetic dyes, much water effluent generated during textile dyeing would pollute the environment as it contains a heavy load of chemicals, where it is estimated that 10 to 15% of the dye is lost in the effluent during the dyeing process (Iqbal and Ashiq, 2007). Thus, renewed international interest has arisen in natural dyes which have gained momentum due to the increased awareness of the environmental and health hazards associated with synthetic dyes (Iqbal and Ashiq, 2007). In addition to their creative potential, excellent fastness, softness, lustrous and bright where appearance is generally regarded as the most important textile attribute for consumer acceptance and endues (Al-Amoudi and Osman, 2009). Natural dyes are classified

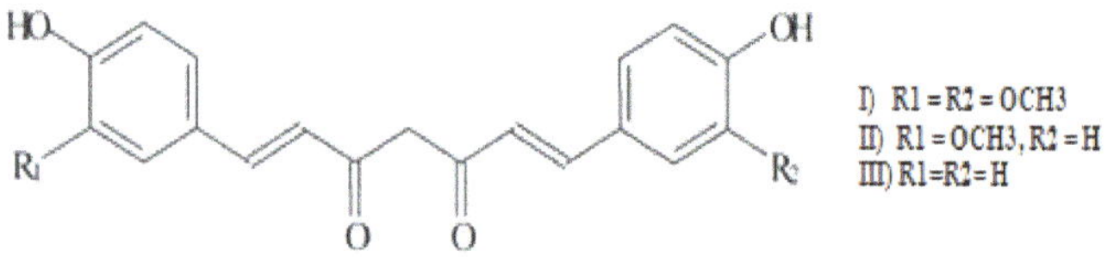

Figure 1. Chemical structure of Curcumin.

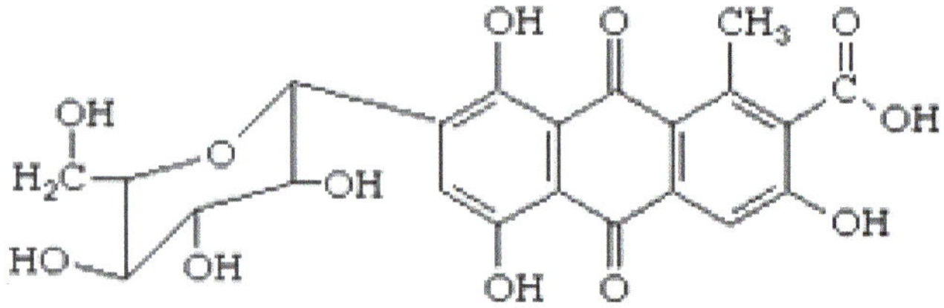

Figure 2. Chemical structure of carminic acid.

according to their chemical structure, sources method of application and color. The most traditional natural colors representing the principle colors are the yellow turmeric, the red cochineal, and the blue indigo.

Turmeric (*Curcuma longa*) is a naturally occurring yellow plant belonging to the ginger family; it has been known to the east Indians for several thousand years, its rhizome is used as a dietary spice coloring agent in foods and textiles. The main pigment compound in turmeric is called curcumin. As shown in Figure 1 (Ferreira et al., 2004), there are three curcuminoids, which make up 2 to 5% of the spice, and are all dicinnamoylmethane derivatives (Balasubramanian, 1990; Stankovic, 2004; Jayaprakasha et al., 2002, 2005): I:diferuloylmethane:1,7-Bis-(4-hydroxy-3-meth oxyphenyl)-hepta-1,6-diene-3,5-dione $C_{21}H_{20}O_6$ C.A.S. number: 458-37-7, Formula weight: 368), II:p-hydroxycinnamoylferuloylmethane:1-(4-Hydroxyphenyl)-7-(4-hydroxy-3-methoxy phenyl)-hepta-1,6-diene-3,5-dione, $C_{20}H_{18}O_5$: C.A.S. number: 33171-16-3, Formula weight: 338) and III: p,p-dihidroxydicinnamoylmethane:1,7-Bis-(4-hydroxyphenyl)-hepta-1,6-diene-3,5-dione.

Curcumin I is the main coloring curcuminoids compound, it is an orange-yellow crystalline powder "natural yellow 3" practically insoluble in water and ether but soluble in ethanol, dimethylsulfoxide, and acetone. The principal coloring components of curcumin exhibit a keto-enol tautomerism and antioxidative properties (Jayaprakasha et al., 2002). It is s identified as [1, 6-heptadiene-3,5-dione-1,7-bis(4-hydroxy-3-ethoxyphenyl)-(1E,6E) or diferuloylmethane ($C_{21}H_{20}O_6$); (Chemical formula: C19H16O4: C.A.S. number: 33171-05-0, Formula weight: 308 (Jiang et al., 2006; Aggarwal et al., 2003).

Most of natural red-color dyes have high solubility in water and act as acid dyes, which is why they bond to protein fibers (Siva, 2007; Kongkachuichaya et al., 2002). Cochineal is one of the species with great economic importance, since ancient times, is the 'American cochineal' *Dactylopius coccus* Costa (which some publications still refer to as *Coccus cacti* L.) because from the dried bodies of the adult females, where the little cochineal bug will give the most color when ground into a fine powder to obtain colors of dark burgundy to bright red to soft lilac and pink based on carmine. Carmine consists of an aluminum or calcium-aluminum lake on an aluminum hydroxide substrate of carminic acid; the coloring principle of cochineal (Marmion, 1991). Carminic acid is a hydroxyanthraquinone with a lateral chain of C-glycosyl and only one position free on the aromatic nucleus; in other words, it is an anthraquinino with other aglycones presents which are kermesic acid (3,5,6,8-tetrahydroxy-1-methyl-2-anthraquinone carboxylic acid), flavokermesic acid and laccaic acid D (3,6,8-trihydroxy-1-methyl-2-anthraquinone carboxylic acid) (Allevi and Tyman, 1998). The composition of carminic acid as shown in Figure 2 was proposed to be $C_{22}H_{20}O_{13}$, and it comprises approximately 10% of cochineal and 2 to 4% of its extract (Allevi and Tyman, 1998). It is used as brilliant red coloring agent (CAS Reg. No.: 1260-17-9; EINECS: 215-023-3; No. Shoultz: 1381; C.I.: light and heat stable of all the colorants and is 75470; Natural Red 4; E120). It is one of the more stable than many synthetic food colors (Siva, 2007).

Carminic acid has been widely used to color foods, beverages, textiles, pharmaceuticals and cosmetics products due its relatively high chemical and biological stability and its apparent lack of mutagenic, carcinogenic or genotoxic effects. This stability and the concomitant absence of toxicity have been attributed to the presence of the C-Glycosyl moiety. Besides, the color attributes, recently, also has been reported to be beneficial to health with potential antibiotic and antitumor properties. At the beginning of the 21st century, it is predicted that many colors will be used for both their additional beneficial functions in the body, as well as, coloring effect (Allevi and Tyman, 1998).

Indigo which has been used for 4000 years is described as the blue of all old textiles. It seems to be the oldest known colorant; C.I. Vat Blue1 of Chemical Formula ($C_{16}H_{10}N_2O_2$) and Chemical Name: 2-(1,3-dihydro-3-oxo-2H-indol-2-ylidene)-1,2-dihydro-3H-indol-3-one. The Indigotin or Indican pigment is shown in Figure 3. It has been made from several different dye plants, such as *Indigofera tinctoria* and *Indigo suffraticosa*. Today, it retains its position as one of the world's most important industrial chemicals (Roessler and Jin, 2003). It is actually used for textile dyeing and printing where small amounts are used for dyeing wool and silk. In 2009, large spills of blue dyes had been reported for downstream of a blue jeans manufacturer in Lesotho. Because of its high value as a trading commodity and low oral toxicity, indigo

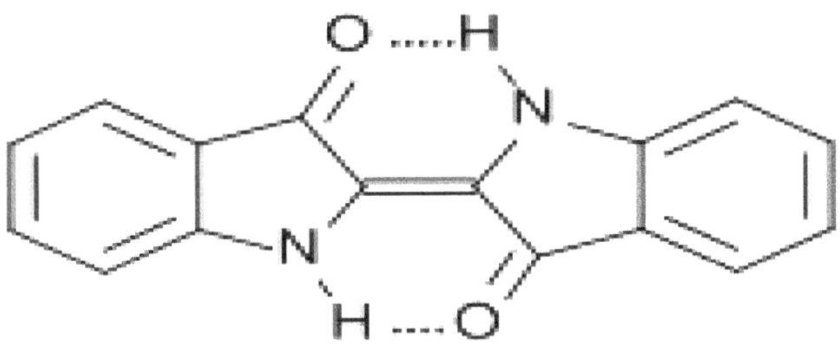

Figure 3. Chemical structure of Indigo.

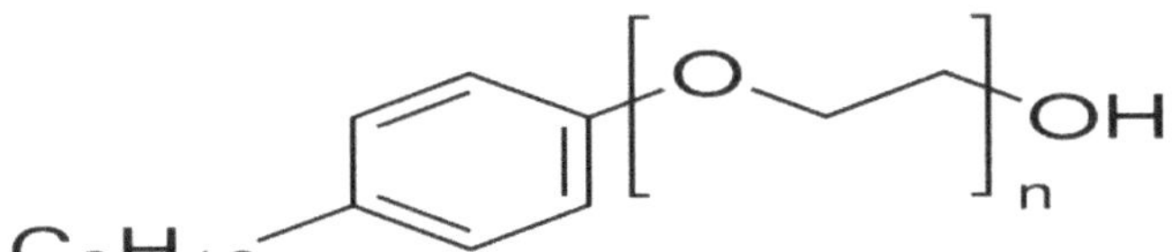

Figure 4. Chemical structure of Nonidet® P 40 Substitute.

was often referred to as Blue Gold. It is a so-called vat dye, which means that it needs to be reduced to its water soluble leuco-form before dyeing (Roessler and Crettenand, 2004). Natural dyes are generally prepared by boiling the crushed powder with water, and sometimes it is left to steep in cold water. Technology for production of natural dyes could vary from simple aqueous to complicated solvent systems to sophisticated supercritical fluid extraction techniques depending on the product and purity required (Siva, 2007). Several researches observed that extraction of natural dyes with co-solvents induced rise to extracts incompletely soluble in water (Heba and Gamal, 2011; Osman et al., 2009).

On the other hand, the use of ultrasonic in the extraction and dyeing methods to increase the yield of colorants is being attempted and represents a promising technique in saving energy, providing effectiveness in the silk treatment, dyeing, and mordanting in comparison with the conventional heating system (Heba and Heffernan, 2009; Heba and Gamal, 2011; Kamel et al., 2007; Vajnhandl and Le Marechal, 2005; Padma et al., 2006; Heba and Heffernan, 2010).

Previous scientific researchers were focused on the application of natural dyes on the different textile fabrics, mordants and mordanting methods, etc, recording their fastness properties and colorimetric data. In the present paper, there is an effort taking place to explore the availability of mixing turmeric, cochineal and indigo powders with different percentages to dye silk that was chosen because it is characterized as the queen of fibers having unique combination of properties as luster and appearance, and composed of different alpha amino acids orienting to form long chain polymer by condensation and polymerization that can be easily attached with natural dyes via ionic and hydrogen bonds to produce renewable aesthetic shades under the influence of pH values. This goal was assisted by water acetone co-solvent that possesses an ecological acceptable for dye extraction and sonic energy due to its considerable saving in processing time and chemical used; on the other hand, it is believed to exhibit only slight toxicity in normal use, and there is no strong evidence of chronic health effects if basic precautions are followed. It is relatively cheap and safe solvent, it has a much higher flashpoint than the alcoholic solvents methanol or ethanol, so is less likely to catch on fire lastly (www.madsci.org/.../1060116066.Ch.r.ht, www.dow.com/productsafety/.../acetone). The renewable colors or shades of the dyed samples have been evaluated on computer color matching system in terms of K/S and L* a* b* color coordinates. As a result, the study will complete the economical point of view together with the aesthetic duty and environmental image.

EXPERIMENTAL

Materials

Degummed and bleached plain Habotain silk fabric (30 g/m^2) purchased from Sherazad Com. NZ was further treated at a liquor ratio of 50:1 with 2 g/L non-ionic surfactant; Nonidet® P 40 Substitute purchased from Sigma-Aldrich NZ Ltd. It is 4-Nonylphenyl-polyethylene glycol as shown in Figure 4. Scouring was carried out for 15 min at 80°C, then thoroughly rinsed and dried at ambient temperature (Kongkachuichaya et al., 2002). Curcumin, indigo, and cochineal powders were purchased from Hands Ashford, Ltd Christchurch-New Zealand. Acetic acid, sodium hydroxide, sodium dithionite and acetone were of laboratory reagent grade.

Extraction and dyeing processes

10% w/v curcumin and indigo powders were mixed together with percentages of 50:50 each of them, and percentage of 75:25 curcumin with indigo. 10% w/v curcumin and cochineal powders were mixed together with percentages of 50:50 each of them in addition to 75:25 curcumin with cochineal. On the other hand, 10% w/v cochineal and indigo powders were mixed together with percentages of 50:50 each of them in addition to 75:25 cochineal with indigo, and 25:75 cochineal with indigo.

Ultrasonic Elmasonic S30 apparatus (80W, 37 kHz) purchased from Elma Hans Schmidbauer GmbH and CO.KG) assisted both the extraction and dyeing processes of the mentioned mixtures percentages in one step, using 20% v.v of acetone: water co-solvent at 50 °C for 1 h at different pH values of 3, 7, and 9, using a material to liquor ratio of 1:100. "In case of indigo dyeing, 10 g/L of sodium dithionite was added to the dye-bath". The dyed samples were washed with 2g/L of each of sodium carbonate, and Nonidet® P 40 Substitute at 60°C for 30 min using a material to liquor ratio of 1:50, then were thoroughly rinsed in warm and cold water, and left to dry at ambient temperature.

Measurement

Color strength of the dyed fabric samples, in terms of the K/S values, was obtained by using the Kubelka Munk equation (Kubelka, 1948; Garland, 1993):

$$F(R) = (1-R)^2/2R = K/S$$

Where K is the absorption coefficient and S is the scattering coefficient for a colorant at a specific wavelength; R is the fractional reflectance value of the dye on the substrate at the λ_{max}. The K/S value at λ_{max} is directly proportional to the concentration of dye on the substrate using Cary 100 UV-Vis Spectrophotometer. The CIE-Lab values of the dyeing were measured on a CM-2600d Konica MINOLTA color meter, and the cylindrical co-ordinates of color were determined after exposure to irradiation from an Xe 150 watt arc lamp. The colors are given as CIE L*a*b* coordinates, as specified by the International Commission on Illumination with L* corresponding to brightness (100 = white, 0 = black), a* to the red-green coordinate (positive sign = red, negative sign = green) and b* to the yellow-blue coordinate (positive sign = yellow, negative sign = blue) (Bechtold et al., 2003).

RESULTS

As shown in Figure 5, on mixing 50% of each of turmeric and indigo together, wavelengths of two colors were observed, representing the yellow color at λ_{max} 423,424 and the bluish green color at λ_{max} 583 - 586. For the yellow, the maximum color strength was achieved in the neutral medium, while for the bluish green; the maximum color strength value was achieved in the alkaline medium.

As shown in Figure 6, on mixing 75% turmeric with 25% indigo, wavelengths of two colors were observed, representing the yellow color at λ_{max} 416 - 453, and the bluish green at λ_{max} 556 - 588. For the yellow color, the maximum color strength was achieved in the neutral medium; on the other hand, the samples exhibited the yellow color in the three mediums, while for the bluish green color, and the maximum color strength value was achieved in both the acidic and neutral mediums.

As shown in Figures 5C and 6C, the pH values of the dye bath have considerable effect on the color co-ordinate values. The lightness L* values increase in the acidic and alkaline mediums, and then these values were decreased when adjusting the pH of dye bath in the neutral medium. This observation relevant to lightness could be attributed with change in increase and decrease of color strength (%) values with change in the pH of the dye bath. The tone of the samples at various pH of dye bath is less greener and more yellower in the acidic and neutral medium as indicated by the values of a* and b* These changes could be attributed with its resonating structures (Moeyes, 1993).

As shown in Figure 7, on mixing 50% of each of turmeric and cochineal together, wavelengths of two colors were observed, representing the yellow color at λ_{max} 419,426,446 and the violet one at λ_{max} 556 - 569, where the maximum color strength values of the two colors were achieved in the acidic medium'.

As shown in Figure 8, on mixing 75% curcumin with 25% cochineal, wave lengths of two colors were observed, representing the yellow color at λ_{max} 419,423,444 and the violet one at λ_{max} 568, where the maximum color strength values of the two colors were achieved in the acidic medium.

As shown in Figures 7C and 8C, the L* (lightness or brightness) has its maximum value at pH 9 in comparison with the other mediums in investigation which reflects the suitability of alkaline medium for obtaining bright color from the two dyes used. The a* and b* values have positive signs, reflecting the red color of cochineal and the yellow color of curcumin respectively, reaching their maximum content at pH 3. Since cochineal dye is characterized by producing colors of dark burgundy to bright red to soft lilac and pink based on carmine[14], it was recorded that silk samples dyed with mixtures of i) 50% of each of curcumin and cochineal, ii) 75 curcumin with 25 cochineal exhibited a greenish yellow color in the acidic medium, yellow color in the alkaline medium and yellowish orange color in the neutral medium with much yellower color in case of 75% curcumin with 25% cochineal.

As shown in Figure 9, on mixing 75% cochineal with 25% indigo, wavelengths of two colors were observed, representing red color in the acidic medium, to violet color in the neutral and alkaline mediums at λ_{max} 521, 558, 581, respectively. The maximum color strength value was achieved in the alkaline medium. The L* (lightness or brightness) has its maximum value at pH 7 in comparison with the other mediums in investigation which reflects the suitability of neutral medium for obtaining bright color from the two dyes. The a* values have positive signs, reflecting the red color of cochineal which achieve its maximum content at pH 3. The b* values have negative signs reflecting the blue color of indigo which achieve its maximum value at pH 9.

As shown in Figure 10, on mixing 50% cochineal with 50% indigo, a violet color was observed at λ_{max} (568, 589, and 581) nm in the acidic, neutral, and alkaline mediums, respectively. The maximum color strength value was achieved in the acidic medium. The L* (lightness or brightness) has its maximum value at pH 7 in comparison with the other mediums in investigation which reflects the suitability of neutral medium for obtaining bright color from the two dyes. The a* values have positive signs, reflecting the red color of cochineal which achieve its maximum content at pH 3. The b* values have negative signs reflecting the blue color of indigo which achieve its maximum value at pH 9.

As shown in Figure 11, on mixing 25% cochineal with 75% indigo, a violet color was observed at λ_{max} 574 nm in the neutral medium and 571 nm in the alkaline and acidic mediums. The maximum color strength value was achieved in the neutral medium. The L* (lightness or brightness) has its maximum value at pH 3 in comparison with the other mediums in investigation which reflects the suitability of acidic medium for obtaining bright color from

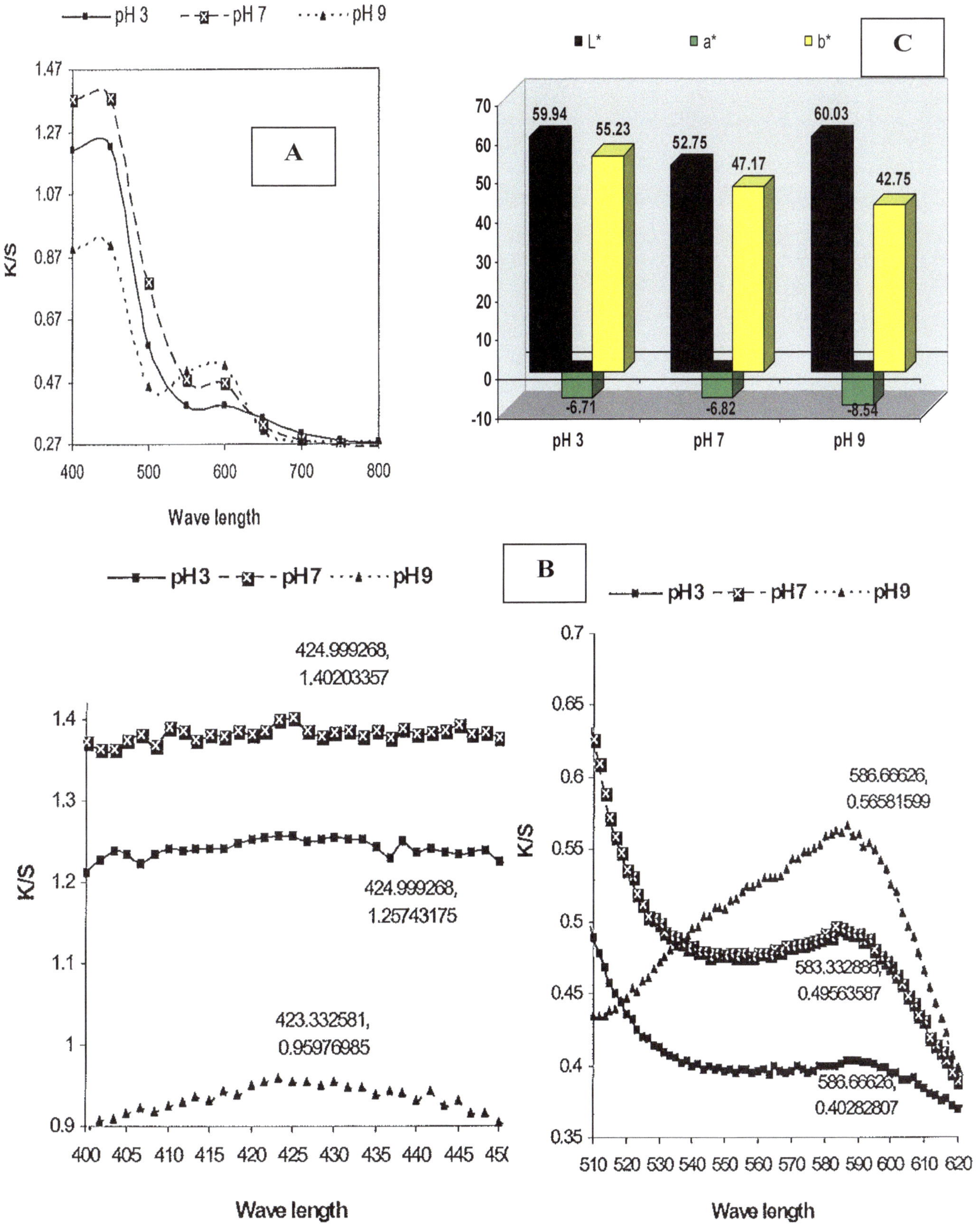

Figure 5. A = Effect of pH values on wave length and K/S of silk dyed with an extract composed of a mixture of 50% turmeric with 50% indigo B = Identification of the highest wave lengths and K/S values according to figure 4A. C = Effect of pH values on the CIE-Lab values of silk dyed with 50% curcumin and 50% indigo extract at different pH values.

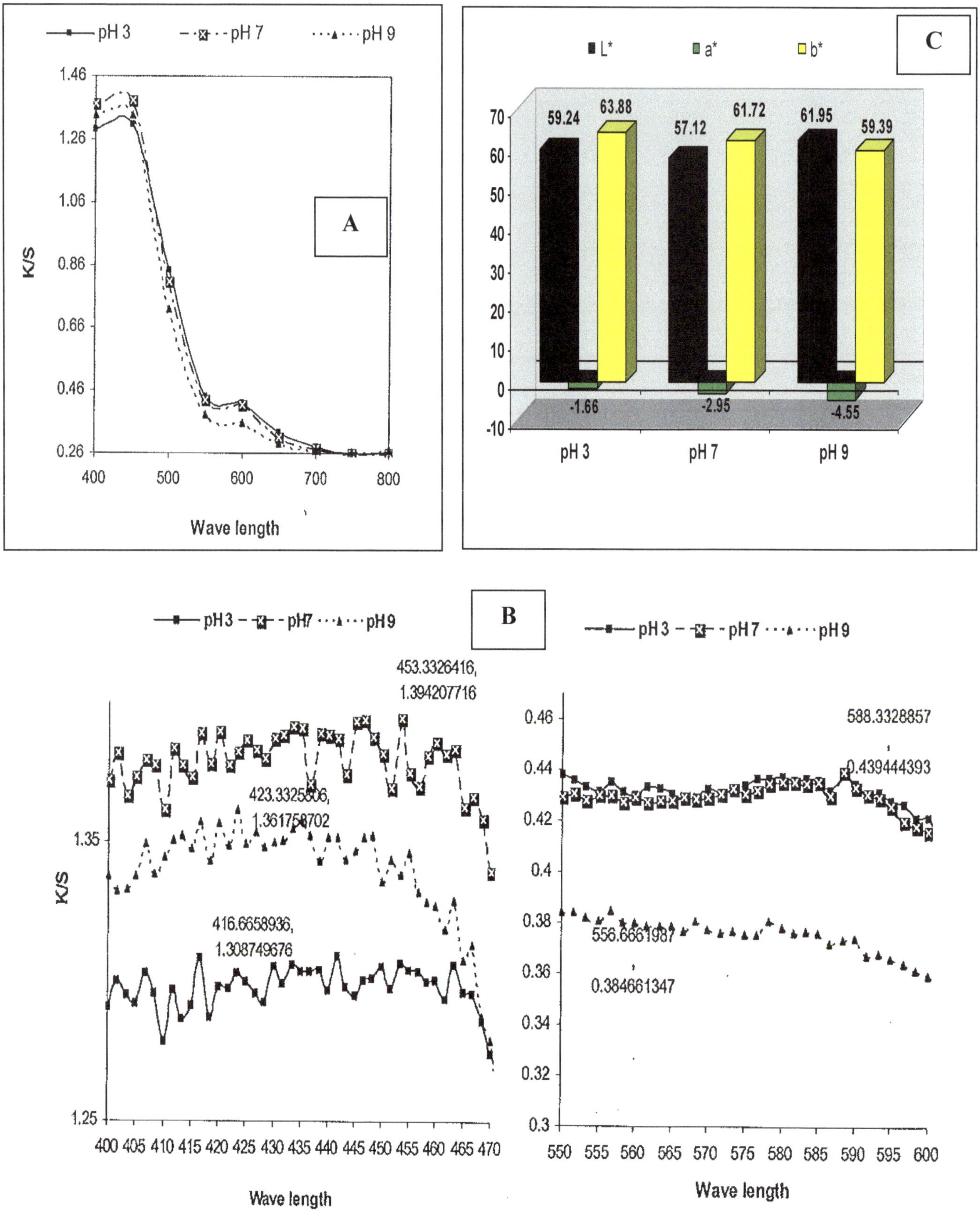

Figure 6. A = Effect of pH values on wave length and K/S of silk dyed with an extract composed of a mixture of 75% turmeric with 25% indigo B = Identification of the highest wave lengths and K/S values according to Figure 5A. C = Effect of pH values on the CIE-Lab values of silk dyed with 75% curcumin and 25% indigo extract at different pH.

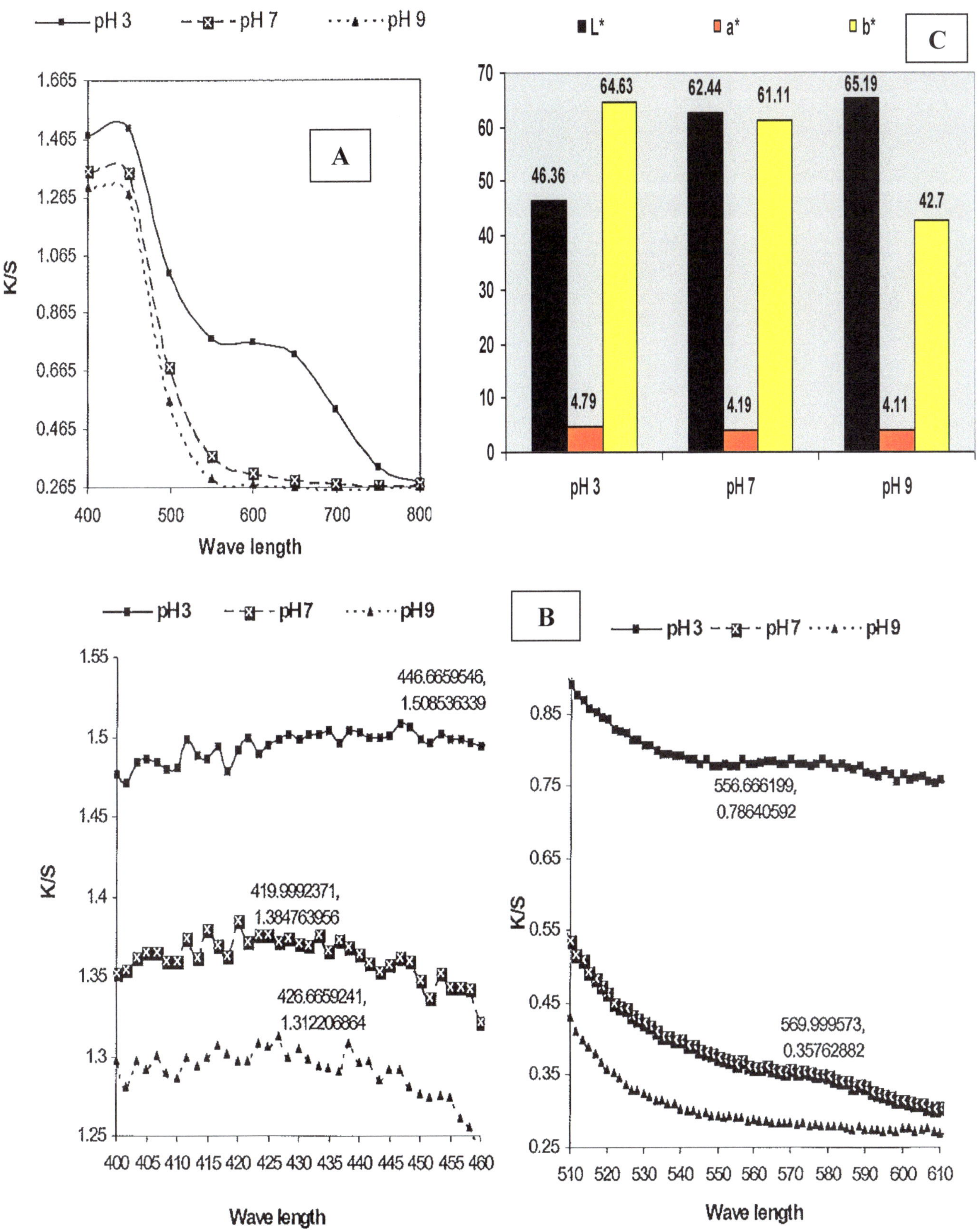

Figure 7. A = Effect of pH values on wave length and K/S of silk dyed with an extract composed of a mixture of 50% curcumin with 50% cochineal B = Identification of the highest wave lengths and K/S values according to Figure 6A. C = Effect of pH values on the CIE-Lab values of silk dyed with 50% curcumin and 50% cochineal extract at different pH values.

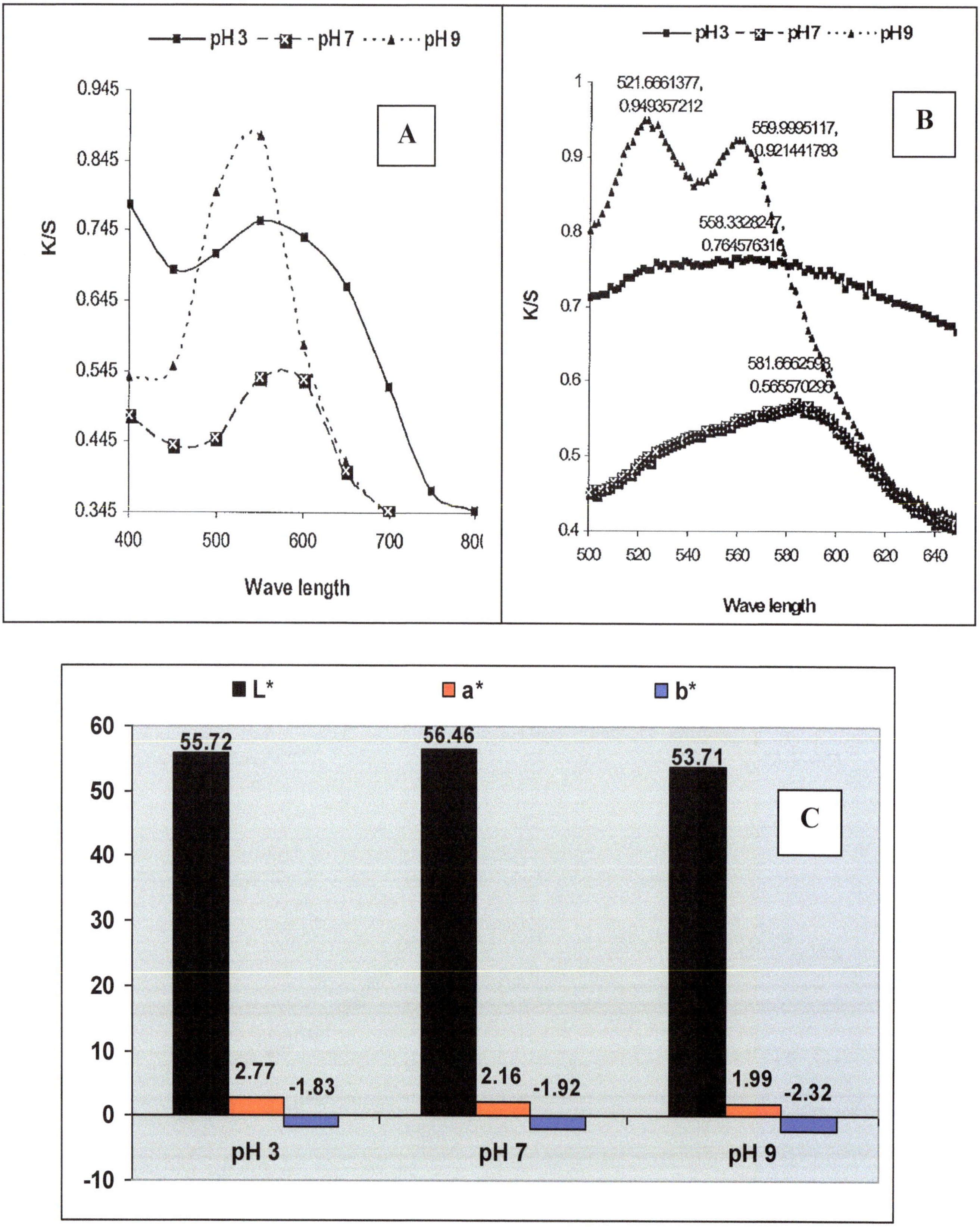

Figure 9. A = Effect of pH values on wave length and K/S of silk dyed with an extract composed of a mixture of 75% cochineal with 25% indigo B = Identification of the highest wave lengths and K/S values according to Figure 8A. C = Effect of pH values on the CIE-Lab values of silk dyed with 50% cochineal and 50% indigo extract at different pH values.

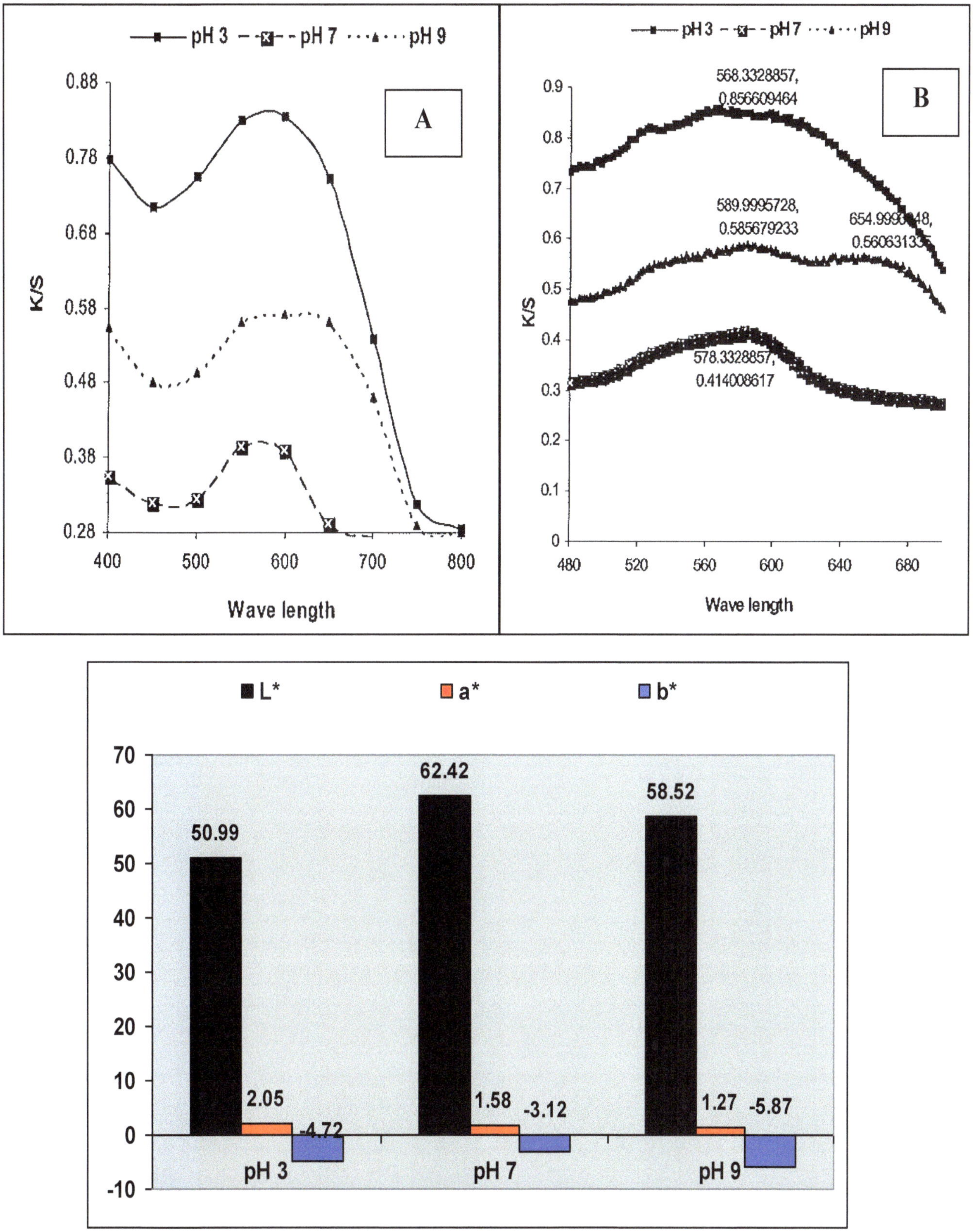

Figure 10. A = Effect of pH values on wave length and K/S of silk dyed with an extract composed of a mixture of 50% cochineal with 50% indigo B = Identification of the highest wave lengths and K/S values according to figure 9A. C = Effect of pH values on the CIE-Lab values of silk dyed with 25% cochineal and 75% indigo extract at different pH values.

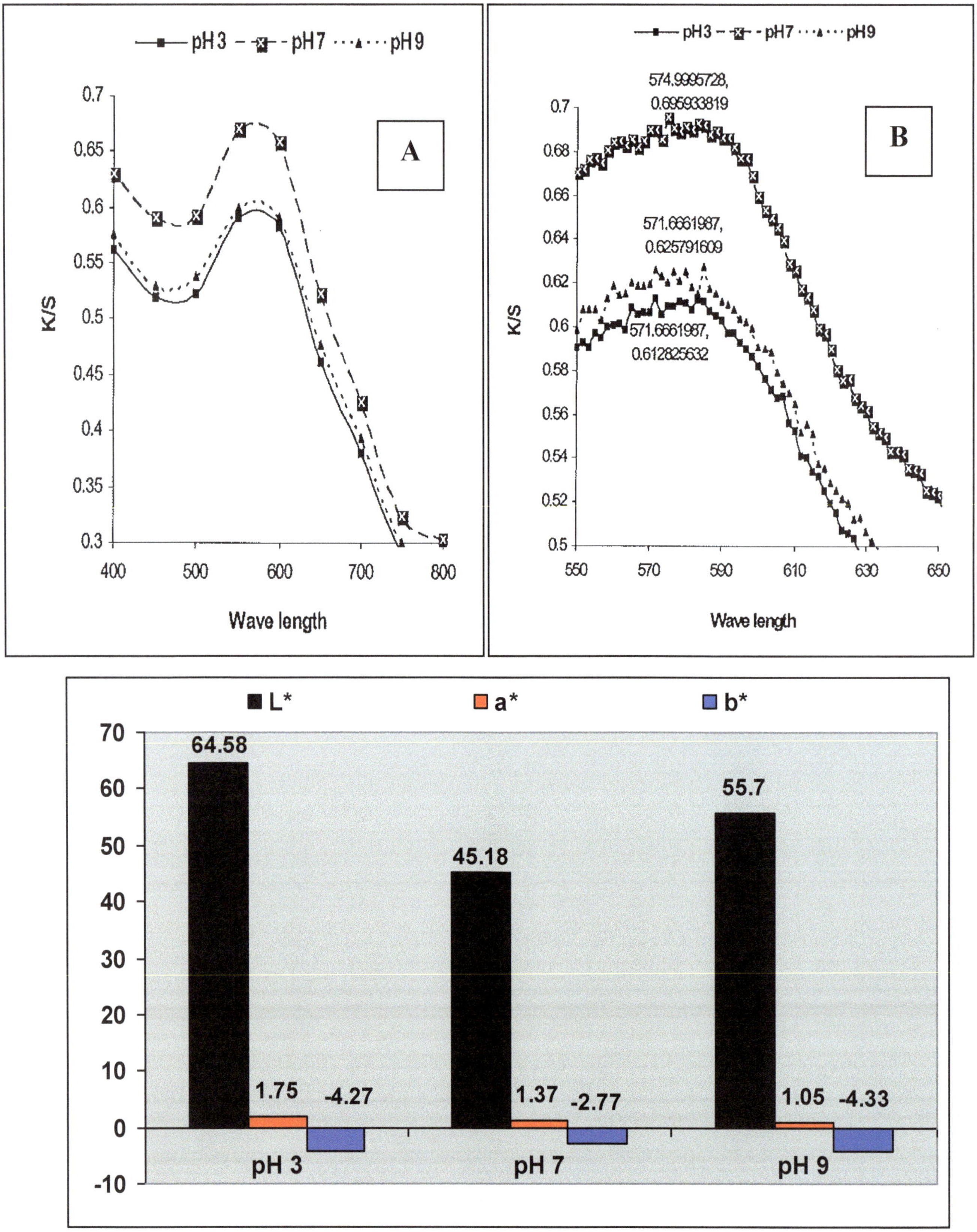

Figure 11. A = Effect of pH values on wave length and K/S of silk dyed with an extract composed of a mixture of 25% cochineal with 75% indigo B = Identification of the highest wave lengths and K/S values according to Figure 10A. C = Effect of pH values on the CIE-Lab values of silk dyed with 75% cochineal and 25% indigo extract at different pH values h.

the applied two dyes. The a* values have positive signs, reflecting the red color of cochineal which achieve its maximum content at pH 3. The b* values have negative signs reflecting the blue color of indigo which achieve its maximum value at pH 9.

DISCUSSION

The pH of the dye solution is one of the most important parameters controlling the adsorption capacity of dye onto silk (Wang et al., 1997; Christie et al., 2000). The effect of pH on the spectroscopy studies of the mixtures of turmeric and indigo onto silk were evaluated in Figures 4 and 5. Results indicated that the maximum color strength values were clearly achieved at acidic medium. From the Greek *chromophorus* (*chromos* color), a chromophore is generally a groups of atoms in a molecule responsible for the interaction with the radiation at certain wavelength; example of common chromophore is conjugated multiple bond, rich in electrons, where the energy of the light which passes through the sample is absorbed by some electrons and they are in that way "excited" to a higher energetic level (state). UV/Visible spectrums of pure curcumin in distilled water showed two well defined λ_{max} at 259 and 405 nm, but it absorbs maximally at 415 to 420 nm in acetone (Stankovic, 2004). The similar λ_{max} were obtained in the spectra of curcumin obtained through extraction from turmeric rhizomes and through dyed fabric in acetone / water solution as well. The presence of the two C=O groups seems to suggest that the 405 nm in the visible region is due to the n-π * transition of conjugated chromophore. The λ_{max} at 259 is due to the π-π* transition that is attributed to alternate double bonds present in the curcumin.

Curcumin is an oil-soluble pigment, practically insoluble in water at acidic and neutral pH, and soluble in alkali. In solutions, the principal coloring components of curcumin exhibit keto-enol tautomerism and, depending on the solvent, up to 95% are in the enol form (Stankovic, 2004). It has a brilliant yellow hue at pH 2.5 to 7 and takes on a red hue at pH > 7, and its decomposition was pH dependent and occurred faster at neutral to basic conditions (Balasubramanian, 1990). At pH <1, aqueous solutions of diferuloylmethane have a red color which indicates the protonated form (H4A+). In the pH range 1 to 7, the majority of diferuloylmethane species are in the neutral form (H3A). Water solubility is very low in pH range 1 to 7 and solutions are yellow. At pH >7.5, the color changes to red. The UV visible spectral data of curcumin reveals two peaks, one in the visible region λ_{max}= (420 nm, log E = 4.77) and the other, λ_{max} = (265 nm with log E = 4.18). The absorption peak in the visible region (λ_{max}= 420 nm) explains the yellow color of the dye. On the other hand, the UV visible spectral data of indigo reveals 580 nm (Stankovic, 2004).

Resonating structure of some natural dyes might be determined via studying curcumin case; Curcumin is an oil-soluble pigment, practically insoluble in water at acidic and neutral pH, and soluble in alkali. In solutions, the principal coloring components of curcumin exhibit keto-enol tautomerism and, depending on the solvent, up to 95% are in the enol form. It has a brilliant yellow hue at pH 2.5 to 7 and takes on a red hue at pH > 7, and its decomposition was pH dependent and occurred faster at neutral to basic conditions (Moeyes, 1993).

At pH <1, aqueous solutions of diferuloylmethane have a red color which indicates the protonated form (H4A+). In the pH range 1 to 7, the majority of diferuloylmethane species are in the neutral form (H3A). Water solubility is very low in pH range 1 to 7 and solutions are yellow. At pH >7.5 the color changes to red. The pKa values for the dissociation of the three acidic protons in Compound 1 (forms H2A-, HA2- and A3-) have been determined to be 7.8, 8.5 and 9.0, respectively as shown in Figure 12 (Stankovic, 2004).

The UV visible spectral data of curcumin reveals two peaks, one in the visible region λ_{max}= (420 nm, log E = 4.77) and the other, λ_{max} = (265 nm with log E = 4.18). The absorption peak in the visible region (λ_{max}= 420 nm) explains the yellow color of the dye. On the other hand, the UV visible spectral data of indigo reveals 580 nm.

A possible mechanism for color change upon addition of OH- to curcumin is shown in Figure 13. The simplest explanation is that phenolic OH is acidic due to the operation of resonance and readily reacts with OH^-. These results in the anion or dianions were represented in Figure 13, the structures with more detailed and quantitative theoretical analysis would, of course, have to be based on quantum mechanical calculations. The presences of certain groups called chromophores are often responsible for color. Curcumin has two C=0 groups separated by CH_2. The C=0 groups are adjacent to alternant array of double bonds. The presence of the two C=0 groups seems to suggest that the 420 nm peak in the visible region is due to the n.... π* transition of the CO chromophore. Thus, the visible transition of curcumin could be attributed to the n... π* transition. The other band with λ_{max} nm (log E = 4.18) should then be due to a π ....π* transition attributed to the altemant set of double bonds present in the neutral curcumin. It was reasonably successful in predicting the λ_{max} of curcumin; it was extended to explain the red shift observed upon addition of OH-. Since the addition of OH- to curcumin results in the anion II, III, or IV the *L* value in expression (I) must be increased due to the delocalization of the negative charge from one end of the model compound to the other end. Alternatively, the predominantly σC-0 bond in phenolic C-OH now attains π character in the anion. Since the C - *0*- bond length is derived from phenolic C - OH, it should be between aromatic C=O and phenolic C-0. It is well known that spectral shift in the n....π* transition due to extension of conjugation is approximately half of the value of $\Delta\lambda_{max}$ (π....π*). This

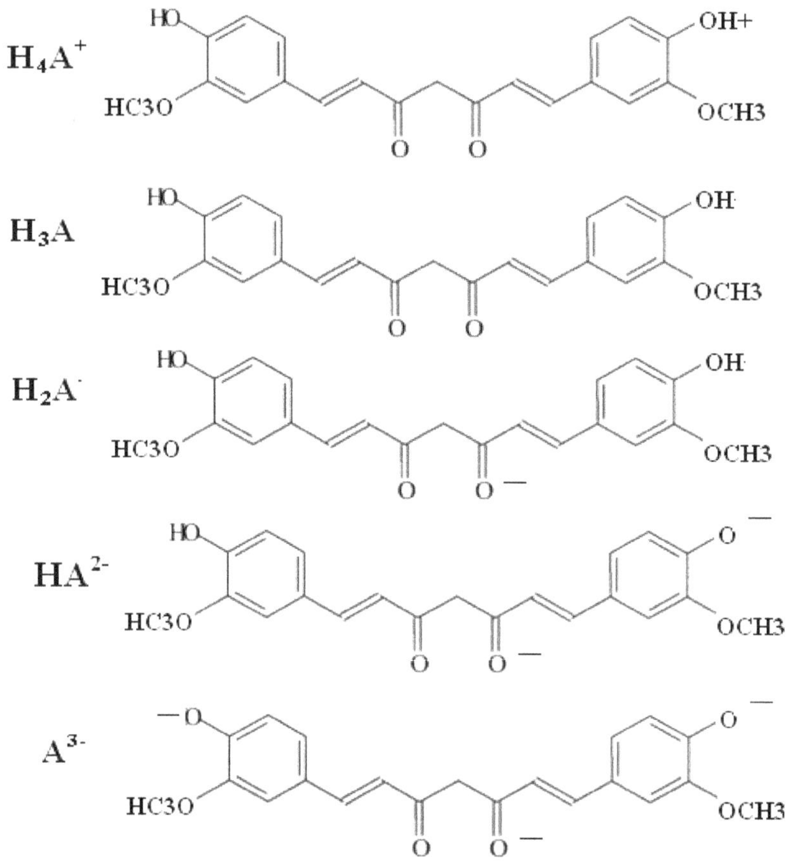

Figure 12. The pKa values for the dissociation of the three acidic protons.

gives λ_{max} for the anion between 479 and 500 nm depending on which parameter is used for the neutral curcumin λ_{max}.. In any case, this falls in the greenish blue to bluish green region of the spectrum, and hence the complement orange-red color is expected for the anionic curcumin. Experimentally, a red color is observed upon addition of OH- to curcumin. The addition of OH- group was modeled by a resonating anionic compound as shown in the top half of Structure II in Figure 5, this was the fact that the π* orbital that is occupied in the anion could be taken into account. This resulted in the lowering of the π* orbital in addition to extension of conjugation. The first π... π* transition for the anion corresponded to the transition between the sixth highest occupied orbital and the lowest unoccupied (seventh orbital) for the anion while the corresponding transition for the neutral curcumin (π... π *) corresponded to a transition between the highest occupied π (fifth) orbital to the lowest unoccupied π* (sixth) orbital. The approximate shift, that is, $\Delta\lambda_{max}$ for this transition was deduced as 820 nm for the UV band and 550 nm for visible band. This changes the $\Delta\lambda_{max}$ of the n ...π* transition of the anionic curcumin in the alkaline medium to approximately 490 nm. This will evidently give a red color to curcumin in alkaline medium.

A sound qualitative explanation of the red shift in the visible band of curcumin is that the *π** is stabilized both by extension of conjugation and its occupancy for Structures II-IV. As a result the transition gap for the n .....π * transition reduces, although the lowest unoccupied π* orbital is different for the neutral curcumin as opposed to the anionic curcumin leading to a red shift. Note that the n....π* transition of neutral curcumin is much stronger (log ε= 4.77). The main reason for the much larger extinction coefficient for curcumin is primarily due to the π-donating alternant system >f conjugated sets of double bonds which can extend the π-bonding through resonance (Stankovic, 2004).

As stated before, since cochineal dye is characterized by producing colors of dark burgundy to bright red to soft lilac and pink based on carmine (Allevi and Tyman, 1998), it was recorded that the silk samples dyed with mixtures of i) 75% cochineal with 25% indigo exhibited violet color in the three mediums, ii) 50% of each of cochineal and indigo exhibited violet color in the alkaline and neutral medium, while grayish blue in the acidic medium and iii) 25% cochineal with 75% indigo exhibited violet color in the neutral medium while grayish blue in the acidic and alkaline mediums. From the mentioned

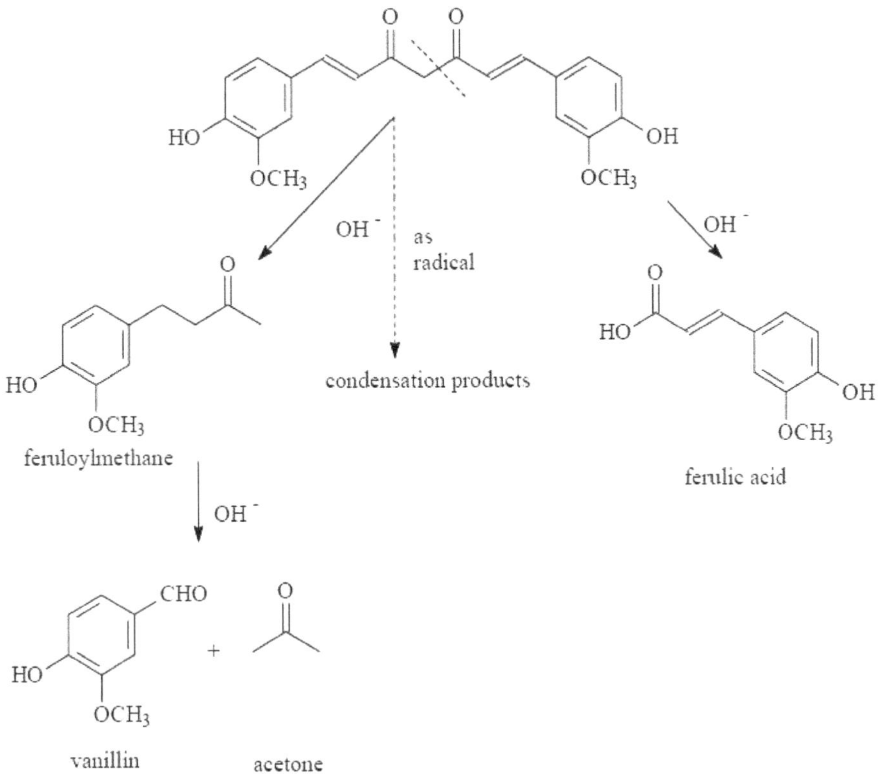

Figure 13. Mechanism for color change upon addition of (OH^- to curcumin).

results, mixtures of dyes exhibited different colors or shades according to a* and b* values which could be attributed with their resonating structures (Montazer and Parvinzadeh, 2004)

Indigo is a vat dye, soluble in alkaline medium but in acidic medium its solubility will be in the least amount. Curcumin acts as direct dye, while cochineal acts as acidic one. The acidic medium is suitable for curcumin and cochineal, but not suitable for indigo. Therefore, the yellow and red color of curcumin and cochineal reached their maximum values in the acidic medium. While the blue color of indigo reached its highest values in the alkaline medium.

The pH of the dye solution is one of the most important parameters controlling the adsorption capacity of dyes onto silk (Moeyes, 1993; Christie et al., 2000). For curcumin and cochineal dyes, the possible mechanisms for the effect of pH on adsorption of dyes are likely to be ionic interactions of the dye anions with the protonated amino groups on the silk fiber. The effect of dye bath pH can be attributed to the correlation between dye structure and silk fibers. Since the dye used is a water-soluble dye containing carboxyl groups, it would interact ionically with the protonated terminal amino groups of silk fibers at acidic pH via ion exchange reaction. The anion of the dye has a complex character, and when it is bound on fiber, further kinds of interactions take place together with ionic forces. This ionic attraction would increase the dyeability of the fiber. At pH > 3, however, the ionic interaction between the carboxylate anion of the dye and silk fibers decreases due to the decreasing number of protonated terminal amino groups of silk and thus lowering its dyeability.

It is worth noting that the lower dyeability at pH > 3 may be attributed to the enhanced desorption of the dye as its ionic bond is getting decreased (Cardon, 2007). In case of indigo dye, different forms of indigo existed in solution, each one having a different substantivity for the fiber. The non-ionic form had the highest, whilst the ionic ones had relatively less substantivity. Silk fibers could best be dyed in the acidic to slightly alkaline pH range. The chemistry of the dye-bath and fiber were considered, in order to explain the effect of pH on color strength. Hydrogen bonding and ionic interactions seemed to be the main reason for the exhaustion of dye onto silk fiber, where hydrogen bonding occurred between the hydroxyl groups and amide groups, while ionic interactions worked at low pH values between the ionized and end-amino groups of the dye and fiber respectively (Christie et al., 2000). On the other hand, ultrasonic represented a promising technique in saving energy by providing additional energy, which was generally attributed to the cavitation phenomenon in a liquid medium, in addition to other mechanical effects, that is, dispersion, degassing, diffusion, and the Intense agitation of the liquid. This phenomenon led to lowering the temperature and reducing the time needed for the extraction, and the dyeing processes (Heba and Heffernan, 2009; Heba and Gamal, 2011; Kamel et al., 2007; Vajnhandl and Le Marechal, 2005; Padma et al., 2006; Heba and Heffernan, 2010) By the way of extraction with acetone, water co-solvent includes the dipole moment, dielectric constant and refractive index values. The solvent polarity can change the position of the absorption or emission band of molecules by solvating a solute molecule or any other molecular species introduced into the solvent matrix (Muhammed et al., 2008; Heba and Gamal, 2011).

Conclusion

Color is one of the most fascinating phenomena which were enticed to chemistry as is well known, full comprehension compound needs knowledge of quantum chemistry and spectroscopy. To encourage the production of eco-friendly aesthetic future demands of renewable shades or colors regarding the environmental awareness towards eco-friendly dyeing and energy save, this article deals on the efficiency of sonic energy in assisting one step extraction and dyeing processes of silk fabric with mixtures of the traditional natural dyes; turmeric, cochineal and indigo powders with different percentages at different pH values in water acetone co-solvent. It was concluded that the color values with respect to *K/S*, and L* a* b* values were found to be influenced by the pH values of the dyeing bath with respect to the chemical structure of the used dyes, where a broad variation in shade and color depth was achieved by applying mixtures of these dyes in various combinations, given that both ultrasonic and water acetone co-solvent are environmentally and ecological acceptable for dyeing technology.

REFERENCES

Aggarwal BB, Kumar A, Bharti AC (2003). Anticancer potential of curcumin: Preclinical and clinical studies. Anticancer Res. 23(1A):363-398.

Al-Amoudi ES, Osman EM (2009). Optimization of Dyeing Performance of an Eco Friendly Natural Dye "*Vervain Barks*" applied to Silk fabrics at different pH values. RJTA 13(1):34-45.

Allevi P, Tyman A (1998). Synthesis of carminic acid, the colorant principle of cochineal. J. Chem. Soc. Perkin Trans. 1(3):575-582.

Balasubramanian K (1990). Two Colorful Applications of the PPP Method. Int. J. Quantum Energy 37(4):449-463.

Bechtold T, Turcanu A, Ganglberger E, Geissler S (2003). Natural dyes in modern textile dye houses -how to combine experiences of two centuries to meet the demands of the future?. J. Cleaner Prod. 11(5):499-509.

Cardon D (2007). Natural Dyes: Sources. Tradition, Technology and Science, (Trad. C. Higgitt). London: Archetype.

Christie RM, Mather RR, Wardman RH (2000). The Chemistry of Color Application, Wiley, Blackwell Sci. Oxford.

El-Nagar K, Sanad SH, Mohamed AS, Ramadan A (2005). Mechanical properties and stability to light exposure for dyed Egyptian cotton fabric with natural and synthetic dyes. Polym. Plast. Technol. Eng. 44(7):1269-1279.

Ferreira ESB, Alison NH, Quye A (2004). The natural constituents of historical textile dyes. Chem. Soc. Rev. 33(6):329-336.

Garland CE (1993). Color technology in textile chemistry. AATCC, In C. Gultekin (Ed.). pp. 107-112.

Heba FM (2009). Environment and energy efficient dyeing of woollen fabric with sticta coronate. Clean Tech. Environ Policy, DOI 10.1007/s10098-009-0267-7.

Heba FM, Gamal AM (2011). Environmental Assessment of Osage orange Extraction and Its Dyeing Properties on Protein Fabrics. Part II: Dyeing Properties. J. Environ. Sci. Technol. 4(4):395-402.

Heba FM, Heffernan S (2010). Environmental aspects on dyeing silk fabric with sticta coronata lichen using ultrasonic energy and mild mordants. Clean Tech. Environ. Policy, Springer-Verlag, DOI 10.1007/s10098-010-0296-2.

Iqbal JM, Ashiq MN (2007). Adsorption of dyes from aqueous solution on activated charcoal. J. Hazard. Mater. 139(1):57-66.

Jayaprakasha GK, Rao JM, Sakariah KK (2002). Improved HPLC for the determination of curcumin, demethoxycurcumin, and bisdemethoxycurcumin. J. Agric. Food Chem. 50(13):3663-3672.

Jayaprakasha GK, Rao JM, Sakariah KK (2005). Chemistry and biological activities of C. longa. Trends Food Sci. Technol. 16(12):533-548.

Jiang H, Timmermann BN, Gang, DR (2006). Use of liquid chromatographyelectrospray ionization tandem mass spectrometry to identify diarylheptanoids in turmeric (*Curcuma longa* L.) rhizome. J. Chromatogr. A. 1111(1):21-31.

Kamel MM, El-Shishtawy RM, Youssef BM, Mashaly H (2007). Ultrasonic assisted dyeing. IV. Dyeing of cationised cotton with lac natural dye. Dyes Pigm. 73(3):279-284.

Kongkachuichaya P, Shitangkoonb A, Chinwongamorna N (2002). Studies on Dyeing of Silk Yarn with Lac Dye: Effects of mordants and dyeing conditions. Sci. Asia. 28:161-166.

Kubelka PI (1948). New contribution to the optics of intensity light-scattering materials-Part I. JOSA. 38:448-451.

Marmion DM (1991). Handbook of US Colorants", Foods, Drugs, Cosmetics and Medical Devices, 3rd edn, John Wiley & Sons, New York.

Moeyes M (1993). Natural dyeing in Thailand, White Lotus, ISBN 974-8495-92-2, Bangkok, Thailand.

Montazer M, Parvinzadeh M (2004). Colorimetric properties of wool dyed with natural dyes after treatment with ammonia. Color Technol. 120(4):161-166.

Muhammed AR, Soliman A, Khattab M (2008). Solvent effect on the Spectral Properties of neutral red. Chem. Central J. DOI:10.1186/1752-153X-2-19.

Osman EM, El-Ebissy AA, Michaell MN (2009). Characterization and Evaluation of the Levelness Parameters of Natural Dyes on Natural Fabrics. RJTA 13(2):61-68.

Padma VS, Mahanta D, Tiwari SC (2006). Ecofriendly sonicator dyeing of cotton with Rubia cordifolia Linn. Using biomordant. Dyes Pigm. 76(1):207-212.

Roessler A, Crettenand D (2004). Direct electrochemical reduction of vat dyes in a fixed bed of graphite granule. Dyes Pigm. 63(1):29-37.

Roessler A, Jin X (2003). State of the art technologies and new electrochemical methods for the reduction of vat dyes. Dyes Pigm. 59(3):223-235.

Siva R (2007). Status of natural dyes and dye-yielding plants in India. Curr. Sci. 92(7):916-925.

Stankovic I (2004). Curcumin, Chemical and Technical Assessment, 61st JECFA (CTA), © FAO, 1(8).

Vajnhandl S, Le Marechal MA (2005). Ultrasound in textile dyeing and the discoloration and mineralization of textile dyes. Dyes Pigm. 65(2):89-101.

Wang YJ, Pan MH, Lin AL, Ho LI., Hsieh YS, Lin JK (1997). Stability of curcumin in buffer solutions and characterization of its degradation products. J. Pharm. Biomed. Anal. 15(12):1867-1876.

www.dow.com/productsafety/.../acetone.

www.madsci.org/.../1060116066.Ch.r.ht

Variation of source distribution of total oxidants: Contributions of oxides of nitrogen, sulphur (IV) oxide emissions and background ozone from Lagos-Nigeria

A. M. O. Abdul Raheem and F. A. Adekola

Department of Chemistry, University of Ilorin, P. M. B. 1515, Ilorin, Nigeria.

Air dispersion modeling system (ADMS)-urban was used as air dispersion tool to model the atmospheric pollutants composition over Lagos air shed. The model runs with single point source showing the variation of concentrations of air pollutants (NO_2, NO_x, O_3, and SO_2) with ambient air dispersion over the city of Lagos, Nigeria. This city is located in the coastal area, and is highly congested with a high intensity of marine, vehicular, and industrial activities. The model was run with input data collected during dry and wet seasons using ADMS-urban. The dispersion pattern of NO_x is during both seasons was significantly different. Whereas, a Gaussian type curve was observed during dry season, a Sigmoid curve was obtained during the wet season. It is noteworthy that the elevation was higher for the dry season than the wet season due to higher photochemical activities. The modelled results obtained in ppb are: $O_{3,}$: 6.13, 23.36; NOx: 2.02, 3.99; SO_2: 9.54, 6.01 during wet and dry seasons, respectively. These predicted concentration values from theoretical model compared fairly with the measured values.

Key words: Seasonal variation, air dispersion modeling system, air pollutants, theoretical and empirical models, photochemical activities.

INTRODUCTION

The temporal and spatial heterogeneity of emission patterns and microclimate combined with the complexity of the urban surface results in complex dispersion pathways at local scales within urban areas. This can lead to strong gradients in vertical and horizontal pollutant concentration (Vakeva et al., 1999). Important oxidants in polluted air are formed within the atmosphere by chemical reactions that occur among the primary pollutants. Among these are peroxy radicals, ozone, nitrogen dioxide and peroxyacetyl nitrate. These, in their gas phase, control the chemistry of the atmosphere and their resultant effects contribute to climate change. These chemical processes affect the current and future composition of the atmosphere while controlling the abundance of the rage of pollutants harmful to health, vegetation and materials (Bloss, 2009).

In order to achieve better understanding of biosphere – atmosphere interactions, climate change simulations and the study of chemistry–climate interactions both of the past and future atmospheres, there is the need for a tool to model the atmospheric composition. Among these models is the dispersion modeling, a typical example of theoretical model. If the physical, chemical and / or biological mechanisms underlying a process are well understood, a steady state or dynamic model can be developed. Steady – state models cannot be used for predicting system responses over time, and therefore, have limited management value. Time–variable models, on the other hand, can handle variable input loads, and can be useful for establishing cause effect relationships. When compared to empirical models, theoretical models are generally more complex. They require a longer period of observation for calibration and the number of variables and parameters to be measured are greater. They also require a significant amount of time for validation. Due to their complexity, and because the understanding of atmospheric systems is incomplete, these types of models are less frequently used than empirical models. Theoretical model also goes along with geographic information system (GIS) as well as geographic positioning system (GPS). The purpose of a dispersion model is to calculate air pollution concentrations according to the pollution emissions data and the nature of the atmosphere (Leksmono et al., 2006). Among many advanced dispersion models, Air dispersion modeling system (ADMS)-urban, AA Quire and Indic Air Viro are designed to model at an urban scale and include point, line and area sources (Chatterton, 2001). However, according to a survey carried out by Beattie (2000) and Woodfield et al. (2006), it was concluded that ADMS-urban is one of the most widely used advanced model tools, hence its attraction and choice for this study.

The ADMS-Urban model is a steady state new generation Gaussian type model which was originally developed (as UK - ADMS) for point source applications (Owen et al., 1999). The model has since been developed as a multi source dispersion modeling system for air quality management applications all over the world.

The behaviour of ADMS-urban has been studied extensively by several evaluations of long-term air pollution concentrations and comparisons with monitored data (Bennet and Hunter, 1997; Carruthers et al., 1996, 1999; Owen and Raper, 2001), while analyses of individual pollution episodes are more rare (Arciszewska and McClatchey, 2001). Local authorities should predict episodes in which pollutant concentrations overcome safety thresholds.

ADMS- urban is an improvement upon ADMS model (Carruthers et al., 1994, 1999) which includes a line source algorithm to enable the calculation of pollutant concentrations from road traffic and industrial sources. ADMS-urban is an urban scale, multi source model allowing up to 3000 grid sources, 1000 road sources and 100 industrial sources to be simultaneously modeled with complete terrain and street canyons. ADMS-urban is also called "new generation" dispersion model which means it uses an approach of boundary layer scaling based on the Monin – Obukhov length and boundary layer depth, rather than the Pasquill stability classes used by many earlier dispersion models (Leksmono et al., 2006). Another feature of ADMS-urban is that in convective conditions, the concentration profiles are assumed to be skewed Gaussian in order to bring material from elevated releases rapidly down to the surface (McHugh et al., 1997). The model is further integrated with a GIS, ArcView 3, and an emission database. This interface is an important visualization tool and also provides a means of manipulating the spatial element of the model input data (Owen et al., 1999).

METHODOLOGY

Study area

The study area for this work is Lagos, former capital of Nigeria. Four sites were used to represent different anthropogenic contributions and one used as control (high % of natural contribution). The coordinates and activities occurring within these sites have been described elsewhere (Abdul Raheem et al., 2009), while Lagos as a Nigeria conurbation and a commercial center have equally been described fully in FRNG (2009).

Emissions data for this study were the measured data generated between 2003 and 2006 (Abdul Raheem, 2007) and recently updated. The meteorological data were collected from Nigeria Meteorological Services (NIMET) Unit of Aviation Ministry, ARIAL, Centre for Energy Research and Development (CERD), Obafemi Awolowo University, Ile – Ife as well as from global Baseline Surface Radiation Network (BSRN) of NASA located in Physics Department, University of Ilorin. The sources in the inventory are spatially described as point and area sources. The inventory is within a GIS allowing aggregation of emission from all sources, sectors up to a grid square basis. A constant value for surface roughness was used for the entire study area (z_o = 1.0 m). This is not representative of the whole study area and may result in inaccuracies in the modeling calculations (CERC, 2004). A surface roughness length of 1.0 m is representative of an urban area such as Lagos but some parts of the study area will be less built – up and a lower value for surface roughness would be more appropriate, this has been previously remarked by Owen et al. (1999) and they commented that such problem is likely to recur with the use of this type of model over a regional scale. The model allows two input values for the surface roughness; one at the source location and the other at the receptor location, but it does not allow for variation in surface roughness for different source locations. However, the algorithm for the evaluation of downwind concentration is displayed in Figure 1.

As shown in Figure 1, the system has a number of distinct features used to achieve the modeling of data (CERC, 2004). These include but not limited to input of meteorological ambient data, measured ambient data, traffic flow and downwind concentration. The modeled results show the concentration – distance line plot (Figures 4 to 11) that reflects the expected three ambient conditions that is, stable, neutral and convective.

RESULTS AND DISCUSSION

Typical examples of the input data used are the mean raw data showing time of the day, pollutants measured, their concentrations, meteorological data and seasonal variations as shown in Tables 1 and 2. The model tool as earlier described is a software (ADMS-urban) licensed to the researcher for a period of one year by Cambridge Environmental Research Consultants, Environmental Software and Services, United Kingdom.

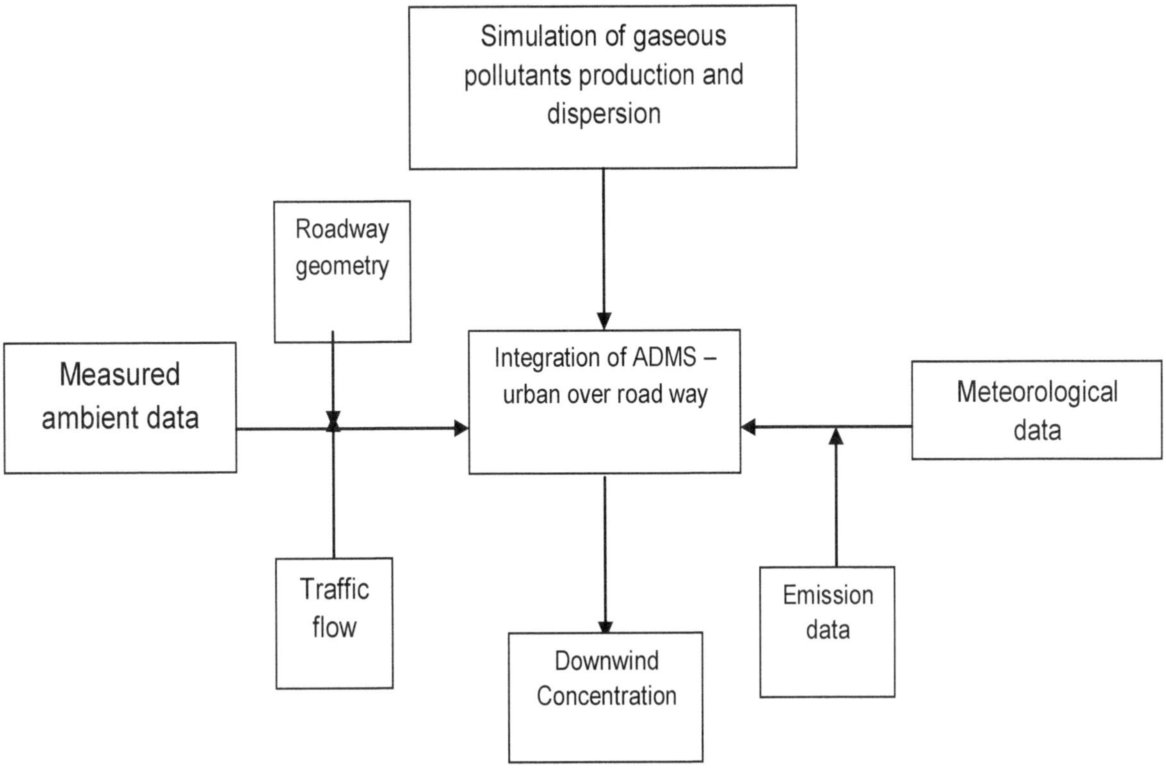

Figure 1. The algorithm for the evaluation of downwind concentration using the ADMS–urban model.

Table 1. Dry season environmental data for Lagos.

Start of sampling	End of sampling	OX (ppb)	NOx (ppb)	SO_2 (ppb)	RELHUM (%)	WND (ms^{-1})	DWND (°C)	AIR temp (°C)	Sun exp (Wm^{-2})
6.30 am	7.30 am	14.26	12.40	10.72	90.58	3.07	107.02	25.99	-1.65
7.45 am	8.45 am	22.92	5.89	7.20	87.58	4.67	156.38	26.38	2.80
9.00 am	10.0 am	28.95	5.39	11.15	73.75	7.09	189.02	29.05	8.90
10.15 am	10.0 am	46.86	5.66	14.82	67.92	7.35	182.63	30.25	11.00
11.30 am	12.30 pm	43.21	6.41	10.51	63.50	8.76	170.67	31.30	12.06
12.45 pm	1.45 pm	85.31	5.68	12.74	60.33	10.11	159.55	32.00	17.30
2.00 pm	3.00 pm	73.77	6.45	16.62	60.08	10.36	155.00	31.98	15.10
3.15 pm	4.15 pm	26.06	6.84	15.47	62.67	10.94	163.79	31.37	13.20
4.30 pm	5.30 pm	12.23	5.72	16.48	67.00	10.21	165.22	30.38	10.70
5.45 pm	6.45 pm	8.58	6.90	19.21	72.75	8.99	166.64	29.20	3.30

Adapted from Abdul Raheem et al. (2009).

After inputting data for dry and rainy seasons (Tables 1 and 2), and the mean traffic flow experienced during the two seasons throughout the sampling periods (Figures 2 and 3) into ADMS-urban as shown in Figure 1, new concentrations were calculated taking into consideration the values of surface roughness, the latitude of the site, Monin – Obukhov length, height of recorded wind, meteorological data, emissions, background concentrations, grid etc. The new concentration calculated is referred to as modeled concentration. Figures 4 to 11 show the dispersion of pollutants from single elevated point source during wet and dry seasons in Lagos. In other words, the concentration – distance line plots. The three ambient stability classes are represented. The deep blue colour represents the unstable condition of the atmosphere; green colour

Table 2. Rainy season environmental data for Lagos.

Start of sampling	End of sampling	OX (ppb)	NOx (ppb)	SO_2 (ppb)	RELHUM (%)	WND (ms^{-1})	DWND (°C)	Air temp (°C)	Sun exp (Wm^{-2})
6.30 am	7.30 am	9.25	3.16	2.30	95.75	4.05	128.38	24.84	-0.13
7.45 am	8.45 am	12.91	3.39	4.46	93.83	5.29	146.58	25.58	1.24
9.00 am	10.0 am	10.36	2.77	6.04	84.5	8.07	191.13	27.52	6.35
10.15 am	10.0 am	9.59	2.45	6.48	80.58	8.79	196.15	28.19	8.10
11.30 am	12.30 pm	9.00	2.25	3.74	78.25	9.72	203.63	28.74	9.25
12.45 pm	1.45 pm	15.03	1.77	4.39	77	10.32	205.68	29.03	14.42
2.00 pm	3.00 pm	16.72	2.87	5.68	76.67	11.45	210.44	28.96	15.74
3.15 pm	4.15 pm	10.78	3.54	7.05	77.58	11.54	213.56	28.58	11.56
4.30 pm	5.30 pm	3.57	3.89	8.13	79.67	11.06	213.20	27.94	7.81
5.45 pm	6.45 pm	1.45	3.89	9.00	83.08	10.05	209.95	27.13	1.65

Adapted from Abdul Raheem et al. (2009).

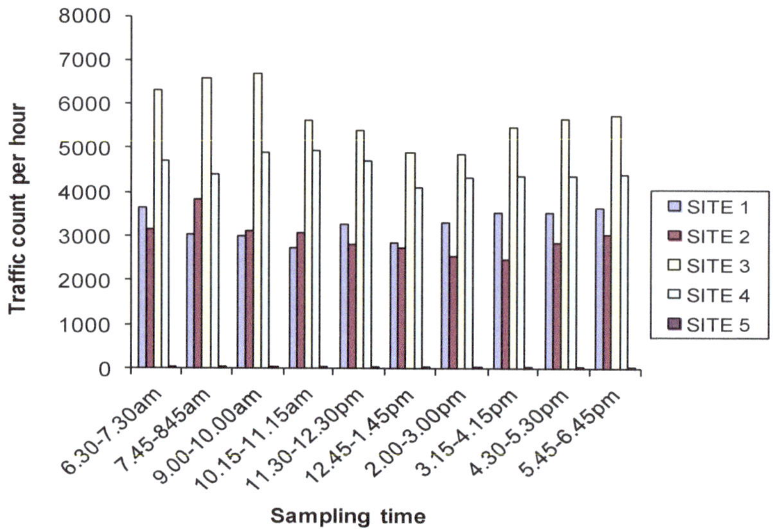

Figure 2. Variation of the average traffic count with the time of the day in Lagos during rainy season.

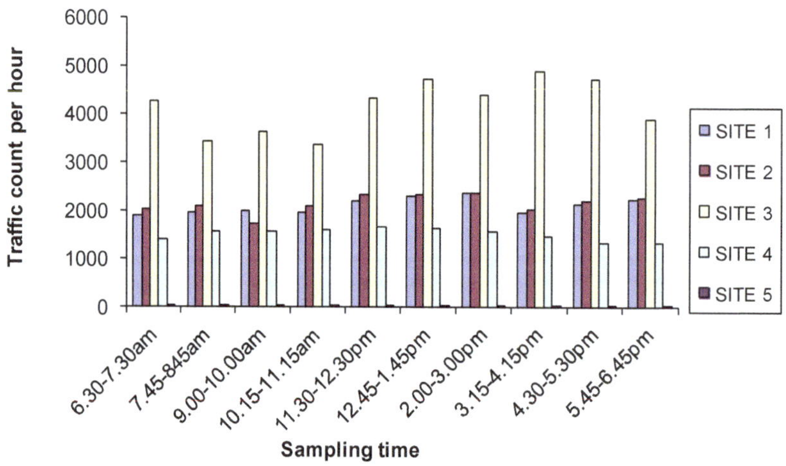

Figure 3. Variation of the average traffic count with the time of the day in Lagos during dry season.

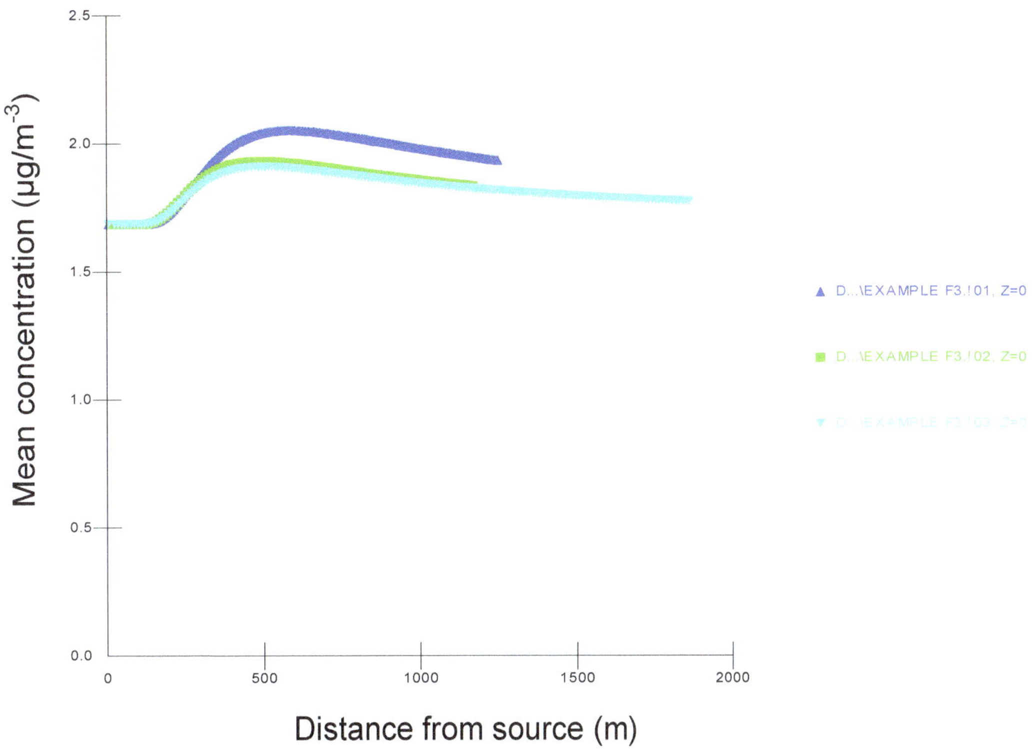

Figure 4. Concentration-distance line plot for no_2 in the rainy season for Lagos.

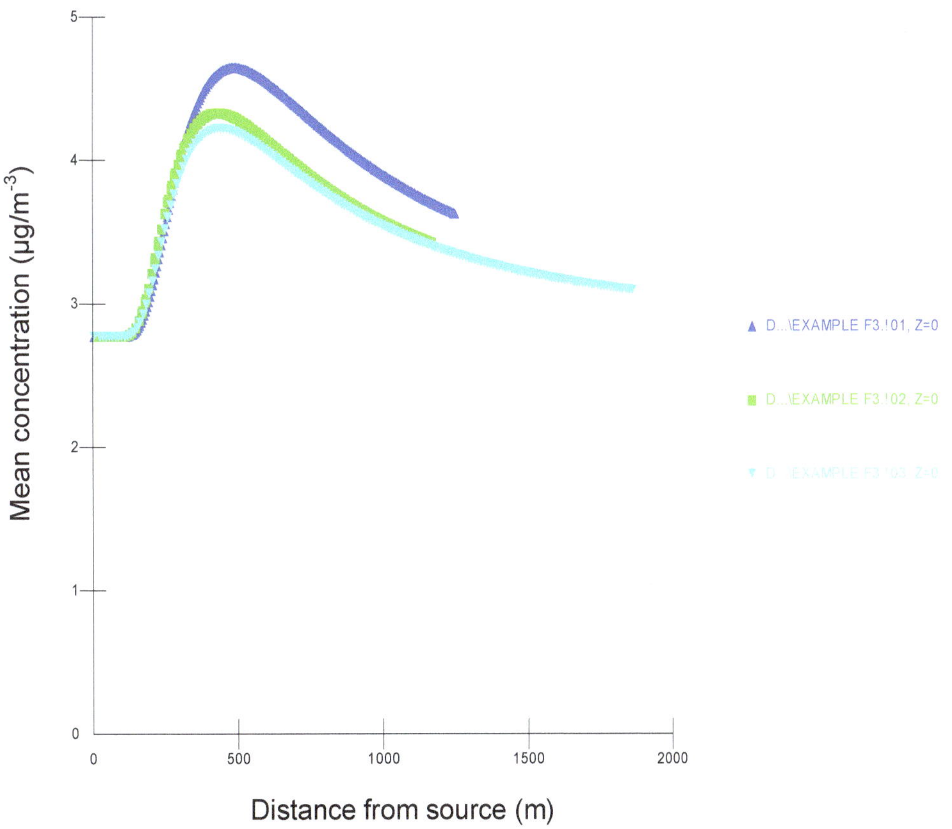

Figure 5. Concentration-distance line plot for no_x in the rainy season for Lagos.

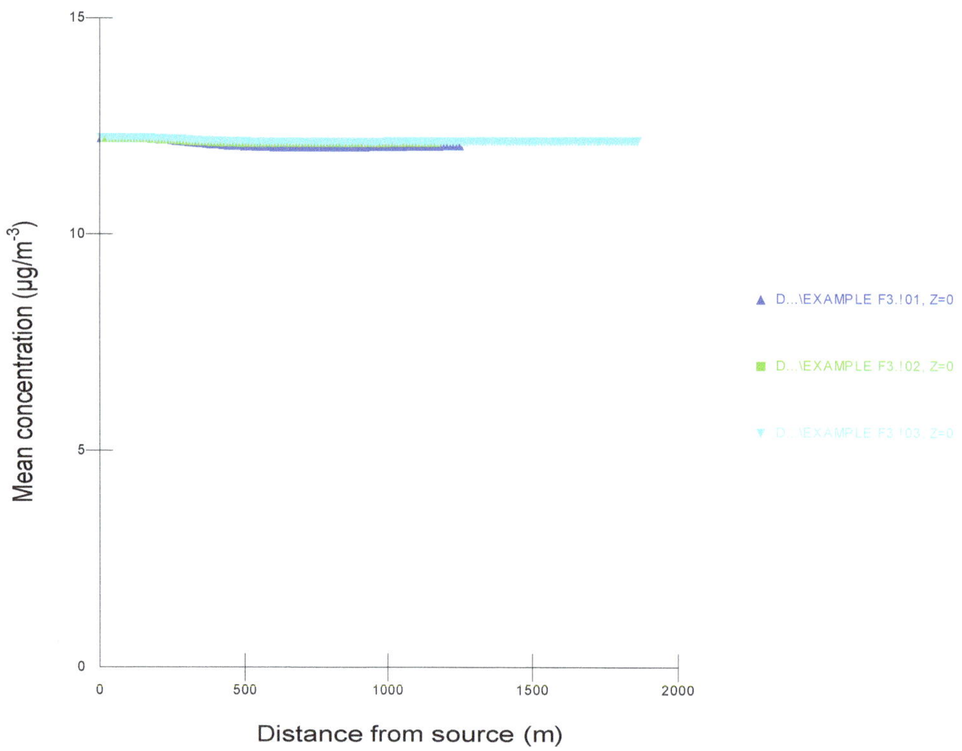

Figure 6. Concentration – distance line plot for o3 in the rainy season for Lagos.

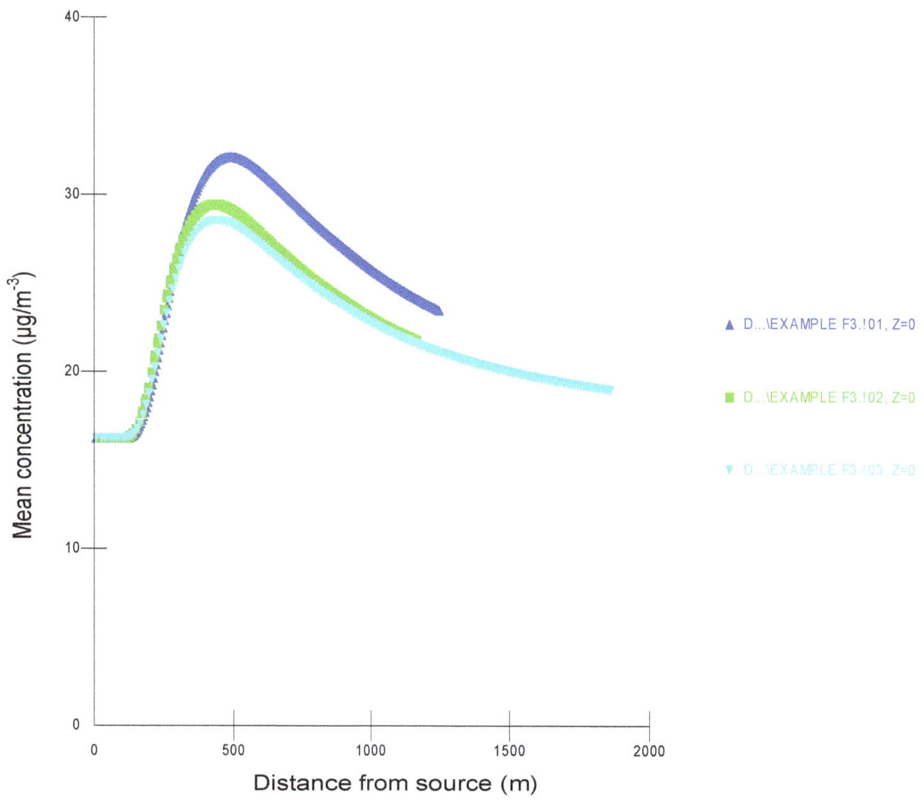

Figure 7. Concentration – distance line plot for so_2 in the rainy season for Lagos.

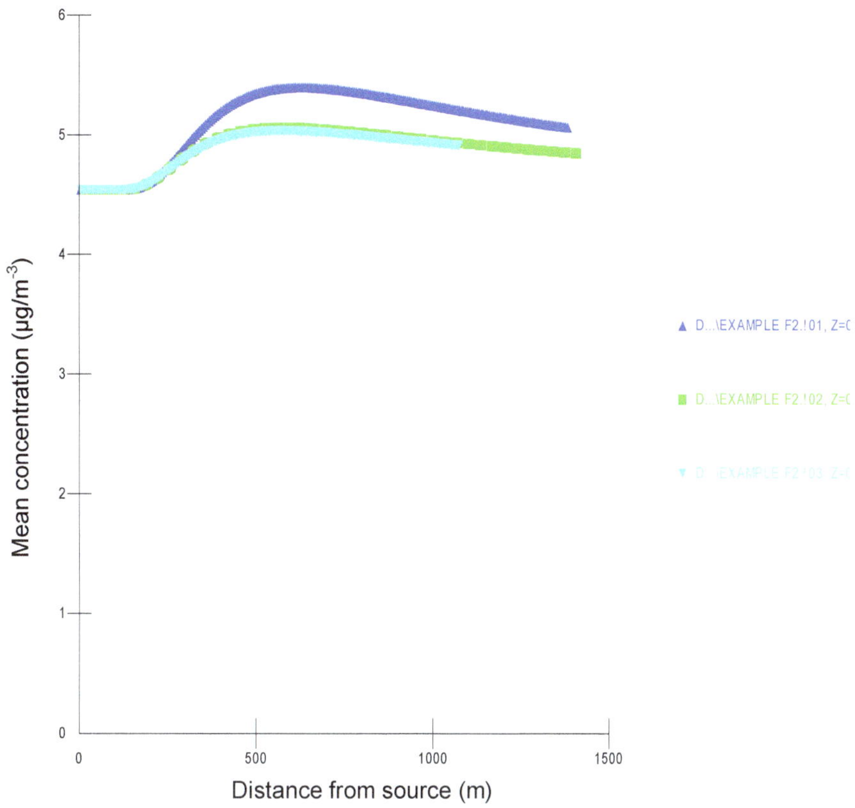

Figure 8. Concentration – distance line plot for no_2 in the dry season for Lagos.

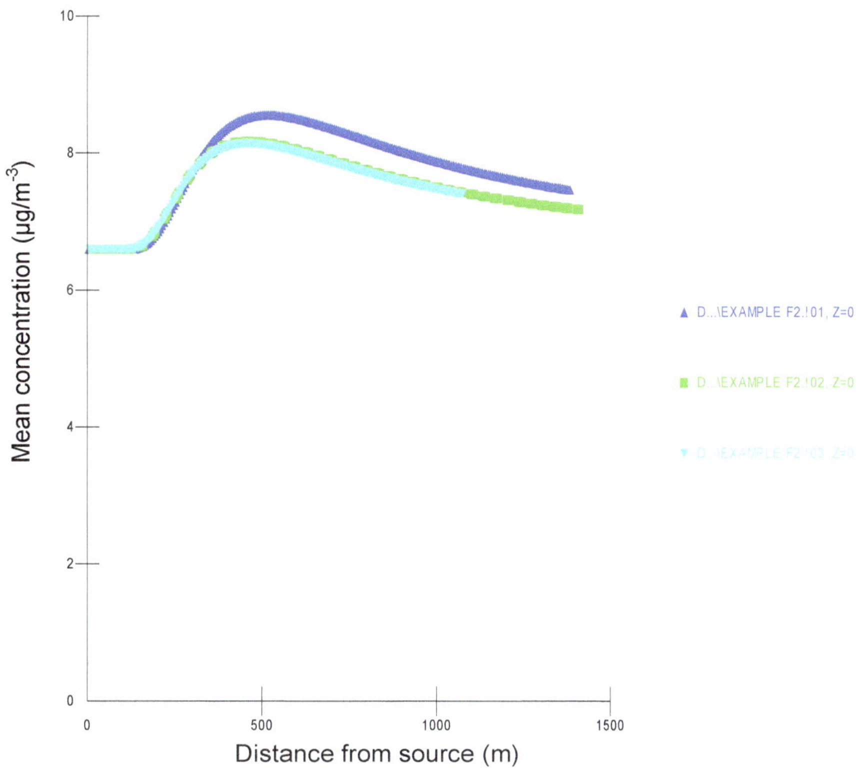

Figure 9. Concentration – distance line plot for no_x in the dry season for Lagos.

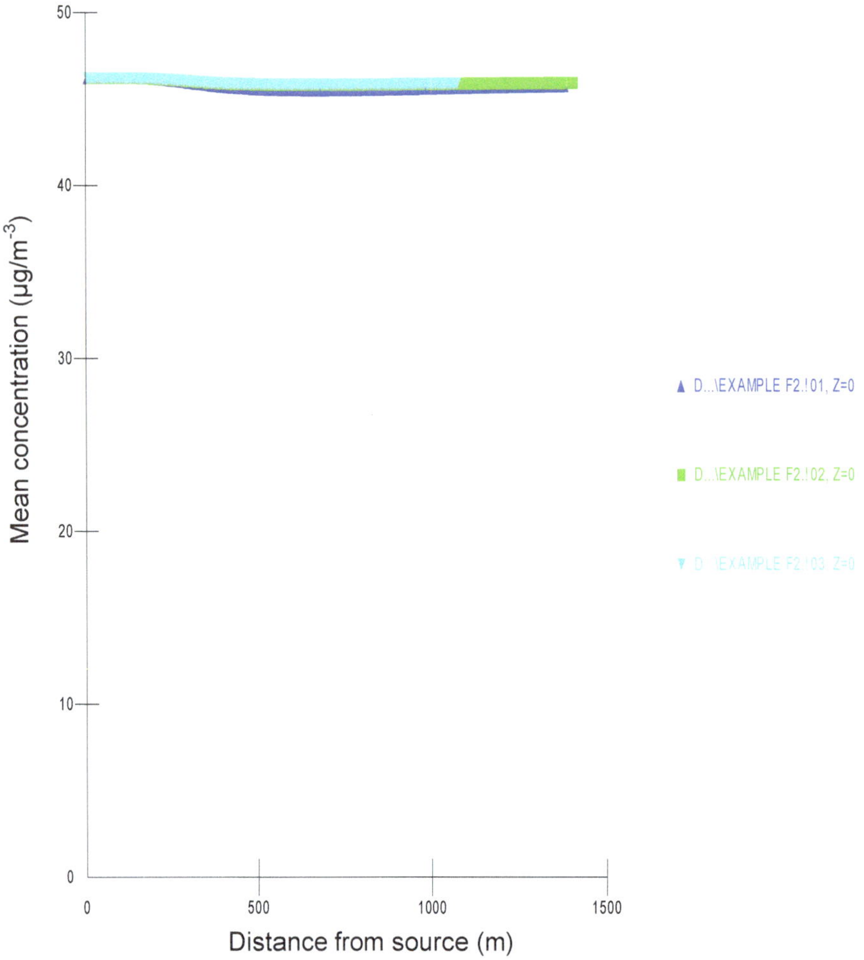

Figure 10. Concentration – distance line plot for o_3 in the dry season for Lagos.

represents the neutral atmospheric condition while the light blue colour represents the stable atmospheric condition as shown from the output of the model.

Figures 8 to 11 show the dispersion of pollutants from single elevated point source during dry season in Lagos. Figures 4 to 11 show the variation of concentrations of NO_2, NO_x, O_3 and SO_2 with distance in Lagos city during both rainy and dry seasons. Figures 6 and 10 show that the ozone concentration is nearly constant during both seasons with concentration range between 12 to 12.5 μgm^{-3} during rain and 47 to 48 μgm^{-3} during dry, from the unstable through neutral to stable conditions of the atmosphere. However, the concentration range shows a very slight decrease before attaining a constant value of 47 – 48 μgm^{-3} beyond a distance of about 500 m in the dry seasons. This may be attributable to its production as a secondary pollutant. The behaviour of NO_2 (1.7 to 2.2 μgm^{-3} in rain to 4.5 to 5.5 μgm^{-3} in dry) and NO_x (2.8 – 4.8 μgm^{-3} in rain and 6.5 to 8.5 μgm^{-3} in dry) are also different for the two seasons. A Gaussian type curve was produced for NO_x during the rainy season while an S – type curve was obtained during the dry season. This shows that the leighton ratio (NO_2) / (NO) which is a daylight factor that depends on ozone concentrations and the sunlight differ for the two seasons. For both seasons NO_2 has an S – type curve. The elevation is also higher for dry than rainy season. This could be attributed to a higher photochemical activity involving the complex chemistry of interconversion of NO_2 or NO_x into O_3 (Chou et al., 2006; Bloss, 2009) in the dry season. There is no visible change for the variation of SO_2 for both dry and rainy season, however, it is the only pollutant that have higher modeled concentration in the wet season compared to the dry season and showing a biogenic source different from the sources of the other two pollutants (Figures 7 and 11).

Comparison of monitored and modeled data

Table 3 shows that the modeled concentrations are generally higher in dry season and this follows a similar

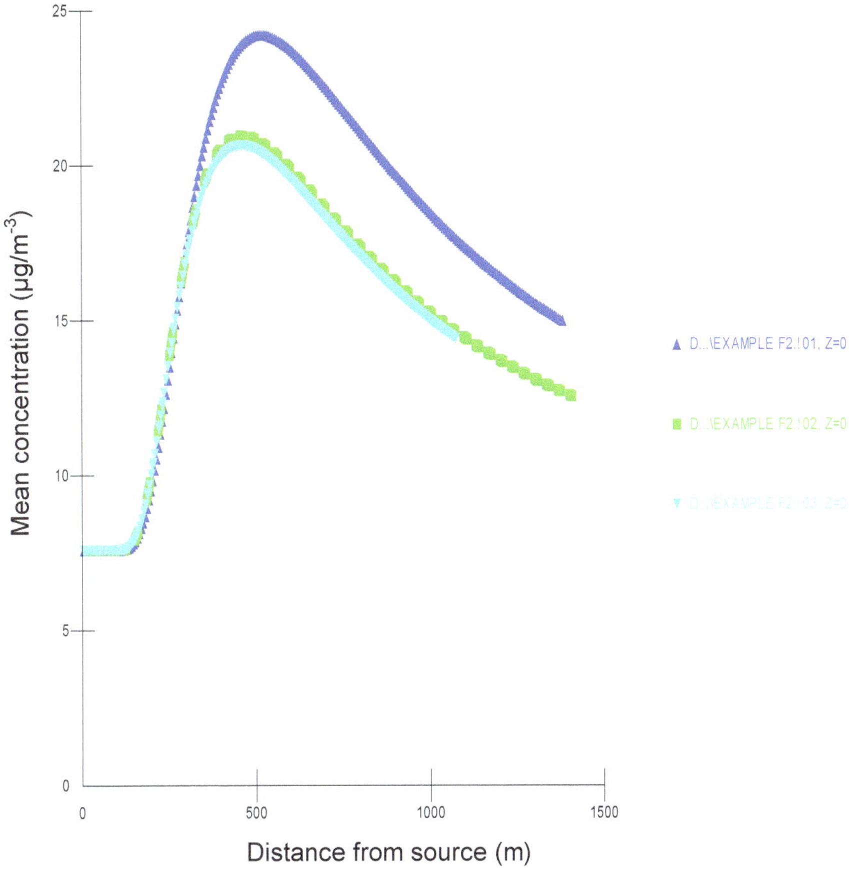

Figure 11. Concentration – distance line plot for so_2 in the dry season for Lagos.

Table 3. The measured and ADMS-urban modeled concentration values of O_3, NO_x and SO_2 in ppb during rainy and dry seasons in Lagos city.

Sites	Measured	O_3 Modeled	Measured	NO_x Modeled	Measured	SO_2 Modeled
Lagos rain	9.87 ± 1.61[a]	6.13	3.00 ± 0.24[a]	2.02	5.73 ± 0.51[b]	9.54
Lagos dry	36.22 ± 3.48[a]	23.36	6.73 ± 0.51 [a]	3.99	13.49 ± 1.72[b]	6.01

[a]Abdul Raheem et al. (2009) and [b]Abdul Raheem et al. (2012).

trend with field measurements. The relatively lower values during wet season have been attributed to the attenuation effect of rain. The modeled result for SO_2 in dry season is however lower by 37.0% than that of rainy season. For comparative purpose, the mean measured concentrations and predicted concentrations by ADMS-urban are summarized in Table 3. The predicted concentration for O_3 is lower than measured value while NO_x and SO_2 are higher during the wet season. This could be due to poor mixing of atmospheric air as a factor of wind speed downwind during the sampling period. The effect of ozone is more felt in rural than urban because of its being a secondary pollutant unlike the oxides of nitrogen and sulphur dioxide. Effect of rain attenuation is more on ozone than the other pollutants. However, in dry season all the modeled values are lower with ozone having the least 35.5% while the other pollutants are greater than 40% lowered than the measured values. Going by the complexity of the theoretical modeling, more information may be required for computation in other to have a good agreement between the predicted concentration and measured concentrations for NO_x, O_3; SO_2.

Though, it is outside the scope of this paper to compare

the urban concentration measurement to that of rural concentration measurement of pollutants, it can be safely concluded that the chemical processes found in the atmosphere within and above cities (such as Lagos-Nigeria) differ from those of the more remote boundary layer as a consequence of the magnitude, diverse and complex nature of local anthropogenic emission sources, the higher concentrations of many trace species found as a result of this, and of the dynamic micro – and macro – environments in the urban landscape (Bloss, 2009).

ACKNOWLEDGEMENTS

We wish to thank Global Change System for Analysis Research and Training (START) USA for sponsoring this work, Cambridge Environmental Research Consultants (CERC) office, United Kingdom for licensing the ADMS – Urban modelling software for a token and the Faculty for the Future (FftF) Schlumberger Foundation for the fellowship award that makes the consolidation of the knowledge of Atmospheric Chemistry a reality.

REFERENCES

Abdul Raheem AMO (2007). Measurement, modelling and analysis of ozone and ozone precursors in the ambient environment of Lagos and Ilorin cities, Nigeria. PhD Dissertation, University of Ilorin, Ilorin, Nigeria.

Abdul Raheem AMO, Adekola FA, Obioh IO (2009). The seasonal variation of the concentrations of ozone, sulphur dioxide and nitrogen oxides in two Nigerian cities. Environ. Model. Assess. 14(4):497-509.

Arciszewska C, McClatchey J (2001). The importance of meteorological data for modelling air pollution using ADMS-Urban. Meteorol. Appl. 8:345-350.

Beattie CI (2000). An Investigation into Urban Air Quality Management (AQM) Practices in England. Summary of Questionnaire Surveys Carried out in 2000, UWE, Bristol.

Bennet M, Hunter GC (1997). Some comparisons of lidar estimates of peak ground – level concentrations with the predictions of UK – ADMS. Atmos. Environ. 31(3):429-439.

Bloss WJ (2009). Atmospheric Chemical Processes of Importance in Cities. In Issues in Envoronmental Science and Technology, 28 Air Quality in Urban Environments. Edited by Hester RE and Harrison RM. Royal Soc. Chem. 2009.

Cambridge Environmental Research Consultants (CERC) (2004). ADMS – Urban Manual.

Carruthers DJ, Dixon P, McHugh CA, Nixon SG, Oates W (1999). Determination of compliance with UK and EU air quality objectives from high resolution pollutant concentration maps calculated using ADMS-Urban. Proc. 6th Int. Conf. Harm. Within Atm. Disp. Models for Regulatory Purposes, Rouen.

Carruthers DJ, Edmunds HA, Bennett M (1996). Validation of the UK – ADMS dispersion model and assessment of the performance relative to R91 and ISC using archive Lidar data. Report Ref. DoE/MW/RR/95/022 London. The UK Environmental Agency.

Carruthers DJ, Holroyd RL, Hunt JCR (1994). UK – ADMS – A new approach to modeling dispersion in the earth's boundary layer. J. Wind Eng. 52:139-153.

Chatterton TJ (2001). Regional and urban scale modeling of particulate matter: can PM_{10} be managed at a local level? PhD Thesis, University of East Anglia.

Chou CCK, Liu SC, Lin CY, Shiu CJ, Chang KH (2006). The trend of surface ozone in Taipei, Taiwan, and its causes: Implications for ozone control strategies. Atmos. Environ. 40:3898-3908.

Leksmono NS, Longhurst JWS, Ling KA, Chatterton TJ, Fisher BEA, Irwin JG (2006). Assessment of the relationship between industrial and traffic sources contributing to air quality objective exceedences: a theoretical modeling exercise. Environ. Model. Softw. 21:494-500.

McHugh CA, Carruthers DJ, Edmunds HA (1997). ADMS and ADMS – Urban. Int. J. Environ. Pollut. 8(3-6):438-440.

Owen B, Edmunds HA, Carruthers DJ, Raper DW (1999). Use of a new generation urban scale dispersion model to estimate the concentration of oxides of nitrogen and sulphur dioxide in a large urban area. Sci. Tot. Env. 235(1-3):277-291.

Vakeva M, Hameri K, Kulmala M, Lahdes R, Ruuskanen J, Laitinen T (1999). Street level versus rooftop concentrations of submicron aerosol particles and gaseous pollutants in an urban street canyon. Atmos. Environ. 33(9):1385-1397.

Woodfield NK, Longhurst JWS, Beattie CI, Chatterton T, Laxen DPH (2006). Regional collaborative urban air quality management: case studies across Great Britain. Environ. Model. Softw. 21(4):595-599.

Federal Republic of Nigeria Official Gazette (FRNG) (2 February 2009). "Legal Notice on Publication of the Details of the Breakdown of the National and State Official Totals 2006 Census" (PDF). Archived from the original on 2007-07-04, Retrieved 2007-06-29. www.population.gov.ng. 96:2.

Toxicity of arsenic (As) on seed germination of sunflower (*Helianthus annuus* L.)

Muhammad Asif Imran[1], Muhammad Nawaz Ch[1], Rass Masood Khan[2], Zulfiqar Ali[1,3] and Tariq Mahmood[3]

[1]College of Earth and Environmental Sciences, University of the Punjab, Lahore, Pakistan.
[2]Department of Botany, University of the Punjab, Lahore, Pakistan.
[3]Nano Science and Catalysis Division, National Centre for Physics, Quaid-i-Azam University, Islamabad 45320, Pakistan.

The effect of different levels (2, 4, 6, 8 and 10 mg/L) of arsenic (as sodium arsenate and sodium arsenite) on germination percentage, mean germination time, days to 50% germination, plumule and radicle length and seedling vigour index of four sunflower cultivars (FH-331, FH-385, FH-405 and FH-415) was determined under laboratory conditions. Germination percentage was gradually decreased with the increase in arsenic concentration in case of all sunflower cultivars. Low concentrations (2 and 4 mg/L arsenic) increased while higher concentrations (6, 8 and 10 mg/L arsenic) posed stress and caused reduction in plumule and radicle length (mm) of seedlings as compared to control. Out of all the four sunflower cultivars, H1 (FH-331) proved the most sensitive one by giving least values for plumule length, radicle length and seedling vigour index as observed during the course of this experiment.

Key words: Sodium arsenate, sodium arsenite, sunflower, seed germination, seedling growth.

INTRODUCTION

Environmental pollution has been proved to be one of by-products of industrial revolution and modern technology strengthened by over population and mismanagement of resources. All metals are natural constituents of our environment and some of them are essentially required as micro-nutrients (Taiz and Zeiger, 2006). Arsenic (As) a ubiquitous, carcinogenic trace metalloid (Pigna et al., 2009) found in virtually all environmental media (Fitz and Wenzel, 2002). Arsenic exists in the environment in various organic and inorganic forms (species) (Mandal et al., 2011). The most important inorganic species are arsenate (As^{5+}) and arsenite (As^{3+}). Its toxicity to plants depends on its valence state. Due to greater cellular uptake, As^{+3} is 60 times more soluble, mobile and toxic than As^{5+} (Giri and Patel, 2012). Arsenic concentration varies in the environment, from less than 1 mg/kg to as high as 100,000 mg/kg in soil (Liu et al., 2002) and one in every 60 people on the planet is living in an area where 50 µg/L or above of arsenic is in ground water (Tibbetts, 2000).

Arsenic contamination may be greater in soils naturally or by human activities through the use of various arsenical pesticides, wood preservatives, industrial wastes and growth promoters for plants and animals (Williams et al., 2005). Irrigation through As-laden ground water in Asia also causes bioaccumulation of As (Dittmar et al., 2010). To evaluate the possible health risk to humans consuming crops irrigated with As contaminated water, information is needed regarding the soil-to-plant transportation of As and to minimize the accumulation of

As in plants consumed directly by humans, farm animals or wildlife (Meharg and Hartley-Whitaker, 2002).

In plants, arsenic is accumulated according to tolerance (Zhong et al., 2011) mainly in the root system, to various degrees in the above ground organs, depending upon its translocation (Mei et al., 2012). The metal tolerance potential of different plant species varies greatly between different species as well as among different genotypes of a same species (Ahmad et al., 2009). Large number of studies indicated that low concentrations of As stimulated the growth of plants but excessive As did harm to plants (Han et al., 2003). Seed germination and seedling growth of wheat were stimulated at low and inhibited at high (5 to 20 mg/kg soil) arsenic concentrations (Zhang et al., 2002).

Sunflower (*Helianthus annuus* L.) being one of newly added crop to human food chain, is cultivated in Pakistan on 32539 ha with average yield of 816.4 kg per ha and average annual production of 612530 tones (GOP, 2012) fulfills the cooking oil requirements of the country population along with other oil-seed crops. The objective of this study was to evaluate the intra-specific/cultivar comparisons for arsenic tolerance in four sunflower hybrids at the seed germination stage which is considered more vulnerable to any stress than latter growth stages of most plant species.

MATERIALS AND METHODS

Seeds and metal salts

Dry, healthy and uniform sized achenes (seeds) of four sunflower cultivars viz. FH-331, FH-385, FH-405 and FH-415 were collected from the Oil Seed Department of Ayub Agriculture Research Institute (AARI) Faisalabad, Pakistan. Seeds were surface sterilized using 5% sodium hypochlorite solution for five minutes and rinsed thrice with distilled water to prevent any kind of fungal contamination. Two salts of arsenic viz. sodium arsenate ($Na_2HAsO_4.7H_2O$) and sodium arsenite ($NaAsO_2$) of Sigma Aldrich, Japan, were used for arsenic treatments application. Twenty seeds were placed in each Petri plate of 9 cm diameter lined with double layered filter paper (Whattman No. 2).

Treatments application

Five levels (2, 4, 6, 8 and 10 mg/L) of both arsenic salts, sodium arsenate ($Na_2HAsO_4.7H_2O$) and sodium arsenite ($NaAsO_2$) were formulated in Hoagland's nutrient solution for the treatment application and one without any arsenic application was considered as control which were provided only nutrient solution. Freshly prepared 10 ml of treatment solution was applied to respective petriplate after wasting the previous solution daily. The petriplates were kept in climatic room under controlled conditions of temperature (25 ºC ± 2 ºC), relative humidity (65%) and continuous white fluorescent light (*PAR* 300 µmol m^{-2} s^{-1}). Number of germinated seeds was counted daily and data were recorded after every 12 h for fourteen days regularly. A seed was considered as germinated when 5 mm of radicle had emerged out of seed coat. Following seed germination parameters were recorded. Germination percentage was measured according to the formula given by Tanveer et al. (2010).

Germination percentage = (Germinated seeds ÷ Total seeds) x 100

Mean germination time was calculated in days according to the equation given by Dezfuli et al. (2008).

$$MGT = \sum Dn \div \sum n$$

Where n is the number of seeds germinated on day D, and D is number of days counted from the beginning of germination.

The time to 50% germination (T_{50}) was calculated according to the following formula modified by Farooq et al. (2005; 2006).

$$T_{50} = ti + [\{(N/2) - ni\}(tj - ti)] \div (nj - ni)$$

Where N is the final number of germinated seeds and ni, nj are cumulative number of seeds germinated by adjacent counts at times ti and tj where $ni < N/2 < nj$.

Total germinated seeds were collected after fourteen days and plumule and radicle were carefully detached for the estimation of plumule length and radicle length in milli meters with the help of a scale. Vigour index for seedling was calculated as the formula given by Cokkizgin and Cokkizgin (2010).

Seedling vigour index (SVI) = [MPL + MRL] × GP

Where MPL = Mean plumule length, MRL = mean radicle length, GP = germination percentage. Experiment was laid out in completely randomized design (CRD) with three replicates. The data regarding all seed germination parameters was subjected to a two-way analysis of variance (ANOVA) with the help of COSTAT computer software (Cohort software, 2003, Monterey, California) and to compare significance of interaction means, least significance difference values (LSD) were also calculated.

RESULTS AND DISCUSSION

Two way analysis of variance (ANOVA) for all parameters revealed significant differences for sunflower hybrids, different treatments (except for Mean Germination Time) and interaction of hybrids and treatments for both sodium arsenate and sodium arsenite used for arsenic treatments application (Table 1).

Germination Percentage (G %age)

Highly significant differences were found among different arsenic treatments and sunflower hybrids (Table 1) for germination percentage. Considerable decrease in germination percentage was recorded with increasing level of arsenic in all the four sunflower cultivars used in this experiment as stated by Ahmad et al. (2009) that higher concentrations of heavy metals suppresses the seed germination parameters. Out of different treatments T5 (10 mg/L As) affected most severely to all seed germination parameters studied. Sodium arsenite posed more stress on germination percentage, as inorganic $As^{(III)}$ is more toxic than $As^{(V)}$ (Winkel et al., 2008) and least value (55%) was calculated in treatment with 10 mg/L arsenic applied as sodium arsenite (Figure 1 and Table 2), whereas with same concentration of sodium

Table 1. Two way analysis of variance (ANOVA) for germination percentage (G %age), mean germination time (MGT), time to 50% germination (T50), plumule length (Pl L), radicle length (Rl L) and seedling vigour index (SVI) of sunflower cultivars.

Source	D F	Mean square											
		Germination percentage		M G T		T 50		Pl L		Rl L		S V I	
		Sodium arsenate	Sodium arsenite	Sodium arsenate	Sodium arsenite	Sodium arsenate	Sodium arsenite	Sodium arsenate	Sodium arsenite	Sodium arsenate	Sodium arsenite	Sodium arsenate	Sodium arsenite
Hybrids (H)	3	497.22**	536.11**	33.93**	26.39**	10.04**	17.41**	307.90**	467.41**	441.56**	491.76**	7533128**	8406116.4**
Treatments (T)	5	554.72**	2064.72**	0.72^{ns}	1.16^{ns}	1.21**	1.13^{ns}	75.22**	137.91**	90.70**	139.52**	4639275.5**	9274712**
H × T Interaction	15	170.83**	181.94**	1.02**	0.54^{ns}	1.04**	1.13*	13.97**	7.12^{ns}	15.93**	12.62^{ns}	770738.76**	628847.1^{ns}
Error	48	53.12	68.75	0.41	0.49	0.31	0.53	5.53	6.48	5.25	9.88	264354.51	340110.31
CV (%)		8.631	10.85	12.825	14.96	13.338	16.92	12.207	12.68	9.909	13.39	14.402	17.33
Mean		84.44	76.39	5.023	4.68	4.158	4.29	19.274	20.07	23.13	23.47	3569.96	3365.39

Significance Level: 0.05, ** Highly significant, ^{ns} Non significant.

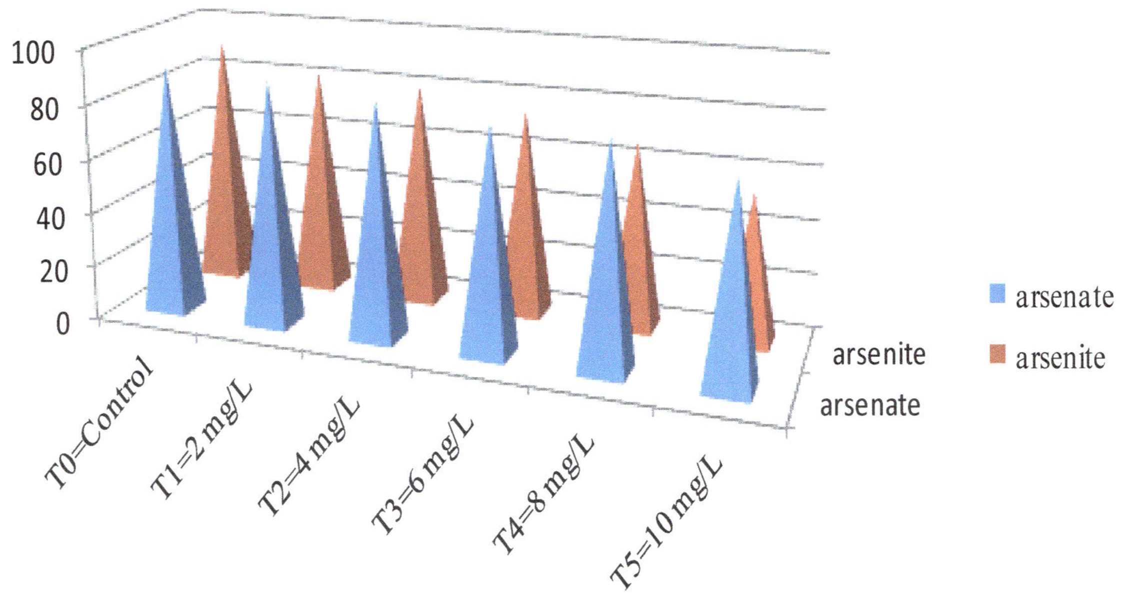

Figure 1. Effect of arsenicals on germination percentage (G %age) of sunflower.

Table 2. Means of arsenical (sodium arsenate and sodium arsenite) treatments for all parameters.

Treatments	Germination percentage		M G T		T 50		Pl L		Rd L		S V I	
	Sodium arsenate	Sodium arsenite	Sodium arsenate	Sodium arsenite	Sodium arsenate	Sodium arsenite	Sodium arsenate	Sodium arsenite	Sodium arsenate	Sodium arsenite	Sodium arsenate	Sodium arsenite
T0- Control	92.08^{a}	92.08^{a}	5.02^{ab}	5.02^{a}	4.50^{a}	4.50^{ab}	17.61^{de}	17.62^{c}	22.5^{c}	22.5^{b}	3670.5^{bc}	3670.5^{bc}
T1- 2 mg/L	90^{a}	84.16^{b}	5.20^{ab}	4.63^{ab}	4.23^{ab}	3.97^{b}	19.55^{bc}	22.97^{a}	23.58^{bc}	26.53^{a}	3908.67^{ab}	4155.3^{a}
T2- 4 mg/L	87.08^{ab}	82.5^{bc}	4.76^{ab}	4.16^{b}	3.73^{c}	3.96^{b}	22.94^{a}	23.07^{a}	26.62^{a}	25.06^{ab}	4296.96^{a}	4098.62^{ab}
T3- 6 mg/L	82.5^{b}	75.83^{c}	4.68^{b}	4.62^{ab}	4.01^{bc}	4.37^{ab}	21.1^{ab}	21.96^{ab}	25.11^{ab}	25.11^{ab}	3732.25^{bc}	3533.96^{c}
T4- 8 mg/L	81.67^{b}	68.75^{d}	5.28^{a}	4.65^{ab}	3.94^{bc}	4.19^{ab}	18.48^{cd}	20.25^{b}	22.37^{c}	22.81^{b}	3336.33^{c}	2909.04^{d}
T5- 10 mg/L	73.33^{c}	55^{e}	5.18^{ab}	4.98^{a}	4.52^{a}	4.74^{a}	15.94^{e}	14.56^{d}	18.61^{d}	17.47^{c}	2475.08^{d}	1824.92^{e}
LSD	5.98	6.81	0.52	0.57	0.37	0.59	1.93	2.08	1.88	2.58	422.04	478.70

Different letters indicate significant differences at 0.05 levels.

Table 3. Means of sunflower cultivars for all parameters

Sunflower hybrids	Germination percentage		M G T		T 50		Pl L		Rd L		S V I	
	Sodium arsenate	Sodium arsenite	Sodium arsenate	Sodium arsenite	Sodium arsenate	Sodium arsenite	Sodium arsenate	Sodium arsenite	Sodium arsenate	Sodium arsenite	Sodium arsenate	Sodium arsenite
H1: FH-331	90.56^{a}	75.83^{b}	6.67^{a}	6.31^{a}	4.41^{b}	5.54^{a}	13.2^{c}	13.65^{d}	15.88^{c}	17.01^{c}	2639.83^{c}	2415.36^{c}
H2: FH-385	83.06^{b}	71.67^{b}	3.46^{d}	3.61^{c}	3.11^{c}	3.31^{c}	22.15^{a}	22.95^{b}	26.94^{a}	27.35^{a}	4120.31^{a}	3623.83^{b}
H3: FH-405	78.06^{c}	73.89^{b}	4.47^{c}	3.94^{c}	4.24^{b}	3.74^{c}	21.59^{ab}	25.17^{a}	25.37^{b}	28.07^{a}	3672.17^{b}	4021.44^{a}
H4: FH-415	86.11^{ab}	84.16^{a}	5.48^{b}	4.86^{b}	4.88^{a}	4.56^{b}	20.15^{b}	18.51^{c}	24.35^{b}	21.45^{b}	3847.56^{ab}	3400.94^{b}
LSD	4.89	5.56	0.43	0.47	0.3	0.48	1.58	1.71	1.54	2.11	344.59	390.86

Different letters represent significant difference at 0.05 levels.

arsenate 73.33% germination percentage value was recorded. Maximum value (92.08%) was noted in control which was devoid of any arsenic application. Sunflower hybrid type H1 (FH-331) showed best germination percentage with 90.56% when arsenic was applied as sodium arsenate while with sodium arsenite germination percentage was reduced to 75.83%. The minimum germination percentage 71.56% was recorded in FH-385(H2). Achenes of H4 (FH-415) responded in a same way (Table 3) with both kinds of arsenic salts in case of germination percentage parameter.

Mean germination time (MGT)

Two way analysis of variance (ANOVA) revealed

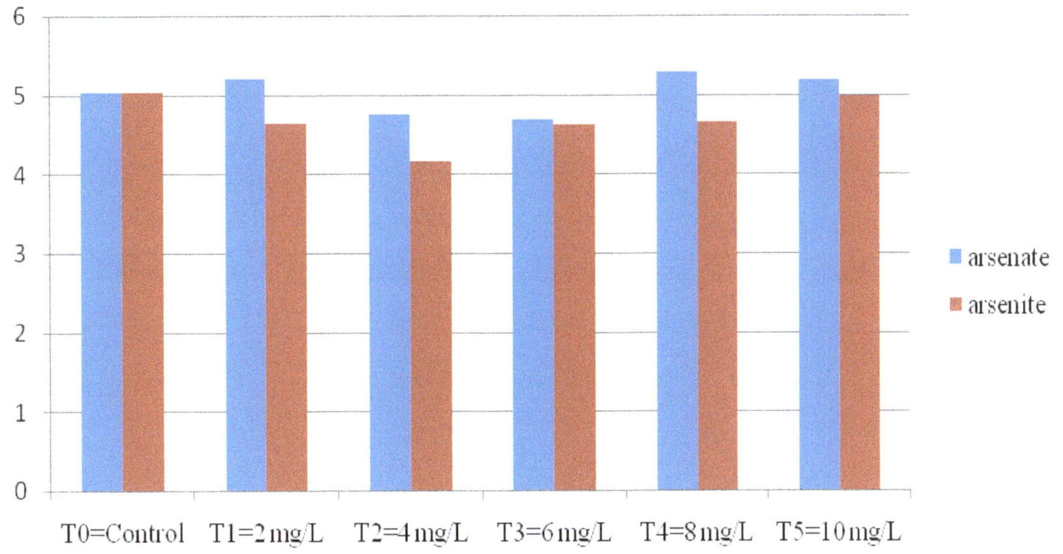

Figure 2. Effect of arsenicals on mean germination time of sunflower.

significant differences for different hybrids in case of both salts of arsenic but there were non-significant differences for various treatments (Table 1 and Figure 2) whereas interaction of hybrids into treatments revealed significant differences for sodium arsenate while non-significant differences for sodium arsenite. Achenes or seeds belonging to treatment T4 (8 mg/L arsenic) showed maximum value (5.28) for mean germination time (MGT) as in Table 2, while out of different cultivars H2 (FH-385) showed minimum MGT value (3.46) and H1(FH-331) showed maximum MGT value (6.67). Both of arsenic salts showed similar effects on mean germination time for sunflower seeds (Tables 2 and 3) as low doses caused the seeds to germinate early as compared to control while high concentration of arsenic salts caused a little bit delay in germination of sunflower seeds (Gulz et al., 2005) with higher MGT value.

Time to 50% germination (T_{50})

Significant differences were found for different sunflower hybrids/cultivars in case of both sodium salts of arsenic whereas in case of treatments of arsenic, sodium arsenite treatments showed non-significant differences. The interaction between treatments and hybrids also showed significant differences (Table 1). Out of different treatments T0 (control) showed same value (4.50 days) as time to 50% germination for both of salts used. Treatment T2 (4 mg/L arsenic) of sodium arsenate gave the least value for days to 50% germination (Table 2 and Figure 3) which shows that low concentration of arsenic boasts up the rate of seed germination (Cheng et al., 2006) but high concentration of arsenic 10 mg/L arsenic in both forms of arsenic slows down the rate of seed germination (Li et al., 2007) by giving maximum value (4.52 and 4.74 for sodium arsenate and arsenite, respectively) for days to 50% germination.

Plumule length (mm)

Different sunflower hybrids and various treatments of arsenic as well as interaction of hybrid into treatments showed significant differences except the interaction value in case of sodium arsenite application as inferred from the analysis of variance (Table 1). Treatment means showed the maximum value (23.07 mm) in case of T2 (4 mg/L arsenic) using sodium arsenite (Table 2 and Figure 4) which is 130.9% of control (17.62 mm). This reveals that low concentrations of arsenic increase the seedling growth up-to certain limit but after that more increase in concentration inhibits the seedling growth (Liu and Zhang, 2007; Geng et al., 2006) as least value (14.56 mm) was recorded in highest concentration (10 mg/L) of sodium arsenite. Sunflower cultivars behaved differently (Table 3) in case of plumule length as out of different sunflower cultivars H3 (FH-405) gave maximum plumule length (25.17 mm) whereas least value (13.2 mm) was observed in H1 (FH-331).

Radicle length (mm)

All the sunflower hybrids and different treatments of arsenic showed significant differences except for the treatment into hybrid interaction values which showed non-significant differences for sodium arsenite (Table 1).

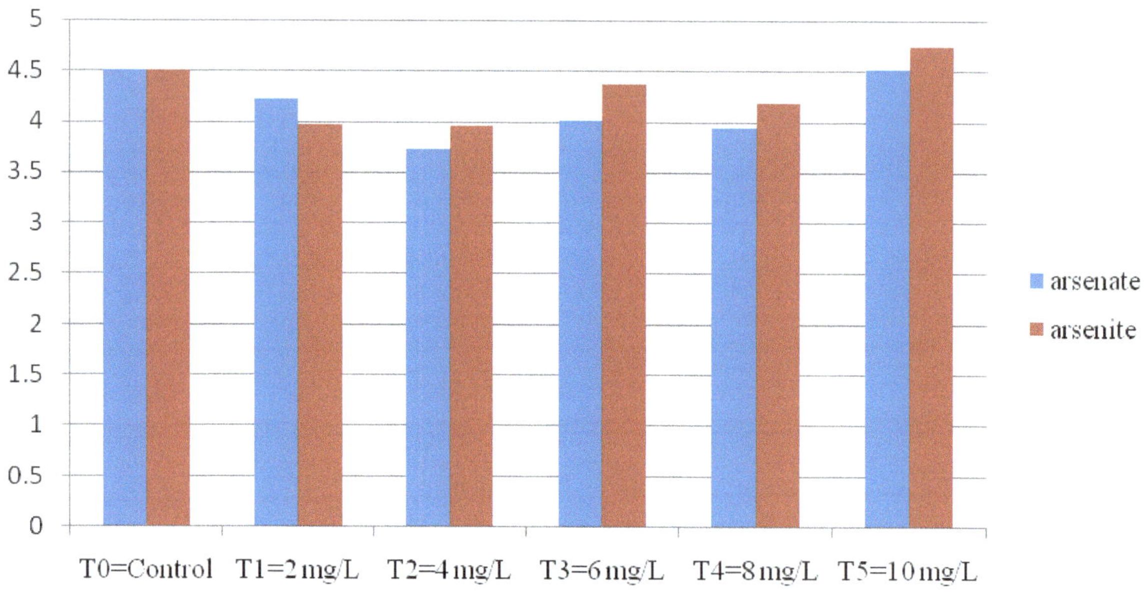

Figure 3. Effect of arsenicals on time to 50% germination of sunflower.

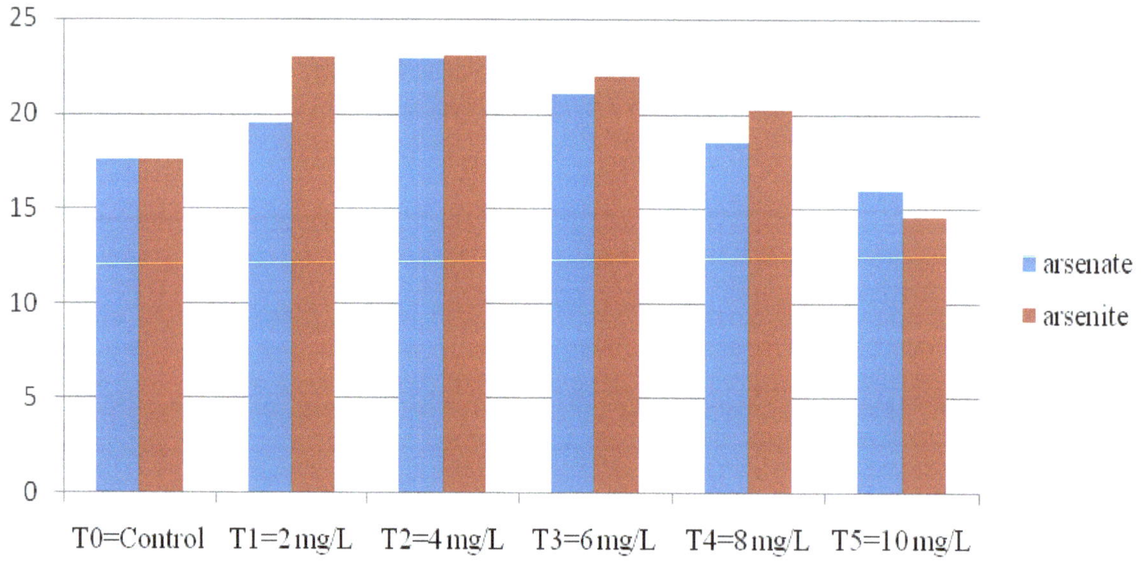

Figure 4. Effect of arsenicals on plumule length of sunflower.

Highest mean value (26.62 mm) of radicle length was recorded in case of T2 (4 mg/L arsenic) as sodium arsenate while the least value (17.47 mm) was observed in case of T5 (10 mg/L arsenic) as sodium arsenite (Table 2, Figure 5), proving that in low concentration arsenic promotes the seed germination but high concentrations of metal inhibits the growth of seedling by suppressing root or radicle growth (Lux et al., 2011). Out of different sunflower cultivars (Table 3) used in this experiment H3 (FH-405) gave maximum radical length (28.07 mm) while H1 (FH-415) gave least value (15.88 mm).

Seedling vigour index (SVI)

Highly significant differences were given by two way analysis of variance (ANOVA) for sunflower cultivars and different arsenic treatments as well as interaction of hybrids into treatments except for interaction value in

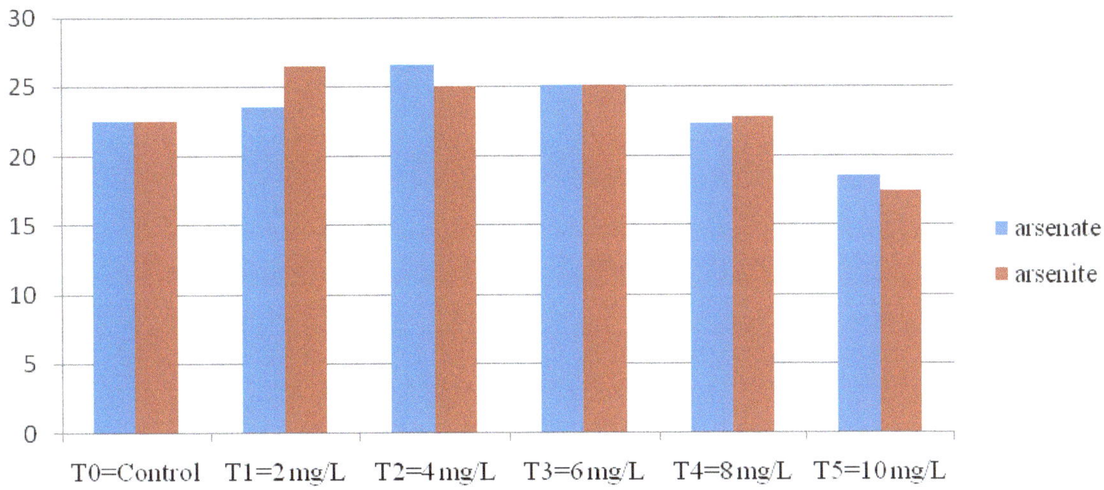

Figure 5. Effect of arsenicals on radicle length of sunflower achenes.

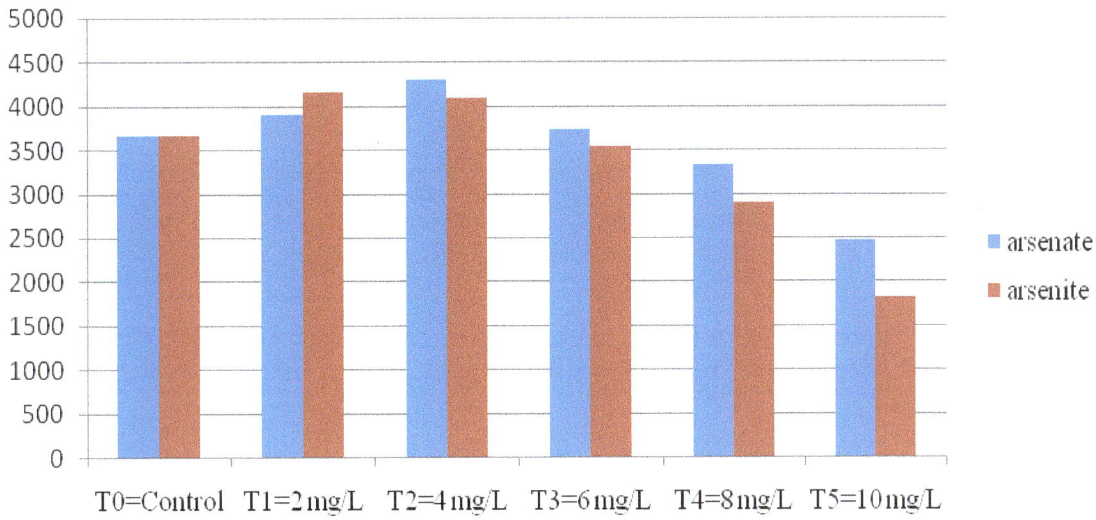

Figure 6. Effect of arsenicals on seedling vigour index (SVI) of sunflower.

case of sodium arsenite which showed non-significant differences (Table 1). Maximum SVI value (4296.96) for treatment means was recorded in case of T2 (4 mg/L arsenic) using sodium arsenate while T10 (10 mg/L arsenic) using sodium arsenite gave least value (1824.92) as shown in Table 2. Out of different sunflower cultivars, H2 (FH-385) gave highest SVI value (4120.31) while H1 (FH-331) gave least value (2415.36) of seedling vigour index (Table 3 and Figure 6).

Conclusion

It can be concluded that arsenic is non-essential to plants and higher concentrations (6, 8 and 10 mg/L arsenic) of both inorganic arsenicals (arsenate as well as arsenite) posed a stress over all germination parameters observed during seed germination of four different sunflower cultivars, but arsenite (As^{3+}) posed more stressful effects than that of arsenate (As^{5+}) contamination. Low concentrations (2 and 4 mg/L arsenic), of both of arsenicals caused increase in plumule and radicle lengths of sunflower seedlings, as compared to control. Gradual decrease in germination percentage was observed with the increase in arsenic concentrations of both arsenic forms in case of all sunflower hybrids. H1 (FH-331) proved to be most sensitive for arsenic application by giving least values for plumule length, radicle length and seedling vigour index. Best seedling vigour index was recorded in sunflower cultivar

H2 (FH-385), showing a little bit resistance to arsenic contamination.

REFERENCES

Ahmad MSA, Hussain M, Ashraf M, Ahmad R, Ashraf MY (2009). Effect of nickel on seed germ inability of some elite sunflower (*Helianthus annuus* L.) cultivars. Pak. J. Bot. 41(4):1871-1882.

Cheng W, Zhang G, Yao H, Wu W, Xu M (2006). Genotypic and Environmental variations in cadmium, chromium, arsenic, nickel and lead concentrations in rice grains. J. Zhejiang Uni. Sci. B. 7:565-571.

Cokkizgin A, Cokkizgin H (2010). Effects of lead ($PbCl_2$) stress on germination of lentil (*Lens culinaris* Medic.) lines. Afr. J. Biotechnol. 9(50):8608-8612.

Dezfuli PM, Sharif-Zadeh F, Janmohammadi M (2008). Influence of priming techniques on seed germination behavior of maize inbred lines (*Zea mays* L.) ARPN. J. Agric. Biol. Sci. 3(3):22-25.

Dittmar J, Voegelin A, Maurer F, Roberts LC, Hug SJ, Saha GC, Ali MA, Badruzzaman ABM, Kretzschmar R (2010). Arsenic in soil and irrigation water affects arsenic uptake by rice: complementary insights from field and pot studies. Environ. Sci. Technol. 44:8842-8848.

Farooq M, Basra SMA, Hafeez K, Ahmad N (2005). Thermal hardening: a new seed vigor enhancement tool in rice. Acta. Botanica Sinica. 47:187-193.

Farooq M, Basra SMA, Hafeez-ur-Rehman, Mehmood T (2006). Germination and early seedling growth as affected by pre-sowing ethanol seed treatments in fine rice. Int. J. Agri. Biol. 8(1):19-22.

Fitz WJ, Wenzel WW (2002). Arsenic transformation in the soil-rhizosphere-plant system: Fundamentals and potential application to phytoremediation. J. Biotechnol. 99:259-278.

Geng CN, Zhu YG, Tong YP, Smith SE, Smith FA (2006). Arsenate (As) uptake by and distribution in two cultivars of winter wheat (*Triticum aestivum* L.). Chemosphere 62:608-615.

Giri AK, Patel RK (2012). Phytoaccumulation potential and toxicity of arsenic ions by *Eichhornia crassipes* in hydroponic system. J. Bioremed. Biodegrad. 3:137.

GOP (Government of Pakistan) (2012). Agriculture Statistics of Pakistan. Ministry of Food, Agriculture and Livestock Division, Economics Wing, Islamabad.

Gulz PA, Gupta S, Schulin R (2005). Arsenic accumulation of common plants from contaminated soils. Plant Soil. 272:337-347.

Han FX, Su Y, Monts DL, Plodinec MJ, Banin A, Triplett GE (2003). Assessment of global industrial-age anthropogenic arsenic contamination. Natur-wissenschaften 90(9):395-401.

Li C, Feng S, Shao Y, Jiang L, Lu X, Hou X (2007). Effect of arsenic on seed germination and physiological activities of wheat seedlings. J. Environ. Sci. 19:725-732.

Liu J, Zheng B, Aposhian HV, Zhou Y, Chen ML, Zhang A, Waalkes MP (2002). Chronic arsenic poisoning from burning high-arsenic-containing coal in Guizhou, China. Environ. Health Perspect. 110:119-122.

Liu X, Zhang S (2007). Intraspecific differences in effects of co-contamination of cadmium and arsenate on early seedling growth and metal uptake by wheat. J. Environ. Sci. 19:1221-1227.

Lux A, Martinka M, Vaculik M, White PJ (2011). Root responses to cadmium in the rhizosphere: a review. J. Exp. Bot. 62(1):21-37.

Mandal A, Lundh D, Nahar N, Bentol H, Bari A, Johnson-Brousseau S, Ghosh S (2011). Modeling arsenic accumulation in plants. 2nd Int. Conf. Emerg. Appl. Info. Tech. doi: 10.1109/EAIT. 2011. P. 91.

Meharg AA, Hartley-Whitaker J (2002). Arsenic uptake and metabolism in arsenic resistant and non resistant plant species. New Phytol. 154:29-43.

Mei L, Wan X, Huang Z, Chen T, Li X, Liu Y (2012). First evidence on different transportation models of arsenic and phosphorus in arsenic hyperaccumulator *Pteris vittata*. Environ. Poll. 161 :1-7.

Pigna M, Cozzolino V, Violante A, Meharg AA (2009). Influence of Phosphate on the Arsenic uptake by Wheat (*Triticum durum* L.) irrigated with Arsenic solutions at three different concentrations. Water Air Soil Pollut. 197:371-380.

Taiz L, Zeiger E (2006). Plant Physiology (4nd Ed.). Sunderland, Massachusetts: Sinauer Associates, Inc. pp. 103-122.

Tanveer A, Rehman A, Javaid MM, Abbas RN, Sibtain M, Ahmad A, Zamir MS, Chaudhary KM, Aziz A (2010). Allelopathic potential of *Euphorbia helioscopia* L, against wheat (*Triticum aestivum* L.), chickpea (*Cicer arietinum* L.) and lentil (*Lens culinaris* Medic.). Turk. J. Agric. For. 34:75-81.

Tibbetts J (2000). Water world. Environ. Health Perspect. 2000. 108:69-73.

Williams PN, Price AH, Raab A, Hossain SA, Feldmann J, Meharg AA (2005). Variation in arsenic speciation and concentration in paddy rice related to dietary exposure. Environ. Sci. Technol. 39:5531-5540.

Winkel L, Berg M, Amini M, Hug SJ, Johnson CA (2008). Predicting ground water arsenic contamination, in South east Asia from surface parameters. Nature Geosci. 1:536-542.

Zhang WH, Cai Y, Tu C, Ma LQ (2002). Arsenic speciation and distribution in an arsenic hyperaccumulating plant. Sci. Total Environ. 300:167-177.

Zhong L, Hu C, Tan Q, Liu J, Sun X (2011). Effects of sulfur application on sulfur and arsenic absorption by rapeseed in arsenic contaminated soil. Plant Soil Environ. 57:429-434.

Lead contamination of traditional hand-dug wells in parts of Kwale County, Kenya

M. W. Chege[1], N. O. Hashim[1], A. S. Merenga[1] and J. Tschiersch[2]

[1]Department of Physics, Kenyatta University, P. O. Box 43844, Nairobi, Kenya.
[2]Institute of Radiation Protection, Helmholtz Zentrum München - German Research Center for Environmental Health (GmbH), Ingolstädter Landstraße 1, 85764 Neuherberg, Germany.

People's health, behavior and intellect to a certain extent are a function of their environment. One important component of this environment is water, a basic necessity that human beings cannot do without. The water available, however, is not always safe for human consumption due to contamination. Among the many water contaminants, lead (Pb) is one of the most dangerous due to its bio-accumulative nature and high toxicity. Children are most vulnerable to Pb toxicity as their main organs such as the brain and the nervous system are still developing. For them, exposure to even relatively low concentrations can result in reduced intelligence quotient (IQ), learning disabilities and attention deficit disorders, behavioral problems such as delinquency and drug abuse, and such conditions as impaired hearing and mental retardation. The main aim of the present study was to determine Pb concentration in water from hand-dug wells in Kwale County in the coastal region of Kenya. Kwale County hosts a number of schools for mentally disabled and hearing impaired persons, while the coastal region in general has among the lowest literacy levels in the country and the highest incidences of drug abuse. Thirty-seven water samples were randomly collected from selected villages and analyzed for Pb content using a flame atomic absorption spectrophotometer (FAAS). Also analyzed were the water pH and conductivity using portable digital meters. Pb concentration in the water samples ranged from below detection limit to 1.397 mg/L with a mean concentration of 0.330 mg/L. Over 62% of the samples had Pb levels above WHO recommended value of 0.01 mg/L. The pH ranged from 5.8 to 10.4 with a mean value of 9.2 while conductivity ranged from 85.3 to 6060 µs/cm with a mean value of 930.351 µs/cm.

Key words: Lead, well water, intelligence quotient (IQ), mental retardation, deafness, Kwale.

INTRODUCTION

Water is essential to life and without it, life as we know it would not be. However, as much as it is precious, it may be a source of diseases and disabilities amongst members of the public. This is because water naturally contains minerals and organisms which may be of a type and/or concentration that is not safe for human consumption. Of particular concern are heavy metals. Heavy metals are non-biodegradable chemical elements that are ubiquitous in the environment and are toxic at relatively low concentrations (Adelekan and Abegunde, 2011; Momodu and Anyakora, 2010; Ghanem et al., 2008). While trace amounts of some metals such as calcium (Ca), chromium (Cr), manganese (Mn), Iron (Fe), cobalt (Co), copper (Cu), zinc (Zn) and molybdenum (Mo) are essential for proper growth and development others such as Pb are non-essential to human beings with no known biological function in the human body (Silveira et al., 2003). Pb is a bio-accumulative and highly toxic element that can cause irreversible damage to body organs such as kidneys and the nervous and reproductive systems (Todd et al., 1996). Pb is particularly toxic to children as their main organs such as the brain and the nervous system are still developing. For them, exposure to even relatively low concentrations can result in such conditions as hearing impairment and mental retardation, reduced intelligence quotient (IQ), learning disabilities and attention deficit disorders, and behavioral problems such as delinquency and drug/substance abuse (Froutan et al., 2011; WHO, 2010; Park et al., 2010; Chen et al., 2007; Norman et al., 2007; Pruss-Ustun et al., 2004; Needleman et al., 2002; Dietrich et al., 2001; Osman et al., 1999; Mendelsohn et al., 1998; Sciarillo et al., 1992; Schwartz, 1992). Chronic exposure to Pb may cause neurological damage and neurobehavioral disorders in the long run (Pabello and Bolivar, 2005; Patrick, 2006). Drugs temporarily restore neurotransmitter functions that are abnormal, and substance abuse may often be crude self-medication in response to the effects of Pb toxicity (Masters, 1999). Pb poisoning can result to iron deficiency anemia, a condition whose prevalence among pregnant women and children in Kenya is very high (Makokha et al., 2008).

In view of the health, intellectual and behavioral impacts of Pb contamination, the present study was carried out in Kwale County in the coastal region of Kenya to determine Pb levels in drinking water from traditional hand-dug wells. Kwale County is home to Mlalani special unit for mentally handicapped children, Kwale School for the mentally handicapped, Kichakamkwaju deaf unit, and Kwale School for the deaf, while the coastal region in general has among the lowest literacy levels in the country and the highest incidences of drug abuse. The literacy level in the region was placed at 62.8% in 2000, the lowest in the country at the time (UNESCO, 2005). Due to continued dismal performance in national examinations, the coastal region was a subject of a documentary tagged *"The last to graduate"* aired in July 2012 on KTN, a local TV channel. In the documentary students as old as 13 years could not read comprehensibly in either English or Kiswahili, the national language in Kenya. The coastal region leads in the use of narcotics (NACADA, 2010) with *Bhangi* (cannabis sativa) being the drug of choice for most young people due to its low price range and relative ease of access (National Assembly Official Report, 2010).

Typical to the coastal region, Kwale County has rich albeit unexploited mineral resources that include rare earth elements such as niobium and titanium (Ministry of Environment and Natural Resources, 1985). Kwale County residents are however, poor village folks who depend mainly on small scale farming for subsistence. Probably due to low education standards, the villagers still ascribe to traditional practices such as witchcraft and more often than not, all sorts of misfortunes ranging from diseases, poor performance in school, poverty and even drought are attributed to sorcery and other supernatural powers (Tinga, 1998). The County generally lacks most basic amenities such as piped water and the residents rely mainly on groundwater from communal traditional hand-dug wells for drinking and other domestic purposes, water that they use without any form of treatment.

Four villages within Kwale County namely Kanana-Shimoni, Munje, Ramisi and Shirazi were chosen for the present study. In these villages, communal hand-dug wells serve as the sole source of drinking water. The wells are indiscriminately sunk without regulation or regard to water quality. The wells are typically shallow and usually no deeper than 20 m. A few of the wells are covered and fitted with a hand pump. Most of the wells however, are open and water is drawn using a tin tied to a rope.

MATERIALS AND METHODS

A total of 37 wells were sampled; 8 in Kanana -Shimoni, 9 in Munje, 13 in Ramisi and 7 in Shirazi. The sample size reflects the population density and thus the total number of wells in each village. Sampling was done during the dry season. 500 ml plastic bottles which had been pre-cleaned and rinsed off with dilute nitric acid, distilled water and sample water, in that order, were used. Water was drawn from the wells in the same way the local residents fetched their water: using a tin tied to a string in the case of open wells and hand pump in the case of closed wells. The samples were tested on site for pH and conductivity using portable digital meters before being transported to the laboratory where they were acidized using nitric acid to a pH<2 for preservation. They were then left to sit for at least 72 h before the commencement of Pb measurement.

The samples were prepared for Pb measurement using the EPA 200.2 digestion method. In this method, 100 ml of thoroughly mixed sample was poured into 150 ml beaker and 2 ml concentrated NO_3 and 1 ml concentrated HCl added. The solution was heated using an electrical burner at a solution temperature of 95 ± 5°C up until the volume was reduced to 20 ml. After cooling, the solute was topped up to 100 ml using distilled water and filtered to rid any particulate matter. The samples were then analyzed for Pb content using a calibrated GF-990 flame atomic absorption spectrophotometer (FAAS) at the School of Pure and Applied Sciences, Pwani University College, Kilifi, Kenya.

RESULTS AND DISCUSSION

Table 1 shows the overall results of the parameters analyzed while Table 2 gives the summary statics. The Pb levels in the water samples ranged from below detection limit (BDL) to 1.397 mg/L with a mean concentration of 0.33 mg/L. As a guideline for lead contamination in drinking water, the World Health Organization (WHO) has set up a reference level of 0.01 mg/L against which remediation measures are required (WHO, 1993). The mean concentration of Pb in the samples was above the WHO reference level, and so

Table 1. Overall results for water samples from Kwale County.

Sample	Type of well	Village	pH	EC (μs/cm)	Pb conc. (mg/L)
S01	Closed	Ramisi	10.1	85	BDL
S02	Closed	Ramisi	7.2	341	BDL
S03	Closed	Ramisi	9.4	394	BDL
S04	Closed	Ramisi	9.9	426	BDL
S05	Closed	Kanana-Shimoni	9.6	506	BDL
S06	Closed	Kanana-Shimoni	10.0	610	BDL
S07	Closed	Kanana-Shimoni	9.7	425	0.014
S08	Closed	Shirazi	9.7	392	0.041
S09	Closed	Shirazi	9.5	2960	0.492
S10	Closed	Shirazi	9.5	472	0.573
S11	Closed	Ramisi	5.8	207	0.743
S12	Closed	Shirazi	9.7	667	0.889
S13	Open	Munje	10.3	98	BDL
S14	Open	Munje	10.0	338	BDL
S15	Open	Ramisi	8.4	360	BDL
S16	Open	Ramisi	10.1	414	BDL
S17	Open	Ramisi	10.2	715	BDL
S18	Open	Ramisi	7.9	949	BDL
S19	Open	Munje	9.8	1106	BDL
S20	Open	Kanana-Shimoni	7.5	5770	BDL
S21	Open	Shirazi	9.5	506	0.015
S22	Open	Ramisi	9.2	2370	0.045
S23	Open	Kanana-Shimoni	9.9	604	0.136
S24	Open	Kanana-Shimoni	9.9	296	0.248
S25	Open	Kanana-Shimoni	10.0	477	0.254
S26	Open	Munje	8.7	576	0.336
S27	Open	Munje	8.3	336	0.365
S28	Open	Munje	8.2	6060	0.421
S29	Open	Kanana-Shimoni	10.4	1000	0.458
S30	Open	Munje	8.3	1078	0.489
S31	Open	Shirazi	10.0	471	0.714
S32	Open	Shirazi	10.1	361	0.777
S33	Open	Ramisi	9.5	721	0.844
S34	Open	Ramisi	9.9	414	0.893
S35	Open	Munje	6.0	659	0.922
S36	Open	Munje	7.5	928	1.146
S37	Open	Ramisi	10.0	331	1.397

were 62.16% of the samples. Pb contamination of groundwater supplies may arise from either anthropogenic sources such as pesticides, fertilizers, industries and mining activities, or from natural deposits in rock and soil. Pb contamination of the well water of Kwale County is most likely mainly due natural sources as there are no active large scale industries, major mining activities or extensive use of fertilizers and pesticides in the region. Moreover, Kwale County like the rest of the Kenyan coast has heavy mineral sand deposits that contain lead-zinc bearing ores (Kariuki, 2002; Njuguna, 2007). The sampling area is also within the Pleistocene coral limestone formations which may serve as a host rock for Pb deposits.

The pH of the water samples was between 5.8 and 10.4 with a mean value of 9.18. WHO has no health based guideline value for pH, but a range of 6.5 to 8.5 is usually recommended (Napacho and Manyele, 2010). 72.79% of the samples had pH values outside this range with 67.57% of them having pH levels above 8.5, an indication of high alkalinity of groundwater in the region. Electrical conductivity (EC) levels ranged between 85.3

Table 2. Summary statistics of the samples analyzed.

Parameter	pH	EC (µs/cm)	Pb conc. (mg/L)
Minimum value	5.8	85.3	BDL
Maximum value	10.4	6060	1.395
Mean	9.18	930.35	0.33
WHO reference value	-	-	0.1
Samples exceeding WHO reference level			62.16%

Table 3. Range mean and standard deviation values for Pb concentrations for each village.

Village	Range (mg/L)	Mean Pb conc. (mg/l)	STDEV (mg/L)
Kanana-Shimoni	BDL - 0.458	0.139	0.172
Munje	BDL - 1.146	0.409	0.406
Ramisi	BDL - 1.397	0.302	0.486
Shirazi	0.015 - 0.889	0.500	0.348

Table 4. Percentage of samples with Pb levels above WHO reference level.

Village	No. of samples	Samples above WHO ref. level	% above WHO ref. level
Kanana-Shimoni	8	5	62.50
Munje	9	6	66.67
Ramisi	13	5	38.46
Shirazi	7	7	100.00

and 6060 µs/cm with a mean value of 930.35 µs/cm. The high conductivity could be due to sea water intrusion bearing in mind the region's proximity to the Indian Ocean. Two of the wells with the highest conductivity (5770 and 6060 µs/cm) had extremely salty water and although the local residents did not use it for drinking, they used it for other domestic purposes like cleaning and cooking.

Out of the 37 wells studied, 12 were covered while 25 were open. The mean Pb concentration in the closed wells was 0.229 mg/L with a standard deviation of 0.342 mg/L while that of closed wells was 0.378 mg/L with a standard deviation of 0.417 mg/L. A bigger proportion of contamination was observed in the open wells (68%) as compared to the closed wells (50%). The higher concentration of Pb in the open wells may be due to siltation. Table 3 shows the range, mean and standard deviation values of Pb in samples from each village while Table 4 looks at the number of wells sampled per village and the percentage of sampled wells with Pb levels above the WHO reference level. Samples from Kanana-Shimoni Village had the lowest mean value of Pb at 0.1 mg/L while those from Shirazi Village had the highest at 0.5 mg/L. Munje and Ramisi Villages had wells with mean Pb concentrations of 0.409 and 0.302 mg/L respectively. Despite registering the lowest mean Pb value, 62.5% of the sampled wells in Kanana-Shimoni Village were Pb contaminated which implies that lead bearing soils are more or less distributed throughout the village. 66.67% of the sampled wells in Munje Village and all the sampled wells in Shirazi Village had Pb levels above WHO action value. Ramisi Village had the least proportion of contaminated sampled wells at 38.46%. Ramisi Village is a settlement center of sorts unlike the other villages which are farmlands in which dwellings and consequently water wells are relatively far between. From information gathered from the villagers, underground streams are alleged to flow in Ramisi Village and this may explain the low percentage of wells that are Pb-contaminated. No prior study is carried out a site before a well is dug and consequently not all wells coincide with points of good water. Again, being a settlement center, the human population in Ramisi Village is high and therefore there exists a likelihood of well contamination resulting from improper disposal of human waste as well as from minor activities such as paint jobs and bicycle repair works. This may explain the high levels of Pb in some of the wells in the village.

Conclusion

There is one question that lingers in the minds of many Kenyans familiar with the prevalent drought situation of

Kwale County: with such a high water table, why is it that the coastal villagers do not harvest the groundwater and use it for irrigation purposes but instead wait upon the almost always unreliable seasonal rainfall? Why the disabilities, the prevalent drug/substance abuse, the low education standards, the poverty, the superstitions? People's health, behavior and intellect to a certain extent are a function of their environment and water is an important aspect of that environment. 62% of the sampled wells in the part of Kwale County studied had Pb contamination levels that can adversely affect the health and life of human beings. In the natural order of things, there is a tendency for living organisms to react intuitively to thwart impending threats, and it is probably for this reason that the Kwale villagers have refrained from using groundwater for irrigation purposes, given its high toxicity evident from the current research. Coincidentally or otherwise, among the villages studied, human settlements were most concentrated in Ramisi Village, the village with the lowest proportion of Pb-contaminated wells.

RECOMMENDATION

As much as it may be of poor quality, water is essential to life and with no access to other sources, Kwale villagers turn to untreated groundwater for drinking, then to superstitions when calamities possibly brought about by the same water strike. Such calamities may be averted by providing quality water to the residents of Kwale County. There is also need for further research to identify and map the wells with good/poor quality water in terms of other chemical components. Such information would be vital in advising the villagers on which wells to use and to avoid as well as form a basis of mass education on the dangers of Pb and other heavy metal contaminants. For comparative analysis, further research on Pb levels in groundwater during the wet season is required.

REFERENCES

Adelekan BA, Abegunde KD (2011). Heavy metals contamination of soil and groundwater at automobile mechanic villages in Ibadan, Nigeria. Int. J. Phys. Sci. 6(5):1045-1058.

Chen A, Cai B, Dietrich KN, Radcliffe J, Rogan WJ (2007). Lead exposure, IQ, and behavior in urban 5-7 year olds: Does lead affect behavior only by lowering IQ? Pediatrics 119(3):650-658.

Dietrich KM, Ris MD, Succop PA, Berger OG, Bornschein RL (2001). Early exposure to lead and juvenile delinquency. Neurotoxicol. Teratol. 23:511-518.

Froutan H, Zadeh AK, Kalani M, Andrabi A (2011). Lead toxicity: a probable cause of abdominal pain in drug abusers. Med. J. Islamic Republic Iran. 25(1):16-20.

Ghanem A, El-Azab SM, Mandour RA, EL-Hamady MS (2008). Relationship between lead levels in drinking water and mothers' breast milk. Mansoura J. Forensic Med. Clin. Toxicol. 16(1):13-24.

Kariuki DK (2002). Natural resources-minerals: a report of the civil society review of the implementation of agenda 21 in Kenya Kenya NGO Earth Summit 2002 Forum.

Makokha AO, Mghweno LR, Magoha HS, Nakajugo A, Wekesa JM (2008). Environmental lead pollution and contamination in food around Lake Victoria, Kisumu, Kenya. Afr. J. Environ. Sci. Technol. 2(10):349-353.

Masters RG (1999). Poisoning the Well: Neurotoxic Metals, Water Treatment, and Human Behavior. Plenary Address to Annual Conference of the Association for Politics and the Life Sciences, Four Seasons Hotel, Atlanta, GA. Sept. 2, 1999.

Mendelsohn AL, Dreyer BP, Fierman AH, Rosen CM, Legano LA, Kruger HA, Lim SW, Courtlandt CD (1998). Low-Level Lead Exposure and Behavior in Early Childhood. Pediatrics. 101(3):10.

Ministry of Environment and Natural Resources (1985). Kwale district environmental assessment report. National Environment Secretariat, Ministry of Environment and Natural Resources, Kenya.

Momodu MA, Anyakora CA (2010). Heavy Metal Contamination of Ground Water: The Surulere Case Study. Res. J. Environ. Earth Sci. 2(1):39-43.

NACADA (2010). Report of Survey on Drug and Substance Abuse in Coast Province Kenya. Main Report. National Campaign Against Drug Abuse Authority, March 2010.

Napacho ZA, Manyele SV (2010). Quality assessment of drinking water in Temeke District (part II): Characterization of chemical parameters. Afr. J. Environ. Sci. Technol. 4(11):775-789.

National Assembly Official Report (2010). Threat of international narcotics trade to Kenya. Prime Ministerial Statement, Wednesday, 24th November, 2010(P).

Needleman HL, McFarland C, Ness RB, Fienberg SE, Tobin MJ (2002). Bone lead levels in adjudicated delinquents; A case control study, Neurotoxicol. Teratol. 24:711-717.

Njuguna FK (2007). Baseline study towards non-toxic environment in Africa. Country Report, Kenya, 08 August 2007.

Norman R, Mathee A, Barnes B, Merwe L, Bradshaw D (2007). Estimating the burden of disease attributable to lead exposure in South Africa in 2000. South Africa Med. 97:773-780.

Osman K, Pawlas K, Schütz A, Gazdzik M, Sokal JA, Vahter M (1999). Lead exposure and hearing effects in children in Katowice, Poland, Environ. Res. 80(1):1-8.

Pabello NG, Bolivar VJ (2005). Young Brains on Lead: Adult Neurological Consequences? Toxicol. Sci. 86(2):211-213.

Park SK, Elmarsafawy S, Mukherjee B, Spiro A, Vokonas PS, Nie H, Marc G, Weisskopf MG, Hu JSW (2010). Cumuative lead exposure and age-related hearing loss: The VA Normative Aging Study, Hearing Exposure. 1-2: 48-55.

Patrick L (2006). Lead Toxicity Part II: The Role of Free Radical Damage and the Use of Antioxidants in the Pathology and Treatment of Lead Toxicity. Altern. Med. Rev. 11(2):124-127.

Pruss-Ustun A, Fewtrell L, Landrigan PJ, Ayuso-Mateos JL (2004). Lead Exposure, pages 1495 – 1552 In: Ezzati, M., A.D. Lopez, A. Rodgers, and C.J.L. Murray. Comparative Quantification of Health Risks: Global and Regional Burden of Disease Attributable to Selected Major Risk Factors. World Health Organization, Geneva.

Schwartz J (1992). Low-level lead exposure and children's IQ: meta-analysis and search for threshold. Environ. Res. 65:42-55.

Sciarillo WG, Alexander G, Farrell KP (1992). Lead exposure and child behavior. Am. J. Public Health 82(10):1356-60.

Silveira MLA, Alleoni LRF, Guilherme LR (2003). Biosolids and heavy metals in soils. Sci. Agric. 60(4):793-806.

Tinga KL (1998). Cultural practice of the Midzichenda at crossroads: Divination, healing, witchcraft and the statutory Law. APP. 55:173-184.

Todd AC, Wetmur JG, Moline JM, Godbold JH, Levin SM, Landrigan PJ (1996). Unraveling the chronic toxicity of lead: an essential priority for environmental health. Environ. Health Perspect. 104 Suppl 1:141-6.

UNESCO (2005). Policy Review Report: Early Childhood Care and Education in Kenya. UNESCO/OECD Early Childhood Policy Review Project. The Section for Early Childhood and Inclusive Education, Division of Basic Education, UNESCO Education Sector.

WHO (1993). Guidelines for drinking-water quality, second edition, volume 1 World Health Organization (WHO), 21993.

WHO (2010). Childhood lead poisoning. World Health Organization (WHO), 2010.

Effect of stabilizers on stiffness modulus of soil layers: A review

Dariush Moazami, Zainuddin Md. Yusoff, Ratnasamy Muniandy and Hussain Hamid

Department of Civil Engineering, University Putra Malaysia, 43400 UPM Serdang, Selangor, Darul Ehsan, Malaysia.

Subgrade layer has a significant contribution in the pavement structure, and the quality of this layer has a great impact on the subsequent pavement design. Pavement design guide as specified in the AASHTO (2002), incorporates engineering properties of unbound and stabilized subgrade soils, since in many cases fatigue or rutting failure in pavements are due to insufficient stiffness modulus in soil layers. Subgrade materials are characterized by common engineering properties including stiffness and strength. Appropriate treatment using different additives is considered an effective way to achieve a more durable pavement structure and also to increase the stiffness modulus. In this paper, the importance of using proper stiffness for underlying layers and the effects of common additives such as lime, cement and fly ash on improving the strength and stiffness modulus of soil layers are highlighted.

Key words: Stiffness modulus, cementitious stabilization, soil properties.

INTRODUCTION

Subgrade in pavement design and construction is a very important layer since it must provide a stable platform for the subsequent top layers. In addition, pavement design is greatly influenced by the quality of this layer. Subgrade materials should be characterized by their engineering properties such as strength and stiffness (stress-strain relationship under traffic loading) in order to control pavements damage. Providing sufficient stiffness for subgrade cannot be easily obtained all the time; hence, reinforcing the subgrade layer is inevitable and it is necessary to improve the existing stiffness modulus by using various stabilizers. Therefore, improving the engineering properties of subgrade soils using cementitious stabilizations is widely used.

Resilient modulus (M_r) and Unconfined Compressive Strength (UCS) are the essential values of stabilized soils in the new Mechanistic Empirical Pavement Design Guide (MEPDG, 2004) when designing a new pavement. However, only limited studies have specified these parameters (Solanki et al., 2009). This paper presents a comprehensive review on the importance of using adequate stiffness for underlying layers and utilizing common additives with their percentages. Furthermore, the associated improvement in the engineering properties of the subgrade is reviewed.

Importance of appropriate subgrade stiffness

Studies conducted on pavement structural design indicates that pavement layer thickness is governed by the stiffness of subgrade and granular layers. Thus information on the stiffness modulus of subgrade and granular layers are required before performing any pavement design. In a pavement structure; base, subbase and subgrade layers with high stiffness values possess higher stress distribution ability than those with low stiffness. Thus the required thickness of the pavement can be reduced when stiffer layer components are used in the pavement structure. Accordingly, considerable reduction in the construction cost can be achieved beside the optimum design.

On the other hand, pavement layers with low stiffness values exhibit weak performance and in order to improve their properties, it may be required to either increase the asphalt final thickness or to improve the stiffness of the soil layers using appropriate additives. Increased stiffness causes the traffic load to spread over a greater area.

Furthermore, numerous studies have indicated that many cases of fatigue or rutting failures in pavements are due to inadequate stiffness of soil layers (Van Zyl and Maree, 1983; Wright and Paquette, 1987; Barksdale and Itani, 1989; Zakaria and Leest, 1996; Jegede, 2000; Giroud and Han, 2004; Mulungye et al., 2007; Xu and Huang, 2011; Cardone et al., 2011).

Barksdale and Itani, (1989) indicated that uncrushed gravels have a lower stiffness modulus than crushed stones making them more susceptible to rutting. In addition, Zakaria and Leest (1996) stated that aggregate type, fines percentage, moisture content, compaction and load applications are influential factors in the induced strains in a pavement structure. Giroud and Han (2004) stated that bearing capacity failure, after repeated traffic loading, in the base or subgrade layer is the main cause of surface rutting.

Xu and Huang (2011) concluded that most of rutting in pavement layers is related to the weakness in the middle and lower layers. Based on studies done by Wright and Paquette, (1987), roughly 25% of the rutting is related to subbase, 50% to the base and 25% to the asphalt concrete surface. Since in the MEPDG, the total rut depth is equal to sum of the rut depths in each layer, the rutting of underlying layers should not be overlooked. In terms of fatigue failure, Mulungye et al. (2007) stated that even in weak soil layers, fatigue cracking normally occurs before rutting. According to the studies done by Cardone et al. (2011), the stiffness of soil and granular layer must be sufficiently high to avoid fatigue cracking.

In order to overcome the problem of rutting in pavement layers, Jegede (2000) stated that stabilization could improve the California Bearing Ratio (CBR) when facing poor soil properties. His suggestion agrees with Van Zyl and Maree (1983) who noted that when the density of soil is increased (increased stiffness), the plastic deformation of soils is reduced significantly. The suggestion of Van Zyl and Maree (1983) clearly indicates that using appropriate stiffness for underlying pavement layers is very important. In addition, since providing sufficient stiffness for subgrade cannot be easily obtained all the time at the site, it is imperative that reinforcing and/or improving the stiffness modulus of subgrade or any of pavement layers using different stabilizers is inevitable.

STIFFNESS MODULUS OF VARIOUS SOILS

The stiffness of unbound pavement materials is characterized by the resilient modulus (M_r). AASHTO (1986) design guide has recommended the resilient modulus for characterizing subgrade stiffness in both flexible and rigid pavements.

In resilient modulus test, repeated axial cyclic stress of a fixed magnitude, loading time and frequency is applied to the top of the cylindrical soil specimen. From the surroundings, it is under a static confining pressure provided by the triaxial pressure chamber. In fact, this test is the triaxial compression test under cyclic stress. Resilient modulus can be estimated from soil classification and soil unit weight as well. For instance, Table 1 summarizes the resilient modulus of different soils based on their classification.

Effect of additives on increasing stiffness modulus

Cement stabilization has been widely used to enhance the performance of subgrade layer in pavement structures (Khoury et al., 2006). Stabilization process consists of mixing stabilizing agents such as lime, fly ash, and cement with raw soil in which cementing compounds are the key elements for strength and stiffness improvement. The degree of effectiveness varies from one stabilizing agent to the other.

Although several studies have been conducted in the past to evaluate the performance of stabilized soils and aggregates in pavement construction (McManis and Arman, 1989; Zaman et al., 1992; Puppala et al., 1996; Misra, 1998; Little, 2000; Miller and Zaman, 2000; Senol et al., 2002; Khoury and Zaman, 2007), only a few studies (Edil et al., 2006; Chauhan et al., 2008; Mohammad and Saaheh, 2008; Li et al., 2009; Solanki et al., 2009; Singh et al., 2010; Solanki and Zaman, 2011; Tastan et al., 2011; Rout et al., 2012), have determined the required inputs for stabilized soils in the new MEPDG (2004). These inputs include the resilient modulus (Mr) and unconfined compressive strength (UCS), when designing a new pavement. Common additives and the associated results of the soil stabilization process are summarized as follows:

Lime

Since lime chemically changes the plasticity properties of the soil, it is frequently used to treat clays. Although lime stabilization is effective to some extent, the strength and stiffness enhancement is moderate.

In a study by Solanki et al. (2009), specimens were prepared by adding 3, 6, and 9% of lime (by dry weight of soil) to the raw soils that were classified as CL-ML with a

Table 1. Default M_R values for unbound granular and subgrade materials at optimum moisture content and density conditions (NCHRP 1-37A, 2004).

AASHTO soil class	Resilient modulus range (psi)	Typical resilient modulus (psi)
A-1-a	38,500 - 42,000	40,000
A-1-b	35,500 - 40,000	38,000
A-2-4	28,000 - 37,500	32,000
A-2-5	24,000 - 33,000	28,000
A-2-6	21,500 - 31,000	26,000
A-2-7	21,500 - 28,000	24,000
A-3	24,500 - 35,500	29,000
A-4	21,500 - 29,000	24,000
A-5	17,000 - 25,500	20,000
A-6	13,500 - 24,000	17,000
A-7-5	8,000 - 17,500	12,000
A-7-6	5,000 - 13,500	8,000

liquid limit (LL) of 27 and a plasticity index (PI) of 5 and CL with LL = 37 and PI = 11. The additive and soil were blended manually to become uniform. After the mixing process, desired amount of water was added to obtain the optimum moisture content (OMC). The mixture was then compacted in a mold with a diameter of 101.6 mm (4.0 in) and a height of 203.2 mm (8.0 in) to obtain approximately 95 to 100% of the maximum dry density (MDD). After compaction, extruded specimens were cured in a moisture-controlled room with 23.0 ± 1.7°C temperature and a relative humidity of approximately 96% for a period of 28 days. AASHTO T 307-99 was used for M_r tests, and the UCS tests were performed in accordance with the ASTM D1633 test method. Their study concluded that there is an increase in M_r value with increase in the percentage of lime content since more cementing compounds were constituted. For the raw soil, the M_r value was found to be 153 MPa, while for samples with 3, 6 and 9% of lime, the M_r values were 5.6, 7.1, 7.3 times more than the control sample, respectively. The UCS was 227 kPa for the raw soil. Addition of 3, 6, and 9% of lime increased the UCS values by a factor of 1.7, 1.7 and 2, compared to the control sample, respectively.

Rout et al. (2012) investigated the resilient modulus (according to AASHTO T-307) behavior of three subgrade soils. Based on unified classification, the tested soils were CH with various plastic properties. The soils were stabilized using lime-cement to evaluate the efficiency in enhancing the resilient characteristics. Furthermore, effects of confining and deviator stresses were investigated. In their study, laboratory specimens were compacted in such a way that the sample's height is two times greater than its diameter (for instance 142 mm in height and 71 mm in diameter). The soil was air dried and mixed with 6% of lime and 3% of cement. The soil specimens were prepared at the moisture content of 3% more than the OMC. After initial blending of soil with lime and cement, the mixture was maintained in the room temperature for at least 72 h before compaction. Both control and treated samples were cured in a humidity room for duration of two and seven days, respectively. Resilient Modulus values for untreated soils ranged between 30 to 50 MPa, whereas for the same treated soil, it was up to three times more especially for the soil with highest plastic property. It was observed that in untreated subgrade soil, increasing the deviator stress resulted in resilient modulus reduction for all the three tested confining pressures (13.8, 27.6, 41.4 KPa). This phenomenon is due to the strain softening behavior in clayey soils. On the other hand, in treaded soils, increasing the deviator stress resulted in an increase in resilient modulus; this could be because of strain hardening behavior of granular materials. This is due to the fact that the treated soils exhibit non-plastic behavior and their modulus depends on confining stresses as expected by granular materials.

Mohammad and Saadeh (2008) investigated stabilization of A-4 clay soil with 10% of lime. Cylindrical soil specimens (D = 71.1 mm and H = 142.2 mm) were compacted with an impact compactor in five layers on which the repeated load triaxial resilient modulus (AASHTO T 307, 2003) and material property tests were conducted. The untreated specimens were immediately tested after the compaction, while the compacted lime-treated samples were covered with polythene bags and kept in the humidity room for a duration of 7, 14, and 28 days for curing. The lowest resilient modulus in untreated subgrade soil was 35 MPa, while for the lime-treated soils it was 2.25 to 3.25 times more based on different cyclic stresses rages from12 to 62 KPa with 41 KPa of confining pressure.

Singh et al. (2010) evaluated the effectiveness of lime in a sulfate containing soil, that was characterized as CL

Table 2. Effect of lime stabilizer on improving M_r and UCS in soils.

Authors' name	Soil type	Additive type	Stabilizer percentage (%)	Moisture (%)	Curing time (days)	Measuring parameter	Initial Value (Mpa)	Rate of increase
Solanki et al. (2009)	CL-ML, CL	Lime	3	OMC	28	M_r	153	5.6
			3			UCS	227	1.7
			6			M_r	153	7.1
			6			UCS	227	1.7
			9			M_r	153	7.3
			9			UCS	227	2.0
Rout et al. (20120	CH	Lime-cement	6-3	OMC+3	7	M_r	30-50	3.0
Mohammad and Saadeh (2008)	A-4	Lime	10	-	28	M_r	35	2.25-3.25
Singh et al. 2010	CL	Lime	5	OMC	28	M_r	58	4.9
Solanki and Zaman 2011	CL-ML	Lime	6	OMC	28	M_{rt}	65	12
	CL	Lime	6	OMC	28	M_{rt}	29	22

type (Unified-based classification) or A – 7 - 6 (AASHTO-based classification). In their study, the optimum lime content was selected based on the strength and pH tests. Harvard miniature compacted specimens with 3, 6, and 9% of lime were cured for 28 days in a moisture-controlled room at 23.0 ± 1.7°C temperature and 96% humidity. Samples were then subjected to UCS test with 0.6 mm/min increment based on the ASTM D 1632-96. The UCS results showed an increase for up to 4% of lime, after which only a minor change was observed in UCS value. Among various tested percentages of lime (0, 1, 3, 5, 7, 9 and 100%); the pH value increased significantly until 5% in the pH test (ASTM D 6276-99). Finally, 5% of lime was chosen as the optimum content. The resilient modulus test (AASHTO T 307-99) was performed on the compacted samples with a diameter of 101.6 mm and a height of 203.2 mm at OMC. The M_r result for the lime stabilized soil improved up to 284 MPa compared to the raw soil with a modulus of 58 MPa when tested at a deviator stress of 42 kPa and confining pressure of 13 kPa; meaning that adding lime had increased the M_r value by a factor of 4.9 compared to the control samples. In their study, the M_r values were based on unsaturated condition, and suction or degree of saturation were not incorporated.

Solanki et al. (2011) evaluated the indirect tensile characteristics of lime stabilized mixtures. According to the USCS, two types of soils including CL - ML (silty clay with sand; LL = 27 and PI = 5) and CL (lean clay; LL = 37 and PI = 11) were selected with 6% of lime. For the resilient modulus in tension mode (M_{rt}), 101.6 mm (4 in) cylindrical specimens were compacted by the use of Superpave gyratory to a height of 63.5 mm (2.5 in). Specimens were cured in a controlled environment (23.0 ± 1.7°C temperature and 96% humidity) for 28 days after compaction, as recommended by the ASTM D 1632 test method. M_{rt} values were improved for the lime stabilized specimens. In addition, the M_{rt} decreased as the stress ratio (Deviator stress/ Maximum Tensile Stress) increased. For instance at a stress ratio of 0.36, 6% of lime resulted in approximately 12 and 22 times increase for (CL - ML) and CL, respectively relative to the initial values of 65 MPa for CL - ML and 29 MPa for CL. Effects of lime stabilizer on improving the engineering properties of raw soil are summarized in Table 2.

Cement

In Solanki et al. (2009), while using the same procedure of stabilization for lime, they added 5, 10, and 15% of Cement kiln Dust (CKD) to CL - ML raw soil. CKD is an industrial waste gathered

Table 3. Effect of cement stabilizer on improving M_r and UCS in soils.

Authors' name	Soil type	Additive type	Stabilizer percentage (%)	Moisture (%)	Curing time (days)	Measuring parameter	Initial value (Mpa)	Rate of increase
Solanki et al. (2009)	CL-ML	CKD	3	OMC	28	M_r	153	4.8
			3			UCS	227	1.3
			6			M_r	153	15.7
			6			UCS	227	4.8
			9			M_r	153	16.7
			9			UCS	227	6.0
Mohammad and Saadeh (2008)	A-4	Cement	8	-	28	M_r	35	10-15

during the production process of Portland cement. Based on their results, the M_r value was 153 MPa for the raw soil, while for the specimens stabilized with 5, 10 and 15% of CKD, the M_r values increased as much as 4.8, 15.7, 16.7 times, respectively. In addition, the UCS value was found to be 227 kPa for the raw soil and increased by a factor of 1.3, 4.8, and 6 for 5, 10, and 15% of CKD, respectively.

Mohammad and Saadeh (2008), using the same procedure of stabilization for lime, added 8% of cement for stabilization of A - 4 clay soil. They concluded that the resilient modulus of cement-treated subgrade soil increased 10 to 15 times (based on different cyclic stresses) relative to the control samples with resilient modulus of 35 MPa. Effects of cement stabilizer on improving the engineering properties of raw soil are summarized in Table 3.

Fly ash

Fly ash is a non-crystalline pozzolanic and slightly cementitious material. It is an environmentally hazardous solid waste, produced mostly from the burning of coal in thermal power plants. Solanki et al. (2009), using the same procedure of stabilization for lime and cement, prepared specimens by adding 5, 10, and 15% of a class C Fly Ash (CFA) to the CL - ML soil. They concluded that, the M_r values was 153 MPa before stabilization and increased by a factor of 1.7, 4.2 and 14.1 when 5, 10, and 15% of CFA stabilizing agent was used. One of the reasons for the higher M_r values is related to the high percentage of pozzolans in CFA, which is responsible for the constitution of cementitious products. The UCS value was 227 kPa for the control sample. Addition of 5, 10 and 15% of CFA increased the UCS by a factor 1.3, 2.4 and 4, respectively as they noted.

Li et al. (2009) evaluated the ability of Fly Ash to increase the unconfined compressive strength and stiffness of the subgrade soils both in the laboratory and in the field. Twelve percent of class C Fly Ash, by dry weight, was mixed with the subgrade soil, which consisted of lean clay (CL), clayey sand (SC), and silty sand (SM) (using Unified Soil Classification) or A-7-6 with a group index (GI) > 10, A - 6 with GI > 2 and A – 2 - 6 with GI < 1 (using AASHTO Soil Classification). The M_r tests (AASHTO T292, 1991) were performed immediately after compaction with optimum water content, or after 14 days of curing while sealed in plastic wrap and kept in a room with 100% humidity. Since, there is no UCS standard method for Fly Ash stabilized materials, in their study a strain rate of 0.21% per min was used for the UCS tests based on the ASTM D 5102 for compacted soil - lime mixtures. In the field, strength and stiffness of the stabilized soils were evaluated by means of Soil Stiffness Gauge (SSG), Dynamic Cone Penetrometer (DCP), after 7 days of curing, and Falling Weight Deflectometer (FWD). Results indicated that initial M_r of the subgrade varied between 34 and 42 MPa (with a mean = 38 MPa) while, the M_r for laboratory - mix Fly Ash stabilized soils ranged from 115 to 167 MPa (with a mean = 139 MPa). On the other hand, the field-mixed Fly Ash stabilized soil had M_r of 60 to 129 MPa (with a mean = 82 MPa). Their results indicated that the addition of 12% of Fly Ash to the raw soil increased the M_r value by a factor of 3.7 in the laboratory and 2.2 in the field, respectively. Their results also showed that by adding Fly Ash to the subgrade UCS increased significantly. The UCS of the subgrade ranged from 92 to 138 kPa (with a mean = 112 kPa), the laboratory stabilized samples had UCS between 360 and 1120 kPa (with a mean = 663 kPa), and the UCS of field stabilized soils ranged from 110 to 410 kPa (with a mean = 183 kPa). Thus it is clearly shown that addition of 12% of type C Fly Ash to the subgrade soil improved the M_r value by a factor of 5.9 in the laboratory and 1.6 in the field, respectively. Furthermore, they reported that mixing condition and type of mixing had caused these differences.

In a study by Edil et al. (2006), seven soft fine-grained soils (six inorganic soils and one organic soil) were mixed with four types of Fly Ashes and subjected to California Bearing Ratio (CBR) and resilient modulus (M_r) tests. The soils were CL, CH, OH (Unified Soil Classification) or A-7-5, A-7-6, A-6 (AASHTO Soil Classification) with the plasticity indices ranging between 15 and 38. Two of the Fly Ashes were selected from high quality class, while the others did not fulfill the requirements of class C or class F based on the ASTM C 618 specification. Tests were conducted on soils and soil–Fly Ash mixtures prepared with optimum water content, 7% more than optimum content as well as a very wet *in situ* condition (9 –18% more than optimum water content). Unsoaked CBR tests were performed both on the raw soils and soil–Fly Ash mixtures according to ASTM D 1883 specification. For the resilient modulus, AASHTO T 292-91 procedure was followed. For compaction procedure, standard proctor was selected after 1 to 2 h delay from mixing. Compacted soil–Fly Ash specimens were cured at 25°C and 100% relative humidity. The CBR tests were conducted after 7 days of curing. However, for resilient modulus test, specimens were cured for 14 and 56 days. Adding 10 and 18% of Fly Ash increased CBR by a factor of 4 and 8, respectively. At 7% wet of optimum, M_r of the raw soil ranged between 3 and 15 MPa. Mixtures with 10 and 18% of Fly Ash had M_r of 12 to 60 MPa and 51 to 106 MPa, respectively. They reported that Fly Ashes with higher CaO/SiO_2 ratio constituted mixtures with higher CBR and M_r. Furthermore, finer grained or more plastic soils have smaller CBR or M_r values before stabilization, while greater CBR ratio and M_r ratio were obtained after stabilization. The resilient modulus did not increase significantly after 7 and 14 days of curing, while the increase was noticeable between 14 and 28 days. After 28 days, little further increase occurred in the resilient modulus. Among the four tested Fly Ashes, type C exhibited the maximum improvement. The effect of Fly Ash on the CBR and M_r was much lower for the organic soil compared to inorganic soils.

Solanki and Zaman (2011), using the same procedure of stabilization for lime, evaluated the indirect tensile properties of soil specimens (CL - ML and CL) stabilized with class C of Fly Ash. Results indicated improvement in M_{rt} values for CFA stabilized specimens. Moreover, increasing the stress ratio (Deviator stress/Maximum Tensile Stress) resulted in M_{rt} decline. For instance, at a stress ratio of 0.36, adding 10% of CFA provided approximately 7.5 and 24.9 times of increase for (CL - ML) and CL, respectively.

Singh et al. (2010), using the same procedure of stabilization for lime, evaluated the effectiveness of CFA (12, 15 and 18%) for CL or A-7-6 type soil. The UCS increased up to 16% of CFA, after which minor change was observed in the UCS strength. Among various percentages of additive (0, 6, 8, 10, 12, 14, 16, 18 and 100%), the pH value had significant increase until 4%. Finally, the use of 16% CFA was recommended for subsequent testing. The M_r results for the CFA stabilized soil indicated 4.5 times increase compared to those of the raw soils which had ranged from 26 to 58 MPa.

Tastan et al. (2011) evaluated the efficiency of different percentages of Fly Ash (up to 30% content) in stabilizing organic soils. Three soft organic soils with different organic contents including silty, sandy peat (Pt, A-8(0)), low plasticity organic sandy clay (OL-OH, A-7-5(32)), and moderately plastic organic clay (OL, A-4(5)), were used in this study. Unconfined compression tests were conducted following ASTM D5102 (ASTM 2009b) with the strain rate of 0.21% per min. The compaction effort was adjusted in such a way that the same impact energy per unit volume as in the standard Proctor effort [ASTM D698 (ASTM 2007)], was experienced. Compacted samples were cured for seven days to simulate the early curing conditions during construction at 100% relative humidity and 25°C. Specimens for the resilient modulus and unconfined compression tests were prepared in a mold with a diameter of 102 mm and a height of 203 mm. Compaction effort was as the same compaction energy per unit volume in the standard Proctor compaction method (600 kN/m^3). They concluded that, for clayey soils with an Organic Content (OC) of less than 10%, UCS value increases noticeably (from 30 kPa without Fly Ash to more than 400 kPa with Fly Ash). On the other hand, for a highly organic sandy silty peat with OC = 27%, lower increases in UCS value were observed (from 15 kPa without Fly Ash to more than100 kPa with Fly Ash). For instabilized organic soils, resilient modulus test could not be performed at wet conditions since the specimens were too soft. Addition of Fly Ash (up to 30%) to OL - OH and OL (OC = 5 and 6%, respectively) resulted in M_r values between 10 to 100 MPa (this depends on the type and percentage of the Fly Ash used). Furthermore, M_r improved up to 120 MPa at Optimum Water Content (OWC). However, for Pt (OC = 27%), the stabilized mixture never produced M_r greater than 30 MPa no matter which Fly Ash type and percentage was used. Finally, they conclude that increase in the organic content decreases the strength of the mixture exponentially. They also reported that the maximum UCS and M_r were obtained when the CaO content was greater than 10% and the CaO/SiO_2 ratio

Table 4. Effect of Fly Ash stabilizer on improving M_r and UCS in soils.

Authors' name	**Soil type**	**Additive type**	**Stabilizer percentage (%)**	**Moisture (%)**	**Curing time (days)**	**Measuring parameter**	**Initial value**	**Rate of increase**
Solanki et al. (2009)	CL-ML	Fly ash	3	OMC	28	M_r	153 Mpa	1.7
			3			UCS	227 KPa	1.3
			6			M_r	153 Mpa	4.2
			6			UCS	227 KPa	2.4
			9			M_r	153 Mpa	14.1
			9			UCS	227 KPa	4.0
Li et al. (2009)	CL,		12	OMC	14	M_r	38 Mpa	3.7
	SC,	Fly ash				UCS	112 KPa	5.9
	SM			-	7	M_r	38 Mpa	2.2
						UCS	112 KPa	1.6
Edil et al. (2006)	CL-CH		10	≥OMC	14	M_r	9 MPa	4
		Fly ash	18					8.7
			10		7	CBR	-	4
			18					8
Solanki and Zaman (2011)	CL-ML	Fly ash	10	OMC	28	M_{rt}	65 MPa	7.5
	CL		10	OMC	28	M_{rt}	29 MPa	24.9
Singh et al. (2010)	CL	Fly ash	16	OMC	28	M_r	26-58 MPa	4.5
Tastan et al. (2011)	OL, OL-OH,PEAT (OC<10)	Fly ash				M_r	10 Mpa	12
						UCS	30	13.4
		Fly ash	Up to 30%	OWC	7	M_r	<1	<30
	OC>27					UCS	15	6.7

ranges from 0.5 to 0.8.

Chauhan et al. (2008) mixed 30% of Fly Ash and 70% of silty sand which yielded the maximum UCS value. They stated that as the number of load cycles and deviator stress increases, the resilient modulus decreases. However, increasing confining pressure resulted in resilient modulus increase. Table 4 summarizes different studies on the Fly Ash stabilizer.

CONCLUSION

A stable subgrade layer is a key component of a pavement structure. Engineering properties such as strength and resilient modulus of subgrade are influential parameters on the overall performance of the pavement structure, since in many cases fatigue or rutting failure in pavements are because of insufficient stiffness modulus in soil layers. The new Mechanistic-Empirical Pavement Design Guide (AASHTO, 2004) incorporates engineering properties of subgrade soils (unbound or stabilized) in the design procedure because of its importance.

Providing sufficient stiffness for subgrade cannot be easily obtained all the time; hence, reinforcing the subgrade layer is inevitable and it is necessary to improve the existing stiffness modulus by using various stabilizers. The following conclusions were derived:

1. In treaded soils, increasing the deviator stress resulted in an increase in resilient modulus since treated soils exhibit non-plastic behavior and their modulus depends on confining stresses as expected by granular materials.
2. M_{rt} values were improved for the lime and Fly Ash stabilized specimens. In addition, the M_{rt} decreased as the stress ratio (Deviator stress/Maximum Tensile Stress) increased.
3. M_r value increases with increase in the percentage of lime content. This is especially true for the soil with higher plastic property.
4. CFA increases the resilient modulus and CBR, the reasons for the higher M_r values is related to the high percentage of pozzolans in CFA. Therefore, Fly Ashes with higher CaO/SiO_2 constituted mixtures with higher CBR and M_r.

Furthermore, the amount of resilient modulus increase differs for laboratory and *in situ* conditions because of differences in mixing conditions and type of mixing. Moreover, finer grained or more plastic soils have smaller CBR or M_r values before stabilization, while greater CBR ratio and M_r ratio were obtained after stabilization as follows:

1. For CFA stabilized soils, curing time is influential in improving the resilient modulus. The resilient modulus did not increase significantly after 7 and 14 days of curing, while the increase was noticeable between 14 and 28 days. After 28 days, little further increase occurred in the resilient modulus. Among Class C, Class F, and non classic Fly Ashes, type C exhibited the maximum improvement.
2. Fly Ash improves the resilient modulus and UCS, while the amount of this improvement reduces with increase in the organic content. The maximum UCS and Mr were obtained when the CaO content was greater than 10% and the CaO/SiO_2 ratio ranges between 0.5 and 0.8.
3. For treated soils, as the number of load cycles and deviator stress increases, the resilient modulus decreases. However, increasing confining pressure resulted in resilient modulus increase.

REFRENCES

AASHTO T 292 (1991). Resilient Modulus of Subgrade Soils and Untreated Base/Subbase Materials.

AASHTO (2002). Standard Method of Test for Resilient Modulus of Subgrade Soils and Untreated Base/Subbase Materials, AASHTO Designation 307, Standard Specifications for Transportation M aterials and Methods of Sampling and Testing, Part II Tests. American Association of State Highway and Transportation Officials ,Washington, DC.

ASTM C618 (2012). Standard Specification for Coal Fly Ash and Raw or Calcined Natural Pozzolan for Use in Concrete.

ASTM D698 (2007). Standard test methods for laboratory compaction characteristics of soil using standard effort.

ASTM D1632 (1996). Standard Practice for Making and Curing Soil-Cement Compression and Flexure Test Specimens in the Laboratory.

ASTM D1633 (2007). Standard Test Methods for Compressive Strength of Molded Soil-Cement Cylinders.

ASTM D1883 (2007). Standard Test Method for CBR (California Bearing Ratio) of Laboratory-Compacted Soils.

ASTM D5102 (1996). Standard Test Method for Unconfined Compressive Strength of Compacted Soil-Lime Mixtures.

ASTM D5102 (2009). Standard test method for unconfined compressive strength of compacted soil-lime mixtures.

ASTM D6276 (1999). Standard Test Method for Using pH to Estimate the Soil-Lime Proportion Requirement for Soil Stabilization.

Barksdale RD, Itani SY (1989). Influence of Aggregate Shape on Base Behavior; In Transportation Research Record 1227, Transportation Research Board, National Research Council, Washington, D. C., pp 173-182.

Chauhan MS, Mittal S, Bijayananda Mohanty B (2008). Performance evaluation of silty sand subgrade reinforced with Fly Ash and fibre. Geotextiles and Geo-membranes 26, ASCE. pp. 429-435.

Cardone F, Cerni G, Virgili A, Camilli S (2011). Characterization of permanent deformation behaviour of unbound granular materials under repeated triaxial loading. J. Constr. Build. Mater. Elsevier 28:79-87.

Edil TB, Acosta HA, Benson CH (2006). Stabilizing Soft Fine-Grained Soils with Fly Ash. J. Mater. Civ. Eng. ASCE pp. 283-294.

Giroud JP, Han J (2004). Design method for geogrid-reinforced unpaved roads. I. Development of design method, II. Calibration and Applications Journal of Geotechnical and Geo-environmental Engineering, American Society of Civil Engineers, ASCE 130:787-797.

Jegede G (2000). Effect of soil properties on pavement failures along the F209 highway at Ado-Ekiti, south-western Nigeria. J. Constr. Build. Mater. 14, Elsevier pp. 311-315.

Khoury N, Zaman M, Laguros J (2006). Semi-quantification of cementing products using X-ray diffraction technique in stabilized aggregate. J. X-Ray Sci. Technol. 14(1):39-54.

Khoury N, Zaman M (2007). Environmental effects on durability of aggregates stabilized with cementitious materials. J. Mater. Civ. Eng. 19(1):41-48.

Lin LM, Tastan O, Benson CH, Edil TB (2009). Field Evaluation of Fly Ash Stabilized Subgrade in US 12 Highway. International Foundation Congress and Equipment Expo. ASCE. pp. 385-392.

Little DL (2000). Evaluation of structural properties of lime stabilized soils and aggregates. Report Prepared for National Lime Association.

McManis KL, Arman A (1989). Class C fly ash as a full or partial replacement for portland cement or lime. Transportation Research Record. 1219, Transportation Research Board, National Research Council, Washington, D. C.

Miller GA, Zaman M (2000). Field and laboratory evaluation of cement kiln dust as a soil stabilizer. Transportation Research Record. 1714,

Transportation Research Board, National Research Council, Washington, D.C.
Misra A (1998). Stabilization characteristics of clays using class C fly ash. Transportation Research Record. 1611, Transportation Research Board, National Research Council, Washington, D.C.
Mohammad L, Saadeh S (2008). Performance Evaluation of Stabilized Base and Subbase Material. Geo Congress, ASCE. pp 1073-1080.
Mulungye RM, Owende PMO, Mellon K (2007). Finite element modeling of flexible pavements on soft soil subgrades, Elsevier J. Mater. Design 28:739-756.
NCHRP (2004). Guide for Mechanistic-Empirical Design of New and Rehabilitated Pavement Structures. Final Report, Project 1-37A,TRB, National Cooperative Highway Research Program, Washington, D.C..
Puppala AJ, Mohammad LN, Allen A (1996). Engineering behavior of lime-treated Louisiana subgrade soil. Transportation Research Record. 1546, Transportation Research Board, National Research Council, Washington, D.C.
Rout RK, Ruttanapormakul P, Valluru S, Puppala AJ (2012). Resilient Moduli Behavior of Lime-Cement Treated Subgrade Soils. Geo Congress, ASCE. pp. 1428-1437.
Senol A, Bin-Shafique Md S, Edil TB, Benson CH (2002). Use of class C fly ash for stabilization of soft subgrade. Proceeding of 5th International Congress on Advances in Civil Engineering, Istanbul Technical University, Turkey.
Singh D, Ghabchi R, Laguros JG, Zaman M (2010). Laboratory Performance Evaluation of Stabilized Sulfate Containing Soil with Lime and Class C Fly Ash. GeoFlorida , Advances in Analysis, Modeling & Design, ASCE. pp. 757-766.
Solanki P, Khoury N, Zaman MM (2009). Engineering Properties and Moisture Susceptibility of Silty Clay Stabilized with Lime, Class C Fly Ash, and Cement Kiln Dust. J. Mater. Civ. Eng. ASCE. pp. 749-757.
Solanki P, Zaman M (2011). Characterization of Lime and Fly Ash-Stabilized Soil By Indirect Tensile Testing. ASCE. Pp. 4438-4448.
Tastan EO, Edil TB, Benson CH, Aydilek AH (2011). Stabilization of Organic Soils with Fly Ash. Journal of Geotechnical and Geo-environmental Engineering, ASCE. pp. 819-833.
Van Zyl NJW, Maree JH (1983). The behaviour of high standard crushed stone base pavement during heavy vehicle simulator test. Civ. Eng. S. Africa 25(7):745-753.
Wright PH, Paquette RJ (1987). Highway Engineering, 5th edition,John Wiley, New York.
Xu T, Huang X (2011). Investigation into causes of in-place rutting in asphalt pavement. J. Constr. Building Mater. Elsevier 28:525-530.
Zakaria M, Leest G (1996). Rutting characteristics of unbound aggregate layers, Constr. Build. Mater. Elsevier 10(3):185-189.
Zaman M, Laguros JG, Sayah AI (1992). Soil stabilization using cement kiln dust. Processing of 7th International Conference on Expansive Soils, Vol. I. Texas Technology University, Lubbock, Texas. pp. 347-351.

Determination of some physical and mechanical properties of Calabrian pine (*Pinus brutia* Ten.) trees grown in the Denizli area of Turkey

Kadir Ozkaya

Pamukkale University, Denizli Vocational School of Technical Sciences, Department of Materials and Materials Processing Technology, 20070, Denizli, Turkey.

The most extensive distribution of Calabrian pine (*Pinus brutia* Ten.) trees in the world is found in Turkey. This species varies depending on the regions in Turkey for growth features, such as climate, soil structure and slope. In this study, the technological properties of Calabrian pine (*Pinus brutia* Ten.) grown in the Denizli area were studied. Within the scope of the study, their physical properties, such as air-dried and oven-dried specific densities, swelling and shrinkage values were determined. Their mechanical properties, such as bending strength, modulus of elasticity in bending, parallel compressive strength to the fibers, parallel tensile strength to the fibers and shear strength in a radial direction were determined. The values obtained were compared with the Calabrian pine (*P. brutia* Ten.) trees grown in different regions.

Key words: Denizli, *Pinus brutia* Ten., physical properties, mechanical properties.

INTRODUCTION

Firstly, the rational and economic use of wood should be provided for meeting the demand for lumber from Turkey's forests, which is supplied by importing from abroad. The most effective way to solve the problem would be to set forth the various properties of Turkey's original tree species and to process the wooden raw materials according to these properties to provide for the manufacturing of products that can ensure the best performance in their place of use. Although, it is known that the properties of Turkey's original tree species have both genetic and ecologic elements and vary depending on the regions, very few studies have been made on this subject (Ilter et al., 2011). This species of Calabrian pine (*Pinus brutia* Ten.), by taking into consideration its distribution around the world and its intensity of distribution in Turkey, has been mentioned in the international literature as "Turkish red pine" and is one of Turkey's original tree species with a high economic value. Calabrian pine has been established in pure forests in the Mediterranean, Aegean and Marmara Regions of Turkey, while it is found sporadically and in groves in the coastal and inner parts of the Western and Central Black Sea regions (Ilter et al., 2011). The first studies related to the technical properties for broadening the areas of usage of this species of Calabrian pine, which is widespread in Turkey, date back to the 1950s (Berkel, 1957). Calabrian pine wood has a porous structure and since it is of medium density, it is appropriate for the production of wood-polymer composites (Yildiz, 1993). Since Calabrian pine has excessive resin in its structure, resin production from Calabrian pine does not cause any significant changes in its physical properties and produces an

Table 1. Geographical and structural information for the trees used in the tests.

Geographical and structural properties of calabrian pine trees			
Regional Directorate	Denizli	Exposure	North
Operations Directorate	Eskere (Beyagac)	Location	Akcatas
Sub-district Directorate	Karacaoren	Soil structure	Serpentine
Division	56	1.30 m diameter (cm)	50
Latitude	37° 12' 04" .4 N	Age	120
Longitude	28° 43' 04" .6 E	Height (m)	20
Altitude (m)	950		

increase in its mechanical properties (Oktem and Sozen, 1996). Calabrian pine can be split open easily. This characteristic of Calabrian pine enables it to be used in cooperage and shipbuilding (Bektas et al., 2003).

The habitat is significant in the formation of the technological properties of trees. Consequently, the properties of Calabrian pine, just as for other tree species, vary according to growth regions (Bektas et al., 2003). For instance, as a result of the anatomical, physical and mechanical studies made on the Calabrian pine grown in the Datça, Marmaris region, it was understood that it could be utilized in producing wire poles, mine poles, constructive materials, yachts and boats, packing cases, fence pickets, cultivation tools, cases, and in the paper and cellulose industry (Bozkurt et al., 1993). Care should be taken in the area where Calabrian pine is grown, especially in places where it is exposed to the bending impact. The Hatay, Adana and Antalya regions could be recommended if Calabrian pine will be used in the areas that require static bending strength, such as mine poles, construction forms and bridges (Ilter et al., 2011).

While it was emphasized in the research studies made that regional differences were significant, a determination has not been made for the trees growing in the Denizli area, the starting point in the western Mediterranean region where the optimal growing conditions for Calabrian pine occur.

In this study, it was aimed to increase the industrial use of Calabrian pine by determining the technological properties of the Calabrian pine species grown in the Denizli area. For this purpose, the trees grown in the areas remaining between Beyağaç – Çameli – Acıpayam, where the most habitats in the Denizli area are found, have been studied.

MATERIALS AND METHODS

Four sample trees were selected by considering the habitat characteristics, such as direction, slope, altitude, diameter, density, etc. in the Eskere – Akçataş area, where the largest distribution of Calabrian pine is found. The areas from which the sample trees were taken and the general properties of tree species were determined according to the TS 4176 (ISO 4471) standards. The geographical and structural properties of the Calabrian pine (*P. brutia* Ten.) selected for the experimental materials used in the study have been given in Table 1.

Preparation of the specimens

After marking the bark on the north side of the trees, they were cut 0.30 m above the soil and their full-length was measured. Wheel-shaped sections 15 cm long were removed from the trunks of each sample tree once every 2 m by starting from 0.30 m high and 2-m. Long trunk sections were removed from each sample tree between 2 and 4 m. The north direction was marked on the transversal section of each piece and they were enumerated according to the sequence of removal. The 2 m long trunk sections were cut to a width of 6 cm in a north-south direction and east-west direction to determine the physical and mechanical properties. The pieces obtained were kept in a dry place for 3 months until they reached equilibrium moisture content (EMC).

Specimens were prepared from the 6 cm wide, 2-m long trunk sections for testing compressive strength, bending strength, tensile strength, shear strength, shrinkage and swelling and 15 cm high specimens from the trunk sections were prepared for testing air-dried and oven-dried density (Figure 1). All the specimens obtained were conditioned at a temperature of 20±2°C and a relative humidity of 65±5% until they reached an EMC.

Analysis of physical properties

Strips 2 cm wide in the north-south and east-west directions were removed from the 15 cm long tree trunk sections obtained from the sample trees for density measurements. Specimens measuring 20×20×30 mm were used according to the TS 2471 (ISO 3130) and TS 2472 (ISO 3131) standards to determine the air-dried (δ_{12}) and oven-dried (δ_o) densities of Calabrian pine.

Specimens prepared from the 1 m long tree trunks of Calabrian pine wood were used for the dimensional tests. Specimens measuring 30×30×15 mm were prepared according to the TS 4083 (ISO 4469) and TS 4084 (ISO 4859) standards to determine the amount of efficiency in the tangential and radial directions.

For the dimensional swelling test (α), the dimensions of the specimens were measured separately in the tangential and radial directions via digital calipers after being kept in a climatization chamber at a temperature of 20±2°C and a relative humidity of 65±5% until they reached the EMC of 12±2% (L_{12}). Later, when the same specimens reached the EMC of 21±2% in a climatization chamber at a temperature of 20±2°C and a relative humidity of 90±5%, they were measured again at the first measurement place (L_{21}) (Sogutlu, 2004).

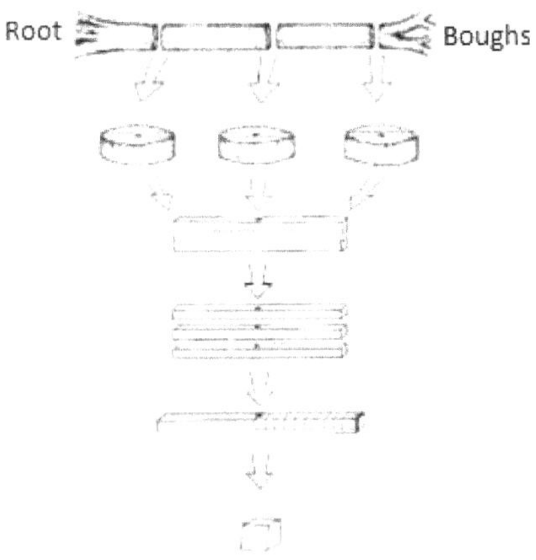

Figure 1. Specimen taking from the sample tree trunk for determining the physical properties (Dogdu, 2006).

The dimensional swelling percentages ($\alpha_{//}$) were calculated according to Formula 1.

$$\alpha // = \frac{L_{21} - L_{12}}{L_{12}} \times 100 \quad (1)$$

The amount of volumetric swelling (α_v) was obtained from the sum of the percentages of swelling in the tangential and radial directions (α_t, α_r) according to Formula 2.

$$\alpha_v = \alpha_t + \alpha_r \quad (2)$$

The amounts of dimensional shrinkage (β) were determined according to the principles stated in the TS 4083. The specimens were kept in a climatization chamber at a temperature of 20±2°C and a relative humidity of 90±5% until they reached the EMC and were measured separately in the tangential and radial directions (L_{21}). Later, when the same specimens reached the EMC of 7±2% at a temperature of 20±2°C and at a relative humidity of 30±5%, they were measured again at the first measurement place (L_7) (Sogutlu, 2004). The dimensional shrinkage percentages ($\beta_{//}$) were calculated according to Formula 3.

$$\beta // = \frac{L_{21} - L_{7}}{L_{21}} \times 100 \quad (3)$$

Separate values were calculated for the percentage of shrinkage in the tangential and radial directions (β_t, β_r). The amounts of volumetric shrinkage (β_v) were obtained from the sum of the percentages of shrinkage in the tangential and radial directions according to Formula 4.

$$B_v = \beta_t + \beta_r \quad (4)$$

A total of 24 specimens were utilized for each test to determine the physical properties. Of these, the results of 16 specimens that fulfilled the homogeneity of normality were evaluated.

Analysis of mechanical properties

Test specimens were prepared from the sample lumber in the dimensions of 6×200 cm for testing the mechanical properties of compressive strength, bending strength, tensile strength and shear strength (Table 2).

All the specimens were able to reach the EMC by being conditioned at a temperature of 20±2°C and a relative humidity of 65±5%. Subsequently, the tests were carried out via the tensile compression testing machine. The pace of the test was adjusted so that the specimens would break in 1.5 to 2 min and the strength at the instant of failure (F_{max}) was measured. A total of 20 specimens were utilized for each test to determine the mechanical properties. Of these, the results of 12 specimens that fulfilled the homogeneity of normality were evaluated.

Statistical analysis

Normality analysis was applied to the results obtained and extreme values were attained. Next, the one-sample T-test was applied for each test implementation separately and the statistical significance of the values was evaluated at an interval of confidence of 95%.

RESULTS AND DISCUSSION

According to the analysis results, all tests were found to be statistically significant ($p<0.05$). The analysis results of the tests carried out to determine the physical properties have been given in Table 3 and to determine the mechanical properties have been given in Table 4.

In Table 5, the values obtained in the study were compared with the values obtained in a study carried out by the Central Anatolia Forestry Research Institute of the Ministry of Forestry of the Republic of Turkey (Ilter et al., 2011).

According to the values of Calabrian pine trees grown in the Denizli area given in Table 5, the air-dried and oven-dried density values (0.567 g/cm^3, 0.532 g/cm^3) were higher than the trees in the Kahramanmaras area, but lower than in the Antalya and Samsun areas. However, the differences among the values were not very significant.

According to the swelling and shrinkage percentages, while the trees grown in Denizli area had the lowest values in a tangential direction, they had the highest values in a radial direction and volumetrically. The longitudinal values could not be compared, since they were not analyzed in the research carried out by the Institute. According to these results, if businesses prefer Calabrian pine, then they could utilize the trees grown in the Denizli area by using the tangential directions on large surfaces for outdoor furniture.

According to the mechanical properties, the Calabrian pine grown in Denizli area had the lowest values in bending strength and shear strength in a radial direction. This situation could result from the anatomical structure, especially from the chemical structures of Calabrian pines. According to the literature, properties of wooden material, such as density, temperature, moisture, knots and fiber direction affect its bending strength (Bozkurt and Erdin, 1997). However, the differences in resin and cellulose-lignin percentages are significant on bending strength (Ilter et al., 2011). The chemical components of

Table 2. Tests and types of specimens for determining the mechanical properties.

Tests	Standards	Type of specimen
Parallel compressive strength to the fibers ($\sigma_{B//}$)	TS 2595 (ISO 3797)	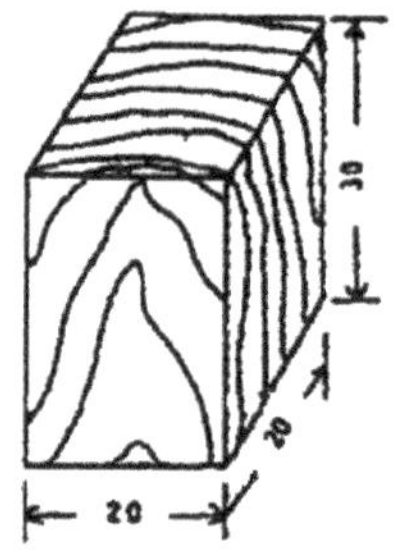
MoR (σ_e) and MoE (E)	TS 2474 (ISO 3133), TS 2478 (ISO 3349)	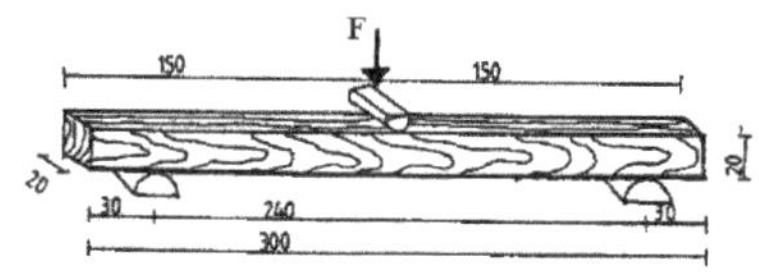
Shear strength in radial direction (σ_m)	TS 3459 (ISO 3347)	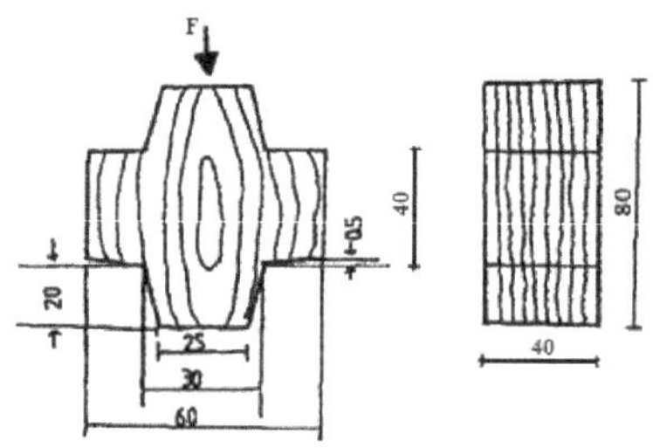
Parallel tensile strength to the fibers ($\sigma_{ç\perp}$)	TS 2475 (ISO 3345)	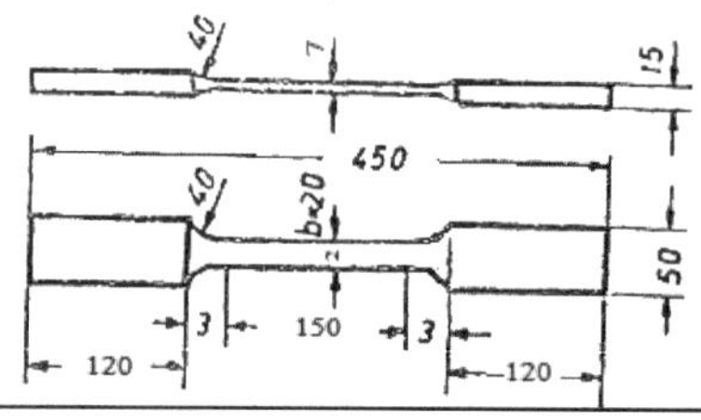

Calabrian pine woods should be analyzed in order to reach a firm conclusion on this issue.

The highest values from the mechanical test results were obtained in parallel compressive strength to the fibers and parallel tensile strength to the fibers. According to this result, the Calabrian pine trees grown in the Denizli area do not have sufficient strength for longitudinal strength to the fibers. If it is preferred as a construction material, especially for building construction and roofs, it should not be preferred for practices in a horizontal direction, such as beams.

Conclusion

In this study, some of the technological properties of Calabrian pine (*P. brutia* Ten.) trees grown in the Denizli area were investigated. Oven-dried and air-dried density values were significant criteria for determining the amount of raw material in its structure (Bozkurt and Erdin, 1997). These values are used to determine the productivity of cellulose in paper production and wood density should be between 0.300 to 0.600 g/cm^3 for commercial wood pulp (Gunduz, 1999). Therefore, the Calabrian pine trees grown in the Denizli area used in this study as the research material were found to be suitable for paper production. As its shrinkage and swelling values were low in a tangential direction but high in a radial direction and volumetrically compared to the other areas, by taking into account the use of direction, Calabrian pine could be preferred for outdoor applications such as arbor, benches and bridges, and for the furniture in damp places, such as bathrooms, saunas and kitchens. In this study, it was found that Calabrian pine trees grown in the Denizli area were less resistant against longitudinal loads to the fibers.

On the other hand, as they obtained better results against loads parallel to the fibers and tensile strength compared to the other areas, the use of Calabrian pine in the right direction should be considered in uses that require resistance to loads, such as prefabricated houses, roofs, wardrobe closets, bookcases and

Table 3. Descriptive statistics for the physical properties.

Properties		Sample size (N)	Arithmetic mean ($\overline{x}$)	Standard deviation (SD)	Variance (V)	Coefficient of variation (CV)	Min. value	Max. value
Air-dried density δ_{12} (g/cm^3)		16	0.567	0.032	0.001	5.643	0.520	0.630
Oven-dried density δ_0 (g/cm^3)		16	0.532	0.033	0.001	6.203	0.490	0.600
Swelling α (%)	α_t	16	7.813	1.059	1.122	13.554	6.330	9.510
	α_r	16	8.866	0.421	0.177	4.748	8.100	9.640
	α_v	16	16.679	0.888	0.788	5.324	15.210	17.910
	α_l	16	0.659	0.175	0.031	26.555	0.410	1.00
Shrinkage β (%)	β_t	16	6.753	0.859	0.738	12.720	5.300	8.65
	β_r	16	6.924	0.475	0.226	6.860	6.180	7.720
	β_v	16	13.677	0.963	0.928	7.041	12.240	15.630
	β_l	16	0.681	0.131	0.017	19.236	0.530	1.040

T, Tangential,; r, Radial; v, Volumetric; *l*, Longitudinal.

Table 4. Descriptive statistics for the mechanical properties.

Properties	N	$\overline{X}$	SD	V	CV	Min	Max
Modulus of elasticity in bending (MoE) (*E*) (N/mm^2)	12	9650.755	577.059	332997.583	5.979	8728.960	10743.720
Bending strength (MoR) (σe) (N/mm^2)	12	95.893	5.516	30.431	5.752	87.110	105.260
Shear strength in radial direction (σ_m) (N/mm^2)	12	6.613	0.525	0.275	7.939	5.623	7.320
Parallel compressive strength to the fibers ($\sigma_{B//}$) (N/mm^2)	12	60.674	3.297	10.872	4.036	56.640	65.570
Parallel tensile strength to the fibers ($\sigma_{ç\perp}$) (N/mm^2)	12	81.697	15.401	237.131	18.851	61.460	113.520

Table 5. Comparison of the properties of Calabrian pine grown in other areas.

Technological properties		Area (City)			
		Denizli	Antalya	Samsun	Kahramanmaraş
Physical properties	δ_{12} (g/cm^3)	0.567	0.580	0.588	0.530
	δ_0 (g/cm^3)	0.532	0.541	0.558	0.500
	α_t (%)	7.813	7.650	8.670	8.120
	α_r (%)	8.866	6.330	6.430	5.590
	α_v (%)	16.679	14.390	15.630	14.170
	α_l (%)	0.659	-	-	-
	β_t (%)	6.753	7.390	7.690	7.720
	β_r (%)	6.924	6.310	6.270	5.230
	β_v (%)	13.677	12.400	12.590	11.750
	β_l (%)	0.681	-	-	-
Mechanical properties	σe (N/mm^2)	95.893	112.306	102.311	99.9740
	σm (N/mm^2)	6.613	9.274	6.831	7.174
	$\sigma_{B//}$ (N/mm^2)	60.674	58.813	53.588	50.607
	$\sigma_{ç\perp}$ (N/mm^2)	81.697	78.103	72.430	77.203

transport boxes. In conclusion, the Calabrian pine (*P. brutia* Ten.) trees grown in the Denizli area have the technological values that could compete with trees grown in other areas.

ACKNOWLEDGEMENT

This study was supported by the Pamukkale University, Scientific Research Project's Unit with Project Code No. 2011BSP008.

REFERENCES

Bektas I, Alma MH, As N, Gundogan R (2003). Relationship between site index and several mechanical properties of Turkish Calabrian pine (*Pinus brutia Ten.*). For. Prod. J. 53(2):27-31.

Berkel A (1957). Kızılçam (*Pinus brutia*) da teknolojik araştırmalar (Technological research studies on red pine {*Pinus brutia*}). İstanbul Üniversitesi Orman Fakültesi Dergisi A. (in Turkish) 7(1):22-68.

Bozkurt Y, Erdin N (1997). Ağaç teknolojisi (Tree technology). İstanbul: İstanbul_Üniversitesi Basımevi, Üniversite Yayın No. 3993, Orman Fakültesi Yayın (in Turkish) 445:372.

Bozkurt Y, Goker Y, Erdin N, As N (1993). Datça kızılçamında anatomik ve teknolojik özellikler (Anatomical and technological properties of red pine in Datça). In: Uluslararası Kızılçam Sempozyumu Bildiriler Kitabı (Proceedings of the International Symposium on Red Pine), 628-635. Ankara: Türkiye Cumhuriyeti, Orman Bakanlığı (in Turkish).

Dogdu YC (2006). Camiyanı karaçamı'nın (*Pinus nigra* Arn. subsp. *pallasiana* var. *pallasiana*) bazı teknolojik özellikleri ve kurutma cetvellerinin oluşturulması (Some technological properties of *Pinus nigra* Arn. subsp. *pallasiana* var. *pallasiana* and determining the kiln drying schedules). Master's thesis, Zonguldak Karaelmas University, Graduate School of Natural and Applied Sciences, Department of Forestry Industrial Engineering, Zonguldak, Turkey (in Turkish).

Gunduz G (1999). Camiyanı karaçamı'nın (*Pinus nigra* Arn. subsp. *pallasiana* var. *pallasiana*) bazı anatomik, teknolojik ve kimyasal özellikleri (Anatomic, technological and chemical properties of Camiyanı black pine {*Pinus nigra* Arn. subsp. *pallasiana* var. *pallasiana*}). PhD dissertation, Zonguldak Karaelmas University, Institute of Science, Bartin, Turkey (in Turkish).

Ilter E, Saracbasi A, Balkiz OD (2011). Bazı kızılçam (*Pinus brutia* Ten.) orijinlerinin teknolojik özelliklerinin belirlenmesi (Determination of technological properties of some red pine {*Pinus brutia* Ten.} origins). Ankara: İç Anadolu Ormancılık Araştırma Enstitüsü Yayınları, Teknik Bülteni No. 291 (in Turkish).

Oktem E, Sozen R (1996). Reçine üretiminin kızılçam (*Pinus brutia* Ten.) odununun fiziksel ve mekaniksel özellikleri üzerine etkisi (Effects of resin production on the physical and mechanical properties of Calabrian cluster pine {*Pinus brutia Ten.*} wood). Ankara: İç Anadolu Ormancılık Araştırma Enstitüsü Yayınları, Teknik Bülten Serisi No. 256 (in Turkish).

Sogutlu C (2004). Bazı yerli ağaç türlerinin kündekârı yapımında kullanım imkânları (Feasibility of using some local wood species in making kundekari {woodwork, cabinet work}). PhD dissertation, Gazi University, Institu. Sci., Ankara, Turkey (in Turkish).

Yildiz UC (1993). Kızılçam odunundan elde edilen odun-polimer kompozitlerinin bazı teknolojik özellikleri (Some technological properties of wood-polymer composites obtained from red pine (*Pinus brutia* Ten.). In: Uluslararası Kızılçam Sempozyumu Bildiriler Kitabı (Proceedings of the International Symposium on Red Pine). Ankara: Türkiye Cumhuriyeti, Orman Bakanlığı (in Turkish).

A preliminary investigation of the signal-to-noise ratio of Toro and Nsukka stations in Nigeria

Kadiri Umar Afegbua[1] and F. O. Ezomo[2]

[1]Department of Earthquake Seismology, Centre for Geodesy and Geodynamics, P. M. B. 11, Toro, Bauchi State, Nigeria.
[2]Department of Physics, Faculty of Physical Sciences, University of Benin Ugbowo, Benin City, Edo State, Nigeria.

Preliminary investigation of Toro and Nsukka stations from the Nigerian National Network of Seismographic Stations (NNNSS) with respect to Signal-to-Noise-Ratios (SNR) has been performed from noise observation and numerically. The results showed that Toro has higher SNR compared to Nsukka. Six months data collected from the stations in 2012 were used for the study. Nsukka Station in the southern part of Nigeria exhibited high noise than its Toro counterpart that is located in the North. This study is intended to investigate the causes of differing SNR at as observed at Toro and Nsukka stations respectively, with a view to adopting measures to forestalling SNR compromise at future stations in Nigeria.

Key words: Seismic noise, Nigerian national network of seismographic stations (NNNSS), signal-to-noise-ratio, measures for signal-to-noise-ratios (SNR) improvement.

INTRODUCTION

One of the primary reasons for siting seismographic stations anywhere in the world is to provide a platform for the monitoring of seismic activities (in the case of regions with low seismicity like Nigeria, United Kingdom, Canada etc); or for routine earthquakes' recording (for active regions like Japan, Indonesia, Turkey, East Afriacn Rift system etc) (Bormann, 1998). By the time one has successfully established seismic stations or network of seismic stations and generating data from them, the next consideration is the appraisal of the quality of data being generated from the stations (Trnkoczy et al., 2002a).

When a seismic trace is completely masked by seismic noise, the signal to noise ratio is in this case, seriously impaired. On the other hand, a station which has a high signal to noise ratio is considered a good station (Bormann et al., 2002). The paramount interest of a seismologist is to obtain data with less noise for research purposes. Therefore, one of the main issues in today's applied seismology is to ensure high signal-to-noise-ratios (SNR) by suitable ways of data acquisition and processing. However, the success of SNR improvement largely depends on our understanding of the ways in which seismic signals and noise differ (Bormann, 2002).

Recorded seismic signals always contain noise and it is important to be aware of both the source of the noise and how to measure it (Havskov and Ottemoller, 2008). Basically, noise observable on seismic trace is either noise generated by the installed seismic equipment or seismic noise from earth vibrations (ambient seismic noise). Sources of ambient noise include tides, water waves striking the coast, turbulent wind, effects of wind on trees and buildings, traffic or human based noise (Boonefoy-Claudet, 2006). Normally, the instrument noise is well below the seismic noise that it is often ignored in most cases (Alguacil and Havskov, 2002). Other sources of seismic noise include running water, surf and volcanic tremor (an almost harmonic noise associated to fluids motion, often lasting hours or days) or background activity as local sources of seismic noise (Alguacil and Havskov, 2002).

Ambient seismic noise can either be cultural or natural. The ambient seismic noise represents the microtremors and microseisms respectively. Cultural noise shows strong diurnal variations and it has characteristic frequencies depending on the distribution of the noise source (McNamara and Buland, 2004).

Installation procedures of seismometers contribute significantly to presence of noise from instrumentation. If seismometers are not properly installed it could lead to generation of self-noise or noise from external influences that cause high noise especially on the horizontal component of the seismometer (Wielandt and Forbriger, 1999). So in the process of selecting a site for permanent seismic stations, the issues of the type of infrastructure to be installed and how to achieve low seismic noise at the site are seriously considered. The noise level depends on the geological situation and on the proximity of sources, some of which are usually associated with the infrastructure. It is expected that a seismograph installed on solid basement rock will fairly insensitive to local disturbances while one sitting on a thick layer of soft sediments will be noisy even in the absence of identifiable sources (Wielandt, 2002). Since the inherent noise is the major determinant of the nature of SNR, avoidance of practices that introduce noise into recorded signal trace must be avoided. This work is aimed at comparing the SNR at both Toro and Nsukka stations from noise observation and computation, infer possible factors that improve or impair data quality and make useful suggestions on how to select a better site where SNR will be enhanced.

NIGERIAN national network ofseismographic stations

The location of the Toro and Nsukka stations are shown in (Figure 1). The installed equipment and geologic foundation at each site are shown in Table 1. The sensor at Nsukka is placed inside a vault of about 6 m deep in a University community with surrounding residential buildings, while that at Toro station is on the surface of a basement in a quiet environment apart from vehicular movements (approximately 30 to 200 m distant) and surrounding few houses and trees. Geographically, the stations at the southern part of Nigeria are located where there are dense vegetations, streams, rivers, and several kilometers away from the Atlantic Ocean, while stations sited in the north are on the basement, no streams, or close by rivers and very far away from the Atlantic Ocean. Toro and Nsukka stations are installed with Entec DR4000 Digitizer with 24-bit high resolution, 132 dB dynamic range, analog and digital alias filtering. The equipment are programmed to operate in both trigger and continuous mode simultaneously (Tables 1 and 2). According to the instrument's manual, the EETEC EP-105 seismometer is based on electrochemical transducer technology. The sensor has a dynamic range of 142 dB and bandwidth of 0.033 to 50 Hz

Abridged geology of Nigeria

Geology plays a significant role when siting a seismic station. So also observance of noise varies from site to site. Apart from being located within the intraplate area, Nigeria's land mass is made of Precambrian to Early Paleozoic crystalline basement rocks, about half of which is covered by sedimentary rocks of Cretaceous to recent age (Osazua, public presentation in Abuja, 2008). About two-thirds of the country's landmass is underlain by the Pre-cambrian basement complex consisting of gneisses, migmatites, schist, and various metamorphic rocks and granites (Figure 2).

These are in places intruded and interspersed by the "Older granites" which originated in the Pan-Africa Orogeny (Olujide and Udoh, 1989). Basement complex rocks outcrop in four main areas of the country: North of Rivers Niger and Benue, covering parts of Kaduna, Plateau, Bauchi, Kano and Sokoto States; southern Nigeria, covering the greater parts of Kwara, Oyo, Ogun; and Ondo States; southeast Nigeria, spanning the northern parts of Cross Rivers State and as far north as Yola; and north of Benue River in Gongola State (Eze et al., 2011). The Basement rocks are overlain by Cretaceous and Tertiary sediments of the seven major sedimentary basins, viz, the Calabar Flank, the Benue Trough, the Chad Basin, lullemmenden (Sokoto) Basin, the Dahomey Basin, and the Niger Delta Basin. Sedimentary successions in these basins are of middle Mesozoic to recent age (Kogbe, 1989). In some cases, the Cretaceous sediments are cut by some major faults which may have been the result of the reactivation of post Pan-African fractures (Merki, 1970) (Table 3).

MEASURES FOR IMPROVING SNR

A high SNR translates to a good performing station. However, if the SNR is low, which of course will compromise the quality of the data, one of the various measures for SNR improvement abound. According to Bormann (1998) and Bormann (2002), these are: frequency filtering or band-pass filtering; velocity filtering and beam forming that frequency filtering cannot achieve; noise prediction-error filtering to determine the characteristics of a given noise field by means of cross- and auto-correlation of array sensor outputs; noise polarization filtering of 3-component recordings which allows one to reconstruct the ground particle motion and to determine its polarization (Shimshoni and Smith, 1964); SNR improvement by recordings in subsurface mines and boreholes and Signal variations due to local site conditions (Bormann, 2002). It has been demonstrated how short-period seismic noise is strongly reduced with the depth of sensor installation in boreholes or mines

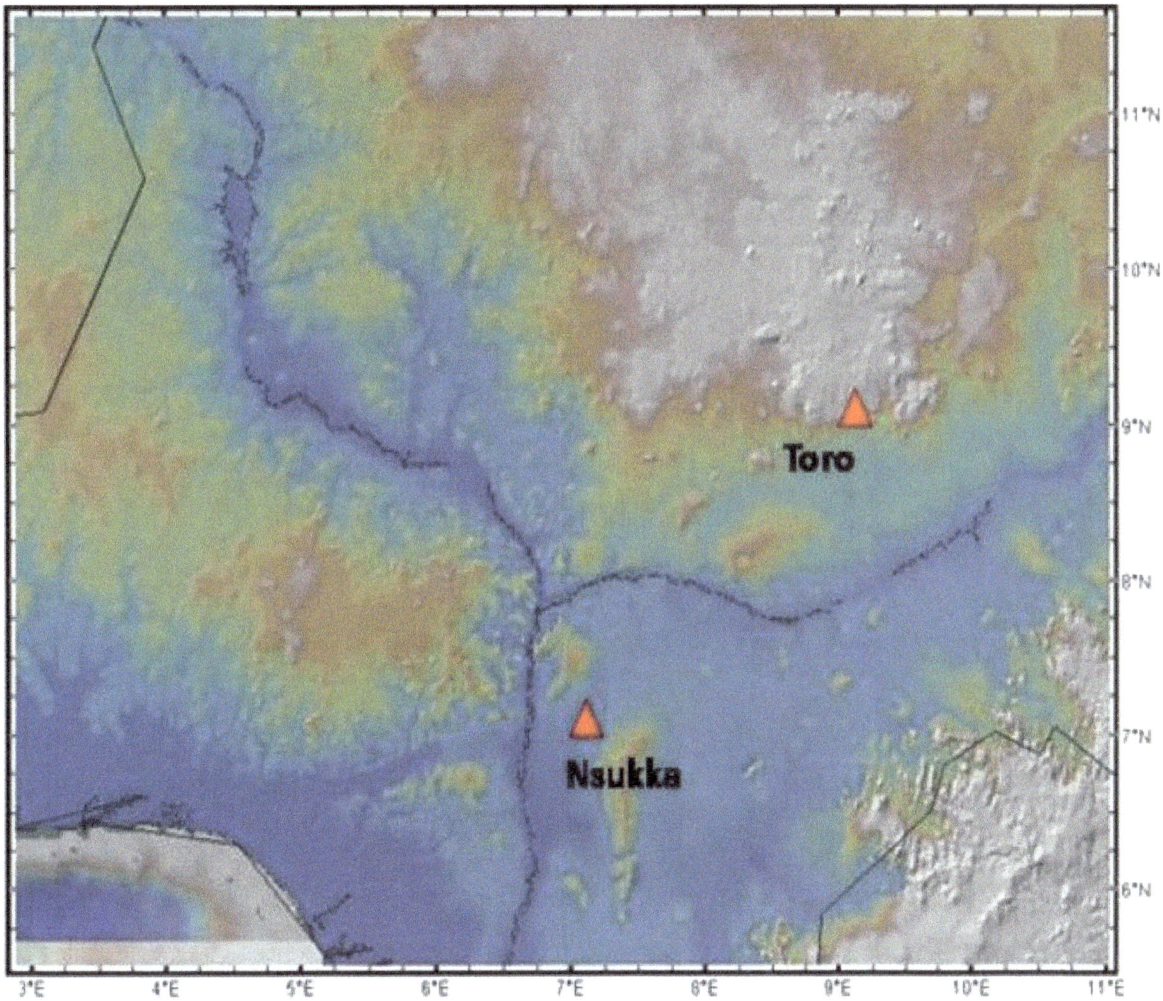

Figure 1. Map of Nigeria showing the locations of Toro and Nsukka stations (courtesy, GeoMapApp).

Table 1. Location of centre for geodesy and geodynamics seismological stations.

S/N	Station name	Latitude	Longitude	Elevation (m)	Geologic foundation	Instrumentation
1	Nsukka (NSU)	06°52.011'N	07°25.045'E	430	Sandstone	Seismograph: DR4000 recorder Seismometer: EP105 broadband seismometer
2	Toro (TOR)	10°03.303N	09°07.089'E	882	Gneiss	Seismograph: DR4000 recorder Seismometer: EP105 broadband seismometer

Table 2. Properties of Toro and Nsukka Stations.

Station	Free period (s)	Damping rate	Generator constant (V/m/s)	Digitizer sensitivity	Sampling rate	Amplifier gain
Nsukka	30	0.7	2000	"	"	"
Toro	60	0.7	2000	"	"	"

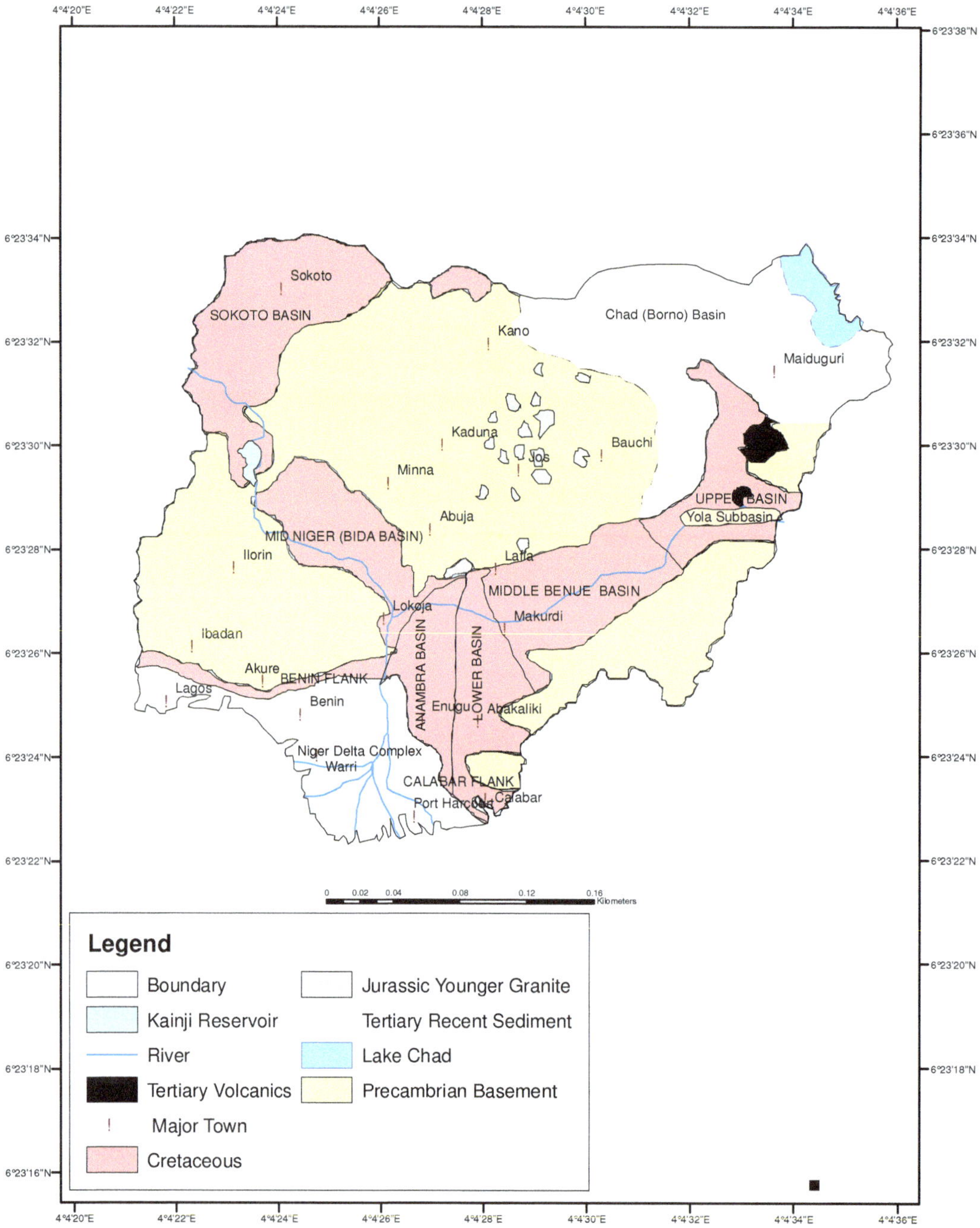

Figure 2. Map of Nigeria showing geology of the country.

Table 3. Classification of different types of outcropping geological formations in quality categories.

S/N	Different types of outcropping geological formations in quality categories
1	Unconsolidated (alluvial)sediments (clay, sands, mud)
2	Consolidated clastic sediment (sandstone, marls)
3	Less compact carbontic rocks, less compact metamorphic rocks, conglomarates, breccias
4	Compact metamorphic rock and carbonic rock
5	Magmatic rocks (granite, basalts)

Grade 5 is the best rock for seismic recordings and grade 1 is the worst according to Vidrih (2001).

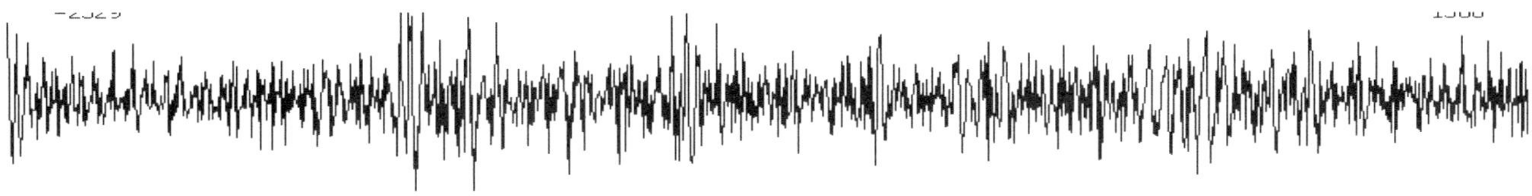

Figure 3. Unfiltered trace from Nsukka Station, very difficult to identify the phases.

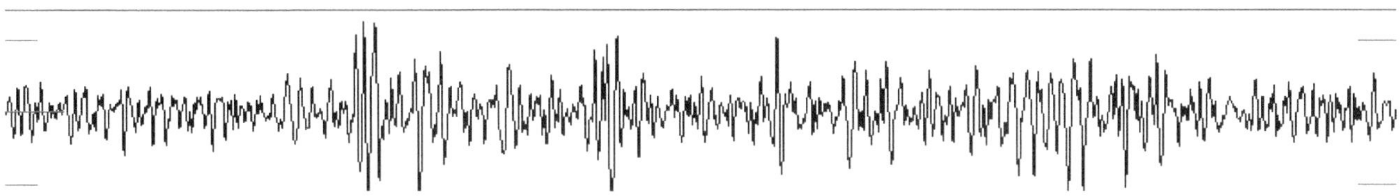

Figure 4. Filtered trace from Nsukka (using 0.01 to 0.1Hz filter band), the signal-to-noise ratio is much improved and later phases are clearly recognizable since the microseisms have been removed by filtering (courtesy: Havscov and Ottemoller, 2008).

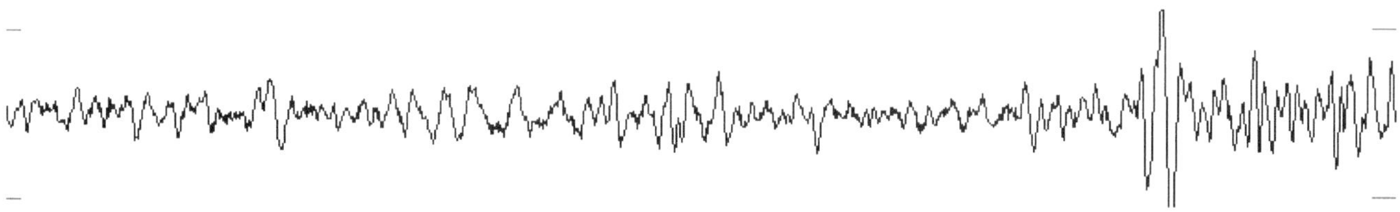

Figure 5. Unfiltered trace from Toro station from 2012-04-11-010-09M.NSN_003.

(NMSOP Chapter 4). However, when installing seismometers at depth, one must also consider effects on the signal, since SNR does not necessarily increase steadily with. This is due to the free-surface effect, peculiarities of the local noise field and geological conditions (Bormann, 2002). Also, while one station of a network may record events rather weakly from a certain source area, the station may do as well as other stations (or even better) for events from another region, azimuth or distance (Bormann, 2002) (Figures 3 to 6).

RESULTS AND DISCUSSION

There are various ways signal-to-noise ratio can be determined. Each method depends on the kind of signal of interest. Since the work is concerned with seismic signal, we can conveniently consider the following equation for SNR computation for both stations (Free encyclopedia):

$$\mathrm{SNR_{dB}} = 10\log_{10}\left(\frac{A_{\mathrm{signal}}}{A_{\mathrm{noise}}}\right)^2 = 20\log_{10}\left(\frac{A_{\mathrm{signal}}}{A_{\mathrm{noise}}}\right).$$

A_{signal} and A_{noise} are the maximum amplitude of the signal and noise of the trace respectively (Russ, 2001; Gonzalez and Woods, 2008). The (log) of the result was finally multiplied by a factor of 10 or 20, depending on the one you are using, to obtain the SRN in logarithmic decibel scale (dB).

If you are measuring sound waves, for example, it is important to know how background noise might interfere with your signal. Understanding the signal-to-noise ratio can give you a better idea about the signal you are interested in. In this study, the above equation was used to compute the maximum amplitude ratio using SEISAN open source seismological software (Havscov and Ottemoller, 1999). The signal to noise ratio of Toro station was found to be 38.7 compared to that of Nsukka station that is 9.8. Other methods can of course, be applied in the future to compare with these results.

Meanwhile, we may not completely right off Nsukka station as a bad station since from standard consideration, a good station has SNR>3. But since the SNR of Toro is almost four times greater than that of Nsukka, it is imperative to take measures that would help to improve the SNR of Nsukka station for better data quality. From Table 1, it is clear that Toro and Nsukka stations are respectively located on basement and sediment respectively, which could be one of the reasons for the sharp SNR differences. SNR is important for magnitude determination. If S:N > 3 the station is considered good otherwise there is a problem.

Figures 7 to 12 were constructed using noise estimation open source software, PQLII, with RAW data obtained from Toro and Nsukka stations at same period in 2012. For space and other considerations, all the figures

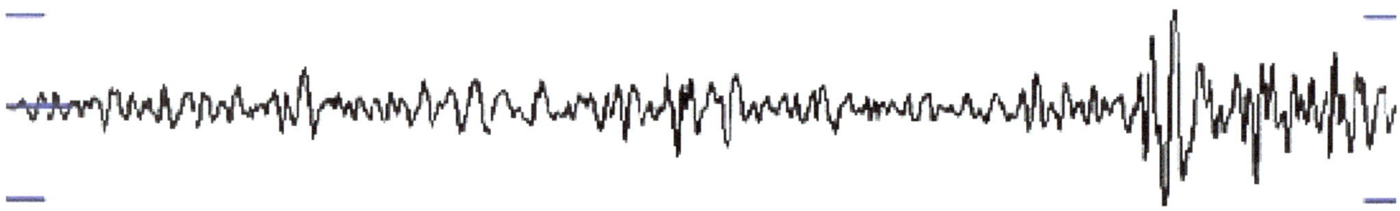

Figure 6. Filtered trace using 0.01 to 0.1Hz band-pass of the Toro station. Good SNR at this station does not give room for a distinct difference between Figures 4 and 5, unlike as observed in Figures 2 and 3.

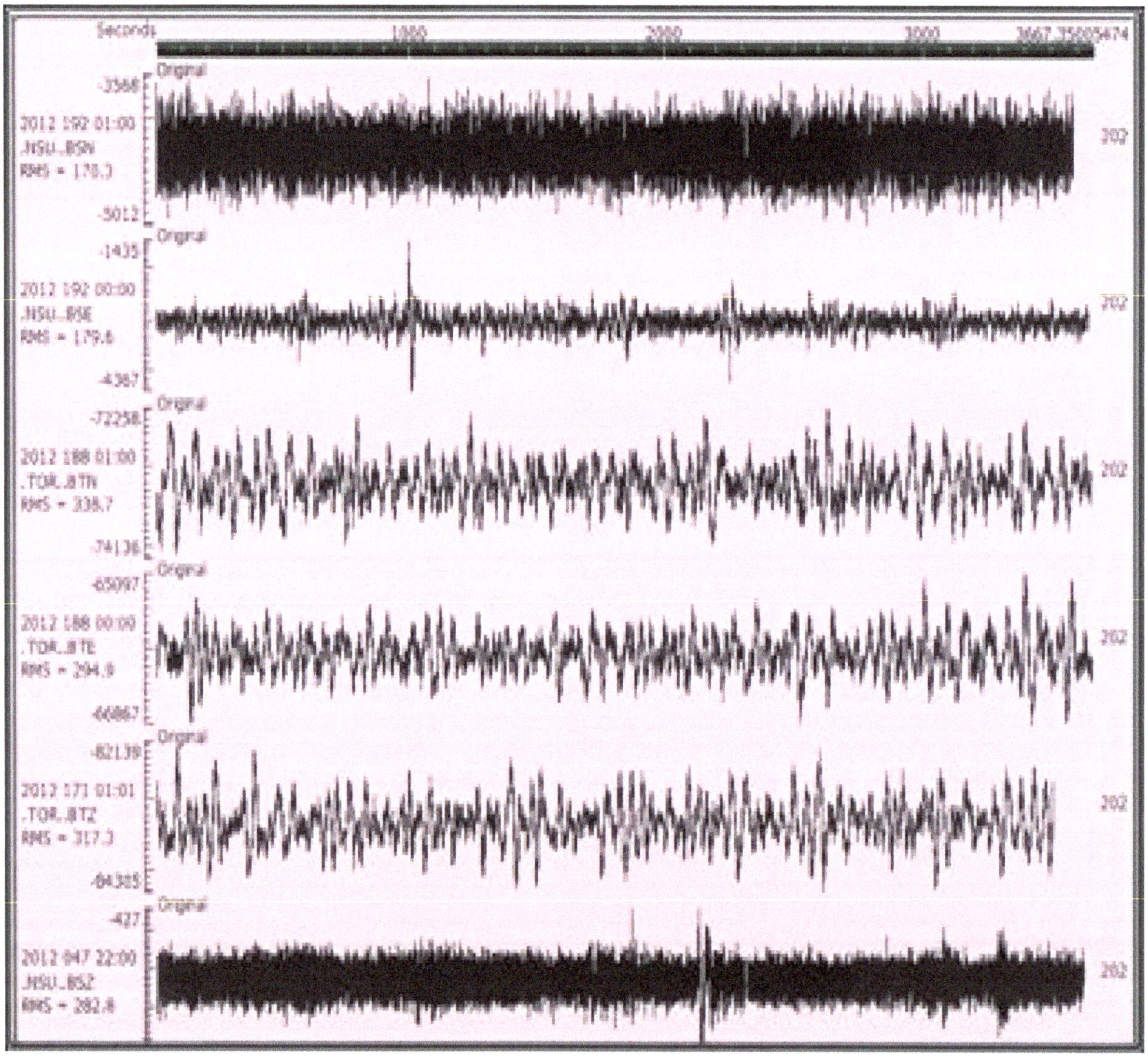

Figure 7. Traces of Z, N, E, components from Toro and Nsukka stations. High noise levels are clearly observed on the components of Nsukka station. If for instance there is an event at the time the seismic traces from Nsukka in this figure were recorded, the event will be completely shrouded by noise, thereby compromising quality of the signal. This is not good for signal to noise ratio of a station.

derived from the noise analysis for six-month period are not presented here. However, Figures 7 to 12 summarily compared the noise levels from both stations. Figures 9 to 12 clearly demonstrated the dominance of

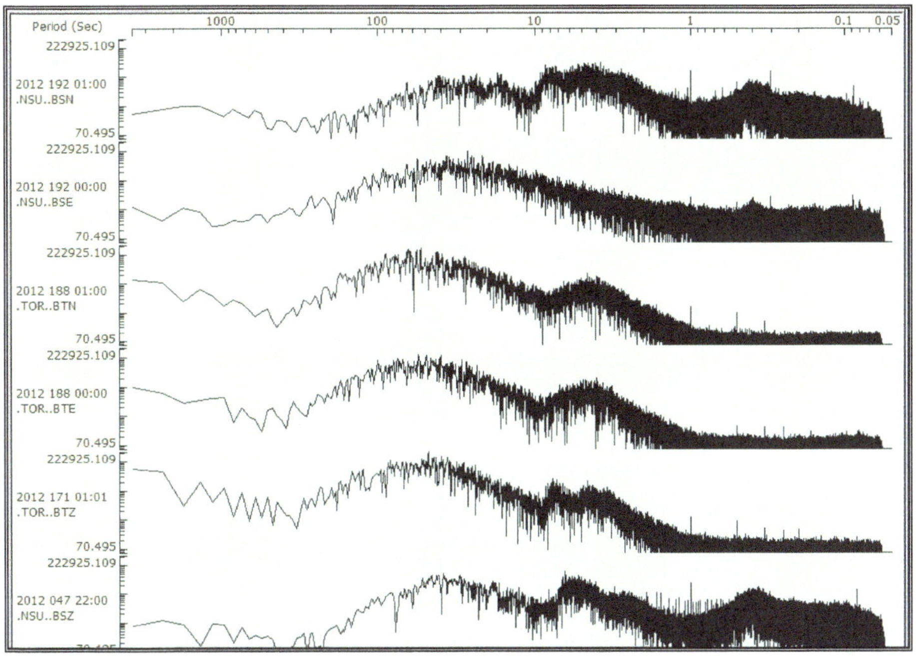

Figure 8. Spectral figures from the three components of Toro and Nsukka stations. The spectra from Nsukka station (first, second and sixth in the figure; especially the first and sixth spectra) exhibited high noise at high frequencies compared to the spectra of Toro station (third, fourth and fifth spectra in the figure). It is likely as a result of the contribution of anthropogenic noise from surrounding human activities and natural noise sources like wind.

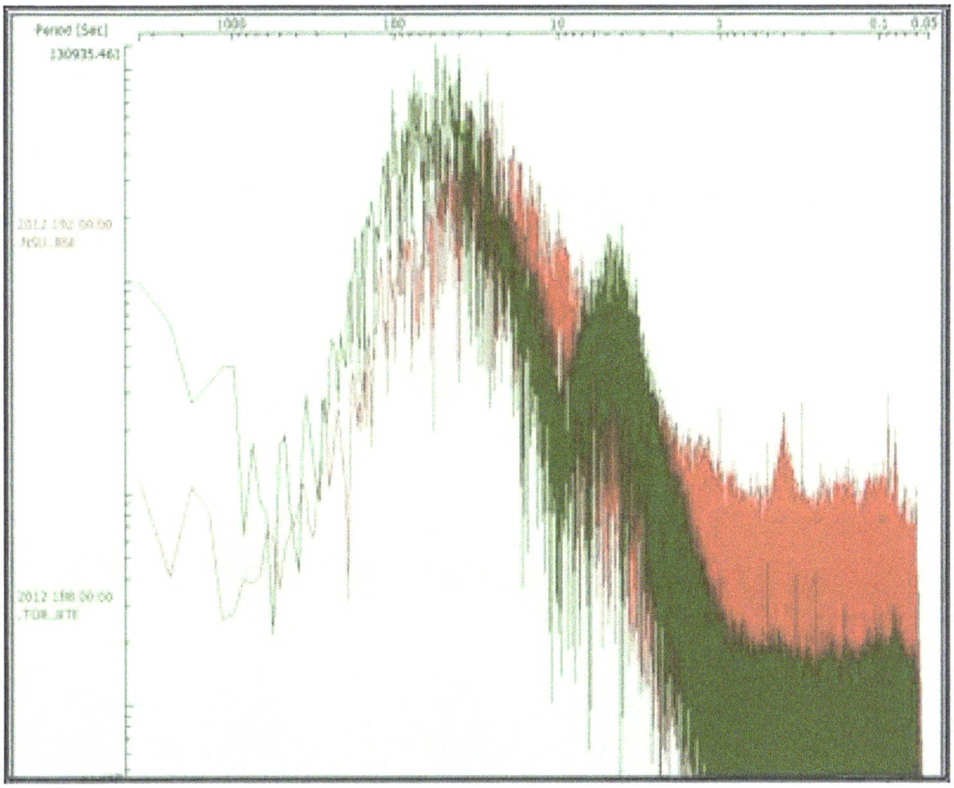

Figure 9. Overly of the Horizontal component of Toro and Nsukka showing higher noise on the Nsukka component (pink colour). The green colour represents trace from Toro.

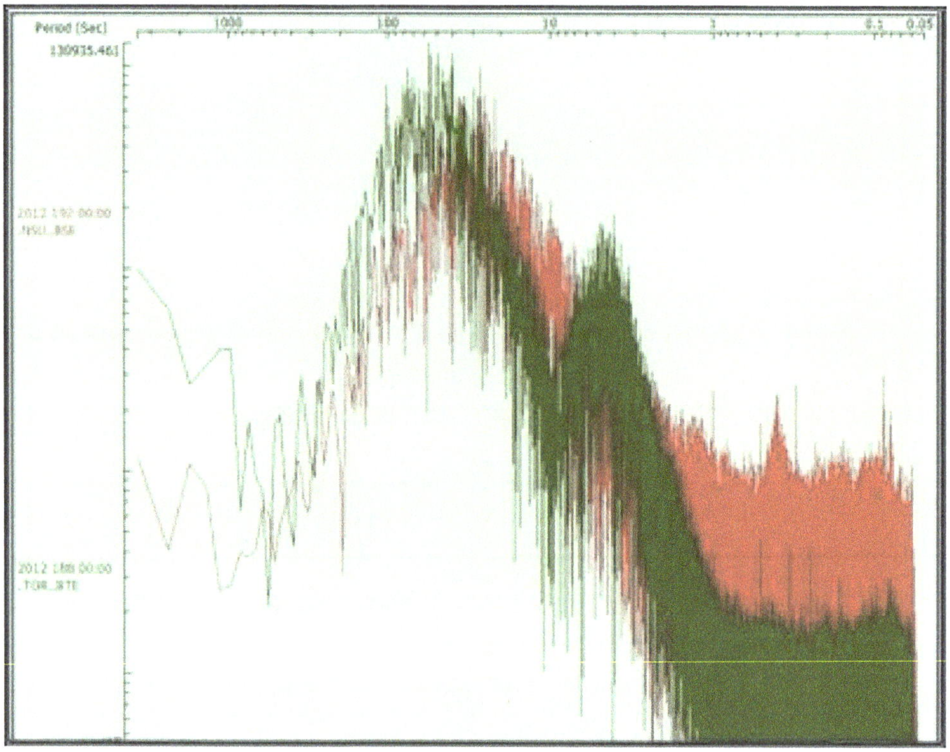

Figure 10. Overly of the vertical (Z) component of Toro and Nsukka showing higher noise on the Nsukka component (green colour). The red colour represents trace from Toro.

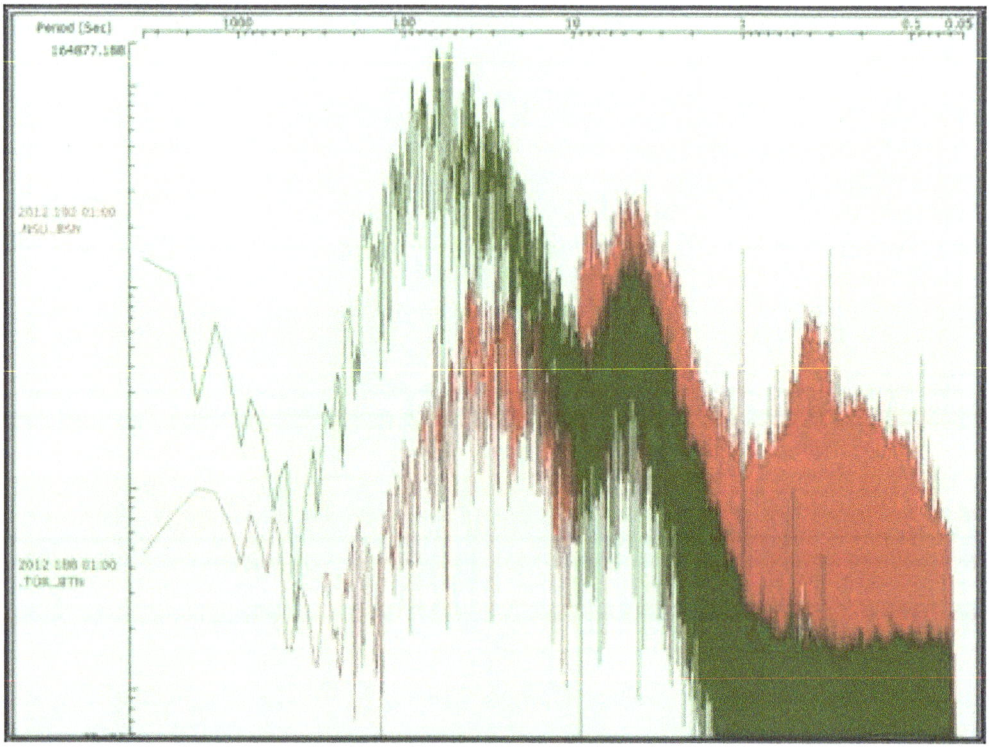

Figure 11. Overly of the N component of Toro and Nsukka showing higher noise on the Nsukka component (pink colour). The Green colour represents trace from Toro.

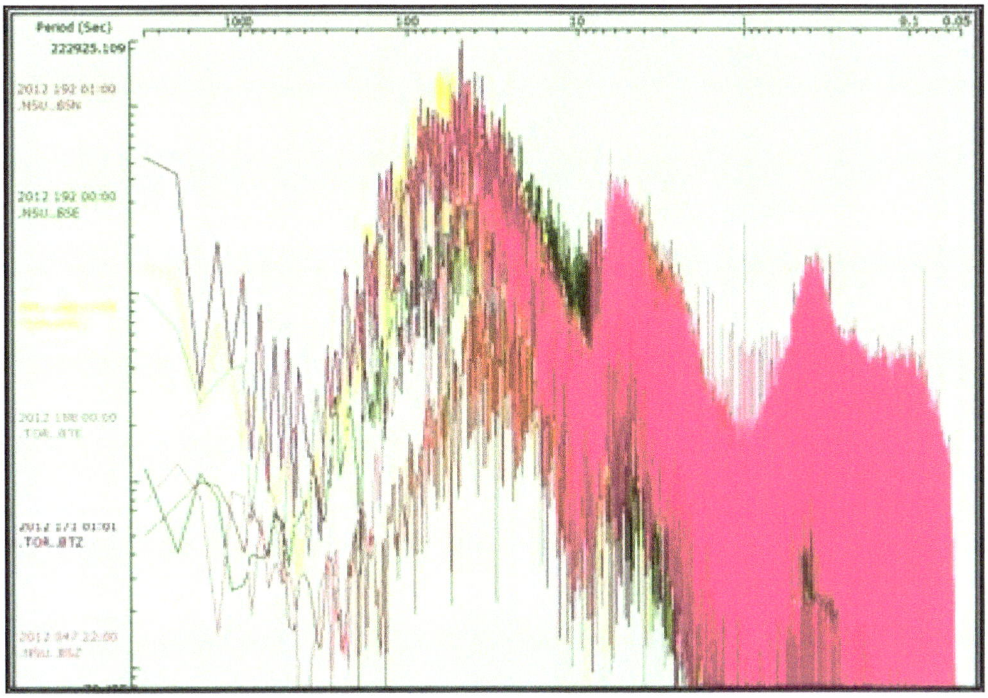

Figure 12. Overly of the three components of Toro with three components of Nsukka showing a remarkable higher noise on the Nsukka component (pink, green and red colours). The yellow, light pink and black colours traces represent Toro station, which are almost covered by the dominance noise from Nsukka.

noise on the Z, E and N components of Nsukka over Toro, using data from the stations and collected at the same time.

Conclusion

The performance of the two Nigerian Seismographic Stations: Toro from the northern part of Nigeria and Nsukka in the south had been compared. The pattern of noise as observed from the stations within a six month (January-July, 2012) revealed a consistent high noise on Nsukka data while data from Toro exhibited a relatively low noise. The low signal to noise ratio as observed at the Nsukka is likely due to the presence of sediments where it is sited, while that of Toro sited on hard rock has high SNR. Another reason that may be responsible for the noise at Nsukka cum low SNR, could be as a result of the low-period sensor (16 s) installed there. Improving data quality translates to improving site quality and this could be achieved at Nsukka if a better site is selected and/or increasing the depth of the vault to a reasonable depth to the bedrock and installing high-period sensor of say, 100 or 120 s (Trnkoczy et al., 2002b). However, the increase in the depth of the vault should not compromise the signal from the station as this would create another problem. This study was carried out within the limit of available tools and techniques. However, there are several ways or methods that could be adopted to check the performance of stations which include the use of Pascal Quick Look Extended (PQLX), Quack, etc.

REFERENCES

Alguacil G, Havskov J (2002). Instrumentation in Earthquake Seismology. pp. 77-85.

Boonefoy-Claudet C (2006). Use of ambient noise for microzonation studies in urban environment; 4th International conference on Earthquake Geotechnical Engineering. June 25-28, 2007, 1580.

Bormann P (1998). Conversion and comparability of data presentation on seismic background noise. J. Seismol. 2:1.

Bormann P, Baumbach M, Bock G, Grosser H, Choy GL, Boatwright J (2002). Seismic Sources and Source Parameters. New Man. Seismol. Observ. Pract. 1:119-124.

Bormann P (2002). Seismic Signal and Noise. New Man. Seismol. Observ. Pract. 1:213-236.

Eze CL, Sunday VN, Ugwu SA, Uko ED, Ngah SA (2011). Mechanical Model for Nigerian Intraplate Earth Tremors. Disaster Management, Earth Observation. http://www.earthzine.org/2011/05/17/mechanical-model-for-nigerian-intraplate-earth-tremors/.

Gonzalez CR, Woods RE (2008). Digital Image Processing. Prentice Hall. 354.

Havskov J, Ottemoller L (2008). Processing Earthquake Data. pp. 64-68.

Havscov J, Ottemöller L (1999). SEISAN earthquake analysis software. Seis. Res. Lett. 70:532-534.

Kogbe CA (1989). Geology of Nigeria. Rockview (Nigeria) Limited. 2:538.

McNamara D, Buland RP (2004). Ambient Noise levels in the continental United States. Bull. Seismol. Soc. Am. 94(40):1517-1527.
Merki PJ (1970). Structural geology of the Cenozoic Niger Delta. African Geology, University of Ibadan Press. pp. 251-268.
Olujide PO, Udoh AN (1989). Preliminary comments on the fracture systems of Nigeria. In Proceedings of the National Seminar on earthquakes in Nigeria (Ed.) Ajakaiye D. E; Ojo S. B. and Daniyan M. A. pp. 97-109.
Russ JC (2001). The Image Processing handbook. CRC Press. ISBN0:8493-7254-2.
Shimshoni M, Smith SW (1964). Seismic signal enhancement with three-component detectors. Geophysic 29(5):664-671.
Trnkoczy A, Bormann P, Hanka W, Holcomb LG, Nigbor RL (2002b). Site selection, preparation and installation of Seismic Stations. New Manual of Seismological Observatory Practice. 1:313-374.
Trnkoczy A, Havskov J, Ottemoller L (2002a). Seismic Networks. New Man. Seismol. Observ. Pract. 1:421-427.
Vidrih R (2001). Seismogeological effects on rocks during the 12 April 1998 upper Soca Territory earthquake (NW Slovenia). Tectonophysics 330(3-4):153-175.
Wielandt E, Forbriger T (1999). Near-field seismic displacement and tilt associated with the explosive activity at Stromboli. Annali Geofisica, 42:407-416.
Wielandt E (2002). Seismic Sensors and their calibration. New Man. Seismol. Observ. Pract. 1:274-275.

Heavy metals concentration levels in selected arable agricultural soils in South Western Nigeria

Olatunde S. Olatunji[1] , Beatrice O. Opeolu[2], Olalekan S. Fatoki[1] and Bhekumusa J. Ximba[1]

[1]Department of Chemistry, Faculty of Applied Sciences, Cape Peninsula University of Technology, Cape Town, South Africa.
[2]Department of Environmental and Occupational Studies, Faculty of Applied Sciences, Cape Peninsula University of Technology, Cape Town, South Africa.

The application of micronutrient supplements such as fertilizer, animal manure and sewage sludge in bid to enhance soil productivity and crop yield, poses the risk of heavy metals elevation and nutrients depletion in arable lands. Arable soils were collected from farmlands in use for the cultivation of food produce in selected farming communities in South Western Nigeria. The soils were air-dried, sieved to give <2 mm grain size and digested according to standard methods. The concentrations of Lead (Pb), Cadmium (Cd), Zinc (Zn) and Copper (Cu) were measured using flame atomic absorption spectrophotometer (FAAS). The concentration of Zn ranged from 12.98 to 38.94 µg/g, and was the highest amongst the measured heavy metals. This was followed by Cu, 6.17 to 20.87 µg/g and then Pb, 0.79 to 8.35 µg/g, while the concentration of Cd, 0.26 to 5.75 µg/g was the least. Geo-statistical analysis on the concentrations of the measured heavy metals revealed a significant ($p < 0.05$) spatial correlation (γ) with the soil pH and organic carbon (OC) (Pb, $\gamma = 0.74$; Cd, $\gamma = 0.67$; Zn $\gamma = 0.57$; Cu, $\gamma = 0.52$) and (Cd, $\gamma = 0.79$; Zn $\gamma = 0.59$; Cu $\gamma = 0.51$), respectively. The OC levels predispose soils to the retention of soil heavy metals and micro/macro-nutrients depending on the nature of OC-metal interaction. The concentrations of the heavy metals were below the European Economic Community (EEC) maximum allowable concentrations (MACs) in agricultural soils, except for Cd. About 56% of the farmland soils contained more than 3 mg/kg of Cd, while about 44% were within the 1 to 3 µg/g EEC MAC for Cd in agricultural soils. Although, the concentration levels of the measured heavy metals in the soils do not appear to be of serious concern, the study results provides evidence of gradual accumulation of Cd in farmlands

Key words: Heavy metals, arable agriculture, farmland, micronutrient supplements, concentration.

INTRODUCTION

Supplementing soil with essential minerals, nutrients and organics fluxes in arable agricultural land is on the increase in a bid to enhance soil productivity and crop yield (Drechsel and Gyiele, 1999; Akinrinde, 2006). This is because nutrient regeneration processes such as weathering of soil minerals, nitrogen fixation, and atmospheric deposition in rain and dust are unable to meet up with the rate of nutrient depletion. Although trace elements and heavy metals occur naturally in all agricultural soils (Alloway, 1990; IPNI, 2008), elevated levels of Lead (Pb), Arsenic (As), Copper (Cu), Cadmium (Cd) and Zinc (Zn) have been reported (Chang and Page, 2000; Alloway, 2004).

Their sources are suspected to be anthropogenic, such as the application of micronutrient fertilizer, animal manure, sewage sludge, pesticides and atmospheric deposition (IPNI, 2008; Tracy and Baker, 2010; Wuana and Okieimen, 2011).

The use of fertilizers, organic manure and other soil nutrients replenishing additives, in order to improve crop yield especially in nutrient depleted soils is on the increase. Too much of fertilizers, manure or sewage sludge can be problematic, while too little means nutrients are not replenished (Akinrinde, 2006). Studies revealed that heavy elements are slowly accumulating in arable soils, especially in soils with depleting trace minerals receiving soil treatments of fertilizer, manure, etc. (Chang and Page, 2000). An estimated over 75% of garden allotment soils in the United Kingdom were reported to be contaminated with potentially toxic metals (Alloway, 2004). Culbard et al. (1988) reported higher mean concentrations of Pb and Zn as a result of pollution, although garden soils tend to contain higher levels than agricultural soils. There is a delicate margin between safe and harmful levels in view of the potential bioavailability, ecosystem redistribution and bio-magnification in higher heterotroughs (Islam et al., 2007).

The build-up of heavy metals in soils occurs at a slow rate, but on large areas (Keller et al., 2001). There may be spatial variability and geographical heterogeneity in heavy metals levels in arable soils, and this may be small and localized in same soil type within an ecosystem community, or regional involving large scale differences (Du Feng et al., 2008). This, however, depends on the physicochemical properties of the farmland soil ecosystems, and their impact on physical and biological factors including soil origin and structure, topography, vegetation cover and soil exposure, soil microclimate, various land use systems and land management (Lipiec et al., 2004; Ithier-Guzman, 2010).

The accumulation and spatial changes in natural soil concentration characteristics of heavy metals is of environmental and health concern. Soils heavy metal accumulation trends may not be detected on time, because metal contamination may not lead to immediate concern in terms of toxicity, except in the case of hot spot pollution (Keller et al., 2001). High concentrations of heavy metals such as Cu and Zn in soils can result in phyto-toxicity, for example, distortion in photosynthetic electron transport, and impairment of carbohydrate metabolism (Tsonev and Lidon, 2005; Mishra and Dubey, 2005). The ubiquitous character and the increase of heavy-metal which flows through the soil system may also cause serious problems for soil fertility, ground water quality, and food chains (Keller et al., 2001). Excess concentration levels of heavy metals have caused the disruption of natural terrestrial ecosystems (Wei et al., 2007; Yadav et al., 2009).

Qualitative and quantitative information about the concentration of heavy metals and trends of heavy metals fluxes, soil withholding capacity/release potential and the stability of heavy metals within soil mineral aggregates in many farmlands in Africa and especially Nigeria is lacking. Adequate and comprehensive information such as these is essential in order to achieve sustainable use, management of heavy metal regimes and preservation of the soil resources against heavy metal build-up, in the long-term. There is need to generate field and laboratory information on the concentration levels, and soil fluxes of heavy metals in arable agricultural soils. This constitutes an important means of access to precise and quantitative information, essential for environmental evaluation of soil quality, risk of soil pollution and retro-gradation of soil characteristics. This study investigated heavy metal concentration levels and fluxes in selected arable agricultural soils in some farming communities of South Western Nigeria, as a basis for the development of flux balancing tool required for the rational management of heavy metal contamination in agricultural soils.

MATERIALS AND METHODS

Study area

The study sites included selected farmlands at Obbo Ile (KW_1), Eleyin (KW_2), Apata Balla (KW_3), Lafiaji (KW_4) and Karati (KW_5) in Kwara State; Ilora-Iseyin Road (OY_1), Ahoro Dada (OY_2), Idere (OY_3), Eruwa (OY_4) and Alagutan Igbeti-Igboho Road (OY_5) in Oyo State; Kajola (OS_1), Odo Iju (OS_2), Majaroko (OS_3), Ikire (OS_4), Ikirun-Iragbiji Road (OS_5) in Osun State and Odeda (OG_1), Ayetoro (OG_2), Idi Iroko (OG_3), Papalanto (OG_4) and Sagamu (OG_5) in Ogun State (Figure 1). The farmlands are all located within the savannah and the rain forest vegetation of the South Western Nigeria.

Field sampling design and sampling

Field investigations and sampling were carried out between November, 2010 and March, 2011. Farmlands in use for the cultivation of food produce were selected in some farming communities in South Western Nigeria covering Ogun, Osun, Oyo and Kwara States (Figure 1). The sampling points (Table 1) were spatially located across each of the selected farmlands in order to capture variability in spatial environmental setting, in the distribution of heavy metals in soils. The soils were visually inspected and top soil and subsurface soils were collected from each sampling points from five arable farmlands in each of the investigated States. Control soil samples were also collected from bush fallows at locations remotely afar off from the farmland.

Soil sample preparation and laboratory analysis

Soil samples were air-dried at room temperature and the dried samples gently crushed with ceramic mortar and pestle after which they were sieved through 2 mm mesh sieves. The less than 2 mm size fraction was stored and sub sampled for the determination of soil parameters.

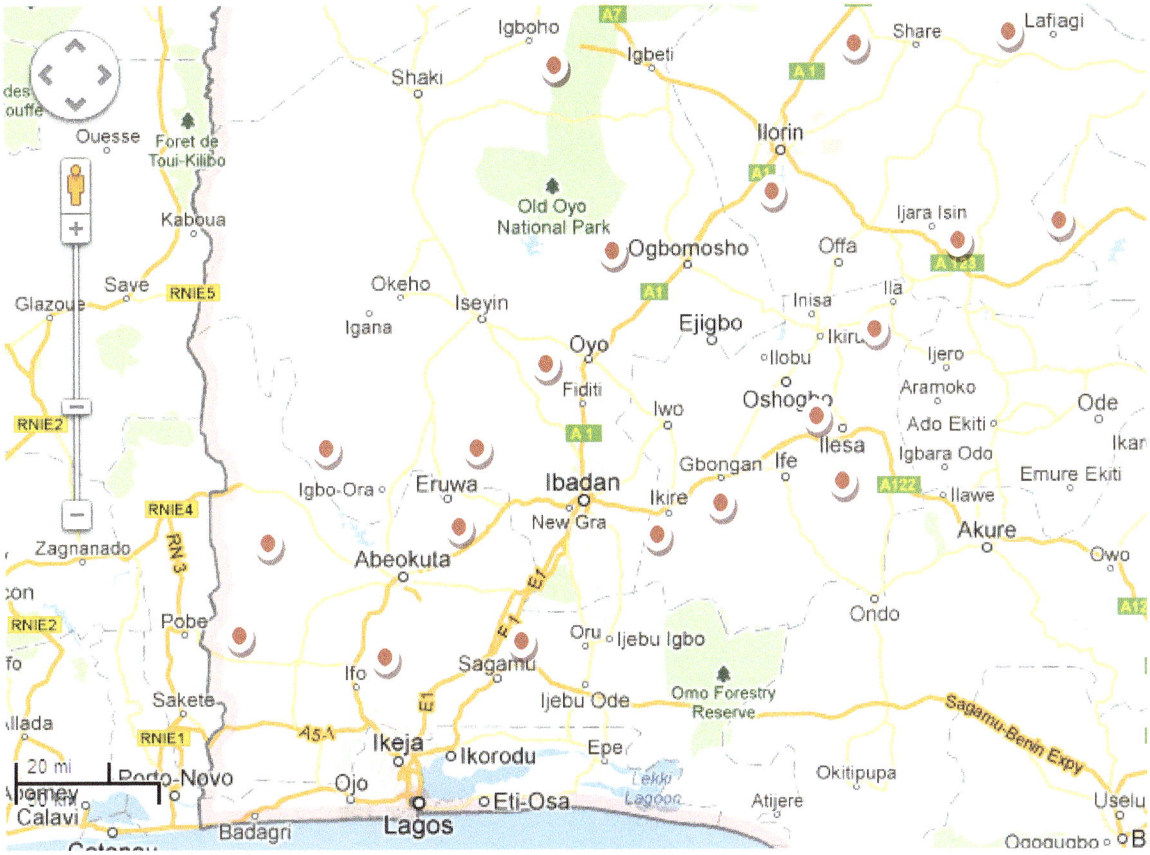

Figure 1. Map of South Western Nigeria showing the farmland study areas in Kwara, Ogun, Osun and Oyo States (Google Maps).

Table 1. Location of selected agricultural farmlands sampling station in some farming communities in Osun, Oyo, Ogun and Kwara States.

Sampling location	Coordinates		Sampling location	Coordinates	
	Northing	Easting		Northing	Easting
OS$_1$ Kajola	07°.721067	004°.636283	OG$_1$ Odeda	07°.344733	003°.647417
OS$_2$ Odo Iju	07°.580951	004°.704250	OG$_2$ Ayetoro	07°.244450	003°.043833
OS$_3$ Majaroko	07°.469752	004°.894567	OG$_3$ Idiroko	06^0.692133	002°.832317
OS$_4$ Ikire	07°.406233	004°.241533	OG$_4$ Papalanto	06°.594317	002°.949733
OS$_5$ Ikirun	07°.874167	004°.659701	OG$_5$ Sagamu	06°. 848321	003°. 604733
OY$_1$ Ilora-Iseyin Rd	07°.803917	003°.905133	KW$_1$ Obbo Ile	08°.071252	005°.302767
OY$_2$ Ahoro Dada	08°.154133	004°.020933	KW$_2$ Eleyin	08°.174167	005°.046583
OY$_3$ Idere	07°.495383	003°.252617	KW$_3$Apata Balla	08°.401067	004°.418133
OY$_4$ Eruwa	07°.552583	003°.445967	KW$_4$ Lafiaji	08°.868330	005°.323401
OY$_5$ Alagutan,	08°.595517	003°.402567	KW$_5$ Karati	08°.878833	004°.900333

Codes: OS, Osun; OY, Oyo; OG, Ogun; KW, Kwara.

Determination of pH and conductivity

The pH of the farmland soils were determined according to the method described by Mclean (1982). A 1:1 mixture of soil and distilled water was prepared and allowed to stand for about 30 min. The pH and conductivity of the suspensions were then measured

Table 2. Results of the determination of pH, OC and concentration levels (μg/g) of some heavy metals in selected agricultural soils in some farming communities in Osun, Oyo, Ogun and Kwara States.

Sampling location	pH	Organic carbon (%)	Pb (μg/g)	Cd (μg/g)	Zn (μg/g)	Cu (μg/g)
Osun (N = 20)						
Minimum value	4.1	1.24	0.19	2.15	21.28	7.91
Maximum value	6.8	2.51	8.35	5.75	38.94	20.87
Mean value	6.05	1.80	3.32	3.92	30.00	15.02
Standard deviation	0.68	0.36	1.99	1.15	6.62	3.88
Oyo (N = 20)						
Minimum value	4.5	1.03	1.04	1.17	15.19	6.34
Maximum value	6.5	1.77	5.23	3.94	27.80	16.74
Mean value	5.65	1.34	3.22	2.29	21.85	12.48
Standard deviation	0.54	0.22	1.34	0.74	4.75	3.25
Ogun (N = 20)						
Minimum value	4.6	0.67	1.24	1.90	17.02	6.48
Maximum value	6.4	1.26	4.79	5.09	30.78	17.11
Mean value	5.68	1.01	3.27	3.47	24.09	13.38
Standard deviation	0.49	0.18	1.03	1.02	5.19	3.46
Kwara (N = 20)						
Minimum value	4.3	0.67	1.01	1.39	12.98	6.17
Maximum value	6.5	1.24	4.70	3.74	23.74	16.28
Mean value	5.42	0.95	2.96	2.55	18.75	12.14
Standard deviation	0.63	0.16	0.82	0.75	3.97	3.10
EEC guideline values (1993)			50 - 300	1 - 3	150 - 300	140 - 300

with pre-calibrated pH meter and conductivity meter electrodes, respectively.

Heavy metals (Cd, Cu, Pb, Zn)

About 2 g each of the <2 mm sieved soil samples were acid digested by the method of Onianwa (2001), and the digests dissolved in dilute HNO_3. The heavy metals (Pb, Cd, Zn, and Cu) in the digests were determined by atomic absorption spectrophotometer (AAS)

Organic carbon (OC)

OC of the farmland soil samples were determined by the rapid dichromate oxidation method described by Walkley and Black (1934) and reviewed by Schumacher (2002).

RESULTS

The digestion and analysis of replicate spiked soil samples (n = 4) showed good recovery between 87.54 and 95.81% and relative standard deviation (RSD) of 4.09 to 15.54%. The recoveries and RSD for the metals were ranged: Pb, 90.58 to 94.75%, RSD, 5.97 to 10.38%; Cd, 91.32 to 93.80%, RSD, 4.01 to 6.54%; Zn, 92.75 to 95.81%, RSD 6.94 to 15.54% and Cu: 87.54 to 90.49%, RSD, 8.46 to 10.72%. The percentage recovery of each metal falls within the generally acceptable recovery 100 ± 20%.

Physical and chemical characteristics

The results of the determination of pH, OC and the concentrations of the heavy metals Cd, Cu, Pb and Zn in soil samples collected from selected agricultural farmlands in Osun, Oyo, Ogun and Kwara States are presented in Table 2.

Soil pH and organic carbon (OC)

The farmland soils were slightly on the acidic to near neutral range (4.1 to 6.8) in surface to deeper 20 cm subsurface soils (Table 2). The pH of the sampled farmland soils in Osun, Oyo, Ogun and Kwara States ranged from 4.1 to 6.8 (6.05 ± 0.68), 4.5 to 6.5 (5.64 ± 0.54), 4.6 to 6.4 (5.68 ± 0.49), and 4.3 to 6.5 (5.42 ± 0.63) µg/g, respectively. OC content of the farmland soil samples are generally low (0.67 to 2.51%). The OC composition content in the soils sampled from the various farmlands in Osun, Oyo, Ogun and Kwara States ranged from 1.24 to 2.51 (1.80 ± 0.36)%, 1.032 to 1.77 (1.34 ± 0.22)%, 0.67 to 1.26(1.01 ± 0.18)%, and 0.67 to 1.24 (0.95 ± 0.16)%, respectively.

The metal content of the soils

Heavy metals occur at measurable concentration in all the agricultural farmlands. The concentrations of Pb, Cd, Zn and Cu in the soils were variable. The distributions of the metals were not of any definite pattern hence, no specific trend was observed for the metal levels in the farmland soils.

The concentrations of Pb in soils of the farmlands in Osun State measured the highest Pb levels which ranged from 0.19 to 8.35 (3.32 ± 1.99) µg/g (Table 2) than in the soils collected from the other States. The least Pb concentrations were measured in farmland soils collected in Kwara State [1.01 to 4.70 (2.96 ± 0.82) µg/g)]. Soil concentrations of Pb in the farmland soils collected in Oyo and Ogun States ranged from 1.04 to 5.23 (3.22 ± 1.34) and 1.24 to 4.79 (3.27 ± 1.03) µg/g, respectively. The concentration of Cd (1.17 to 5.75 µg/g) was the least among the measured metals. Cd concentrations in Osun State farmland soils ranged 2.15 to 5.75 (3.92 ± 1.15) µg/g. This was followed by farmland soils in Ogun State which ranged from 1.90 to 5.09 (3.47 ± 1.02) µg/g, and then Kwara State 1.39 to 3.74 (2.55 ± 0.75) µg/g. The least Cd concentration (0.26 to 5.75 (2.29 ± 0.74) µg/g) were detected in the soils from Oyo State.

Zn levels in agricultural farmlands soils of Osun, Oyo, Ogun and Kwara States were ranged from 21.28 to 38.94 (30.33 ± 6.62) µg/g, 15.19 to 27.80 (21.85 ± 4.75) µg/g; 17.02 to 30.78 (24.09±5.19) *µ*g/g and 12.98 to 23.74 (18.75±3.97) *µ*g/g, respectively (Table 2). Farmland soils of Osun State had the highest Zn concentration, while farmland soils of Kwara State had the least Zn concentration. Cu levels in all the agricultural farmland soils were ranged from 7.91 to 20.87 (15.02 ± 3.88) µg/g, 6.48 to 17.11 (13.38 ± 3.46) µg/g, 6.34 to 16.74 (12.48 ± 3.25) µg/g, and 6.17 to 16.28 (12.14 ± 3.10) µg/g in Osun, Ogun, Oyo and Kwara States, respectively.

DISCUSSION

Agricultural soils are complex environmental systems influenced by natural and anthropogenic activities (Prieler and Anderberg, 1996). Visual inspection of the soils in the study areas showed that the farmland soils are porous, aerated, well drained, moderately weathered and ferruginized. Their texture transits typically between sandy loam - clay loam on the surface, and clayey in the subsoil horizons except in farmland of Kwara State which are mostly loamy to sandy loam on the surface.

The concentration of heavy metal available in soils is largely controlled by adsorption and desorption processes occurring in soils and this depends on soil mineral assemblage, organic matter (OM) and chemistry. Lewis (2005) reported that the mechanism of adsorption and desorption is influenced by the net electrostatic charges on the heavy metals and soil particulates, concentration of the heavy metals, soil pH, redox potential and other complex set of interacting soil characteristics. The magnitude of OC in soils and the accumulation of OM within soil is a balance between the returns or addition of plant residues and their subsequent loss due to the decay of these residues by microorganisms. OC levels in most of the farmland soils were lower than the 2.5% considered adequate for arable land crop production in South Western Nigeria. Eludoyin and Wokocha (2011) reported a mean of 0.96% OC in plot under repeated maize cropping and 2.02% OC in forest soils around Ibadan. About 26% of the soils contain less than 1.00% OC, while about 45% contains between 1.10 and 1.50% OC. About 21 and 7.5% of the soils had between 1.51 to 2.00% OC and greater than 2.00% OC, respectively with the highest level of 2.51% measured in Osun State. In general, OC levels in the farmland soils are the order Osun > Oyo > Ogun > Kwara States. There were significant ($p < 0.05$) spatial variation in soil OC levels and land use types with strong positive correlations above the least correlation value (γ_{min}) 0.64, at a 0.01 significant level in the different land uses.

Soil accumulation and availability of heavy metals are affected by soil OM, soil pH, grown crop species and crop rotation (IPNI, 2008). The concentration of Zn was the highest among the measured heavy metals in the agricultural farmland soils investigated, and ranged between 12.98 and 38.94 µg/g. This was followed by those of Cu which ranged from 6.17 to 20.87 µg/g and then Pb, 0.79 to 8.35 µg/g. The concentration of Cd, 0.26 to 5.75 µg/g was the least in all the sampled farmland soils. Geo-statistical analysis on the concentration levels of the measured heavy metals revealed spatial correlation with the soils pH and OC levels and characteristics, and this defines soils' holding capacity for heavy metals and other mineral nutrients. There is a linear threshold between the concentration levels of some heavy metals and the OC in the farmland soils, with significant correlation ($p < 0.05$) between OC levels and the heavy metal content in the soils (Pb, $\gamma = 0.74$; Cd, $\gamma = 0.67$; Zn $\gamma = 0.57$; Cu, $\gamma = 0.52$). The OC levels predispose soils to the retention of soil heavy metals and micro/macro-nutrients, although this depends on the nature of interaction between them.

Soil pH regulates almost all biological and chemical

reactions in soil (Brady and Weil, 2002; Kabata-Pendias, 2001), thus the distribution of soil pH may provide a useful index for the potential soil holding capacity for heavy metals, nutrients and fertility of soil types. The observed mean pH of the farmland soils were in the acidic: 6.05 ± 0.68, 5.64 ± 0.54, 5.68 ± 0.49 and 5.42 ± 0.63 μg/g (Table 2) in Osun, Oyo, Ogun and Kwara States, respectively. Solomon (2008) reported that pH value of less than 5.5 is considered as a problematic for most microbial activities, and this affects the availability soil heavy metals and nutrients, consequently, there may be less heavy metals in the farmland soils collected in Kwara. The soil pH threshold variance showed significant correlation ($p < 0.05$) with the heavy metal levels in soil (Cd, γ = 0.79; Zn, γ = 0.59; Cu, γ = 0.51) except for Pb (Pb, γ = 0.39). This suggests that, the pH of soils may determine the solubility's of metals in soil solution and their erosion pattern in the farmland soils.

Spatial information about the prevalence of heavy metals according to the soil parent material in most arable agricultural farmland soils in Nigeria is scarce. The detected concentration levels were geochemically indexed in comparison to the European Economic Community (EEC, 1993) recommended maximum allowable concentration (MAC) threshold in agricultural soils, in order to ascertain whether the investigated soils contain metal levels that may be of concern. The measured concentration levels of the heavy metals were below the EEC MACs, except for Cd. Over 56% of the farmland soils contained more than 3 mg/kg of Cd, while about 44% were within the range of 1 to 3 mg/kg EEC MAC for Cd in agricultural soils. The concentration levels of the metals in the surface and subsurface farmland soils were within the normal range of values in unpolluted soils as reported for different countries of the world, except for Cd (Alloway, 1990; EEC, 1993; Brady and Weil, 2002). Furthermore, there was no significant difference ($p < 0.05$) between the values reported for the surface and subsurface soils, although values recorded in top soil were slightly higher than that recorded in subsurface soils. Thus, it can be suggested that the heavy metal concentrations of soil samples from the study areas do not indicate heavy metal contamination of the soils except for Cd, which were 1.2 to 1.9-fold higher than the EEC threshold, especially in Osun and Ogun States. Interpolating the metals concentration data and their spatial distribution on the farmlands, most soils in Osun State appeared to contain higher levels of the metals compared to other states, which may be associated with organic sequestration. According to Fennessy and Mitsch (2001), total nutrient of soil may decrease due to prolonged usage, hence the need to enhance the essential nutrients and minerals quality of arable farmlands. Although, the use of fertilizer on the farmlands was common, the observed levels were apparently a product of the natural background and inputs from diffused sources which are not readily ascertained. The prevention and/or rational management of heavy metal contamination through the application of fertilizer, sewage sludge, manures and other soil nutrient supplement in agricultural soils, requires the use of flux balancing technique.

Conclusion

The concentration levels of heavy metals in the evaluated farmland soils do not appear to be of serious concern for now, except for Cd. The study results provide evidence of gradual accumulation of Cd in arable farmlands, while the concentration levels of the heavy metals in the evaluated farmland soils do not appear to be of serious concern. The sources of the metals in farmlands are suspected to be either from the use of nutrient replenishing materials (such as fertilizer), traffic and aerial deposition.

REFERENCES

Akinrinde EA (2006). Issues of Optimum Nutrient Supply for Sustainable Crop Production in Tropical Developing Countries, Pak. J. Nutri. 5(4):387-397.

Alloway BJ (1990). Heavy metals in soils. Blackie Academic & professional, Glasgow. P. 339.

Alloway BJ (2004). Contamination of soils in domestic gardens and allotments: a brief overview. Land Contam. Reclam. 12(3):179-187.

Brady N, Weil R (2002). The nature and properties of soils, 13th Edition. Prentice Hall. Upper Saddle River, New Jersey, USA.

Chang AC, Page AL (2000). Trace element slowly accumulating, depleting in soils. Calif. Agric. 54(2):49-55.

Culbard EB, Thorntonn I, Watt J, Wheatley M, Moorcroft S, Thompson M (1988). Metal contamination in British urban dusts and soils. J. Environ. Qual. 17:226-234.

Du Feng LZ, XuXuexuan ZX, Shan L (2008). Spatial heterogeneity of soil nutrients and above ground biomass in abandoned old-fields of Loess Hilly region in Northern Shaanxi, China. Acta Ecol. Sinica. 28(1):13-22.

Drechsel P, Gyiele LA (1999). The economic assessment of soil nutrient depletion, Analytical issues for framework development. International Board for Soil Research and Management. Issues in Sustainable Land Management No. 7. Bangkok:I BS RAM.

EEC (1993). Agricultural soil and the environment. Archieved from www.ec.europa.eu/agriculture/publi/fact/envir/2003_en pdf on July 26, 2004.

Eludoyin OS, Wokocha CC (2011). Soil dynamics under continous monocropping of maize (*Zea mays*) on a forest alfisol in South-Western Nigeria. Asian J. Agric. Sci. 3(2):58-62.

Fennessy MS, Mitsch WJ (2001). Effects of hydrology on spatial patterns of soil development in created riparian wetlands. Wetl. Eco. Manag. 9:103-120.

IPNI (2008). Crop fertilization and heavy metal accumulation in soils. Enviro-brief, of the Scientific Staff of the International Plant Nutrition Institute (IPNI), No 7. Archived from www.ipni.net/publication/envb.nsf/FILE/09062-07-Enviro-Brief-07.pdf on 24 Sept, 2012.

Islam E, Yang X, He Z, Mahmood Q (2007). Assessing potential dietary toxicity of heavy metals in selected vegetables and food crops. J. Zhejiang Univ. Sci. B. 8(1):1-13.

Ithier-Guzman (2010). Assessing the Ability of Soils and Sediment to Adsorb and Retain Cs-137 in Puerto Rico. A dissertation submitted to the College of Marine Science University of South Florida.

Kabata-Pendias AP (2001). Trace element in soils and plants. CRC Press, Boca Raton, FL. USA.

Keller A, von Steiger B, van der Zee SEATM, Schulin R (2001). A stochastic empirical model for regional heavy-metal balances in agro ecosystems. J. Environ. Qual. 30:1976-1989.
Lewis BG (2005). Soil chemistry. Environmental and ecological chemistry. Volume II. Encyclopedia of life support system (EOLSS). Soil Chemistry, Northwestern University, Evanston, Illinois archieved from www.eolss.net/Sample-Chapter/C06/E6-13-03.00.pdf on August 09, 2012.
Lipiec J, Walczak R, Józefaciuk G (2004). Plant growth in relation to soil physical conditions. Institute of Agrophysics, Pol. Acad. Sci. Lublin. 97: 239-246.
McLean EO (1982). Soil pH and lime requirement. In Page AL, RH Miller and DR Keeney (eds.) Methods of soil analysis. Part 2 - Chemical and microbiological properties. (2nd Ed.). Agronomy. Soil Sci. Soc. Am. J. 9:199-223.
Mishra S, Dubey RS (2005). Heavy metal toxicity induced alterations photosynthetic metabolism in plants. Taylor and Francis Group, LLC. Archived from http://www.psi.cz/ftp/ola/Handbook on 29 Oct., 2012.
Onianwa PC (2001). Roadside topsoil concentration of lead and other heavy metals in Ibadan, Nigeria, Soil Sed. Contam. 10(6):577-591.
Prieler S, Anderberg S (1996). Assessment of long term impacts of cadmium and lead to agricultural soils in the upper Elbe and Oder River Basins. Interim report of the International Institute for Applied Systems Analysis (IIASA), A2361.
Schumacher BA (2002). Methods for the determination of total organic carbon (TOC) in soils and sediments. Ecological risk assessment support centre office of research and development US EPA.
Solomon D (2008). Presentation On the Relationships Existing in Minerals Soil Between pH On the One Hand and the Activity of Microorganisms and the Availability of Plant Nutrients On the Other. Bahir Dar University, Bahir Dar, Ethiopia.
Tracy D, Baker B (2010). Heavy Metals in fertilizers used in organic production. Archived from http://www.horticulture.umn.edu/amarkhar/courses/pdf on 12 Aug, 2012.
Tsonev T, Lidon FJC (2005). Zinc in plants - An overview. Emir. J. Food Agric. 24(4):322-333.
Walkley A, Black IA (1934). An examination of the Degtjareff method for determining organic carbon in soils: Effect of variations in digestion conditions and inorganic soil constituents. Soil Sci. 63:251-263.
Wei Z, Xi B, Zhao Y, Wang S, Liu H, Jiang Y (2007). Effect of inoculating microbes in municipal solid waste composting on characteristics of humic acid. Chemosphere 68:368-374.
Wuana RA, Okieimen FE (2011). Heavy metals in contaminated soils: A review of sources, chemistry, risks and best available strategies for remediation. International Scholarly Research Network ISRN Ecology, Volume 2011, Article ID 402647, 20 pages doi:10.5402/2011/402647.
Yadav SK, Juwarkar AA, Kumar GP, Thawale PR, Singh SK, Chakrabarti T (2009). Bioaccumulation and phyto-translocation of arsenic chromium and zinc by *Jatropha curcas* L.: Impact of diary sludge and biofertilizer. Biores. Technol. 100(20):4616-4622.

Adsorption study of *Nymphaea alba* for the removal of manganese from industrial waste water

Shazia Khurshid, Shaheen Kausar, Sadia Afzal, Ahmed Adnan, Arshad Mehmood, Muhammad Arshad and Zafer Ahmad Bukhari

Department of Chemistry, G. C. University Lahore, Pakistan.

The application of adsorbents obtained from *Nymphaea alba* as a replacement for costly conventional methods of removing heavy metal ions from waste water has been reviewed. Detailed adsorption study of manganese on *N. alba* was investigated. Batch adsorption study was carried out as a function of amount of adsorbent, contact time, thermal treatment, pH and agitation speed. The adsorbent to solution ratio and the metal ion concentration in the solution affects the degree of metal ion removal. Instrumentation employed was atomic absorption spectrometer. *N. alba* was an excellent adsorbent as compared to number of other low cost adsorbents.

Key words: Manganese, low cost adsorbent, industrial waste water, *Nymphaea alba*.

INTRODUCTION

Pollution is the introduction of contaminants into the natural environment that causes adverse change. Water is contaminated by the discharge of industrial waste, untreated domestic sewage into surface waters. Point and non point sources are causes of pollution to our water resources due to tremendous population growth (Franklin, 1991).

Millions of people worldwide are suffering with the shortage of fresh and clean drinking water. Rapid industrialization, population expansion and unplanned urbanization have largely contributed to the severe pollution to water reservoirs and surrounding soils. It is well known that 70-80% of all illnesses in developing countries are related to water contamination. Pollutants discharged in wastewaters are toxic to aquatic life thus disturbing natural bio balance of water body in which the waste is being disposed (Sundarrajan et al., 2000, Ali, 2010).

Heavy metals are the major contaminant in industrial wastewater. The metals commonly include cadmium (Cd), lead (Pb), copper (Cu), zinc (Zn), nickel (Ni) and chromium (Cr). These metals deposit in human bodies either through direct intake or through food chains (Argun and Dursun, 2008). Polluted water can be treated by different remediation techniques, for example ion exchange, oxidation, electrodialysis, reverse osmosis, electrolysis and adsorption. The basic principles of these techniques used for the removal of pollutants are based on their chemical, electrical, physical, biological and thermal properties. Treatment costs of these technologies ranges from 10-450 US dollars per cubic meter of treated water except adsorption technique. The cost of water treatment using adsorption is 5.0-200 US dollars per cubic meter of water (Gupta et al., 2012). Adsorption is considered to be the best wastewater treatment technique due to its universality, ease of handling and inexpensiveness. Adsorption can remove soluble and insoluble pollutants. The removal capacity of this method is up to 99.9%. Therefore adsorption has been used for the removal of a variety of pollutants from different contaminated water sources (Ali et al., 2012).

Manganese is the part of natural environment. Asyndrome related to manganese chronic exposure to higher levels of manganese may result in the symptoms of lethargy, mask like face, tremors and psychological disturbances. Respiratory effects have also been reported in workers chronically exposed by inhalation (Flynn and Susi, 2009).

Water containing Mn has rust colour, that is the cause of staining of faucets, sinks, or laundry, and has an off-taste or odour. Manganese levels vary in different places in a water body. Infants should not drink water that is above the health advisory level of 300 µg/L (Agency for Toxic Substances and Disease Registry (ATSDR) 1997).

Due to over population and industrialization, demand for water is increasing day by day. So, alternate sources of clean water are needed and wastewater treatment may serve this purpose. Among various water treatment techniques, adsorption on activated carbon is on the top due to its universal nature. Activated carbon is the best adsorbent able to adsorb inorganic as well as organic pollutants that contaminate water resources. In addition to activated carbon, other adsorbents can also be used to remove toxic metal ion from waste water (Ali, 2010). Dead biomasses of different origin constitute cheap adsorbent for inorganic pollutants as well as organic contaminants in the industrial waste water (Panayotova and Velikov, 2008; Pesavento, 2003).

Plant wastes are inexpensive as they have no or very low economic value. Most of the adsorption studies have been focused on untreated plant wastes such as papaya wood, maize leaf, tea leaf powder, lalang leaf powder, rubber leaf powder, peanut hull pellets, sago waste, salt bush leaves tree fern, rice husk ash and Neem bark, grape stalk wastes, etc. Some of the advantages of using plant wastes for wastewater treatment include simplest of the technique, little processing, good adsorption capacity, selective adsorption of heavy metal ions, low cost, free availability and easy regeneration (Burkel and Stoll, 1999).

Nilofar flower grows abundantly near or in water bodies in normal climatic conditions of Pakistan (Figure 1). *Nymphaea alba* is large, cup-shaped, white flowers up to 20 cm across. *N. alba* (Nilofar flower) is easily available in Pakistan. It is present in ponds and lakes. The present paper describes the use of flowers of *N. alba* as an adsorbent from industrial waste effluents. It is highly economic, naturally occurring and effective adsorbent. It can be used not only for manganese but for other heavy metals like copper, cadmium, chromium, nickel, lead, iron, cobalt, zinc and mercury.

Figure 1. Nilofar flower in water bodies in normal climatic conditions of Pakistan.

METHODOLOGY

All the chemicals and solvents used in experimental work were highly purified and of analytical grade (Merck). Double distilled water was used for preparation and dilution of solutions. Atomic Absorption Spectrophotometer (Shimadzu, Japan AA7000F) was used for the determination of manganese contents. Fuel was acetylene and oxidant was air with the ratio of 1:3, pH meter was used to check pH ranges of solutions, Technico10S-207 Incubator Shaker was used for continuous shaking, Gallenkampe oven was used for heating purpose, Distillation assembly with water vacuum pump.

Preparation of adsorbent

N. alba from different ponds was washed with double distilled water. First it was dried in sun and then in oven at 50°C for one hour. After complete drying it was powdered using a mechanical grinder, sieved and stored to save it from moisture and contamination.

Adsorption experiment

Waste water samples were collected in plastic bottles from different industries and were refrigerated. The samples were doubly filtered before the treatment. Dilutions were made for each sample by diluting 10 ml of each sample to 100 ml. 50 ml of these effluents were transferred to different conical flasks. Atomic absorption spectrophotometer was used for the determination of Mn. Chemical compounds were ionized into free analyte by burners that provide the heat energy. Standard solutions of different concentrations were prepared and 50 ml from each solution was transferred to different conical flasks. Solutions were analysed for concentration of metals after filtration.

Various parameters that were used to check the adsorption behaviour of *N. alba* were:

1. Effect of concentration of adsorbent.
2. Effect of contact time.
3. Effect of temperature.
4. Effect of pH.
5. Effect of agitation speed.

A small portion of waste water was filtered off through membrane filter (0.45 µm). Filtered samples (50 ml) were transferred in Erlenmeyer conical flask. Total concentration of manganese in the samples was determined by means of atomic absorption spectrometer. Adsorption experiments were made at room temperature after using 50 ml of known concentration of

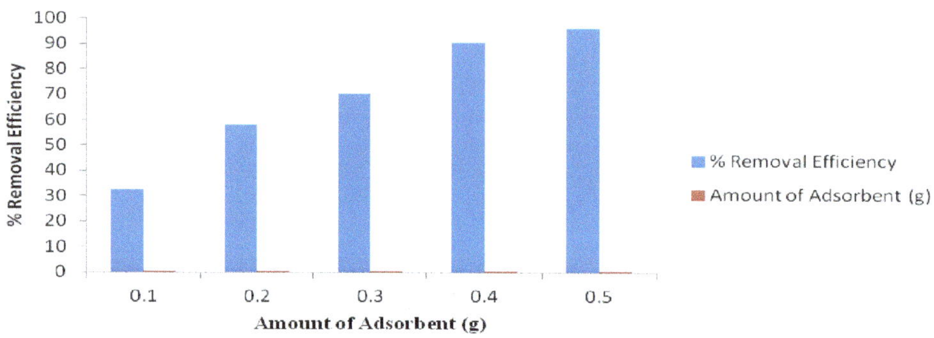

Figure 2. Effect of amount of adsorbent on adsorption of Mn.

manganese solutions in different flasks. Various amounts of adsorbent were put in each flask and these were shaken at 100 rpm Biosorption was monitored by measuring the decline in concentration of Mn in the samples by atomic absorption spectrometer. The effect of various other parameters on adsorption was also studied.

The effect of concentration of adsorbent

50 ml of 50 ppm solutions were taken in conical flasks. Different amounts of adsorbent (0.1 g, 0.2 g up to 0.7 g) were added in the solutions and were agitated at a speed of 100 rpm for 30 min. The solutions were double filtered and analysed by atomic absorption spectrophotometer.

The effect of contact time

50 ml of 50 ppm solutions were taken in conical flasks. 0.5 g of adsorbent was added in each flask and was shaken at 150 rpm for various time intervals (10, 20, 30, 40 up to 100 min). Then solutions were doubly filtered off and were analysed by atomic absorption spectrophotometer.

The effect of thermal treatment on Mn adsorption

50 ml of 50 ppm solution was taken in conical flasks. 0.5 g of adsorbent was added to the flasks. Flasks were agitated at 150 rpm for 100 min at various temperature ranges (10, 20 to 90°C). Solutions were filtered and analysed by atomic absorption spectrophotometer.

The effect of pH on Mn adsorption

50 ml of 50 ppm solutions were taken in conical flasks and pH of flasks was adjusted over a pH range of 1-10 by means of 0.1M NaOH/HCl solutions. 0.5 g adsorbent was added in each flask. Flasks were shaken in shaker at 150 rpm for 100 min at 70°C. Solutions were double filtered and analysed by atomic absorption spectrophotometer.

The effect of agitation speed on Mn adsorption

Adsorbent (0.5 g) was added in 50 ml of 50 ppm solution and pH was adjusted at 7. Flasks were agitated at 70°C for 100 min at various agitation speeds as 25, 50, and 75, 100 up to 225 rpm. Solutions were double filtered and analysed by atomic absorption spectrophotometer.

Removal of manganese from chemical industry

Waste effluent collected from a chemical industry near Muridke (District Sheikhpura) was treated using same method as used for manganese standard solutions. 50 ppm of diluted waste water was double filtered. Its concentration was measured by atomic absorption spectrometer.

RESULTS AND DISCUSSION

Removal of manganese (II) from industrial waste water through *N. alba* (Nilofar flower) is an efficient technique. *N. alba* is locally present in lakes and ponds in Pakistan. The use of *N. alba* for the removal of toxic heavy metal like copper, cadmium, chromium and manganese from different industrial effluent is very easy, efficient and adoptable method. Method adopted was to find the initial concentration of Manganese (II) in the solutions and then determining the concentration after treatment. The difference showed the removal of manganese metal from waste water.

Effect of various parameters on adsorption of manganese

Effect of amount of adsorbent

A 50 ml aliquot of sample solution containing 24.26 ppm Manganese (II) was transferred to each flask after filtration. Different amounts of adsorbent ranging from 0.2 to 0.7 g were added to 50 ml waste water. Solutions were shaken at 175 rpm for 100 min at 60°C. Samples were analysed by atomic absorption spectrometer. With varying amounts of the adsorbent, adsorption varied and the maximum adsorption of 96% was attained with 0.5 g of adsorbent per flask (Figure 2).

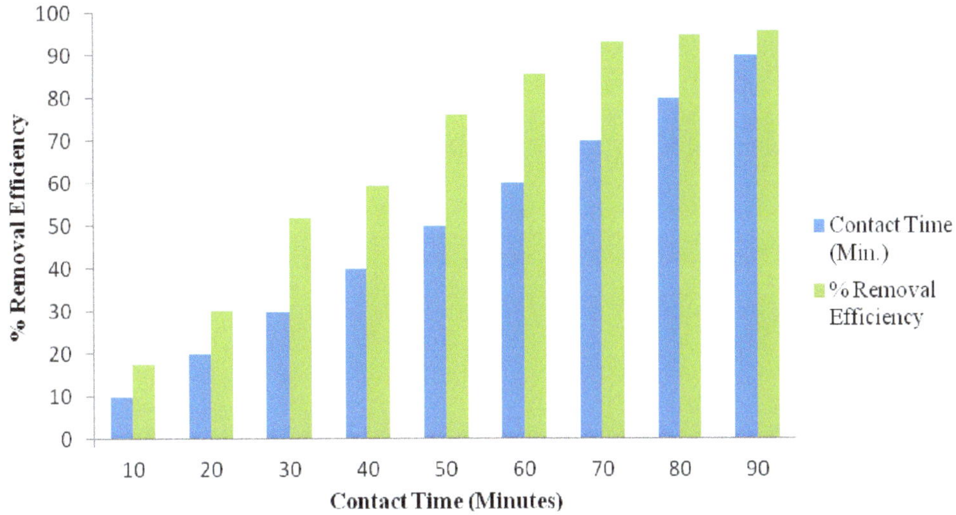

Figure 3. Effect of amount of contact time on % removal of Mn.

The effect of contact time on manganese adsorption

Effect of length of contact between the adsorbent and the solution was studied by keeping other parameters constant. Time duration varied from 20 to 140 min. Results revealed that more contact time resulted in better removal of metal ions from solution. The maximum removal of 96% was observed when contact time of 100 min after which there was insignificant change (Figure 3).

Effect of temperature on manganese adsorption

The effect of temperature was investigated using a temperature range of 10 to 90°C keeping all other parameters constant. To counter the evaporation at elevated temperatures (above 40°C), deionized water was added into flasks at regular short intervals to keep uniform volume through the experiment. Adsorption of Manganese (II) ions from waste water increased with increasing temperature and the maximum removal of 96% was observed at 70°C. A further increase in temperature did not enhance the adsorption (Figure 4).

Effect of pH on manganese adsorption

Waste water samples were treated at various pH ranges like 1 to 10. Maximum removal of 96% was seen at pH 7 (Figure 5).

Effect of agitation speed on adsorption of manganese

By shaking adsorbent and metal ions mix well and have more contact to interact. Waste water samples were shaken in incubator shaker at various speeds and it was found that the speed of adsorption increased by increasing the shaking and reached the maximum value at agitation speed of 150 rpm. At this speed, the maximum adsorption of 96% was observed (Figure 6 and Table 1).

Conclusion

N. alba can act as good adsorbent not only for the removal of manganese ions but also for various other toxic metals like lead, mercury, copper, chromium, cadmium, cobalt, nickel and iron. The removal of manganese (II) from industrial waste water through *N. alba* (Nilofar flower) was found to be an efficient technique. It was an efficient strategy and its performance was the best at temperature of 70°C and pH 7.0. The technique employed was very simple and environment friendly as no manmade chemicals were involved. This method can be adopted by various industries for the waste water treatment. The experimental parameters at temperature of 70°C, pH 7 and minimum amount of adsorbent level of 1.0% were considered to be 96% efficient in removing manganese (II) from industrial waste water. So industries can apply this methodology for removal of toxic metals from industrial waste water effluents, thus preventing land and water pollution.

ACKNOWLEDGEMENTS

Authors wish to thank Miss Nabila Aslam (GCU Lahore)

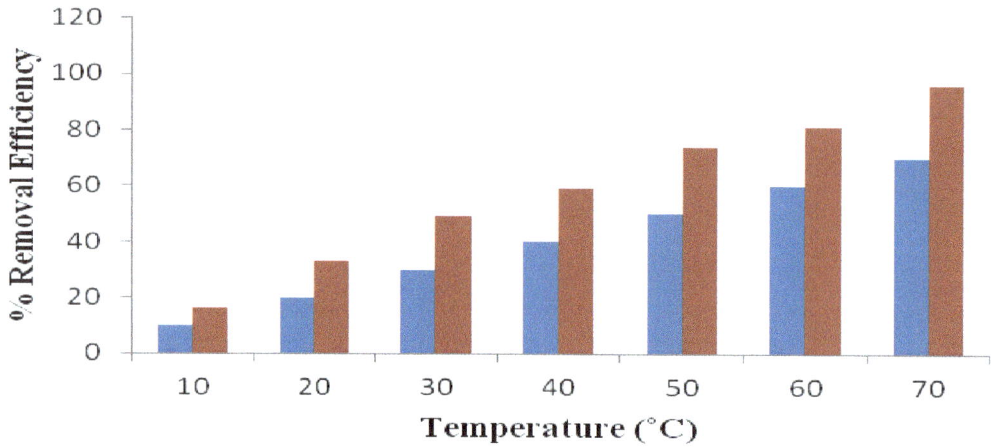

Figure 4. Effect of temperature on % removal efficiency of Mn.

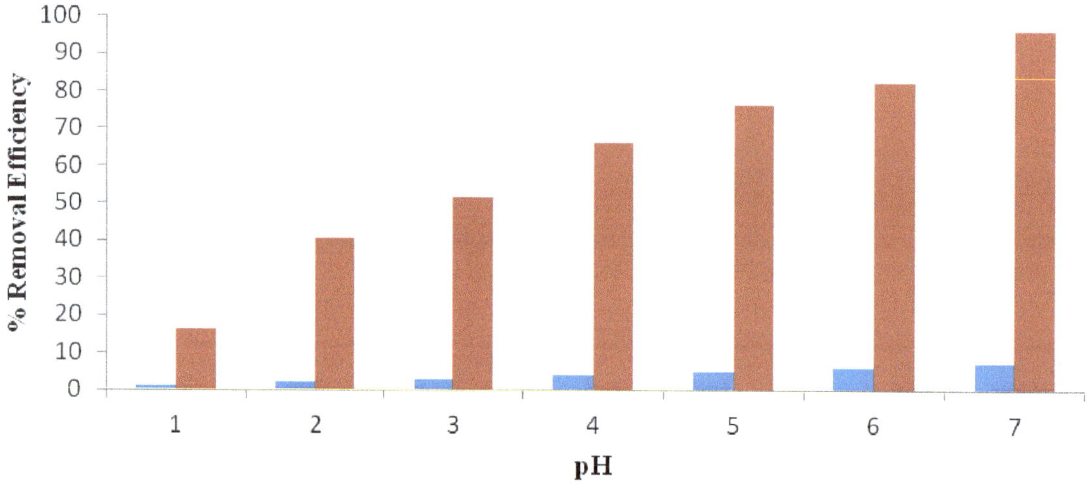

Figure 5. Effect of pH on removal of Mn.

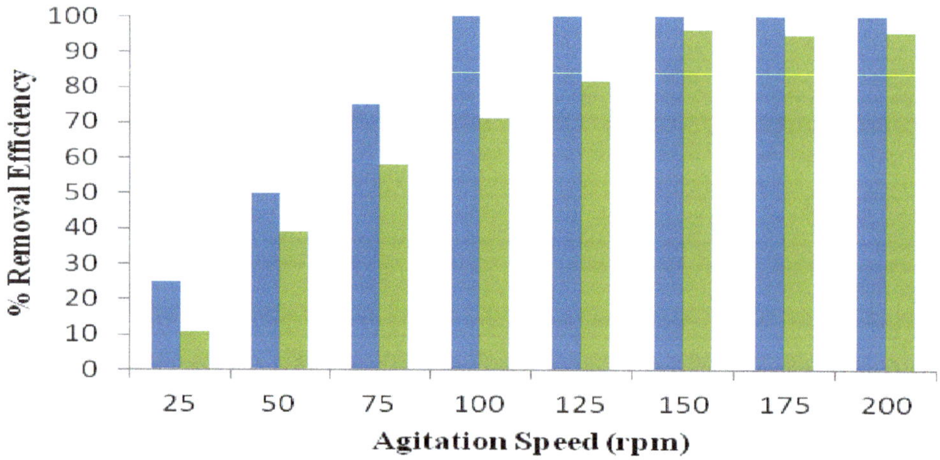

Figure 5. Effect of agitation speed on the removal of Mn.

Table 1. Concentration of manganese in different industrial waste water.

Industries	Concentration of Manganese (ppm.)
I	0.019
II	0.135
III	0.322
IV	5.445
V	18.325
VI	19.464
VII	0.039
VIII	24.268
IX	0.468
X	0.623

and Mr. Salman (PU Lahore) for providing instrumental analysis support.

REFERENCES

Agency for Toxic Substances and Disease Registry (ATSDR). (1997) Toxicological Profile for Manganese. U.S. Department of Health and Human Services, Atlanta, GA.

Ali I, Mohd A, Tabrez AK (2012). Low cost adsorbents for the removal of organic pollutants from waste water. J. Environ. Manage. 113:170-183.

Ali I (2010). The quest for active carbon adsorbents substitutes: Inexpensive adsorbent for toxic metal ions removal from wastewater, separation and purification reviews, 39:95-171.

Argun ME, Dursun S (2008). A new approach to modification of natural adsorbent for heavy metal adsorption. Bioresour. Technol. 99:2516-2527.

Burkel RS, Stoll RC (1999). Naturally occurring arsenic in sandstone aquifer water supply wells of North Eastern Wisconsin. Ground Water Monit. Remed.19:114-121.

Flynn MR, Susi P (2009). Neurological risks associated with manganese exposure from welding operations. Int. J. Hyg. Environ. Heal. 212:459-469.

Franklin LB (1991). Low cost adsorbents for the removal of organic pollutants from wastewater. Waste water engineering: treatment disposal and reuse.

Gupta VK, Ali I, Saleh TA, Nayak A, Agarwal S (2012). Chemical treatment technologies for waste-water recycling, *RSC.* 2:6380-6388.

Panayotova M, Velikov B (2008). Influence of zeolite transformation in a homoionic form on the removal of some heavy metal ions fromwastewater. J. Environ. Sci. Health. 38:545-554.

Pesavento M (2003). Adsorption of lead (II) and copper (II) on activated carbon by complexation with surface functional groups. Anal. Chem. Acta. 480:171-180.

Sundarrajan M, Fernandis AZ, Subrahmanyam G, Prabhudesai S, Krishnamurthy S, Rao K (2000). Over expression of G1/S cyclins and PCNA and their relationship to tyrosine phosphorylation and dephosphorylation during tumour promotion by metanil yellow and malachite green. Toxicol. Lett.116:119-130.

Geoelectrical investigation of aquifer problems in Gosa area of Abuja, North Central, Nigeria

Omeje M. [1] **, Husin W.** [1]**, Noorddin I.** [2]**, Oha I. A.** [3]**, Onwuka O.S** [3]**, Ugwuoke P. E.** [4] **and Meludu O.** [5]

[1]Department of Physics, Faculty of Science, Universiti Teknologi Malaysia, 81310, Johor Bahru, Malaysia.
[2]Centre for Research and Innovation Management, National Defence University, Malaysia.
[3]Department of Geology, University of Nigeria Nsukka, Enugu State, Nigeria.
[4]Centre for Energy Research and Development, University of Nigeria, Nsukka, Enugu State, Nigeria.
[5]Department of Physics, School of Pure and Applied Sciences, Federal University of Technology, Yola, Nigeria.

The study of aquifer problems in Gosa area of Abuja, North-central Nigeria, has been undertaken. Abuja, the capital city of Nigeria, is underlain by Precambrian basement rocks. Vertical electrical sounding (VES) was carried out at twelve locations within and around the Gosa area. The interpretation of the data obtained from the sounding revealed that six non-aquiferous geoelectrical layers overlie the unfractured basement. The first layer with average thickness of about 0.5 m and resistivity of 83 Ohm-m. The thickness of the second layer is about 0.7 m and resistivity of 438 Ohm-m. The third layer is about 3.4 m thick and resistivity of about 63 Ohm-m. The fourth layer with average thickness of about 16.0 m and resistivity, 236 Ohm-m. The fifth layer with average thickness of 42.7 m and resistivity of about 68 Ohm-m. The thickness of sixth layer varies from 42.7 m to infinity and resistivity of about 2067 Ohm-m. Eight boreholes of varying depths, 45 to 70 m were drilled. Other areas were drilled with good yield of static water level of about 3.7 m, whereas Gosa aquifers were found dried. It could be attributed to basaltic intrusion that shattered the aquifer bearing rocks thereby devoid water in the zone after pronouncing from geophysical investigation.

Key words: Geologic map, accessibility map, vertical electrical sounding (VES) profiles.

INTRODUCTION

Basement complex rocks do not inherently make good aquifers. The hydrogeologic characteristics of basement rocks are only enhanced when the rocks are fractured and/or when they are weathered. The conditions are better enhanced when the rocks are overlain by thick overburden. Groundwater potentials of a basement complex area are often determined by geophysical means, which determines the thickness of the overburden and the network of fractures that may exist in the area. Geophysical surveys are also important for groundwater investigation in basement areas in view of the discontinuous (localized) nature of basement aquifers (Satpatty and Kanugo, 1976). The use of the vertical electrical sounding (VES) method is popular for groundwater investigation in both soft rock (sedimentary) and hard rock (igneous and metamorphic) terrains (Barongo and Palacky, 1989; De Beer and Blume, 1985; Mbonu et al., 1991; Shemang, 1993). In the Basement Complex of Nigeria, extensive application of geoelectrical method for groundwater investigation has been reported (Pulawski and Kurth, 1977; Acworth, 1987; Olorunfemi and Okankune, 1992; Olorunfemi and Fasuyi, 1993;

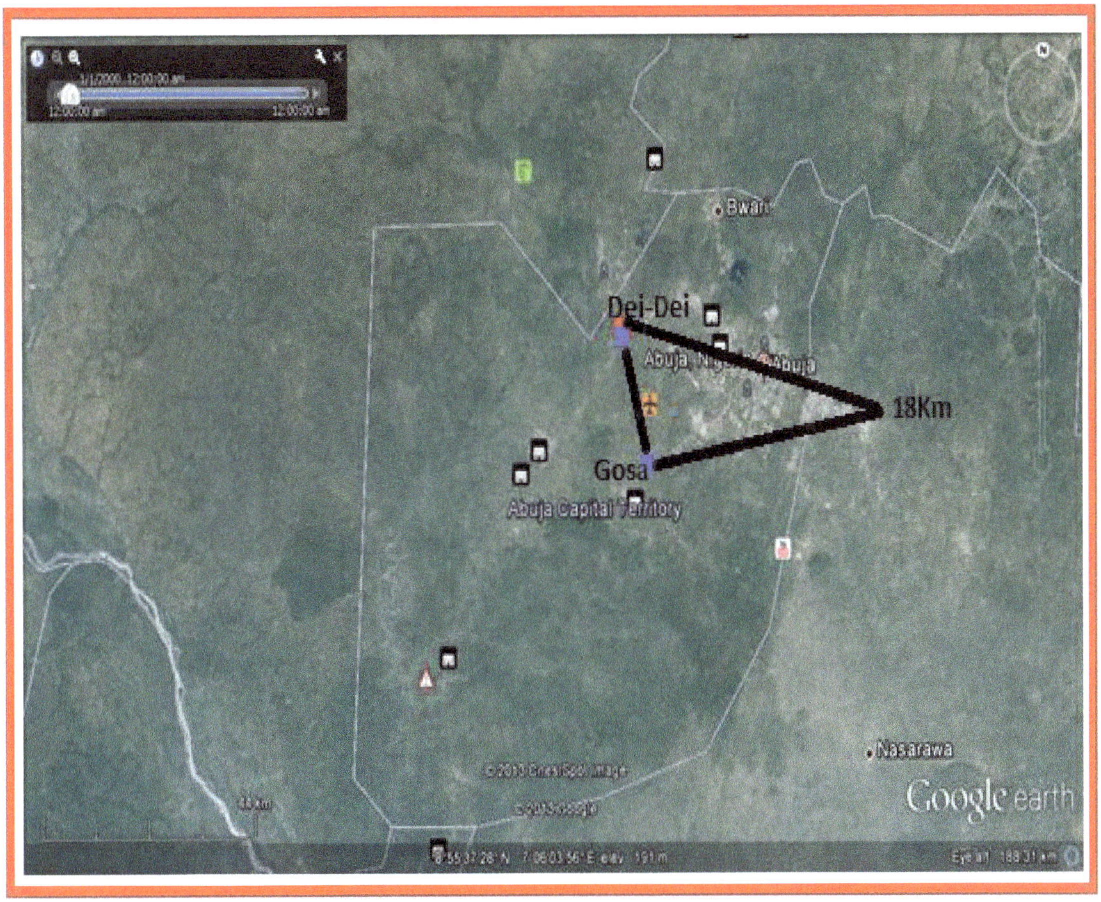

Figure 1. The Google earth map of the study area showing the study arean with blue dot.

Edet and Okereke 1997; Nur and Ayuni, 2004).

The area of study includes most parts of Abuja main city, and western surburbs. It includes areas around Maitama, Asokoro, Wuse, Garki, Gwarimpa, Kado, Jabi, Kubwa, Dei-Dei, Lugbe and Gosa. It is bounded by Latitudes 8° 53`N to 9° 13`N and longitudes 7° 00`E to 7° 30`E. Gosa has become important because of the increasing population of the suburbs of Abuja. Many residents of the area embark on the development of private boreholes to augment public water supplies which are inadequate. But most boreholes here are abortive, and so counter the intentions of the owners, in addition to loss of capital. This work is a first-hand study aimed at establishing the groundwater conditions in the Gosa area of Abuja, Nigeria, which will help in adequate planning and drilling of boreholes in the area.

Geology and hydrogeology of the area

The area of study forms part of the Basement Complex of north central Nigeria; with lithologic units falling under three main categories, which include (1) Undifferentiated migmatite complex of Proterozoic to Archean origin, (2) Metavolcano-Sedimentary rocks of late Proterozoic age and (3) older granite complex of late Precambrian - Lower Paleozoic age, also known as Pan-African granites. All these rocks have been affected and deformed by the Pan-African thermotectonic event. Detailed reports of the lithological description, age, history, structure and geochemistry of the Basement Complex of Nigeria are given in Oyawoye (1972); Black et al. (1979); Ajibade et al. (1987); Rahaman (1988); Caby (1989) and Dada (2008). Figure 1 shows the Google earth map of the location. Figure 2 is the geologic map of Nigeria showing the study area with red arrow.

In the study area, all the three major rock categories mentioned above are well represented (Figure 3). The rocks are generally weathered into reddish micaceous sandy clay to clay materials, capped by laterite. The hydrogeology of basement areas is simple since there is an inherent limitation to the occurrence of groundwater. However, where the regolith is thick, and there is a dense network of fractures, the potentials for the accumulation of groundwater in basement complex rocks may increase.

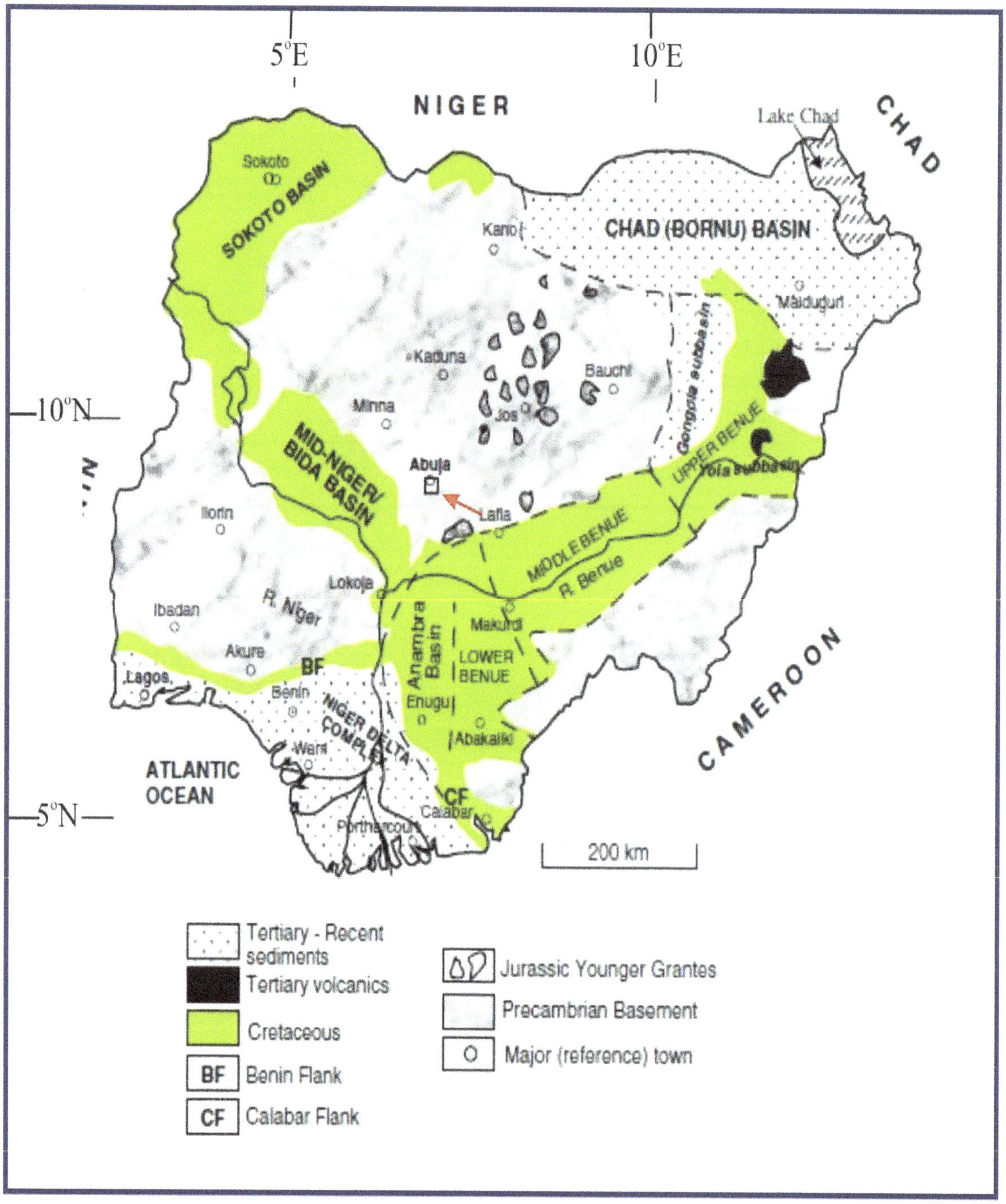

Figure 2. Geological Map of Nigeria, showing the position of Abuja (red arrow) in the Basement Complex of North central Nigeria (modified from Obaje, 2009).

MATERIALS AND METHODS

Geophysical investigation

The search for groundwater in Gosa and it's environ was carried out through the use of electrical methods of geophysical survey, (Figure 4). The Schlumberger configuration in VES was used to obtain field data. Vertical electrical sounding probes the vertical variation in resistivity of the subsurface, thereby indicating the presence of fluid and ionic concentration in the subsurface materials. It is also applied to determine the depth to bedrock, delineate the various units that constitute the overburden (regolith), determine the degree of fracturing of the bedrock; all of which would help in making the choice for a feasible site for constructing a successful borehole. VES data for this work were obtained using the Allied Omega C2 Terrameter. The field data obtained was presented as curve of apparent resistivity values against half of the current electrode separation (AB/2) in metres on a log-log scale.

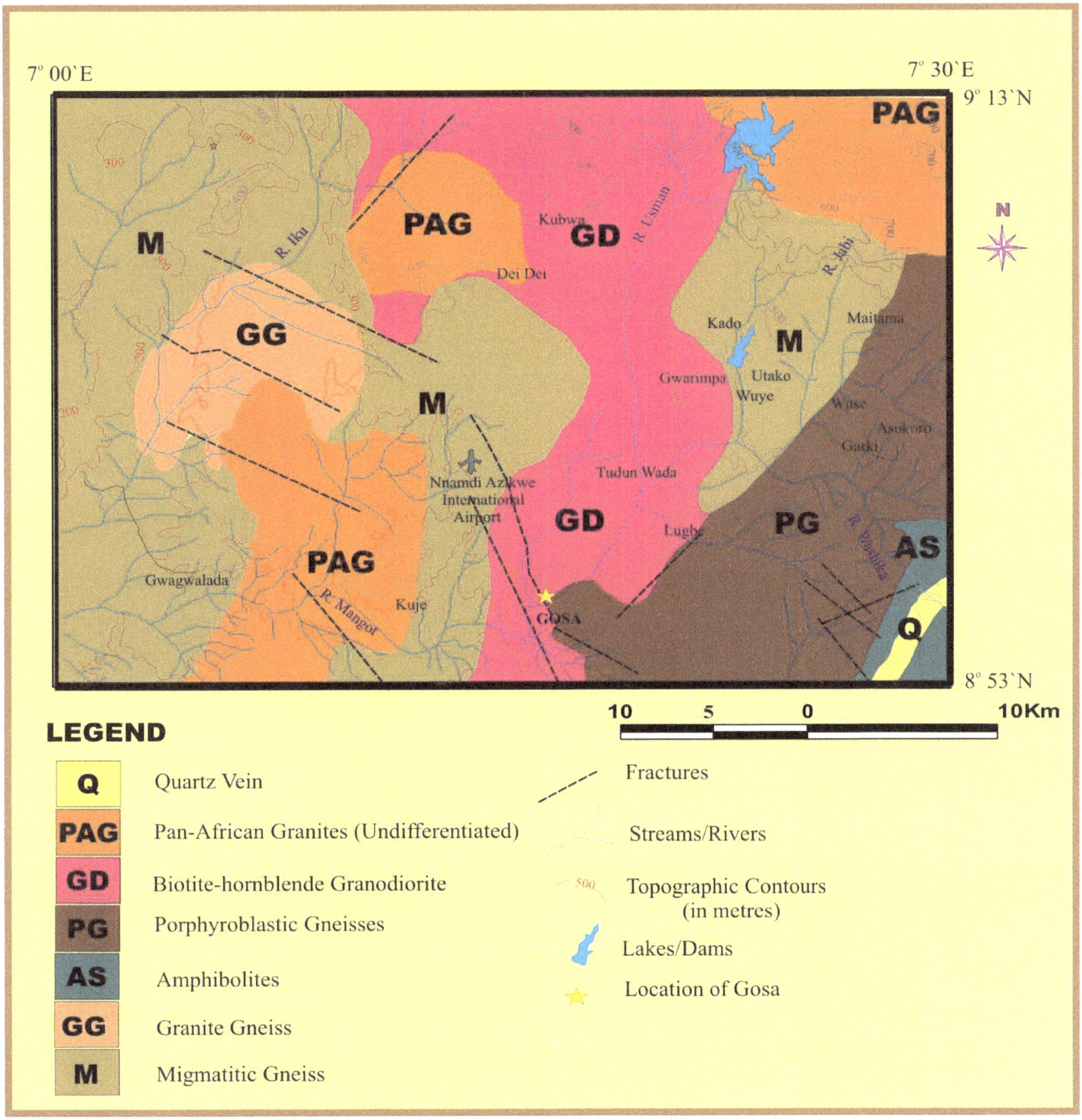

Figure 3. Geological map of the study area.

A 3-mode filter was employed to jettison associated noise due mostly to electrical signals and interference from adjacent power transmission lines. The VES data obtained were analyzed using the WINRESIST 2004 Version computer software to improve the quality of the interpretation by iteration and modelling to goodness of fit. The profiles plots of apparent resistivity against electrode spacing are shown in Figures 5, 6, 7, 8, 9 and 10.

Drilling of boreholes

Drilling boreholes

Among the drilled boreholes, two monitoring boreholes for this work were drilled at Dei-Dei and Gosa for proper investigation of

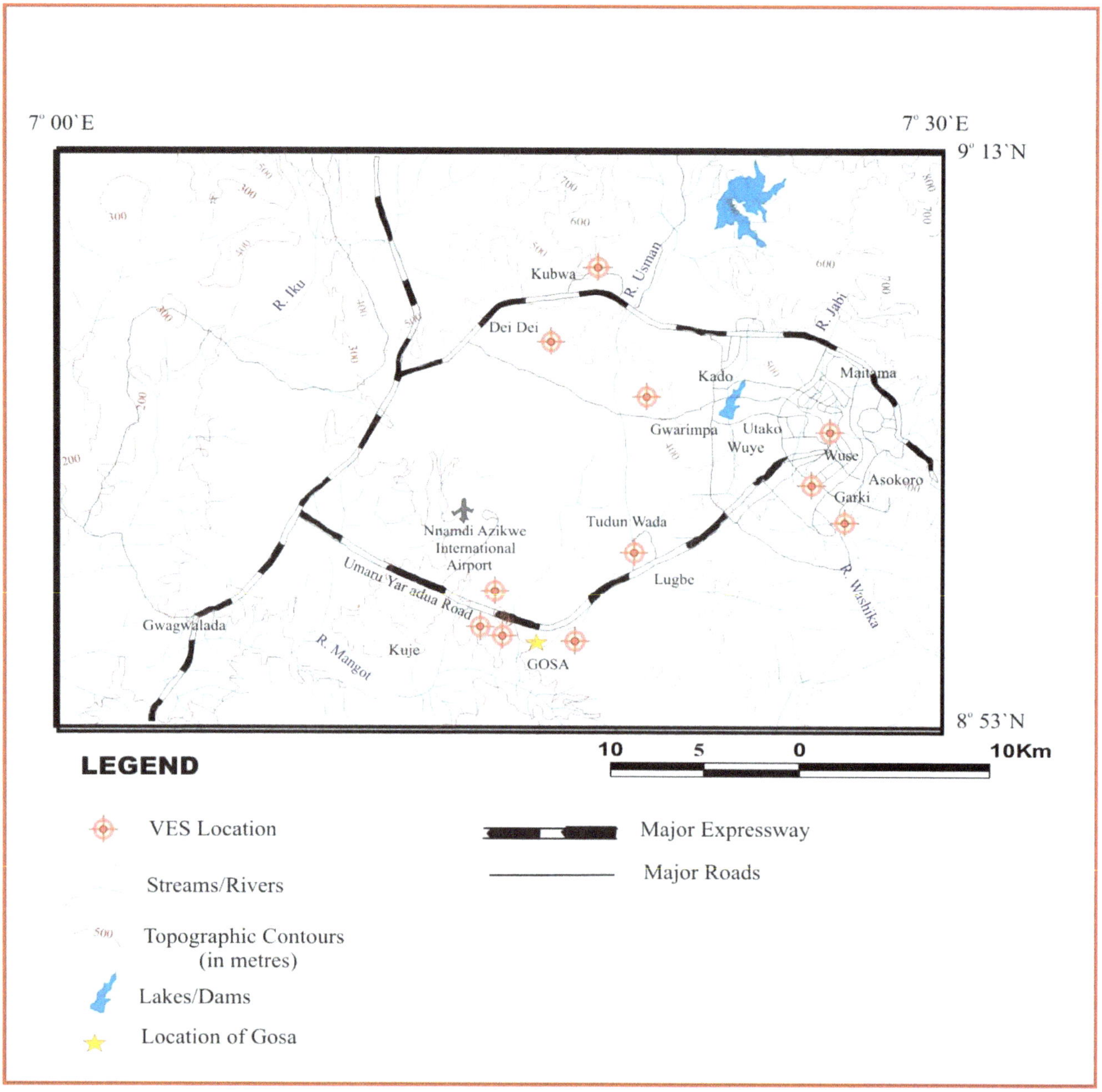

Figure 4. Accessibility map of the study area, showing positions of VES points.

subsurface step-by-step information with well log and GPS for the coordinates. The boreholes were drilled with the help of 30 ton capacity rig machine with compressor made of INGERSOL. The rocks were clay and sand, and the medium was heterogeneous as the thickness of the different layers differed in the boreholes.

Sample inventory

The identification of boundaries between layers with noticeably different particle sizes using visual manual logging method, record the thickness when the layer changes. Layer thickness change may range from less than one metre to tens of metres. After boundary of distinct layers have been clearly marked an indelible felt-tipped pen, using a single entry for each layer:

(a) Record the date the sample is logged and the initials of the logger,
(b) Record the designate geologically logged cut segment,
(c) Determine and record the depth interval for each layer,
(d) Photograph each layer using digital camera.

The two sites were Dei-Dei (about 60 m), and Gosa (49 to 50 m), Figure 3 shows the drilling points.

Site one borehole (S1): Details about the logging are given in Table 2.

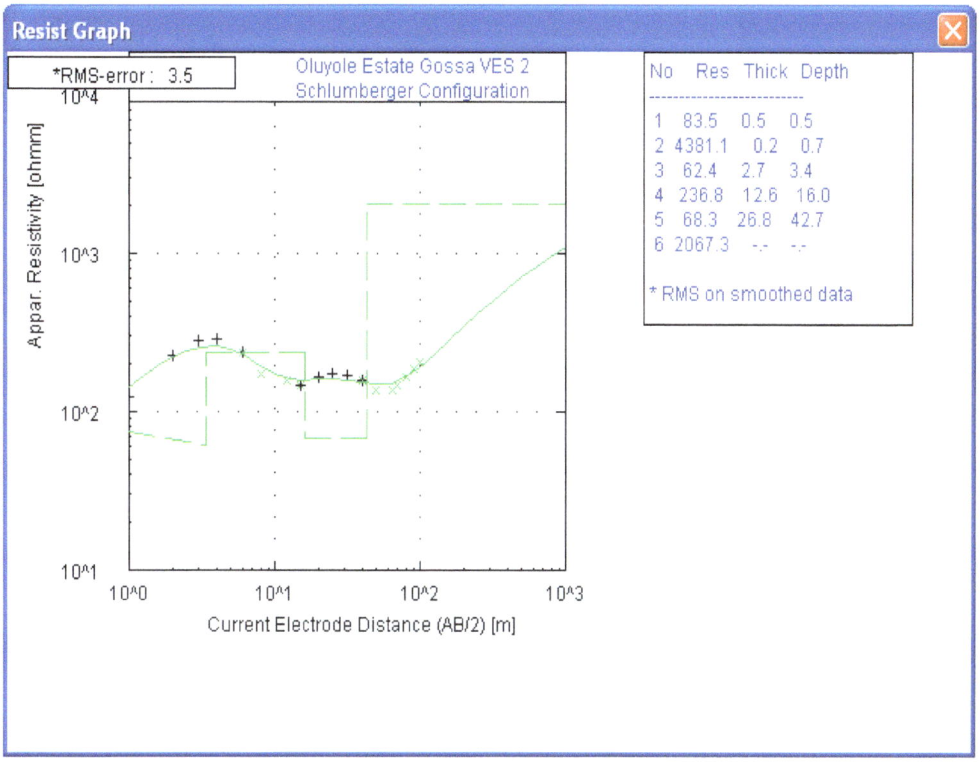

Figure 5. Gosa.

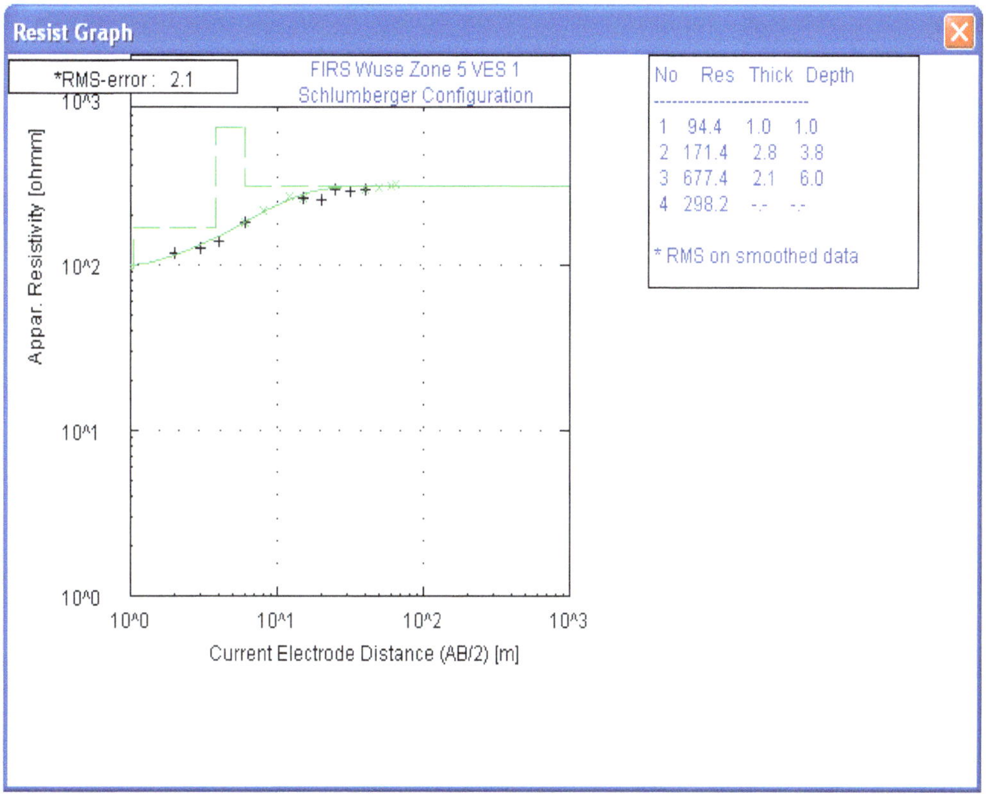

Figure 6. Wuse Zone 5.

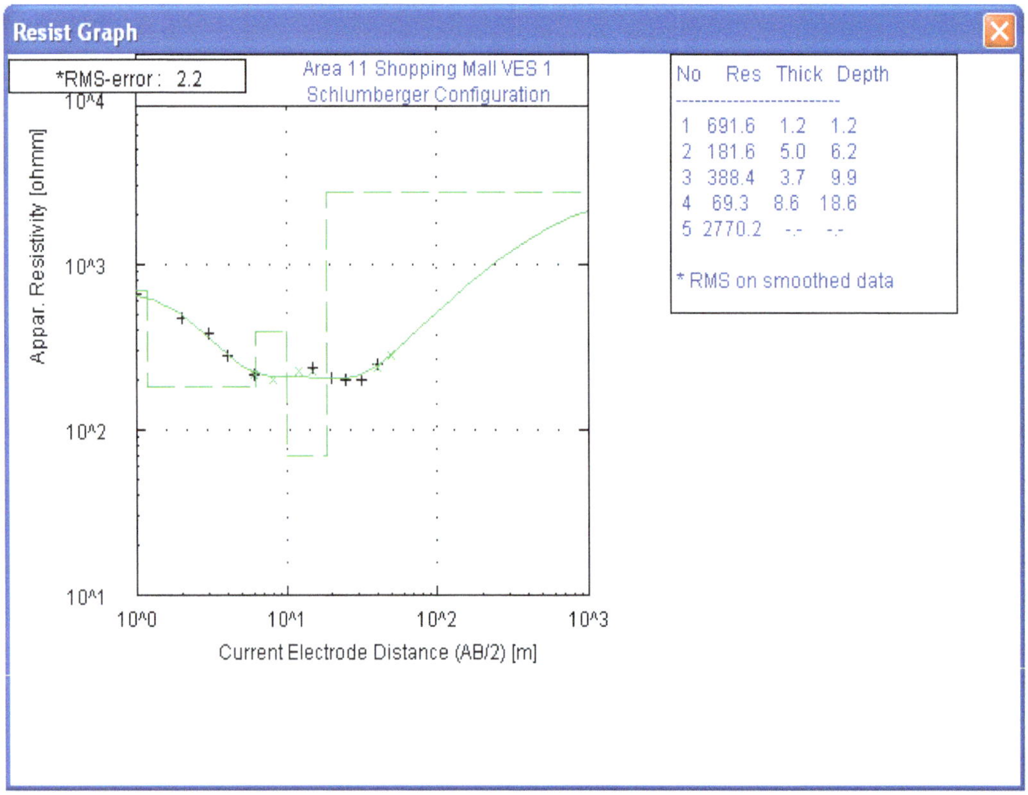

Figure 7. Area 11.

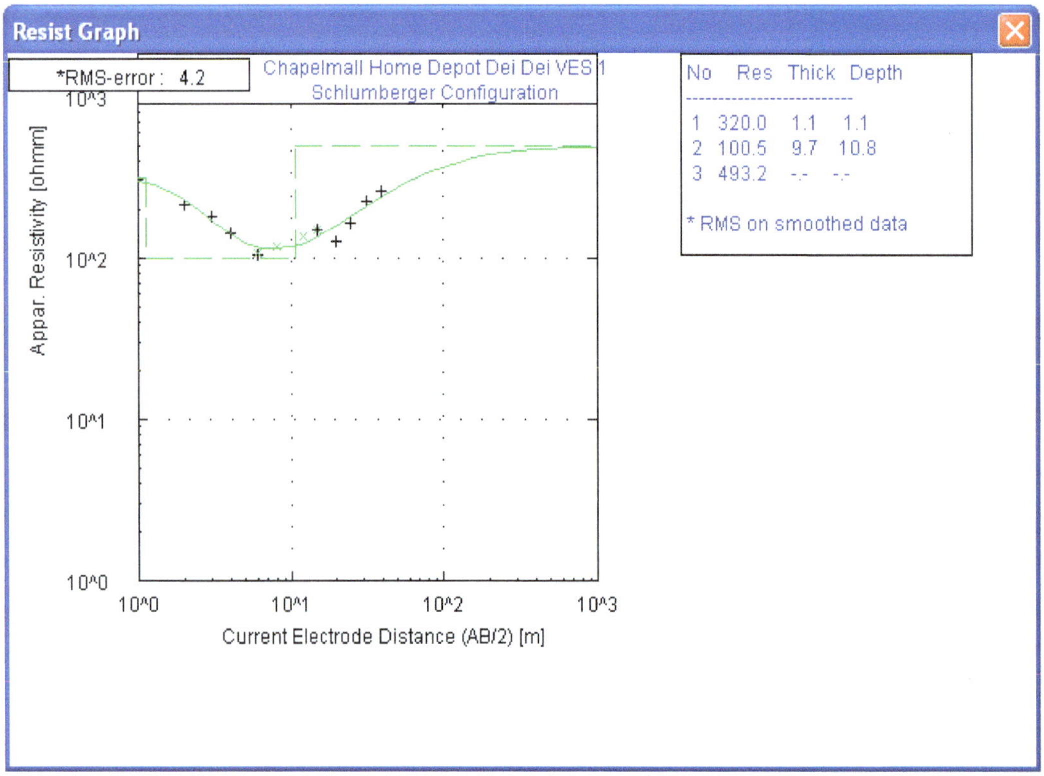

Figure 8. Dei-Dei.

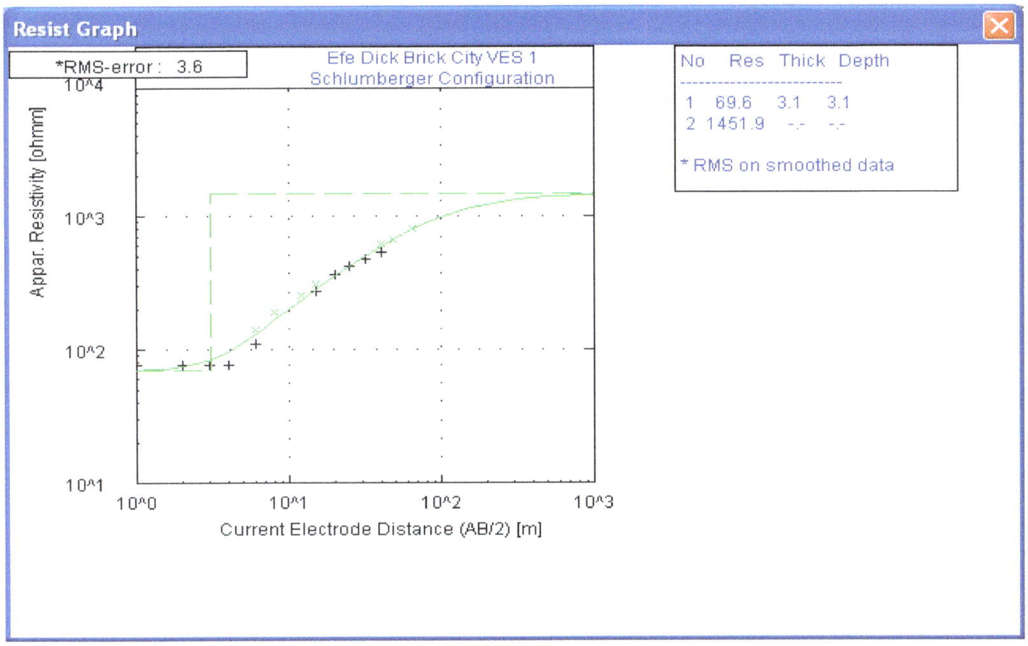

Figure 9. Brick City.

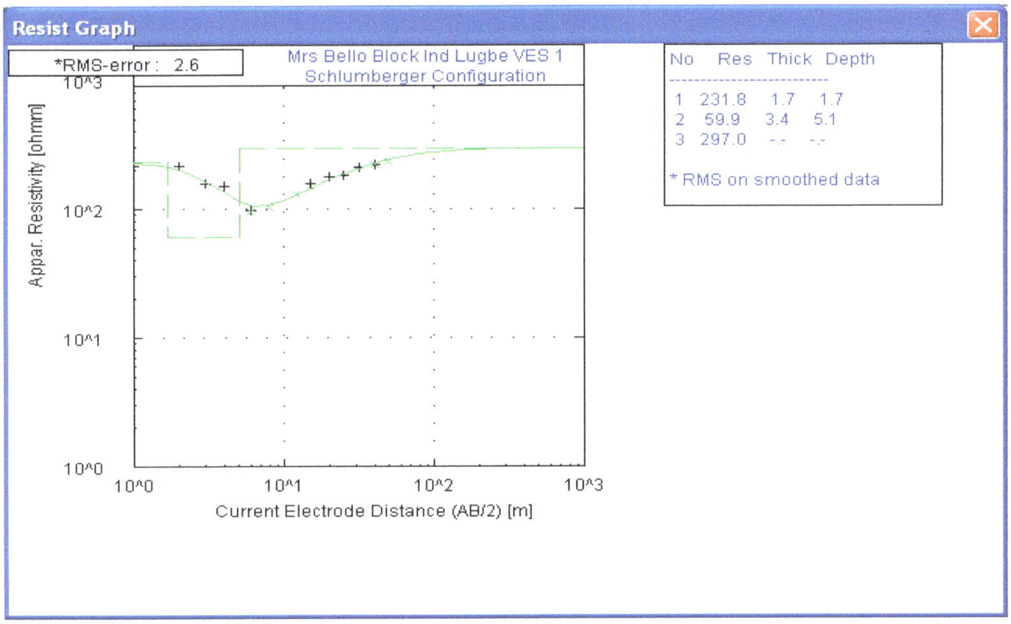

Figure 10. Lugbe.

Site two borehole (S2): Details about the logging are given in Table 3.

DATA INTERPRETATION AND RESULTS

VES data analysis and interpretation

Twelve VES points were studied in this work; five from areas around Gosa and seven from other parts of Abuja, (Figure 4). The five points in Gosa area were close to abortive borehole sites. Table 1 shows a summary of the results of the interpretations. The table shows that at the location of VES 1, there is a thick lateritic topsoil of up to 3.4 m, underlain by a thick micaceous sandy clay layer of weathered basement rock that cannot sustain boreholes. This is directly underlain by unweathered, unfractured

Table 1. Summary of results obtained from the computer output of the eight (12) VES in the area.

S/N	VES No.	Location	Resistivities(ohm-m) of geoelectric layers						Thickness (m) of geoelectric layers						Lithology	Remarks
			ρ1	ρ2	ρ3	ρ4	ρ5	ρ6	h1	h2	h3	h4	h5	h6		
1.	VES 1	Gosa	83,	438,	62,	236,	68,	2067	3.4,	39.3,	∞				Lateritic topsoil, weathered basement, fresh basement	Dry
2.	VES 2	Gosa	410.6,	130.5,	47.72,	4869,	164.4		0.44,	1.22,	3.33,	4.88,	∞		Sandy clayey topsoil, fresh basement	Dry
3.	VES 3	Gosa	211.2,	82,	122,	54.29,	1140,	82043	0.6,	2.03,	2.86,	5.97,	12.48,	∞	Sandy topsoil, weathered basement, fresh basement	Dry
4.	VES 4	Gosa	1580,	332.8,	47.57,	19369,	72.7		0.49,	1.33,	4.44,	6.86,	∞		Sandy topsoil, clayey weathered zone, fresh basement	Dry
5.	VES 5	Gosa	184.3,	1104,	37.87,	1473			0.22,	2.46,	12.95				Sandy topsoil, weathered basement, fresh basement	Dry
6.	VES 6	Wuse zone 5	94,	171,	677,	298			1.0,	2.8,	2.2,	∞			Sandy topsoil, clayey sand, weathered basement	Wet
7.	VES 7	Lugbe	231,	59,	297				1.7,	3.4,	∞				Sandy clayey topsoil, weathered and fractured basement	Wet
8.	VES 8	Area 11	691,	181,	388,	69,	2770		1.2,	5.0,	3.7,	8.7,	∞		Lateritic topsoil, fractured basement	Wet
9.	VES 9	Dei-Dei	320,	100,	493				1.1,	9.7,	∞				Sandy clay topsoil, weathered basement	Wet
10.	VES 10	Phase 4	42,	1323,	217,	39,	444		0.9,	0.3,	4.6,	16.4,	∞		Lateritic clay topsoil, weathered and fractured basement	Wet
11.	VES 11	Kubwa	257,	1219,	530,	100000			0.8,	0.9,	13.9,	∞			Sandy clay topsoil, weathered basement	Wet
12.	VES 12	Brick City Estate	69,	1451					3.1,	∞					Sandy clay topsoil, fractured basement	Wet

Table 2. Depth and Lithologic Unit of Borehole at Dei-Dei. The drilling point coordinate, Lat. 9° 6'52"N and Long. 7° 15'39"E, using GPS, Model: Extrex high sensitivity 2000 to 2007 Garmin Ltd.

Sample ID	Depth (m)	Thickness (m)	Lithology description
S1L1	0-4	4	Sandstone, brownish and ferruginous, interbedded with quartz feldspar
S1L2	4-10	6	Coarse sand with clay, bright red.
S1L3	10-11.3	1.3	Slightly micaceous Sandy clay, brownish pebbly, fine to coarse feldspar.
S1L4	11.3-18.5	7.2	Fin to coarse sandy clayey and gravel
S1L5	18.5-24	5.5	Sand. Brown, clayey at the top, fine to coarse
S1L6	24-33	9	Light grey coarse sand, granite gravel
S1L7	33-45	12	Silty sand feldspar, blackish to grey
S1L8	45-49	4	Grey silty sand, low grade
S1L9	49-57	8	Greyish to Purple silty sand
S1L10	57-64.5	7.5	Sand, fine to coarse, pebbly blackish to grey

Table 3. Depth and Lithologic Unit of Gosa Site Borehole. Coordinate (Lat: 8° 56' 45.6" N and Long: 7° 13' 26.2" E, GPS- Model: Extrex high sensitivity 2000-2007 Garmin Ltd) was used for coordinate.

Sample ID	Depth (m)	Thickness (m)	Lithology description
S2L1	0-5.6	5.6	Sandy clay, reddish brown laterite top soil.
S2L2	5.6-15.6	8	Sandy clay, fine to medium, brownish to yellow.
S2L3	13.6-24.9	11.3	Clay sandy feldspar Yellowish brown pebbly
S2L4	24.9-32.3	7.4	Micaceous clayey, grey to black
S2L5	32.3-41.8	9.5	Sandy shinny greyish to black feldspar.
S2L6	41.8-50.8	9	Fine medium shinny, qartz interbed, greyish ash feldspar

basement. VES 2 shows an area with thicker lateritic topsoil underlying fresh basement at a very shallow depth. At the location for VES 3, the unweathered basement is encountered at a deeper depth; yet the area does not sustain a borehole. The location for VES 4 and 5, the fresh basement is encountered at shallow depth. We are of the opinion that part of the problem in the Gosa area is the occurrence of micaceous clayey sediment in addition to the existence of shallow unfractured basement rocks. The lithology around the other VES points is similar to that of Gosa area described above; however, all the wells are productive.

Borehole log interpretation

It is clear from the above log that in Dei-Dei, 7.2 m thickness of the basalt is buried about 4 m of clay capped by a top laterite bed. The basalt lie directly a thin gravely and sandy layer marked by deep brownish laterite surface. The gravely layers, though highly permeable produces good yield, deep brown in colour, the brown colour resulting from high iron concentration derived from the intensive ferruginisation associated with the ground surface before the volcanic eruption. The fractured bottom layer of the basalt provides less contaminated water.

In Gosa, the coarse granites weather into water bearing sandy residue, the syenitic rock types, with the predominance of unstable minerals eg. Feldspars, decompose into plastic or soft clay and other argillites which behave only as aquitards or aquicludes. Generally, only small amount of water can be obtained or dry well in the freshly unweathered bedrock below the weathered layers. Even when fractured and prevented water from being transmitted into the borehole.

Conclusion

Results from the VES surveys and well log carried out in the study area indicated the presence of lateritic or sandy topsoil, weathered basement and fractured basement. The similarity in the geoelectric sections suggests a homogenous aquiferous setting in the area. However, this is not the case as boreholes drilled in the Gosa area were dry, even when drilled to depths of over 120 m. Based on the findings stated above, the following conclusions can be made.

(1) In Gosa, the area is inextensive fracturing in younger granite, and not often coincide with the broad depression of valley within the formation. In nature, it situated in an elongated depression simulating one of such situation but

the hydrogeological information which are the deeply dissected by fractures within the rock formations are not interconnected.

(2) In Dei-Dei, the rock compositions are made up of coarse granitic material, gave relatively higher yields in the apparently fractured or faulted zones which could be the evidence of volcanic activity marked by the occurrence of flat toped lateritised basalt.

In this study, it suggests that most of the successful boreholes in the basement rocks should terminate at the depth of about 60 m in deeply weathered and fractured zones that are interconnected.

ACKNOWLEDGEMENTS

The authors would like to thank the ministry of Higher Education (MOHE) for their funding through Universiti Teknologi Malaysia Research Grant Scheme Project. The authors will gratefully acknowledge the Nigerian Geological Survey Agency and Federal Ministry of Water Resources for their support in this work. Thanks to SYB Sinyoung Borehole Limited for providing the Rig machine and Compressor used in drilling the boreholes. Also, to Maxico Hydrosolution consult for providing Campus Ohmega for geosurvey.

REFERENCES

Acworth RI (1987). The development of crystalline basement aquifers in a tropical Environment. Quart. J. Eng. Geol. 20:265-272.

Ajibade AC, Woakes M, Rahaman MA (1987). Proterozoic crustal development in the Pan-African regime of Nigeria. In Proterozoic lithospheric evolution geodynamics. 17:259-271.

Barongo JO, Palacky GD (1989). Investigation of electrical properties of weather layers in the Yale area, Western Kenya, using resistivity soundings. Geophysics 56:133-138.

Black RR, Caby R, Moussine-Pouchkine A, Bayer R, Bertrand JM, Boullier AM, Fabre J, Lesquer A (1979). Evidence for Late Precambrian plate tectonics in West Africa. Nature 278:223-227.

Caby R (1989). Precambrian terrains of Benin, Nigeria and Northeast Brazil and the Late Proterozoic South Atlantic fit. Geol. Soc. Am. Special Paper. 230:145-158.

Dada SS (2008). Proterozoic evolution of the Nigeria – Boborema province. Geologic Soceity London special publications. 294:121-136.

De Beer JH, Blume J (1985). Geophysical and hydrogeological investigation of the groundwater resources of western Hereroland, southwest African/Namibia. Trans. Geol. Soc. S. Afr. 88:483-493.

Edet AE, Okereke CS (1997). Assessment of hydrogeological conditions in basement aquifers of the Precambrian Oban Massif, Southeastern Nigeria. J. Appl. Geophys. 36:95-204.

Mbonu PDC, Ebeniro JO, Ofoegbu CO, Ekine AS (1991). Geoelectrical Sounding for the determination of aquifer Characteristics in part of Umuahia area of Nigeria. Geophysics 56:284-291.

Nur A, Ayuni NK (2004). Hydrologeolectrical study in Jalingo Metropolis and environs of Taraba State, Northeastern Nigeria. Global J. Geol. Sci. 2:101-109.

Obaje NG (2009). Geology and mineral resources of Nigeria. Lecture Notes Earth Sci. pp. 120-221.

Olorunfemi MO, Fasuyi SA (1993). Aquifer types and geoelectrical/hydrogeologic characteristics of the central basement terrain of Nigeria. J. Afr. Earth Sci. 16:309-317.

Olorunfemi MO, Okakune ET (1992). Hydrogeological and Geological significance of Geoelectrical survey of Ile-Ife. Nig. J. Mining Geol. 28:221-222.

Oyawoye MO (1972). The basement complex of Nigeria. In T.F.J. Dessauvagie and A.J. Whiteman (eds). Afr. Geol. Ibadan pp. 66-102.

Pulawski B, Kurtn K (1977). Combined use of resistivity and seismic refraction methods in groundwater prospecting in crystalline areas. Study project. Kenya. DANIDA pp. 5-33.

Rahaman MA (1988). Recent advances in the study of the Basement Complex of Nigeria. Precambrian Geology of Nigeria. Geol. Survey Nig. Pub. pp. 11-43.

Satpatty BN, Kanugo BN (1976). Groundwater Exporation in Hard rock terrain: A case study. Geophys. Prospect. 24:725-736.

Shemang EN (1993). Groundwater potentials of Kubami River Bassin, Zaria, Nigeria, from D. C Resistivity study. Water Res. 2:36-41.

The application of some biodiversity indices in the Tortum Stream, Erzurum, Turkey

Ozden FAKIOGLU, Mine KOKTURK and Muhammed ATAMANALP

Faculty of Fisheries, Ataturk University, 25240 Erzurum, Turkey.

The aim of the present study was to investigate biodiversity index, dominancy index and equivalence index of diatom community in the Tortum Stream and its tributaries. Diatom samples were collected from nine stations on the main river and tributaries between July 2012 and May 2013. A total of 36 taxa of Bacillariophyceae were identified. Dominant species was *Cocconeis placentula* in all stations. Biodiversity index was lower on the upstream river than downstream and tributaries. Not only domestic and agriculture pollution but also dam construction and stone pit affected Tortum Stream negatively. In conclusion, species richness decreased while organic pollution tolerant species increased.

Key words: Diatom, biodiversity index, dominant index, evenness index, Tortum Stream.

INTRODUCTION

Benthic algal communities have developed in virtually all substrates that receive light, whether in small streams or large rivers. These communities have generated important part of water ecosystems due to their role in food chain and their quick response and direct to a lot of change in environmental parameters. Cell structure of diatom is directly connected to physicochemical parameters of water, so diatom is an indicator for identification of river water quality. Therefore, it is used as a tool to make comparison between aquatic ecosystems (Allan and Castillo, 2007).

The use of phytoplankton communities or other aquatic organisms as bioindicators goes back to very ancient times. For the first time, in 1954, benthic organisms were used as bioindicators by Patric due to their easy adaption to changes in the chemical and physical displacement (Thunmark, 1945; Nygaard, 1949; Lepistö and Rosenström, 1998).

In Turkey, the earliest study in which diatoms were used to evaluate the water quality changes was carried out by Kalyoncu and Barlas (1997) in which he gave more information about that study (saprobic index etc.). Organic pollution is usually closely related to enhance nutrient concentrations. Based on the significant correlation with nutrients and organic pollution variables, this study suggests that Saprobic Index (SI), Trophic Index (TI) and Swiss Diatom Index (DI-CH) are integrating the effects of enrichment of organic pollution. In another study, among the observed diatom indices, TI and DI-CH better reflect the changes in water quality than the SI (Kalyoncu et al., 2009). The diatom community is closely related to water quality and bio-invitational methods can be used in the river monitoring system in Turkey (Solak et al., 2012). Tortum River Basin has about 1900 km^2 area covering both Tortum and Uzundere towns and it is formed by Tortum Stream and its tributaries (Pırlak, 1993).

The aim of the present study was to evaluate the use of epilithic diatoms as indicators of water quality in the Tortum Stream and its tributaries.

MATERIALS AND METHODS

Study area

The Tortum Stream is located in the north part of the Eastarn

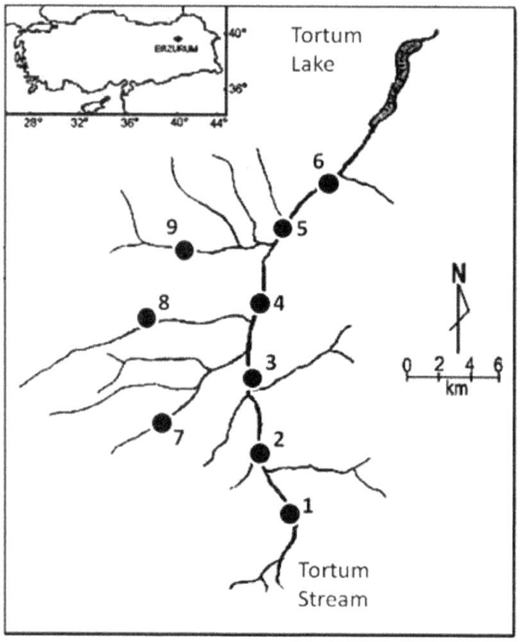

Figure 1. Location and the sampling stations of the Tortum Stream and its tributaries.

Anatolia, Turkey and its basin lies within the boundaries of Erzurum province. The stream form originates from town Tortum, forms the Tortum and the Tortum Waterfall, and then drains into the Coruh River (Figure 1). It has an average of 50 km length (Kıvrak and Gürbüz, 2010).

Diatom identification and counting

The diatom samples were collected from 9 stations monthly between June 2012 and May 2013. During the study period, the diatoms were sampled by scraping the 25 cm^2 upper surface of the epilithon, with a stiff toothbrush and collected in 250 ml sample bottle. The samples were fixed using Lugol's iodine. The composition and relative abundant of diatoms was estimated at 1000x magnification by binocular microscope and diatoms were identified after cleaning with a sulphuric and nitric acid mixture. The common taxonomic literatures were used (Krieger, 1932; Round, 1953; Cramer, 1991; Cox, 1996; Kelly, 1997; John et al., 2002). Three slides were prepared from each side and approximately 200 valves were enumerated in each slide to determine the relative abundance of each taxa (Kalyoncu et al., 2009).

Biodiversity indices

Five diversity indices (Shannon-Wiener, Simpson, Margalef, Menhinick, and McIntosh), two evenness indices and one dominant index were studied. MINITAB 15 (Cluster Observation) Software was used to interpret the relationship between the indices.

Shannon-Wiener diversity index (H')

This index is applied to biological systems which is derived from a mathematical formula by Shannon in 1948 (Türkmen and Kazancı, 2010):

$$H' = -\sum_{i=1}^{S} p_i \log_e p_i \,, \quad p_i\text{: } n_i/n$$

Where s is the total number of species and p_i is the number of individuals belonging to i species (n_i) / total number of individuals (n) (Hill, 1973; Krebs, 1998; Kwak and Peterson, 2007; James and Aderaje, 2010).

Simpson diversity index (D)

$$1 - D = \left[\sum ni(ni-1)\right] / N / (N-1)$$

Where ni is the number of individuals belonging to i species and N is the total number of species (Hill, 1973; Krebs, 1998; Kwak and Peterson, 2007; James and Aderaje, 2010).

Simpson dominance index (C)

$$C = \sum_{i=1}^{s} (p_i^{\,2}),$$

Where s is the number of species and p_i is the relative abundance of species i (Hill, 1973; Krebs, 1998; Kwak and Peterson, 2007; James and Aderaje, 2010).

Margalef diversity index (Dmg)

$$D_{mg} = S - 1 / \mathrm{Log}N$$

Where S is the number of species and N signifies the number of individual in a sample (James and Aderaje, 2010).

Menhinick diversity index (D_{mn})

$$D_{mn} = S / \sqrt{N}$$

Where S is the number of species and N signifies the number of individual in a sample (Menhinick, 1964; James and Aderaje, 2010).

McIntosh diversity index (M_c)

$$M_c = \left[N - \sqrt{\left(\sum n_i^{\,2}\right)}\right] / \left[N - \sqrt{N}\right]$$

Where ni is the number of individuals belonging to i species and N is the number of species (Türkmen and Kazancı, 2010).

Pielou evenness index (J')

$$J' = H' / \log_e S$$

Where S is the number of species and H' is the Shannon-Wiener Diversity Index (Hill, 1973; Krebs, 1998; Kwak and Peterson, 2007; James and Aderaje, 2010).

Table 1. The list of diatoms in Tortum Stream.

Phyllum: Heterokontophyta	Order: Rhopalodiales
Classis: Bacillariophyceae	Family: Rhopalodiaceae
Order: Achantes	*Epithemia adnata* (Kützing) Brebisson
Family: Cocconeidaceae	*E. sorex* Kützing
Cocconeis placentula Ehrenberg	Order: Thalassşophyceae
Order: Bacillariales	Family: Catenulaceae
Family: Bacillariaceae	*Amphora ovalis* (Kützing) Kützing
Hantzshia amphioxys(Ehrenberg) Grunow	Classis: Coscinodiscophyceae
Nitzschia amphibian Grunow	Order: Aulacoseirales
Family: Surirellaceae	Family: Aulacoseiraceae
Suriella brebissonii Krammer & Lange-Bertalot	*Aulacoseria ambigua (Grunow)Simansen*
Order: Cymbellaceae	*A. granulata (Ehrenberg) Simansen*
Family: Cymbellaceae	Order: Thalassiosirales
Cymbella affinis Kützing	Family: Stephanodiscaceae
C. lensolata Cuspidata Pantocsek	Genus: Cyclotella
C. turgidula Grunow	*Cyclotella bodanica* Eulenstein ex Grunow
Encyonema minutum Hilse	*C. ocellata* Pantocsek
Family: Gomphonemataceae	Genus: Stephanodiscus
Didymosphenia geminata (Lyngbye) M. Schmidt	*Stephanodiscus rotula* (Kützing) Hendey
Gomphonema angustadum Kützing	*S. hantzschii* Grunow
G. capitatum Ehrenberg	Order: Melosirales
G.olivaceum (Hornemann) Brébisson	Family: Melosiraceae
G. parvalu (Kützing) Kützing	Genus: Melosira
Family: Rhoicospheniaceae	*Melosira varians* C.Agardh
Rhoicosphenia curuata (Kützing) Grunow	Classis: Fragilariophyceae
Order: Naviculales	Order: Fragilariales
Family: Naviculaceae	Family: Fragilariaceae
Caloneis bacillum (Grunow) Cleve	*Diatoma vulgaris* Bory de Sain-Vicent
Navicula cryptocephala Kützing	*Fragilaria arcus* (Ehrenberg) Cleve
Family: Pleurosigmataceae	*F.capucina* Desmazieres
Gyrosigma acuminatum (Kützing) Rabenhorst	*Meridion circulare* (Graville) C.Agardh
Family:Stauroneidaceae	*Ulnaria capidata* (Ehrenberg) P.Compere
Craticula cuspidata (Kützing) D.G.Mann	*U. ulna* (Nitzsch) P.Compere

McIntosh evenness index (M_cE)

$$M_cE = \left[N - \sqrt{\left(\sum n_i^2\right)}\right] / \left[N - \left(N/\sqrt{S}\right)\right]$$

Where ni is the i. türe ait birey sayısını, S is the number of species and Nis the number of individual (Türkmen and Kazancı, 2010).

RESULTS

In this study, a total of 3 classes, 10 orders, 15 families and 36 taxa, were identified belonging Bacillariophyceae (Table 1). While low species diversity was identified in summer, high species diversity was found between winter and spring (Table 2). During the study period, the highest total number of individuals was found in station 1 and the lowest total number in station 6. Total number of individuals was about 50% higher in the downstream sampling stations than in the upstream sampling stations. The number of species varied depending on station with an average of 24 species (Figure 2). The values of Shannon-Wiener Diversity Index ranged from 0.46 to 1.92. The lowest value was for station 2 and the highest for station 7. The values of Simpson Diversity Index were between 24.26 and 24.86. The lowest value was for station 2 and the highest value for station 7 (Figure 3).

In this study, *Cocconeis placentula* reached a significantly high abundance level in the all of the stations. In addition, both Shannon-Wiener Diversity and Simpson Diversity indexes were shown to be of eutrophication assessment in the stream ecosystem (Table 3). The values of Margalef Diversity Index ranged from 4.01 to 4.88. The lowest value was for station 2 and

Table 2. Monthly variation in species diversity ('+' indicates species, E not encountered).

Species	Months										
	Jun	**Jul**	**Sep**	**Oct**	**Nov**	**Dec**	**Jan**	**Feb**	**Mar**	**Apr**	**May**
Aulacoseria ambigua (Grunow)Simansen	E	E	E	E	E	E	E	E	E	+	E
Caloneis bacillum (Grunow) Cleve	E	E	E	E	E	+	+	+	+	+	E
Cocconeis placentula Ehrenberg	+	+	+	+	+	+	+	+	+	+	+
Craticula cuspidata (Kützing) D.G.Mann	E	+	+	E	E	+	+	+	+	+	+
Cyclotella bodanica Eulenstein ex Grunow	E	+	E	E	E	E	+	+	E	E	E
C. ocellata Pantocsek	E	E	E	E	E	E	E	+	E	+	+
Cymbella affinis Kützing	E	+	+	+	E	E	+	+	E	+	+
C. lensolata Cuspidata Pantocsek	E	E	E	E	E	E	E	E	E	+	+
C. turgidula Grunow	E	E	E	E	+	E	+	+	+	+	+
Diatoma vulgaris Bory de Sain-Vicent	E	+	+	+	E	+	+	+	+	+	+
Didymosphenia geminata (Lyngbye) M. Schmidt	E	E	E	+	E	+	+	+	E	+	+
Encyonema minutum Hilse	E	E	E	E	E	E	E	+	E	E	E
Epithemia adnata (Kützing) Brebisson	E	E	E	E	E	E	+	E	E	+	+
E. sorex Kützing	E	E	E	E	E	E	E	E	E	E	E
Fragilaria arcus (Ehrenberg) Cleve	E	E	E	E	+	+	+	+	+	+	+
F.capucina Desmazieres	E	E	E	E	E	+	+	+		+	
Gomphonema angustadum Kützing	E	+	+	+	E	+	+	+	+	+	+
G. capitatum Ehrenberg	E	E	E	E	E	E	E	E	E	E	E
G. parvalu (Kützing) Kützing	E	E	E	E	E	E	E	E	E	E	E
G.olivaceum (Hornemann) Brébisson	E	E	E	E	E	E	+	E	E	E	E
Gyrosigma acuminatum (Kützing) Rabenhorst	E	E	E	E	E	E	+	E	E	E	+
Hantzshia amphioxys (Ehrenberg) Grunow	E	E	E	E	+	+	+	+	+	+	+
Melosira varians C.Agardh	E	+	E	E	E	E	+	+	+	+	+
Meridion circulare (Graville) C.Agardh	E		E	E	E	E		+			
Navicula cryptocephala Kützing	E	+	+	+	+	+	+	+	+	+	+
Nitzschia amphibian Grunow	E	+	+	+				+		+	+
Rhoicosphenia curuata (Kützing) Grunow	E	+	+	E	E	+	+	+	+	+	+
S. hantzschii Grunow	E	E	E	E	E	E	+	E	E	+	
Stephanodiscus rotula (Kützing) Hendey	E	E	E	E	E	E	+	E	E	+	
Suriella brebissonii Krammer & Lange-Bertalot	E	E	E	E	+	E	E	E	E	E	E
Ulnaria capidata (Ehrenberg) P.Compere	E	E	+	+	+	E	+	E	E	+	+
U. ulna (Nitzsch) P.Compere	E	E	+	+	+	E	+	+	+	+	+

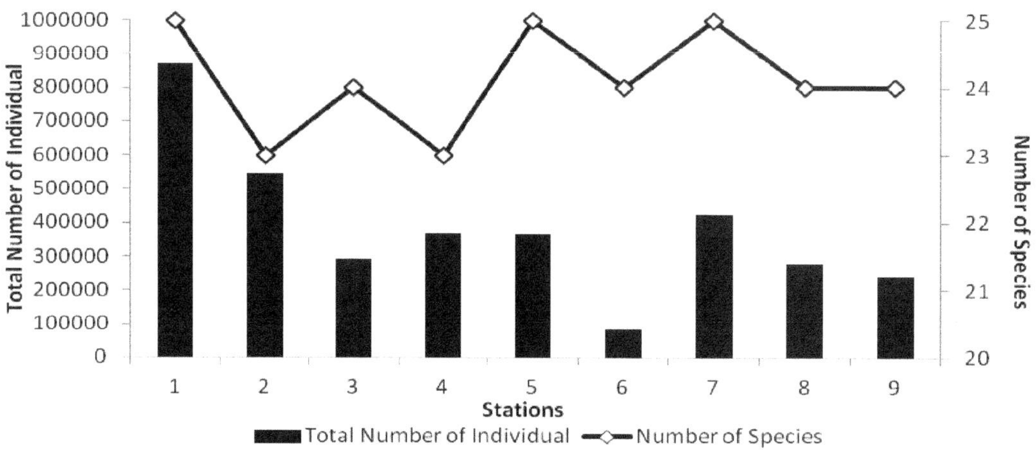

Figure 2. Total number of individual and species for each station.

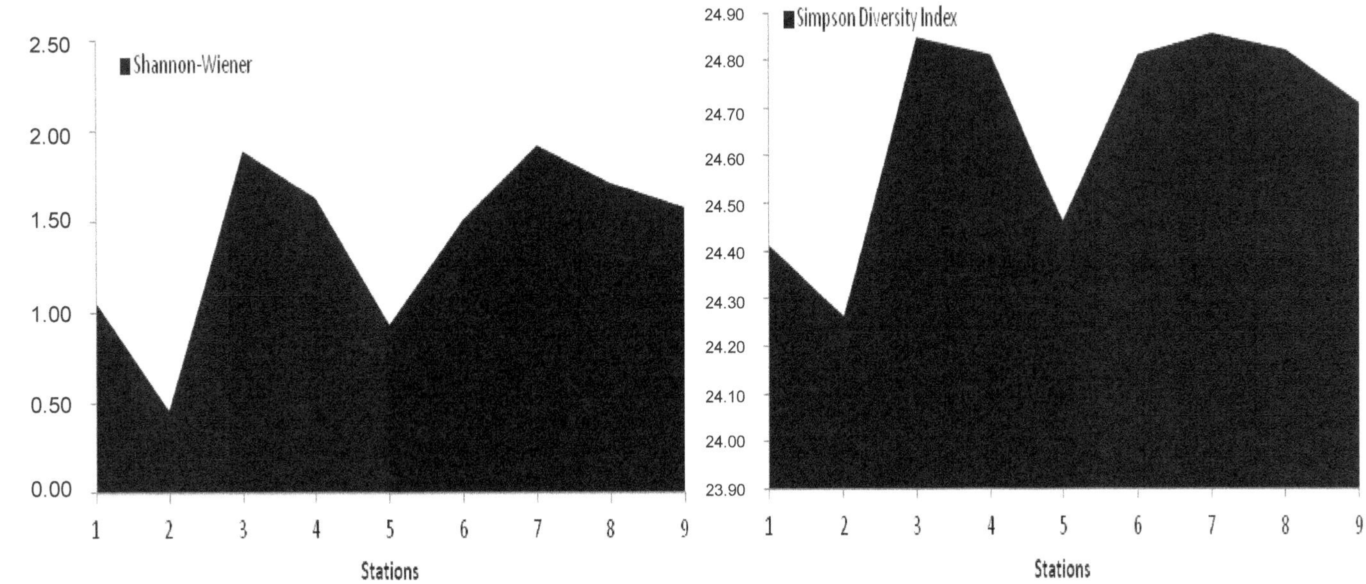

Figure 3. The variation of Shannon-Wiener Diversity and Simpson Diversity connect to station.

Table 3. Changes in dominant species to sample stations.

Station	Species	C
1	*Cocconeis placentula*	0.577
2	*Cocconeis placentula*	0.731
3	*Cocconeis placentula*	0.075
	Caloneis bacillum	0.044
4	*Cocconeis placentula*	0.121
	Navicula cryptocephala	0.047
5	*Cocconeis placentula*	0.527
6	*Cocconeis placentula*	0.136
	Navicula cryptocephala	0.015
	Cymbella affinis	0.025
7	*Cocconeis placentula*	0.029
	Diatoma vulgaris	0.034
	Ulnaria capidata	0.056
8	*Cocconeis placentula*	0.110
	Craticula cuspidata	0.533
9	*Cocconeis placentula*	0.266
	Ulnaria ulna	0.011

the highest value for station 6. These stations did not have a domestic land but they were under the stresses of dams conclusion and quarries. The values of Menhinick Diversity Index were between 0.03 and 0.08. The lowest value was for station 1 and 2 and the highest value for station 6. The values of McIntosh Diversity Index were between 288.295 and 736.85. The lowest value was calculated for station 6 and the highest value for station 2 (Figure 4).

Pielou Evenness Index was between 0.15 and 0.60. The lowest value was station 2 and the highest value was for stations 3 and 6. Both of them found similar value due to dam construction. The McIntosh Evenness Index was between 0.25 and 1.26. The lowest value was station for 2, 3, 4, 8 and 9 and the highest value was for station 5 and 7 (Figure 5). In this study, Shannon-Wiener and Simpson Diversity Indices, Margalef and Menhinick Diversity Indices were similar to each other (Figure 6). The number of species was similar for stations 1, 2, 5 and for stations 3, 8, 9, 7, 4 in the Tortum Stream and its tributaries. However, station 6 was not similar with the other stations (Figure 7).

DISCUSSION

The present study identified 36 species from Bacillariophyceae, whereas Kıvrak and Gürbüz (2010) found 113 taxa in the same habitat during the period of 2005-2006. Our results show that diatom species could be under the stress of dam construction. According to Simson et al. (2003), natural disasters for example, erosion are in charge of increase of sediment concentration in this circumstance which is a trigger of water quality change. In parallel with these changes are nagative effects diatom assemblages and species diversity.

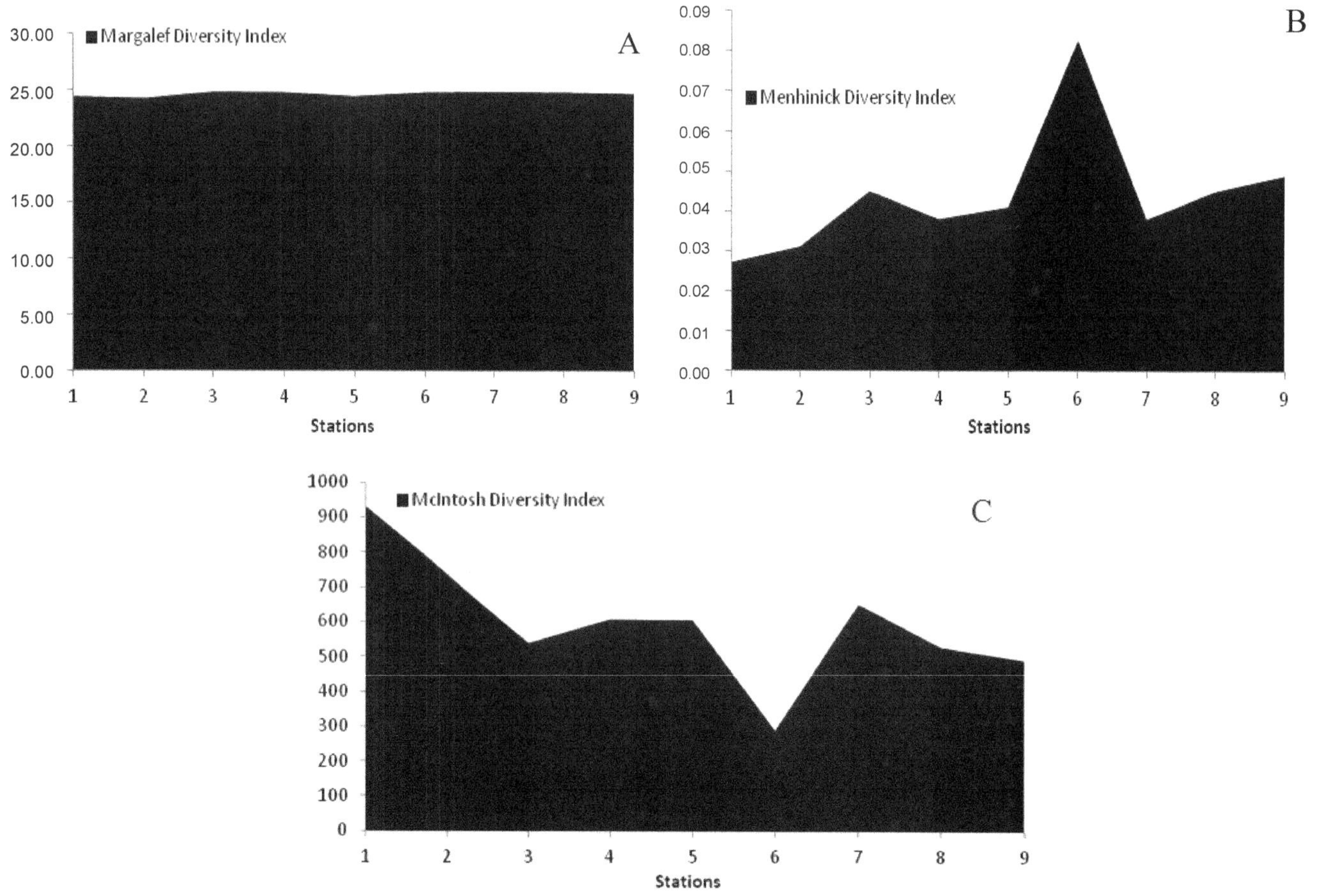

Figure 4. The variation of (A) Margalef Diversity, (B) Menhinick Diversity Index and (C) McIntosh Diversity Index.

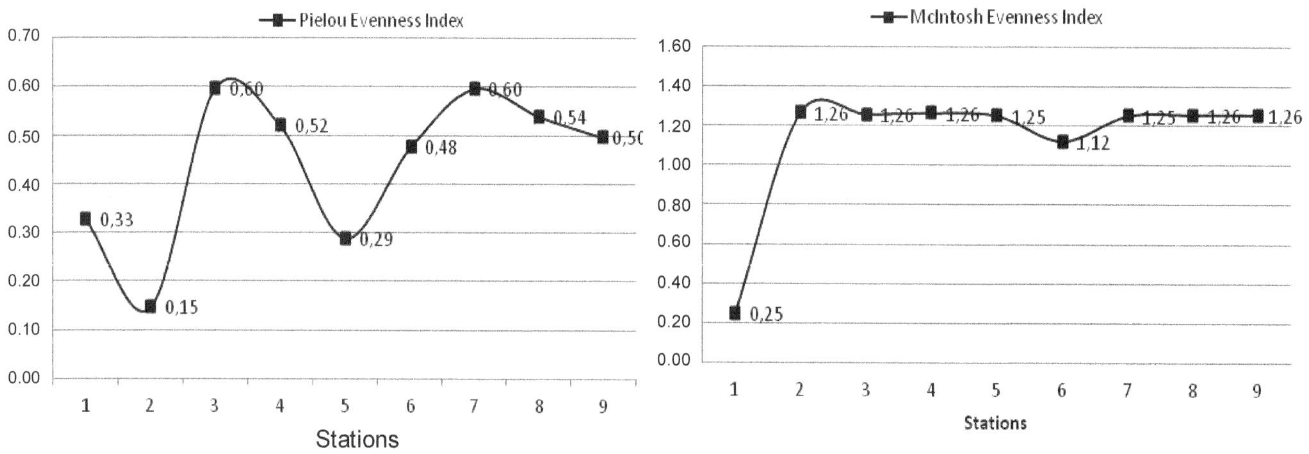

Figure 5. The variation of Pielou evenness index and the McIntosh connect to station.

In our study, *C. placentula* was the dominant taxa in all the stations. Kıvrak and Gürbüz (2010) identified *C. plasentula var. euglypta*. These taxa were adapted organic pollution (Soininen, 2002). *Caloneis bacillum* was found commonly in the station 3, *Navicula cryptocephala* was in the station 4, *Cymbella affinis* was in station 6, *Diatoma vulgaris* and *Ulnaria capidata* were in the station 7, *Craticula cuspidata* was in the station 8 and *Ulnaria*

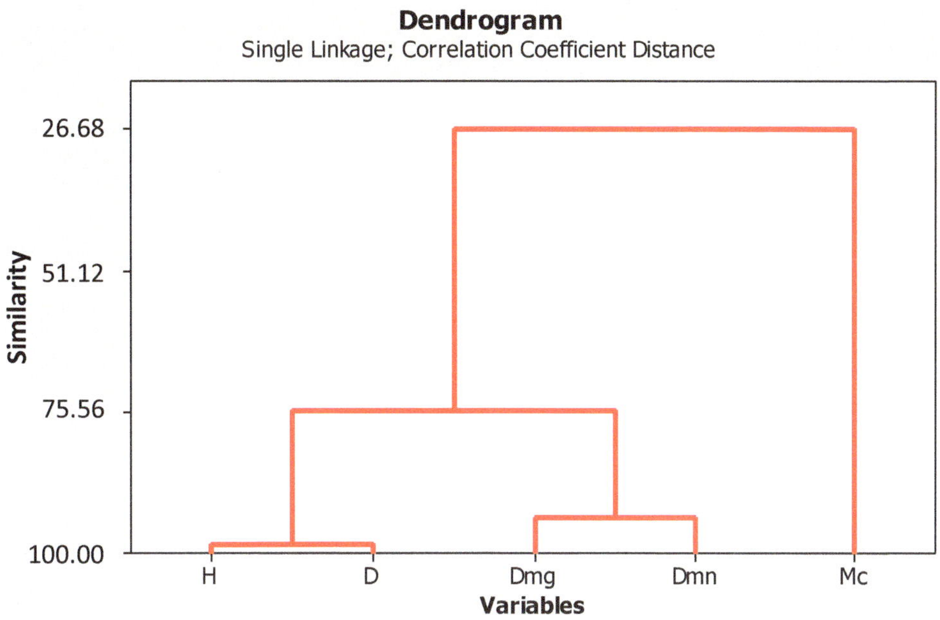

Figure 6. Similarity of different diversity indices.

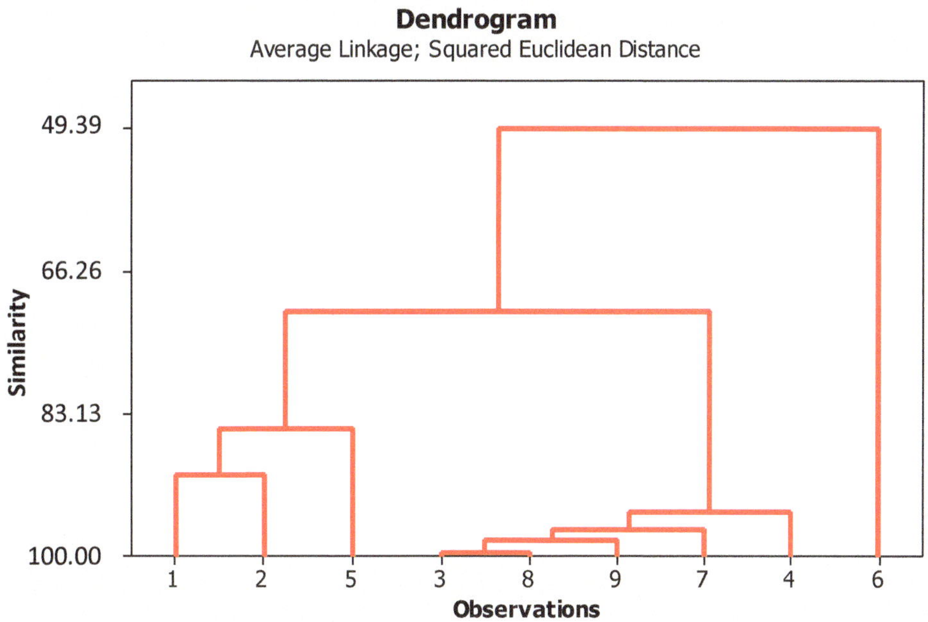

Figure 7. Similarity of different stations on the Tortum Stream and its tributaries.

ulna was in the station 9. This study showed that organic pollution-tolerant species were found in the stations which were located around the settlement and cultivation areas. On the other hand, dam construction and fast flow rate influence species diversity in the Tortum Stream.

In the present study, species diversity increased according to biodiversity indices in part of downstream and Tortum Stream's tributaries. The result of site 7 point out that around the station has been stressed both domestic and waste water pollutions. The structure of the ecosystem of rivers, streams on biodiversity, hydrological diversity and human activities are effective (Allan and Castillo, 2007). Pei and Liu (2011) suggested that dominant species were the upper more than middle part of Niyang River and they reported that useful indices can determine the ecological characteristics of lotic systems.

Conclusion

In conclusion, Tortum Stream and its tributaries were affected by domestic waste, agricultural waste, dam constructions and quarries. They created pressure on the ecosystem of the stream to reduce organic pollution-tolerant species. We suggest that precaution such as sewer system, demolition waste management (for example, dams and quarries) and control of illegal quarries should be taken in order to reduce detrimental effects of lotic ecosystem.

ACKNOWLEDGEMENT

Thanks are due to Murat ARSLAN (Faculty of Fisheries, Ataturk University, Turkey) for manuscript revision and suggestion.

REFERENCES

Allan JD, Castillo MM (2007). Stream Ecology Structure And Function Of Runnin Waters. Cluwer academic publishers Second Edition. London. P.429.

Cox EJ (1996). Identification of freshwater diatoms from live material. Chapman and Hall, London. P.158.

Cramer J (1991). Bibliotheca Phycologia. Desmides de Madagascar (Chlorophyta, Zygophyceae) par P. Bourrelly and A. Coute. Berlin Stuttgart. P. 346.

Hill MO (1973). Diversity ve Evenness: A Unifying Notation and Its Consequences. Ecology. pp. 427-432.

James BK, Adejare LI (2010). Nutrients and Phytoplankton Production Dynamics of a Tropical Harbor in Relation to Water Quality Indices. J. Am. Sci. pp. 261-275.

John PM, Whitton BA, Brook AJ (2002). The Freshwater algal flora of the British Isles. Cambridge Univ. Press, Cambridge. P. 498.

Kalyoncu H, Barlas M (1997). Isparta Stream epilithic diatoms determined depending on the water quality of seasonal changes. In: IX. National Fisheries Symposium, 17-19 Sept. 1997.

Kalyoncu H, Çiçek NL, Akköz C, Yorulmaz B (2009). Comparative performance of diatom indices in aquatic pollution assessment. Afr. J. Agric. Res. pp. 1032-1040.

Kelly MG (1997). Use of Trophic Diatom Index to Monitor Eutrophication in Rivers. Water Research. pp. 236-242.

Kivrak E, Gürbüz H (2010). Tortum Stream (Erzurum) Epipelic Diatom and its Relationship to Some Physicochemical Properties, Ekoloji. pp. 102-109.

Krebs CJ (1998). Ecological methodology, 2nd edition. Benjamin/Cummings, Menlo Park, California. P. 624.

Krieger WV (1932). Die Desmidiaceen der Deutschen Limnologischen Sunda-Expedition.

Kwak TJ, Peterson JT (2007). Community indices, parameters, and comparisons. *in* C.S. Guy and M.L. Brown, Editors. pp. 677-763. Analysis and interpretation of freshwater fisheries data. American Fisheries Society, Bethesda, Maryland. ISBN-13: 978-1-888569-77-3.

Lepistö L, Rosenström U (1998). The most typical phytoplankton taxa in four lakes in Finland. Hydrobiologia, pp. 89-97.

Menhinick EF (1964). A Comparison of Some Species-Individuals Diversity Indices Applied to Samples of Field Insects. Ecology, pp. 859-861.

Nygaard G (1949). Hidrobiological studies on some Danish ponds and lakes. Part II: The quotient hypothesis and some little known plankton organisms. Kongelige Danske Videnskabernes Selskab Biologiske Skrifter. P. 293.

Pei G, Liu G (2011). Distribution Patterns of Benthic Diatoms During Summer in The Niyang River, Tibet, China. Chinese J. Oceanol. Limnol. 29(6):1192-1198.

Pirlak L (1993). Uzundere, residue and grows naturally in Oltu County of Cranberry (*Cornus mas* L.) A Study on Propagation. Ataturk University Natural and Application Science Institu. Turkey, PhD Thesis. P. 164.

Round FD (1953). An investigation of two benthic algal communities in Malham Tarn, Yorkshire. J. Ecol. pp.174-197.

Simson KS, Benfield EF, Macko SA (2003). Food web structure and the role of epilithic biofilms in cave streams. Ecol. 84:2395-2406.

Soininen J (2002). Responses of epilithic diatom communities to environmental gradients in some Finnish Rivers. Int. Rev. Hydrobiol. 87:11-24.

Solak CN, Barinova S, Ács E, Dayioğlu H (2012). Diversity and ecology of diatoms from Felent Creek (Sakarya river basin), Turkey. Turk. J. Bot. pp. 191-203.

Thunmark S (1945). Zur Soziologie des Süsswasserplanktons. Eine methodisch ökologische Studie. Folia Limnologica Skandinavica, pp. 1-66.

Türkmen G, Kazancı N (2010). Applications of various biodiversity indices to benthic macroinvertebrate assemblages in streams of a national park in Turkey. Review of Hydrobiology. pp. 111-125.

Permissions

All chapters in this book were first published in IJPS, by Academic Journals; hereby published with permission under the Creative Commons Attribution License or equivalent. Every chapter published in this book has been scrutinized by our experts. Their significance has been extensively debated. The topics covered herein carry significant findings which will fuel the growth of the discipline. They may even be implemented as practical applications or may be referred to as a beginning point for another development.

The contributors of this book come from diverse backgrounds, making this book a truly international effort. This book will bring forth new frontiers with its revolutionizing research information and detailed analysis of the nascent developments around the world.

We would like to thank all the contributing authors for lending their expertise to make the book truly unique. They have played a crucial role in the development of this book. Without their invaluable contributions this book wouldn't have been possible. They have made vital efforts to compile up to date information on the varied aspects of this subject to make this book a valuable addition to the collection of many professionals and students.

This book was conceptualized with the vision of imparting up-to-date information and advanced data in this field. To ensure the same, a matchless editorial board was set up. Every individual on the board went through rigorous rounds of assessment to prove their worth. After which they invested a large part of their time researching and compiling the most relevant data for our readers.

The editorial board has been involved in producing this book since its inception. They have spent rigorous hours researching and exploring the diverse topics which have resulted in the successful publishing of this book. They have passed on their knowledge of decades through this book. To expedite this challenging task, the publisher supported the team at every step. A small team of assistant editors was also appointed to further simplify the editing procedure and attain best results for the readers.

Apart from the editorial board, the designing team has also invested a significant amount of their time in understanding the subject and creating the most relevant covers. They scrutinized every image to scout for the most suitable representation of the subject and create an appropriate cover for the book.

The publishing team has been an ardent support to the editorial, designing and production team. Their endless efforts to recruit the best for this project, has resulted in the accomplishment of this book. They are a veteran in the field of academics and their pool of knowledge is as vast as their experience in printing. Their expertise and guidance has proved useful at every step. Their uncompromising quality standards have made this book an exceptional effort. Their encouragement from time to time has been an inspiration for everyone.

The publisher and the editorial board hope that this book will prove to be a valuable piece of knowledge for researchers, students, practitioners and scholars across the globe.

List of Contributors

Agba Hemen Emmanuel
Department of Physics, Benue State University, Makurdi, Nigeria

Kungur Samuel Terhemba
Department of Physics, College of Education, Katsina-Ala, Benue, Nigeria

Akaagerger Nguvan Becky
Department of Physics, Benue State University, Makurdi, Nigeria

G. A. Agbo
Department of Industrial Physics, Ebonyi State University, P. M. B. 053, Abakaliki, Nigeria

O. N. Okoro
Department of Industrial Physics, Ebonyi State University, P. M. B. 053, Abakaliki, Nigeria

J. E. Ekpe
Department of Industrial Physics, Ebonyi State University, P. M. B. 053, Abakaliki, Nigeria

O. Akinsoji
Department of Chemistry, Cape Peninsula University of Technology, Bellville, South Africa

O. S. Fatoki
Department of Chemistry, Cape Peninsula University of Technology, Bellville, South Africa

B. J. Ximba
Department of Chemistry, Cape Peninsula University of Technology, Bellville, South Africa

B. O. Opeolu
Department of Environmental and Occupational Studies, Cape Peninsula University of Technology, Cape Town, South Africa

O. S. Olatunji
Department of Chemistry, Cape Peninsula University of Technology, Bellville, South Africa

M. L. Akinyemi
Department of Physics, Covenant University, P. M. B. 1023, Ota, Ogun State, Nigeria

M. R. Usikalu
Department of Physics, Covenant University, P. M. B. 1023, Ota, Ogun State, Nigeria

Cecily Nwokocha
Department of Physics, Alvan Ikoku Federal College of Education, P. M. B. 1033 Owerri, Imo State, Nigeria

Raymond Kasei
Centre for Development Research, University of Bonn, Germany

Urias Goll
Department of Environmental Studies, University of Liberia, Liberia

Mohammad Reza Beheshti
Department of Water Science and Engineering, Science and Research Branch, Islamic Azad University, Tehran, Iran

Amir Khosrojerdi
Department of Water Science and Engineering, Science and Research Branch, Islamic Azad University, Tehran, Iran

Seyed Mahmud Borghei
Department of Civil Engineering, Sharif University of Technology, Tehran, Iran

Mumtaz Shah
Department of Civil and Environmental Engineering, University of Engineering and Technology (UET) Taxila, Pakistan

Hashim Nisar Hashmi
Department of Civil Engineering, University of Engineering and Technology (UET) Taxila, Pakistan

M. O. Adeniyi
Department of Physics, University of Ibadan, Ibadan, Nigeria

E. F. Nymphas
Department of Physics, University of Ibadan, Ibadan, Nigeria

E. O. Oladiran
Department of Physics, University of Ibadan, Ibadan, Nigeria

Chike Luke Orjiekwe
Department of Chemical Sciences, Igbinedion University, Okada, Edo State Nigeria

Kehinde Oguniran
Department of Chemical Sciences, Igbinedion University, Okada, Edo State Nigeria

Solola Saheed Abiodun
Department of Chemical Sciences, Igbinedion University, Okada, Edo State Nigeria

G. A. Agbo
Department of Industrial Physics, Ebonyi State University, Abakaliki, Nigeria

B. Alfa
Department of Physics, Ibrahim Badamasi Babangida University, Lapai, Nigeria

G. F. Ibeh
Department of Industrial Physics, Ebonyi State University, Abakaliki, Nigeria

I. S. Adamu
Department of Physics, Ibrahim Badamasi Babangida University, Lapai, Nigeria

Feza Karaer
Department of Environmental Engineering, Görükle Campus, 16059, Uludag University, Faculty of Engineering-Architecture, Bursa, Turkey

Aslıhan Katip
Department of Environmental Engineering, Görükle Campus, 16059, Uludag University, Faculty of Engineering-Architecture, Bursa, Turkey

Saadet İleri
Department of Environmental Engineering, Görükle Campus, 16059, Uludag University, Faculty of Engineering-Architecture, Bursa, Turkey

Sonay Sarmaşik
Department of Environmental Engineering, Görükle Campus, 16059, Uludag University, Faculty of Engineering-Architecture, Bursa, Turkey

Nurcan Aydoğan
Bursa Metropolitan Municipality-General Directorate of Water and Sewerage Administration (BUSKİ), Bursa Eastern Wastewater Treatment Plant, Küçük Balıklı, Osmangazi, Bursa,Turkey

Osamu Takagi
Information and Research Center, International Research Institute (IRI), 1108-2 Sonno, Inage, Chiba 263-0051, Japan

Masamichi Sakamoto
Aquavision Academy, 1228-3 Tsubuura, Narita, Chiba 287-0236, Japan

Hideyuki Kokubo
Information and Research Center, International Research Institute (IRI), 1108-2 Sonno, Inage, Chiba 263-0051, Japan

Hideo Yoichi
Information and Research Center, International Research Institute (IRI), 1108-2 Sonno, Inage, Chiba 263-0051, Japan

Kimiko Kawano
Information and Research Center, International Research Institute (IRI), 1108-2 Sonno, Inage, Chiba 263-0051, Japan

Mikio Yamamoto
Information and Research Center, International Research Institute (IRI), 1108-2 Sonno, Inage, Chiba 263-0051, Japan

Alemayehu Kiros
Department of Physics, Faculty of Science, Addis Ababa University, P. O. Box -1176, Addis Ababa, Ethiopia

A. V. Gholap
Department of Physics, Faculty of Science, Addis Ababa University, P. O. Box -1176, Addis Ababa, Ethiopia

Giovanni E. Gigante
La Sapienza University, Rome, Italy

Khalid Mahmood
Department of Space Science, University of the Punjab, Lahore, Pakistan

Syeda Adila Batool
Department of Space Science, University of the Punjab, Lahore, Pakistan

Asim Daud Rana
Department of Space Science, University of the Punjab, Lahore, Pakistan

Salman Tariq
Department of Space Science, University of the Punjab, Lahore, Pakistan

Zulfiqar Ali
College of Earth and Environmental Sciences, University of the Punjab, Lahore, Pakistan
Nano Science and Catalysis Division, National Centre for Physics, Q.A.U Campus, Islamabad, Pakistan

Muhammad Nawaz Chaudhry
Nano Science and Catalysis Division, National Centre for Physics, Q.A.U Campus, Islamabad, Pakistan

Nwachukwu R. Ekere
Department of Pure and Industrial Chemistry, University of Nigeria, Nsukka, Nigeria

Okiri P. Ukoha
Department of Pure and Industrial Chemistry, University of Nigeria, Nsukka, Nigeria

Nasibeh Sadafi
Department of Architecture, Faculty of Engineering, National University of Malaysia, 43600, Bangi, Selangor, Malaysia

M. F. M. Zain
Department of Architecture, Faculty of Engineering, National University of Malaysia, 43600, Bangi, Selangor, Malaysia

M. Jamil
Department of Architecture, Faculty of Engineering, National University of Malaysia, 43600, Bangi, Selangor, Malaysia

A. C. Marcus
Department of Chemistry, Ignatius Ajuru University of Education, P. M. B. 5047, Rumuolumeni, Port Harcourt, Rivers State, Nigeria

C. O. B. Okoye
Department of Pure and Industrial Chemistry, University of Nigeria, Nsukka, Enugu State, Nigeria

C. N. Ibeto
National Centre for Energy Research and Development, University of Nigeria Nsukka, Enugu State, Nigeria

Yildiz Cabuk
Department of Forest Industrial Engineering, Faculty of Forestry, Bartin University, 74100 Bartin-Turkey

Selman Karayilmazlar
Department of Forest Industrial Engineering, Faculty of Forestry, Bartin University, 74100 Bartin-Turkey

S. Murat Onat
Department of Forest Industrial Engineering, Faculty of Forestry, Bartin University, 74100 Bartin-Turkey

Rifat Kurt
Department of Forest Industrial Engineering, Faculty of Forestry, Bartin University, 74100 Bartin-Turkey

I. Emmanuel
Department of Physics, Federal University of Technology, P. M. B. 704, Akure, Ondo State, Nigeria

B. Adeyemi
Department of Physics, Federal University of Technology, P. M. B. 704, Akure, Ondo State, Nigeria

K. D. Adedayo
Department of Physics, Federal University of Technology, P. M. B. 704, Akure, Ondo State, Nigeria

G. E. Kifanyi
Department of Civil Engineering, Tshwane University of Technology, Private Bag X680, 0001 Pretoria, South Africa

B. M. B. Shayo
Department of Civil Engineering, Mbeya University of Science and Technology, P. O. Box 131 Mbeya, Tanzania

J. M. Ndambuki
Department of Civil Engineering, Tshwane University of Technology, Private Bag X680, 0001 Pretoria, South Africa

Q. Khadijah
Department of Textile Printing, Dyeing and Finishing, Faculty of Applied Arts, Helwan University, Egypt

M. Heba
Department of Textile Printing, Dyeing and Finishing, Faculty of Applied Arts, Helwan University, Egypt
Department of Fashion Design, Faculty of Art and Design, King Abdul Aziz University, Saudi Arabia

A. M. O. Abdul Raheem
Department of Chemistry, University of Ilorin, P. M. B. 1515, Ilorin, Nigeria

F. A. Adekola
Department of Chemistry, University of Ilorin, P. M. B. 1515, Ilorin, Nigeria

Muhammad Asif Imran
College of Earth and Environmental Sciences, University of the Punjab, Lahore, Pakistan

Muhammad Nawaz Ch
College of Earth and Environmental Sciences, University of the Punjab, Lahore, Pakistan

Rass Masood Khan
Department of Botany, University of the Punjab, Lahore, Pakistan

Zulfiqar Ali
College of Earth and Environmental Sciences, University of the Punjab, Lahore, Pakistan
Nano Science and Catalysis Division, National Centre for Physics, Quaid-i-Azam University, Islamabad 45320, Pakistan

Tariq Mahmood
Nano Science and Catalysis Division, National Centre for Physics, Quaid-i-Azam University, Islamabad 45320, Pakistan

M. W. Chege
Department of Physics, Kenyatta University, P. O. Box 43844, Nairobi, Kenya

N. O. Hashim
Department of Physics, Kenyatta University, P. O. Box 43844, Nairobi, Kenya

A. S. Merenga
Department of Physics, Kenyatta University, P. O. Box 43844, Nairobi, Kenya

J. Tschiersch
Institute of Radiation Protection, Helmholtz Zentrum München - German Research Center for Environmental Health (GmbH), Ingolstädter Landstraße 1, 85764 Neuherberg, Germany

Dariush Moazami
Department of Civil Engineering, University Putra Malaysia, 43400 UPM Serdang, Selangor, Darul Ehsan, Malaysia

Zainuddin Md. Yusoff
Department of Civil Engineering, University Putra Malaysia, 43400 UPM Serdang, Selangor, Darul Ehsan, Malaysia

Ratnasamy Muniandy
Department of Civil Engineering, University Putra Malaysia, 43400 UPM Serdang, Selangor, Darul Ehsan, Malaysia

Hussain Hamid
Department of Civil Engineering, University Putra Malaysia, 43400 UPM Serdang, Selangor, Darul Ehsan, Malaysia

Kadir Ozkaya
Pamukkale University, Denizli Vocational School of Technical Sciences, Department of Materials and Materials Processing Technology, 20070, Denizli, Turkey

Kadiri Umar Afegbua
Department of Earthquake Seismology, Centre for Geodesy and Geodynamics, P. M. B. 11, Toro, Bauchi State, Nigeria

F. O. Ezomo
Department of Physics, Faculty of Physical Sciences, University of Benin Ugbowo, Benin City, Edo State, Nigeria

Olatunde S. Olatunji
Department of Chemistry, Faculty of Applied Sciences, Cape Peninsula University of Technology, Cape Town, South Africa

Beatrice O. Opeolu
Department of Environmental and Occupational Studies, Faculty of Applied Sciences, Cape Peninsula University of Technology, Cape Town, South Africa

Olalekan S. Fatoki
Department of Chemistry, Faculty of Applied Sciences, Cape Peninsula University of Technology, Cape Town, South Africa

Bhekumusa J. Ximba
Department of Chemistry, Faculty of Applied Sciences, Cape Peninsula University of Technology, Cape Town, South Africa

Shazia Khurshid
Department of Chemistry, G. C. University Lahore, Pakistan

Shaheen Kausar
Department of Chemistry, G. C. University Lahore, Pakistan

Sadia Afzal
Department of Chemistry, G. C. University Lahore, Pakistan

Ahmed Adnan
Department of Chemistry, G. C. University Lahore, Pakistan

Arshad Mehmood
Department of Chemistry, G. C. University Lahore, Pakistan

Muhammad Arshad
Department of Chemistry, G. C. University Lahore, Pakistan

Zafer Ahmad Bukhari
Department of Chemistry, G. C. University Lahore, Pakistan

M. Omeje
Department of Physics, Faculty of Science, Universiti Teknologi Malaysia, 81310, Johor Bahru, Malaysia

W. Husin
Department of Physics, Faculty of Science, Universiti Teknologi Malaysia, 81310, Johor Bahru, Malaysia

I. Noorddin
Centre for Research and Innovation Management, National Defence University, Malaysia

I. A. Oha
Department of Geology, University of Nigeria Nsukka, Enugu State, Nigeria

O.S Onwuka
Department of Geology, University of Nigeria Nsukka, Enugu State, Nigeria

P. E. Ugwuoke
Centre for Energy Research and Development, University of Nigeria, Nsukka, Enugu State, Nigeria

O. Meludu
Department of Physics, School of Pure and Applied Sciences, Federal University of Technology, Yola, Nigeria

Ozden FAKIOGLU
Faculty of Fisheries, Ataturk University, 25240 Erzurum, Turkey

Mine KOKTURK
Faculty of Fisheries, Ataturk University, 25240 Erzurum, Turkey

Muhammed ATAMANALP
Faculty of Fisheries, Ataturk University, 25240 Erzurum, Turkey